Palaeomagnetism and the Continental Crust

Palaeomagnetism and the Continental Crust

J. D. A. Piper

Department of Geological Sciences,
University of Liverpool

Open University Press
Milton Keynes

and
Halsted Press
John Wiley & Sons
New York — Toronto

First published 1987 by
Open University Press
Open University Educational Enterprises Limited
12 Cofferidge Close
Stony Stratford
Milton Keynes MK11 1BY, England

Copyright © J. D. A. Piper 1987

All rights reserved. No part of this work may be reproduced in any form by mimeograph or by any other means, without permission in writing from the publisher.

British Library Cataloguing in Publication Data

Piper, J. D. A.
  Palaeomagnetism and the continental crust.
  1. Plate tectonics   2. Palaeomagnetism
  I. Title
  551.1'36   QE511.4

  ISBN 0-335-15201-5

Published in the U.S.A., Canada and Latin America by
Halsted Press, a Division of John Wiley & Sons, Inc.,
New York

Library of Congress Cataloging-in-Publication Data
Piper, J. D. A.
  Palaeomagnetism and the continental crust.

  1. Palaeomagnetism.   2. Earth—Crust.   3. Continents.
I. Title.
OE501.4.P35P54 1986   538'.7   86-16297

  ISBN 0-470-20743-4

Text design by Clarke Williams

Printed in Great Britain

Contents

Palaeomagnetic Database		x
Acknowledgements		xi
Introduction		1
1	**Basic principles of rock and mineral magnetism**	**5**
1.1	*Definition of magnetic behaviour, diamagnetism, paramagnetism and ferromagnetism*	5
1.2	*Types of ferromagnetism*	10
1.3	*Domain structure*	11
1.4	*The acquisition of magnetism by rocks*	15
1.5	*Total magnetisation and Köenigsberger ratio*	17
2	**The magnetic minerals**	**19**
2.1	*Introduction*	19
2.2	*The magnetite-ulvöspinel series*	22
2.3	*The hematite-ilmenite series*	25
2.4	*Maghemite*	27
2.5	*Goethite*	28
2.6	*Pyrrhotite*	29
2.7	*Other natural magnetic materials*	30
3	**Outline theory of rock and mineral magnetism**	**31**
3.1	*Introduction: the properties of superparamagnetic, single-domain and multi-domain particles*	31
3.2	*The demagnetising factor, magnetic forces and relaxation times*	35

	3.3	Single-domain behaviour: the theoretical approach	37
	3.4	Coercivity	39
	3.5	Temperature dependence of unblocking of magnetic remanence	39
	3.6	Thermoremanent magnetisation	40
	3.7	Chemical remanence	43
	3.8	Susceptibility	44
	3.9	Anisotropy of susceptibility	46
	3.10	Magnetisation during slow heating and cooling	49

4 The magnetisations in rocks   53

    4.1 Igneous rocks   54
           The volcanic and hypabyssal environments
           Ophiolites and serpentinisation
           The plutonic environment
    4.2 Sedimentary rocks   58
           The depositional environment: depositional remanent magnetisations and post-depositional detrital remanent magnetisations
           The diagenetic and weathering environments: chemical remanent magnetisations
           Red sediments
           Limestones
           Banded ironstone formations
    4.3 Metamorphic rocks   72

5 Field and laboratory methods   75

    5.1 Objectives   75
    5.2 Field sampling   76
    5.3 Magnetometers   77
           Astatic and parastatic magnetometers
           Spinner magnetometers
           Cryogenic magnetometers
    5.4 Susceptibility meters   82
    5.5 Demagnetisation techniques   84
           Alternating field (a.f.) demagnetisation
           Thermal demagnetisation
           Chemical demagnetisation
    5.6 Laboratory tests for magnetic minerals and domain structure   90
           Domain structure
           Isothermal remanence tests
           Curie point (thermomagnetic) determinations
           Low temperature behaviour
           The Lowrie-Fuller test

6 Palaeomagnetic directions, palaeomagnetic poles and apparent polar wander — 99

- 6.1 *Palaeomagnetic directions and their presentation* — 99
  Spherical projections
  Orthogonal projections
- 6.2 *Combining directions of magnetisation* — 103
  Fisher statistics
  Comparison of mean directions
  Site and group mean directions of magnetisation
  Fisher and Bingham distributions
- 6.3 *Identification and separation of magnetic components* — 110
  Vector subtraction
  Remagnetisation circles
  Use of vector plots
  Multicomponent analysis by the Hoffman-Day method and graphical techniques
  Other methods
- 6.4 *Dating components of magnetisation* — 120
  Geochronological dating
  The fold test
  The conglomerate test
  The contact test
  The unconformity test
- 6.5 *The geomagnetic field and palaeomagnetic poles* — 128
  The dipole field and axial geocentric dipole model
  The axial geocentric dipole — A valid assumption?
  Reversals and excursions of the geomagnetic field
- 6.6 *Apparent polar wander (APW)* — 135
  Definition and presentation
  Interpretation of APW: the rate of continental movement
  Interpretation of APW: tectonic models
  The palaeomagnetic data base

7 Archaean and Proterozoic palaeomagnetism — 151

- 7.1 *Introduction* — 151
- 7.2 *Archaean–early Proterozoic APW (2900–2200 Ma)* — 154
- 7.3 *The 2200–2000 Ma APW path (Fig. 7.3)* — 161
- 7.4 *The 2000–1750 Ma APW path (Fig. 7.4)* — 165
- 7.5 *The 1750–1600 Ma APW path (Fig. 7.5)* — 169
- 7.6 *The 1600–1350 Ma APW path (Fig. 7.6)* — 172
- 7.7 *The 1350–1150 Ma APW path (Fig. 7.7)* — 175
- 7.8 *The 1150–950 Ma APW path (Fig. 7.8)* — 178
- 7.9 *Reversal asymmetry and APW* — 185
- 7.10 *Tectonic implications* — 187
- 7.11 *Precambrian APW and rates of continental movement* — 189

8 Growth and consolidation of the continental crust in Archaean and Proterozoic times — 193

    8.1 *Introduction* — 193
    8.2 *Geochemical and isotopic signatures* — 194
    8.3 *The sedimentary record* — 197
    8.4 *Greenstone belts and straight belts* — 198
    8.5 *Mobile belts and accretionary orogenic belts* — 203
    8.6 *Proterozoic anorogenic magmatism and crustal consolidation* — 209
    8.7 *Mineralisation in the continental crust* — 209

9 Late Proterozoic–Cambrian palaeomagnetism: breakup and dispersal of the continental crust — 215

    9.1 *Introduction* — 215
    9.2 *The 900–700 Ma APW path (Fig. 9.2)* — 218
    9.3 *The 700–570 Ma APW path (Fig. 9.3)* — 221
    9.4 *The Cambrian–Ordovician APW Paths (Fig. 9.4)* — 226
           Gondwanaland
           Laurentia
           Siberia
           Baltica
    9.5 *Palaeomagnetism of late Precambrian–early Palaeozoic microplates and accretionary terranes* — 233
           The Armorican microplate
           The Bohemian microplate
           The Anglo-Welsh microplate
    9.6 *Geological implications* — 239
           The late Precambrian glaciations
           Passive margins and the timing of breakup
           The Vendian–Lower Cambrian alkaline province
           Environmental changes and marine transgression
           The isotopic and chemical signature
           The palaeogeography of Vendian–Cambrian times

10 Palaeozoic palaeomagnetism and the formation of Pangaea — 253

    10.1 *Introduction* — 253
    10.2 *Gondwanaland* — 254
    10.3 *Palaeozoic palaeomagnetism of Western Europe* — 259
           The British region
           Palaeomagnetism and the Caledonian orogeny
           Palaeozoic palaeomagnetism of the remainder of Europe
           Palaeomagnetism and the Hercynian orogeny
    10.4 *Palaeozoic palaeomagnetism and the remagnetisation hypothesis* — 276

10.5	*Palaeozoic palaeomagnetism of the remaining continental plates*	280
	Laurentia	
	The European plate (USSR sector)	
	Siberia	
	Khazakhstania and the China plates	
10.6	*Palaeomagnetism and the reconstruction of Pangaea*	289
10.7	*The palaeogeography of Palaeozoic times*	291

11 Mesozoic and Cenozoic palaeomagnetism and the breakup of Pangaea 299

11.1	*Introduction*	299
11.2	*Laurentia and Eurasia*	300
11.3	*Gondwanaland*	305
11.4	*Sub-plates, microplates and suspect terranes*	311
	The Mediterranean region	
	The Japanese arc	
	North and South China, Kolyma, Sikote Alin and Korea	
	South East Asia	
	West Antarctica	
	The Andean margin of South America	
	Mexico	
	The Cordilleran margin of North America	

12 The geomagnetic field, continental movements and configurations and mantle convection since Archaean times 329

12.1	*Introduction*	329
12.2	*The polarity time scale: Cenozoic and Mesozoic times*	
	Palaeozoic magnetostratigraphy	330
12.3	*The geomagnetic polarity history from the palaeomagnetic record*	339
12.4	*The polarity history of Precambrian times*	341
12.5	*The strength of the geomagnetic field*	343
12.6	*Palaeosecular variation*	345
12.7	*APW: continental drift or true polar wander?*	349
12.8	*Geomagnetic intervals, continental velocities and geological cycles*	353
12.9	*Palaeomagnetism and mantle convection*	360
12.10	*Supercontinents and the geoid*	363
12.11	*The palaeomagnetic signature of global behaviour in Archaean and Proterozoic times*	370

References	375
Appendix	420
Index	422

Announcement

Palaeomagnetic Database

J D A Piper   ISBN 0 335 15211 2

A catalogue of palaeomagnetic data compiled by the author is available from the publishers as a separate publication. It contains a complete listing of palaeomagnetic results from investigations in every part of the globe from Archaean to Cenozoic times. Since the last such compilations were published over a decade ago the amount of information has grown rapidly. This fully undated and expanded database provides an essential reference tool for workers in this field throughout the world.

Acknowledgements

I am indebted to all of those people who have contributed in some way to this book. I have particularly appreciated the encouragement and advice which Rod Wilson has provided over a number of years; Rod painstakingly read and corrected much of this text and suggested many improvements. The late Professor Janet Watson critically commented on much of the content of Chapter 8, and I trust that the final version does some measure of justice to the understanding of Precambrian geology which she imparted to myself and many others. Pat Brenchley, Peter Crimes, Stephen Drury and Nick Kusznir also gave valuable advice on specific aspects of geological and geophysical interpretation. My former research students Jonathan Stearn, Robert Smith, Graeme Taylor and Timothy Poppleton have contributed valuable discussions and ideas and I am grateful to them and to Neil Roberts for allowing me to mention unpublished results. Robert Simbalist and Alan Smith kindly made available the ULTRAMAP program and Alan McCormack gave invaluable help with data and word processing; Alan also wrote most of the computer programs used here. Pauline Lybert and Phyllis Kon efficiently undertook the typing of the more exacting parts of the text and saw to it that I met my deadlines. I am particularly grateful to Joe Lynch for drafting many of the diagrams, and to Joan Dean both for her assistance with copying and sorting and for the enthusiastic way in which she has assisted with my palaeomagnetic studies over a number of years. Former Liverpool students have contributed considerably to the collection and processing of palaeomagnetic and other data included here, notably Simon Acklam, Susan Alexander, Graham Cox, David Norris and Theresa Strange; without their help this book and its complementary Directory of Palaeomagnetic results could not have been completed within a realistic period of time, and I am most grateful for their assistance. Finally, my thanks go to Julie and Adrian for their forbearance and encouragement.

(Proverbs 2:2–6)

The author and publishers are very grateful to the following for permission to

reproduce copyright material indicated:
Figure 2.1(b), The Royal Society of London (reproduced and relabelled from *Phil. Trans. R. Soc. Lond.*, A268, 507-550 (1971); Figure 2.5, The Terra Scientific Publishing Company (reproduced from *J. Geomag. Geoelect.*, 24, 69-90 (1972)); Table 3.1, Elsevier Scientific Publishing Company (reproduced with minor additions from *Phys. Earth Planet. Int.*, 26, 1-26 (1981)); Figure 3.1, Elsevier Scientific Publishing Company (reproduced and partly redrawn from *Phys. Earth Planet. Int*, 26, 1-26 (1981)); Figures 3.3 and 3.4, The Terra Scientific Publishing Company (redrawn from *J. Geomag. Geoelect.*, 21, 757-773 (1969)); Figure 3.6, Elsevier Scientific Publishing Company (reproduced from *Phys. Earth Planet. Int.*, 26, 1-26 (1981) and relabelled); Figure 3.7, Elsevier Scientific Publishing Company (reproduced from Earth Planet Sci. Lett., 28, 133-143 (1975) and relabelled); Figure 4.1, Chapman and Hall (redrawn from Figure 4.5 of *Palaeomagnetism* by D. H. Tarling (1983)); Figure 4.2, Elsevier Scientific Publishing Company (reproduced from *Phys. Earth Planet. Int.*, 26, 1-26 (1981) and relabelled); Figure 4.5, Chapman and Hall (redrawn and modified from Figure 9.12 in *Palaeomagnetism* by D. H. Tarling (1983) from an original figure in *Continental Red Beds* by P. Turner, published by the Elsevier Scientific Publishing Company); Figure 5.4, Chapman and Hall (reproduced and relabelled from Figure 2.4 in *Methods in Palaeomagnetism and Rock Magnetism* by D. W. Collinson (1983); Figure 5.5, Elsevier Publishing Company (reproduced and relabelled from Figures 2, page 382 and 3, page 384, in *Methods in Palaeomagnetism* (eds. D. W. Collinson, K. M. Creer and S. K. Runcorn (1967)); Figure 5.8, Royal Astronomical Society (redrawn from *Geophys. J. R. astr. Soc.*, 67, 395-413 (1981)); Figure 5.11, Elsevier Scientific Publishing Company (redrawn from *Phys. Earth Planet. Int.*, 26, 47-55 (1981)); Figure 5.13, Elsevier Scientific Publishing Company (reproduced and relabelled from *Earth Planet. Sci. Lett.*, 63, 353-367 (1983); Figure 6.1, Chapman and Hall (reproduced and labelled from Figure 12.2 of *Methods in Palaeomagnetism and Rock Magnetism* by D. W. Collinson (1983); Figure 6.5, Elsevier Scientific Publishing Company (reproduced and relabelled from *Earth Planet. Sci. Lett.*, 40, 433-438 (1978)); Figure 6.6, Royal Astronomical Society (reproduced and partially relabelled from *Geophys. J. R. astr. Soc.*, 45, 297-304 (1976)); Figure 6.7, Elsevier Scientific Publishing Company (reproduced and partially relabelled from *Phys. Earth Planet. Int.*, 20, 12-24 (1979)); Figure 6.8, Royal Astronomical Society (reproduced from *Geophys. J. R. astr. Soc.*, 62, 699-718 (1980)); Figure 6.9, Chapman and Hall (reproduced and relabelled from Figure 5.3(c) in *Palaeomagnetism* by D. H. Tarling (1983)); Figure 6.10, Royal Astronomical Society (reproduced from *Geophys. J. R. astr. Soc.*, 72, 165-172 (1983)); Figure 6.11, Geological Society of America (reproduced from Figure 48 of *Palaeomagnetism and Plate Tectonics* by M. W. McElhinny and copied directly from *Geol. Soc. Amer. Bull.*, 71, 645-768 (1960, not copyrighted)); Figure 6.13, Elsevier Scientific Publishing Company (reproduced from *Earth Planet. Sci. Lett.*, 40, 91-100 (1978)); Figure 6.18, The University of Chicago Press and Dr. C. R. Scotese (reproduced and slightly modified from *Jour. Geol.*, 87, 217-277 (1979)); Figures 8.2 and 8.3, Royal Astronomical Society (reproduced and modified (Figure 8.2) or simplified (Figure 8.3) from *Geophys. J. R. astr. Soc.*, 74, 163-197 (1983)); Figures 8.4 and 8.5, Geological Society of America (redrawn from *Geol. Soc. Amer. Mem.*, 161, 11-34 (1983)); Figure 9.9, Elsevier Scientific Publishing Com-

pany (reproduced from *Earth Planet. Sci. Lett.*, 70, 325–345 (1984)); Figure 9.11, John Wiley and Sons (reproduced from *The Caledonide Orogen: Scandinavia and Related Areas* (eds. D. G. Gee and B. A. Sturt) 19–34 (1985)); Figure 10.6, D. Reidel Publishing Company (reproduced and partially relabelled from *Regional Trends in the Geology of the Appalachian-Caledonian-Hercynian-Mauretainide Orogen* (ed. P. E. Schenk) 11–26 (1983)); Figure 11.4(b), inset, Martinus Nijhoff/Dr. W. Junk Publishers (reproduced from *Geol. Mijnb.*, 64 281–295 (1985)); Figure 12.2, Cambridge University Press (reproduced in part from Figure 4.2 of *A Geologic Time Scale* by W. B. Harland, A. V. Cox, P. G. Llewellyn, C. A. G. Pickton, A. G. Smith and R. Walters (1982)); Figure 12.10, Elsevier Scientific Publishing Company (reproduced from *Earth Planet. Sci. Lett.*, 62, 314–320 (1983) with additions); Figure 12.17, Elsevier Scientific Publishing Company (reproduced from *Earth Planet. Sci. Lett.*, 67, 123–135 (1984)); Figure 20.21, Royal Astronomical Society (reproduced with modifications from *Geophys. J. R. astr. Soc.*, 74 163–197 (1983)). The copyright of the following figures is held by the American Geophysical Union: Figure 3.8, (*J. Geophys. Res*, 85., 2625–2637 (1980)); Figure 5.12 (reproduced and relabelled from *Rev. Geophys. Space Phys.*, 70, 171–192; Figure 6.15 (*J. Geophys. Res.*, 84, 5480–5486 (1979)); Figure 6.20 (*J. Geophys. Res.*, 85, 3659–3669 (1980)); Figure 8.6, (redrawn from *Geodynamics Series*, Volume 5, 120–140 (1981)); Figures 11.4 and 11.6, (reproduced with some modifications or additions from *Rev. Geophys. and Space Phys.*, 18, 455–482 (1980)); Figure 11.8, (reproduced from *J. Geophys. Res.*, 87, 3697–3707 (1982)); Figure 12.13, (redrawn and modified from *Trans Amer. Geophys. Un.*, (EOS) 65, 18–19 (1984)); Figure 12.1 is directly redrawn from a figure published by the American Geophysical Union (*J. Geophys. Res.*, 84, 615–676 (1979)).

Department of Geological Sciences
University of Liverpool
March 1986.

Introduction

Palaeomagnetism is the study of the ancient magnetism in rocks, and has the ultimate objective of recovering the direction and strength of the geomagnetic field over the whole of geological time. Although broad constraints can now be placed on the magnitude of the Earth's gravitational field at some points on the geological time scale, palaeomagnetism remains the only important component of palaeogeophysics, because the ancient magnetism of the Earth is the only physical property which can be precisely and routinely recovered. When reasonable assumptions are made about the form and continuity of the geomagnetic field in the past, palaeomagnetism provides the only quantitative information for constraining the ancient positions and configurations of the crustal plates. This book concerns palaeomagnetism and what its results have told us about the continental crust. The subject of *archaeomagnetism*, which aims to describe and utilise the geomagnetic field during historical times, is not discussed here. *Geomagnetism*, the discipline concerned with the description and analysis of the Earth's magnetic field, is discussed only insofar as it impinges on the subjects of recovering the ancient geometry of the continental crust and interpreting the past behaviour of the crust and mantle.

The first six chapters describe the theory and practice of the subject, and are set at a level which will, I trust, be useful and understandable to undergraduate students and research workers in the Earth Sciences alike. Although the subjects of archaeomagnetism and geomagnetism are not specifically covered here, the groundwork material in these chapters may be of interest to workers in these fields. Much more detailed advanced treatises on the instrumentation of palaeomagnetism and the magnetic minerals can be found in Collinson's *Methods in Palaeomagnetism* (Chapman and Hall, 1983) and O'Reilly's *Rock and Mineral Magnetism*, (Blackie, 1983) respectively.

Chapters 7, 9 10 and 11 analyse the palaeomagnetic record over the whole of geological time, beginning with the Archaean and working forward to the Cenozoic. Since this is also a book about the continental crust, the implications of the past geometries and movements of the crust described by the palaeomagnetic evidence are examined at each stage. The results from the ocean basins (which form a significant part of the data base covering the past 190 Ma) and the total geometry of contemporary plate tectonics are not covered here. Indeed, this narrowing of the field which can reasonably be surveyed in a single text, compared with McElhinny's *Palaeomagnetism and Plate Tectonics*, reflects the great expansion of available information since 1973. My approach has been to present a general global survey which incorporates all of the continental data prior to Mesozoic times, and then treat the youngest part of the record in terms of mean polar paths. The palaeomagnetic literature is replete with discussions of the primary, secondary or mixed nature of the remanences in rock collections and it would have been easy to become embroiled in these detailed discussions without contributing significantly to the original literature. This would both have diminished the breadth of the material covered here and obscured the overview which this part of the book is aiming to provide for a general Earth Science readership. Accordingly, the discussion of palaeomagnetic data in Chapters 7, 9, 10 and 11 is addressed in general rather than specific terms. I have tried to avoid discussions of individual pole positions, since I believe that it is crucial that no interpretation should be based on individual poles, or even a small number of data points. Whilst the interpretation of some individual results utilised in this analysis will certainly change in the context of future palaeomagnetic and geochronological studies, the significance of these changes will be lost within the much larger body of data.

In spite of its undisputed importance, palaeomagnetism has enjoyed a somewhat chequered reputation with the Earth Science community at large. This reputation has too often been coloured by tectonic interpretations, which may fit small palaeomagnetic data sets perfectly, but satisfy neither the geological and geophysical constraints, nor the longer-term palaeomagnetic data base. It is my belief that any palaeomagnetic models must survived or be discarded on the basis of how well, or otherwise, they satisfy the geological and geophysical evidence. It is for this reason that I have aimed to bring together the subjects of palaeomagnetism and the continental crust into a single book. Hence the large-scale aspects of the geological and geophysical evidence, which are critical to the testing of the palaeomagnetic models, are discussed in general terms in the last six chapters.

I make no apologies for considering only the single-continent model for the Proterozoic interpretation in Chapter 7. How thorough the case for this model is, has not been widely appreciated. In part, this has resulted from the lack of an up-to-date listing of the global palaeomagnetic data, and a compilation of such a listing has gone hand-in-hand with the writing of this book. Originally it was hoped to include these listings in abbreviated form as an appendix to this text, in a similar form to the one which appeared with McElhinny's book. However, it became clear that the data base has now outgrown this possibility. These compilations are accordingly appearing as a separate Directory of Global Palaeomagnetic Data in the form of computer listings, which can be conveniently corrected and updated at regular intervals. Prior to the beginning of the present decade the apparent polar wander paths for Precambrian times were severely underconstrained. The data base has now, however, reached a level at which it can be used to test the most tightly-constrained model for the pre-Phanerozoic crust and, as I demonstrate in Chapters 7 and 9, this model can now be shown to be the correct one. If this had not been the case, very many more data would have been required to evaluate the alternative, and less tightly-constrained, models for the evolution of the crust. Thus, the palaeomagnetic solution to the nature of the primary continental crust has been derived piecemeal from an heterogeneous data set of very variable quality. This may not be aesthetically pleasing, but it does yield a solution which is eminently testable. As I show in Chapter 8, this solution is outstandingly well-founded in the geological record.

Chapter 9 also brings together the several geological and geophysical facets of the continental breakup and dispersal in Cambrian times. Although this event has long been inferred from the geological record, the separate analyses of the palaeomagnetic data and the subsidence history of the early Palaeozoic sea-boards have recently identified this episode as one of the most dramatic events in the Earth's history. The analysis of the Palaeozoic palaeomagnetic record in Chapter 10 has taken a quite different form from the one I originally envisaged. A survey of the global data both convinced me that the problem of remagnetisation in Palaeozoic sedimentary successions is a profound one, and that a sufficient primary record is preserved to define a true record of apparent polar wander over most of Palaeozoic times. Indeed it is now possible to demonstrate a conformity between orogenic episodes and palaeomagnetically-defined continental suturing. The Palaeozoic record has an important lesson to teach us: the concept of fixed continents, recording a constant field direction over long periods of time is an erroneous one. Clearly, it is now quite meaningless to talk about, say, the Devonian pole for Europe. We are dealing with a dynamic Earth and the apparent polar wander paths are seen to migrate more or less continuously over the whole of geological time.

The discussion of Mesozoic and Cenozoic palaeomagnetism in Chapter 11 is a fairly brief assessment of a time interval which has generally been studied in much greater detail than that of pre-Mesozoic times. In part this is a conscious attempt to present a more balanced perspective of the whole of geological time. In addition, this youngest evidence is best integrated with results from oceanic regions and analyses of marine magnetic anomalies to describe contemporary plate tectonics. However, whilst the analytical Chapters 7 to 12 are concerned mostly with the large-scale implications of palaeomagnetism, it seems likely that the major developments in this subject area over the next decade will come from applications to medium- and small-scale tectonic problems. Thus the major part of Chapter 11 is devoted to results from orogenic terranes, and in particular to regions which are tectonically active at the present time.

From early Proterozoic times onwards I follow the recent geological time scale of Harland *et al.* (1982). The eras and periods of this scale are summarised in the Appendix, while the epoch and age names may be consulted in Fig. 12.2 (Mesozoic and Cenozoic times) and Fig. 12.3 (Palaeozoic times). The proper position of the Archaean–Proterozoic boundary is still a subject of active debate, and rather than take it as the timing of the termination of greenstone-style tectonics (which certainly varies from place to place), I have elected to consider this boundary to be at 2700 Ma, which is when continental coherence can first be demonstrated from the palaeomagnetic record.

Most texts on geomagnetism concern themselves only with the recorded field over historical times. However, the concluding Chapter 12 considers what palaeomagnetism has to tell us about the geomagnetic field over the *whole* of geological time. This subject is also relevant to the long-term nature of mantle convection and these subjects are integrated here with the topics of geological cycles, crustal velocities and the geometry of supercontinents. Thus, Chapter 12 explores mainly the geophysical implications of the palaeomagnetic evidence, while Chapters 7 to 11 explore mainly the geological implications. Recently, much interest has been shown in the possibility of cycles in the behaviour of the Earth. The background to the vast literature on this subject is summarised in Williams's *Megacycles* (1981). In Section 12.8 I attempt to bring together the palaeomagnetic and geological evidence for cyclic behaviour. Whether or not this evidence allows us to recognise a 'beat' within the Earth is still very controversial, but the most important outcome of this discussion is to show that heat release from the Earth's interior, and its consequent expression in geological events, has been a pulsatory rather than a uniform phenomenon. This observation should radically alter our conception of palaeogeography and tectonics. The ensuing discussion of mantle convection (Section 12.9) is an elementary view of a large and complex field, but it leads into a discussion of the geoid. This in turn is extended to a consideration of one of the most exciting developments of the last two or three years, namely a recognition of the link between the geoid and the growth and shape of supercontinents. This evidence is discussed in some detail in Section 12.10 and has the potential to recognise some order in the vast range of continental movements defined by the palaeomagnetic evidence.

The palaeomagnetic data base has now grown very large (although there are still some geological periods and many regions which are poorly covered). It is expanding so rapidly that this may be the last attempt to synthesise the global data for the whole of geological time. With more than a thousand research workers concerned directly or indirectly with palaeomagnetic studies, and more than half of these residing in the Eastern Bloc, is is quite impossible to do justice to the vast number of individual studies. Hence the text represents very much a personal view and I have only been able to highlight those studies which have seemed to me to be particularly important; I can only offer my sincere apologies to those workers whose studies I have overlooked or misrepresented. In addition, there has had to be some economy in the fields covered. I have said virtually nothing about the history of palaeomagnetism, but the reader will find this subject well described in Tarling's *Palaeomagnetism* (Chapman & Hall, 1983). Furthermore no mention is made of the expanding Earth controversy. Although more VLBI (very long baseline interferometry) and low-orbit satellite results will be required to finally settle this issue, it is implicit in the analysis developed here that the diameter of the Earth is now approximately the same as it was in Archaean times.

1
Basic principles of rock and mineral magnetism

1.1 Definition of magnetic behaviour, diamagnetism, paramagnetism and ferromagnetism

A magnetic field is caused by the movement of electric charge. At an atomic level it can result both from the spin of the electrons about their axes, (this produces *spin dipole moments*) and by motions of the electrons in their orbits about the atomic nuclei (this produces *orbital dipole moments*). In the absence of external influences these magnets would normally be randomly oriented, but they respond to an external field in a way that depends upon the electronic configuration of the atoms and on the atomic structure of the substance in which they occur. The magnetic flux is related to the applied field H by the relationship $B = \mu_0 H$ in the SI system (see caption to Table 1.1 for an explanation of these units) where μ_0 is the permeability of free space. All material substances acquire a magnetic moment when placed in a magnetic field and this is their *magnetic susceptibility*. The magnetic moment per unit volume of a substance placed in the magnetic field M (the *intensity of magnetisation*, or simply the *magnetisation*), is related to the applied field H by the relationship:

$$M = kH = \frac{kB}{\mu_0} \tag{1.1}$$

where k is the *volume susceptibility*.

The magnetisation per unit mass, J, is related to the applied field H by the relationship:

$$J = \chi H = \frac{\chi B}{\mu_0} \tag{1.2}$$

where χ is the *mass susceptibility*.

Table 1.1 Magnetic susceptibilities of minerals and rocks

	Magnetic behaviour	Volume susceptibility (k) $\times 10^{-8}$ SI units*			Volume susceptibility (k) $\times 10^{-8}$ SI units*
1. Minerals			**2. Rocks**		
1.1 Oxides, hydroxides and sulphides			*2.1 Sedimentary rocks*		
Magnetite	FI	125,700–2,010,900 (620,000)	Coal		2·4
Hematite	CA + F	50–3,800 (680)	Sandstones		0–2,000 (40)
Pyrrhotite	F	125–628,400 (157,000)	Red sediments		0·5–5,000
Maghemite	FI	300,000	Shales		6–1,860 (60)
Ilmenite	P	3,140–377,000 (188,000)	Limestones		3–350 (31)
Chromite	P	300–11,800 (750)	Dolomites		0–94 (13)
Rutile	P	0·4			
Arsenopyrite	P	300	*2.2 Igneous rocks*		
Goethite (limonite)	A	80–280	Rhyolites		25–3,770 (1,120)
Pyrite	P	5–530 (160)	Granites		0–5,030 (250–470)
Galena	D	−3.3	Basalts		25–18,200 (7,500)
Cassiterite	P	110	Submarine Basalts		40–230
Sphalerite	P	1–70	Dolerites		100–16,300 (1,800–5,700)
Chalcopyrite	P	40	Gabbros		88–9,000 (7,200)
Cuprite	D	−1.1	Dunite		(513)
Quartz	D	−1.6	Pyroxenite		(650–13,200)
			Peridotite		9,600–19,600 (1,630)
1.2 Others			Andesite		(1,700)
Gold	D	−3·5	Diorite		57–12,600 (8,800)
Silver	D	−2·7			
Graphite	D	−12·8	*2.3 Metamorphic rocks*		
Rock salt	D	−1·4	Slates		0–3800 (50–600)

Anhydrite	D	−1·5
Gypsum	D	−1·5
Sulphur	D	−1·2
Calcite	D	−1·3
Rhodochrosite	P	464
Siderite	P	487
Magnesite	D	−1·4
Water	D	−0·9
Kaolinite	D	−4·2
Ice	D	−0·8
Illite	P	4·2
Montmorillonite	P	6
Chamosite	P	231
Fe-montmorillonite (nontronite)	P	176
Cordierites	P	20–110
Garnets	P	148–860
Biotite	P	229–333
Muscovite	P	30–200
Amphiboles	P	49–320
Pyroxenes	P	180–325
Fayalite (Fe olivine)	P	553
Epidote	P	70–100
Phyllites		(100)
Greenschists		9–230 (90)
Amphibolites		7–1,250 (80–150)
Granulites		30–1,500 (320)
Gneisses		12–2,500
Serpentinites		310–1,800 (360)

The susceptibility values are the range of recorded values from many sources in the literature, and the value in brackets is the average from one or more studies. The reported susceptibility values show a variation which depends primarily on the content of the first five minerals listed in this table; the variable values quoted for some mineral species are probably caused by small amounts of impurities of these minerals.
* In the SI (Système International) units used in this book, magnetic forces are measured in terms of the force between two poles in free space. The magnetisation (M) due to a material, and its intensity of magnetisation per unit volume (J) are both measured in amperes per metre (Am^{-1}); they are equivalent to the gauss and oersted in the unrationalised c.g.s. or M.K.S. systems of units. M and J are related to magnetic induction (B) in the magnetised medium by $B = \mu_0 (H + J)$ where μ_0 is the permeability of free space and is equal to 4×10^{-7} henries m^{-1}, B is measured in tesla (1 tesla = 10^4 gauss and 10^9 gammas, and 1 gamma is equivalent to one nanotesla). Hence magnetic fields are expressed in tesla (T), and J is expressed either as the magnetisation per unit volume (Am^{-1}) or the magnetisation per unit mass ($Am^2 kg^{-1}$). The volume susceptibility, k, which is dimensionless in the SI system, has a magnitude 4π times greater than in the c.g.s. and M.K.S. systems in which it is measured in gauss/oersted. Quartz with a mass susceptibility of -0.6×10^{-6} e.m.u./g, and a volume susceptibility of -1.6×10^{-6} e.m.u. in the c.g.s. system, has a mass susceptibility of -0.6×10^{-8} $Am^2 kg^{-1}$, and a volume susceptibility of -1.6×10^{-8} in SI units.

The external magnetic field exerts a force on each of the orbiting electrons which makes the orbit precess. The net result of this is to create a magnetic moment which tends to oppose the applied field. The acquired magnetisation thus has a negative value and the susceptibility is negative. This is the phenomenon of *diamagnetism*, and it is present in all substances, although it tends to be obscured by other, larger, effects in some materials. These latter effects are referred to as *paramagnetism* and *ferromagnetism*.

In addition to the orbital motion of the electrons, they spin on their own axes and have magnetic dipole moments. Although the strengths of these dipoles are not altered when an external field is applied, they tend to line up in the direction of the external field and produce an increase in the magnetism. Only substances with incomplete electron shells show this *paramagnetic* effect. If there are equal numbers of oppositely-directed electron moments, the spin magnetisms completely cancel and only the orbital magnetism remains important to produce diamagnetic behaviour. However, if the atom has an odd number of electrons the spin dipoles do not cancel and the paramagnetic effect dominates. Thus atoms with an even number of electrons tend to be diamagnetic and those with an odd number are paramagnetic. Table 1.1 summarises susceptibility data for common natural diamagnetic and paramagnetic materials: the paramagnetic effect is usually much larger than the diamagnetic one, but only a few common ions have appreciable paramagnetic properties and these belong to the transition series of elements. They owe their particular properties to the small differences in the energy levels of the two outer electron shells, so that varying numbers of electrons can be involved in forming chemical bonds. One or more electrons can be released from the penultimate electron shell in addition to the outermost electrons, so that up to five unpaired electrons can occur within single atoms of Mn^{2+}, Fe^{2+} and Fe^{3+} and contribute to a strong paramagnetic effect. Thus the paramagnetism, and hence the susceptibility, of rock-forming minerals is primarily a function of the iron and manganese content (Table 1.1). The other transition elements are much rarer in nature. The susceptibilities of paramagnetic and diamagnetic materials are practically independent of the applied field magnitude but paramagnetism is strongly temperature-dependent because thermal fluctuations tend to disorient the alignment with the applied field. The *Curie Law* states that paramagnetic susceptibility is inversely proportional to absolute temperature.

Paramagnetism and diamagnetism are types of magnetism which exist only in the presence of an external field: once that field is removed the magnetism disappears. A few materials belong to a third class and exhibit *ferromagnetic* behaviour. They have a much larger susceptibility which is strongly dependent on the strength of the applied field (Table 1.1), and they are of primary interest in the study of palaeomagnetism because they retain a memory of the applied field after it is removed. The magnetic behaviour of a ferromagnetic material is defined by the *hysteresis loop* (Fig. 1.1). At low fields the behaviour is reversible and if the field is removed the magnetism is lost. The *initial susceptibility* is then defined as:

$$\left(\frac{dJ}{dH}\right)_{H=0} = \chi \qquad (1.3)$$

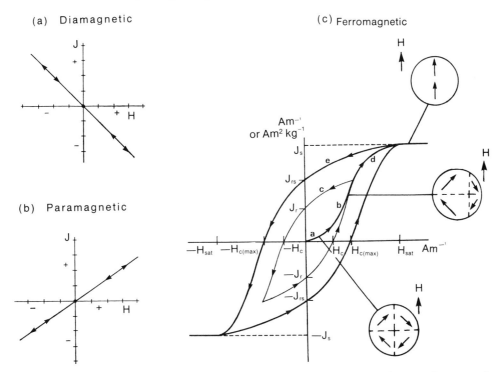

Figure 1.1 Magnetisation (J) as a function of applied field (H) for (a) diamagnetic, (b) paramagnetic and (c) ferromagnetic material. In ferromagnetic materials, J initially increases linearly with applied field; this part of the curve is reversible so that if H is reduced to zero, J also falls to zero. If H is increased further the slope of the curve decreases and if H is then reduced to zero, J does not fall to zero, but follows the path acquiring an isothermal remanence J. A field $-H_c$ in the opposite direction is required to reduce J to zero. An increase in the applied field beyond a value H_{sat} produces no further increase in J which has then reached its saturation value. The circles show a schematic representation of the domain structure: initially magnetisation increases through the movement of domain boundary walls so that domains parallel to H grow at the expense of anti-parallel domains; ultimately the magnetisations of whole domains are rotated.

When the field is increased, an irreversible change in the magnetisation takes place: removal of the field is not accompanied by disappearance of all the magnetisation, and the spontaneous magnetisation reaches a saturated value J_s, (the *saturation magnetisation*) at a finite value of the applied field H. If the applied field is reversed, J decreases along a different curve so that a finite amount of magnetism, J_r (the *remanent* or *residual* magnetism) is retained when H = 0. An increasing magnetic field in the opposite direction reduces this residual magnetisation until J becomes zero when $H = -H_c$, the *coercive force*.

Diamagnetic and paramagnetic effects are explicable in terms of the behaviour of isolated atoms. Ferromagnetic behaviour is due to the interactions between atoms within the framework of a crystal lattice. Again these effects are most important in the elements of the transition series (of which iron is the first and most important member) because these elements have the inner orbital electrons much closer together than the outer valency electrons. The latter electrons result in several unbalanced spins within each atom; they can move freely through the material and interact strongly from atom to atom. These interactions are coupled either between adjacent atoms by *direct exchange*, or via an intermediate anion by *superexchange*. The potential energy of the interaction is termed the *exchange energy*. It will normally be minimised when the atoms are linked up so that the spins of adjacent atoms are antiparallel and their magnetic moments cancel each other out. However, in a few substances, this energy can be minimised by an alignment of the atoms with their dipoles parallel and additive. They then retain a strong resultant magnetisation or *remanence* when the external magnetic field is removed. This is *ferromagnetism*.

1.2 Types of ferromagnetism

Although the transition elements iron, cobalt and nickel can be true ferromagnetics in the elemental form, the natural materials which carry a magnetic remanence are not simple ferromagnetic substances. Instead, the ordering of the spin dipoles is complex, and both positive and negative exchange effects may exist in the same crystal structure. If there is a small overlap between electron orbits, the coupling between them is antiparallel with the result that the lattice is divided into two parts magnetised in opposite directions. In *antiferromagnetic* materials the magnetisations of these two lattices are exactly balanced and there is no net external magnetic field (Fig. 1.2). Sometimes the equal magnetic moments of the two sub-lattices are not exactly antiparallel and then there is a small resultant magnetisation in a direction bisecting the two dipole directions. This is described

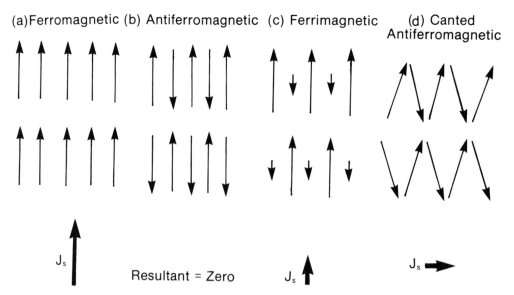

Figure 1.2 The coupling of spin magnetisations in materials carrying a permanent remanence. J_s is the resultant (spontaneous) external magnetic field.

as *canted antiferromagnetism*. If the atomic moments of the two sub-lattices are unequal there is a resultant magnetic moment which is effectively a weak ferromagnetism, and this is known as *ferrimagnetism* (Fig 1.2).

Ferromagnetic, antiferromagnetic and ferrimagnetic behaviours are strongly dependent on interatomic spacing. As the temperature is increased the lattice expands and eventually the critical interatomic distances at which exchange and superexchange couplings can take place are exceeded. The spontaneous magnetisation then reduces to zero at the *Curie temperature*, although the material continues to exhibit paramagnetism and diamagnetism at higher temperatures. The critical temperature at which coupling between the opposing sub-lattices in an antiferromagnetic material takes place is called the *Néel temperature*.

1.3 Domain structure

The alignment of magnetic dipoles in a ferromagnetic material has been attributed already to exchange forces between the unpaired electrons. In reality, this is only one of the sources of magnetic energy within a magnetised crystal structure. Energy is also required to maintain the orientations of dipoles; it depends on the crystal structure of the material and on the shape of the particle. This *magnetostatic energy* (Table 1.2) is another energy required to maintain the distribution of magnetic dipoles in the absence of an external field. The balance of all the forces operating within a magnetic grain may cause it to become subdivided into smaller volumes or *domains* of uniform magnetisation, such that the total overall energy has the minimum attainable value. Adjacent domains have contrasting directions of magnetisation and are separated by narrow *domain walls* from each other. In pure crystalline materials the domains may be several μm in size but in the imperfect grains found in natural materials they are usually much smaller. Actual domain sizes depend not only on impurities and defects in the lattice, but also on the shape of the grain; they are of the order of $0\cdot1-1$ μm in magnetite and $10-100$ μm in hematite, and the domain walls between are estimated to have a thickness of ca $0\cdot01$ to $0\cdot1$ μm.

In any crystal structure there will be one or more preferred directions along which the magnetisations may be stably aligned. These are the *easy directions* of magnetisation and are controlled by the crystal symmetry. Spins prefer to be oriented along them because they then have a minimum *magnetocrystalline energy* (Table 1.2). In the domain walls separating the adjacent regions of uniform magnetisation from one another the spins are gradually canted around from one direction to the other over a distance of about 1000 atoms. Since the spin magnetic moments within the wall do not lie in these preferred directions, they require additional energy to overcome the crystalline anisotropy. Thus it requires a definite amount of energy (the *wall energy*) to form a domain wall, and the subdivision of a grain only continues as long as the energy required for the formation of an additional boundary wall is less than the consequent reduction in the magnetostatic and magnetocrystalline energies (Fig. 1.3).

The hysteresis loop (Fig. 1.1) can now be considered in the context of domain structure. Initially, the domains will arrange themselves so that the internal field is essentially zero. When a field is applied, domains which are oriented close to

Table 1.2 The energies present in a magnetised crystal lattice

Type of Energy	Cause	Method(s) by which the energy can be minimised
1. Exchange	Interaction between unpaired valency electrons of adjoining atoms, in reality an interaction between the wave functions of the ions.	By alignment of the ionic spins antiparallel to one another.
2. Magnetostatic	A *shape anisotropy* effect sometimes called the energy of self demagnetisation. The organisation of the elementary dipoles in a material will, depending on the shape of the body, be distributed to create an internal field opposing the external field.	By reducing the size of the magnetised volume or by increasing the length:width ratio of the magnetised volume in the direction of the applied field.
3. Magnetocrystalline anisotropy	A *crystalline anisotropy* effect. It is easy to magnetise materials along some crystallographic axes and hard along others. The energy is the difference between the work done in magnetising the crystal along the hard and easy directions.	By aligning the domain magnetisations along the easiest axis of magnetisation.

4. Domain wall	The change in direction of magnetisation between one domain and the next occurs across a finite zone in which the spins swing around progressively from one alignment to the other. These spins require extra energy to overcome the magnetocrystalline effect.	By minimising the wall area
5. Magnetostrictive	Magnetic interactions between adjoining spins usually produce a mechanical deformation, and hence a strain, called *magnetostriction*. In addition, impurities in the lattice cause dislocations and produce local strains which act as barriers to changes in magnetisation.	By removing internal strains or impurities and by removing external strains.

(Modified after Strangway (1970)).

the applied field direction, grow at the expense of those with magnetisations at a large angle to the field, and this displaces the boundaries between the walls. Along the initial part of the hysteresis curve, the magnetisation proceeds by small and reversible boundary displacements. The magnetisation which appears in each grain will be just sufficient to produce an internal field equal and opposite to the external field; this maintains the free energy at a constant minimal amount. Along the steeper part of the path, the magnetisation is due to larger irreversible boundary displacement: as the walls are forced over barriers (imperfections or inclusions) in the lattice they cannot return to their original positions when the field is removed. In the highest part of the curve, magnetisation proceeds by the rotation of all the domains into the applied field by overcoming both the magnetostatic and magnetocrystalline energies; large fields are required for this, and the increase in magnetisation is relatively slow. In magnetically 'hard' (high-coercivity) materials the movement of domain walls is difficult and usually inhibited by lattice impurities or imperfections. In magnetically 'soft' (low-coercivity) materials, the domain walls are easily moved, a condition which typically applies to pure and ordered crystal lattices.

Any particle smaller than the size of a domain must be *single domain* (SD) since it is not big enough to contain walls. At some critical size the particle can subdivide into two or more domains. The particle is then *multi-domain* (MD). In a simple structure with parallel domain walls (Fig. 1.3) the lattice distortions are equal and oppositely directed and the *magnetostrictive energy* (Table 1.2) is zero. However, the magnetostatic energy is then large, and a domain structure with

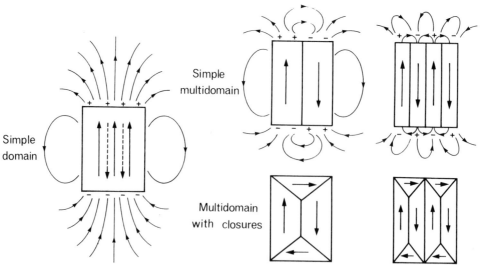

Figure 1.3 The subdivision of a ferromagnetic grain into domains. The domain magnetisations are indicated by solid, straight arrows, the dashed arrows indicate the demagnetising field and the curved arrowed field is the external field. This subdivision permits alignment of the spins in adjacent atoms everywhere except across the boundary walls and prevents the formation of strong magnetic fields in and around the grain. The structure actually achieved is a compromise between minimum exchange coupling energy (uniform alignment everywhere) and minimum magnetostatic energy (random alignment).

closures (Fig. 1.3) is more likely to occur than one with a simple subdivision by parallel walls, because the sum of all the energies is smaller. In fact, in large MD grains the domain walls can attain positions in which the magnetostatic energy is zero. Such grains can be readily demagnetised to have no overall magnetic moment and the hysteresis effect is small.

In general, the hysteresis effect increases with decreasing grain size. The magnetisation induced by a given external field in small grains is less than would be induced in large grains of equivalent concentration but when the field is removed a stronger residual magnetism remains. In SD grains the movement of domain walls no longer plays a part in the magnetisation phenomena: changes can only occur by rotation or inversion of the grain moments into alignment with the direction of an applied field. Since a large energy barrier exists between the directions in which the magnetisation is stable, a strong field is required to do this. The magnetisation of an SD grain can only be destroyed by heating it above its Curie point, although the magnetisation can be reversed, provided that a large opposing field is directed near to the preferred axis of magnetisation. In grains with only few domains, the walls are few in number and have their possible movements restricted by the size and shape of the grain; they tend to behave in a similar way to SD grains and are referred to as *pseudo-single domain* (PSD) grains.

If particles are very small then even at temperatures below the Curie temperature, the thermal vibrations of the lattice have energies comparable to the total magnetic energy. In some cases it is not possible for the lattice to retain a stable magnetisation even at room temperature and the particles are described as *superparamagnetic*. This effect is proportional to temperature and inversely proportional to particle volume. The critical volume (at a given temperature) below which thermal agitation destroys the magnetism in SD particles is the *superparamagnetic threshold*. At this point a small increase in volume or a small decrease in temperature can change a particle from superparamagnetic to SD and a remanence is then *blocked* into the particle. A distinction should be recognised between this threshold temperature and the Curie temperature: in reality unblocking/blocking happens at a temperature which is usually well below the Curie point of the material, and although the material behaves like a paramagnetic material above its blocking temperature(s), it is much more sensitive to external magnetic fields (Section 3.3).

Since large multi-domain grains are so readily demagnetised and since very small superparamagnetic grains can retain no permanent moment, it is probable that the magnetic remanence in rocks of interest in palaeomagnetism resides in the medium-sized SD and PSD grains. This prediction is examined in more detail in Chapters 3 and 4.

1.4 The acquisition of magnetism by rocks

The permanent magnetism present in a rock is termed the natural remanent magnetisation (NRM). It is usually the vector resultant of the *primary magnetisation* acquired when the rock was formed and the *secondary magnetisation* acquired during subsequent geological time. The secondary magnetisation may itself include several components which are generally imposed on the rock by later thermal and tectonic events. The major part of all palaeomagnetic investigations involves the recognition and separation of these magnetic components (Chapter 6). Unfortunately, to describe a magnetisation as 'primary' implies that it was acquired when the rock was formed. This is often a point of uncertainty and many workers prefer to assign the term *characteristic magnetisation* to a remanence found throughout a geological formation, because it avoids genetic connotations. In igneous rocks a primary magnetisation may be acquired by cooling from high temperatures through the blocking temperatures of the magnetic minerals, and it is then described as a *thermoremanent magnetisation* (TRM). A TRM is not all acquired at the Curie point, but continues to be acquired from the Curie point down to ambient temperatures. Fractions of the total remanence can thus be considered to be blocked between temperature intervals. If the rock is heated at a later stage to some temperature, T, below the Curie point, T_c, the fraction of the remanence originally acquired between T and the ambient temperature is unblocked and a new *partial thermal remanence* (PTRM) is acquired when the rock cools again; this new component will coexist with the residual component acquired between T_c and T. PTRMs are additive so that the PTRM acquired between say T_1 and T_3 is equal to the sum of the PTRMs acquired between T_1 and T_2 and T_2 and T_3, where $T_1 < T_2 < T_3$, but the PTRMs acquired within a given temperature interval are not the same for all temperatures: MD grains acquire most of their TRM over a range of a few tens of degrees immediately below the Curie point, while SD grains tend to acquire their TRM over a fairly broad temperature range. When magnetic materials are created at elevated temperatures but below their Curie points — such as magnetite in serpentinites and hematite produced by hydrothermal activity — they acquire a magnetism which has similar properties to a TRM; this is referred to as a *thermochemical remanence* (TCRM).

The magnetisation acquired when chemical changes produce new magnetic materials (or change them from one form to another) at low temperatures is termed *chemical remanent magnetisation* (CRM). Grains with a primary CRM or TRM may both form detrital fragments as a result of weathering, and these grains will tend to be aligned into the Earth's magnetic field during sedimentation. This effect is most pronounced when the particle size is small so that the settling time is long. The sediment can then acquire a *detrital remanent magnetisation* (DRM).

When a rock is subjected to a local magnetic field it will acquire a magnetisation defined by the hysteresis behaviour, which may reach the maximum (saturation) remanence if the local field is large enough. This is described as an *isothermal remanent magnetisation* (IRM) and is simply the residual magnetism left after an external field is applied and removed from a magnetic material. When lightning strikes the ground it essentially represents a strong line current, and creates a circular field about the point of impact which imparts a large IRM over a local area. Rocks with lightning-induced IRMs are recognised by their large, and usually unstable, low-coercivity magnetisations; the effects are confined to no more than a few square metres of outcrop area and can normally be excluded from the palaeomagnetic analysis.

Over geological time, rocks are exposed to the small, ambient magnetic field of the Earth in which they can acquire a time-dependent IRM called a *viscous remanent magnetisation* (VRM). Thermal vibrations of the lattice occur spontaneously, and they cause the domain walls to shift slightly and irreversibly; the net result is to expand domains oriented close to the ambient field direction, at the expense of those with a direction remote from the field. At any fixed temperature the effect is a logarithmic function of time, but as temperature is increased the acquisition of VRM is speeded up. The grain-sizes most sensitive to VRM acquisition are those near the superparamagnetic threshold. If a VRM is an important component of the remanence in a rock, it will be aligned close to the present geomagnetic field direction, and since the field was last reversed 700,000 years ago (Section 12.2), the VRM must then have been acquired over a timescale shorter than this. The ability of a rock to acquire VRM can be tested for by measuring the rock samples after intervals of laboratory storage in a controlled field direction. Most commonly samples from a single rock unit possess variable amounts of VRM compounded with a characteristic remanence, and the resultant NRMs have directions smeared between the present field direction and the characteristic remanence direction. A direction close to the characteristic remanence direction indicates that VRM is unimportant and the rock is a promising recorder of ancient remanence.

1.5 Total magnetisation and Köenigsberger ratio

In the natural situation the rock magnetisation, $\bar{J}$, is the vector resultant of the remanent magnetisation, $\bar{J}_r$ (which, as we have seen, may itself comprise several components), and the induced magnetisation, $\bar{J}_i$, which is the component induced in the applied field and is present only as long as that field is present:

$$\bar{J} = \bar{J}_r + \bar{J}_i \tag{1.4}$$

The components of J cannot be separated by magnetic surveys but require sampling and laboratory measurement of the susceptibility and the NRM.

The ratio of the remanent to the induced magnetisations is the *Koenigsberger ratio*, Q:

$$Q = \frac{\bar{J}_r}{\bar{J}_i} = \frac{J_{NRM}}{\chi H} \tag{1.5}$$

Rocks with high Q values ($>0 \cdot 1-1$) tend to be magnetically stable and good recorders of the ancient geomagnetic field, while rocks with low Q ($<0 \cdot 1$) values are unstable. This is a general reflection of domain structure: in general, rocks with $Q > 0 \cdot 5$ are dominated by SD grains, and rocks with $Q < 0 \cdot 5$ are dominated by MD grains, although this relationship is only an approximate one (Stacey, 1963).

It will be apparent from Table 1.1 that the intensities, susceptibilities (and hence Q ratios) of each rock type may vary by up to several orders of magnitude; each rock group, however, tends to be characterised by a distribution which approximately reflects the content of ferromagnetic minerals. Because of the wide distribution of values, and because only positive values occur, the Gaussian distribution is a poor fit to these data, and arithmetic mean values are not meaningful. A logarithmic normal distribution provides a much better fit (Irving, Molyneux & Runcorn, 1966); values from a single collection usually show a single peak when plotted on a logarithmic abscissa scale and normal curves can be fitted to the data to estimate the mean and standard deviation for the magnetic properties of the whole collection.

2
The magnetic minerals

2.1 Introduction

The Earth contains about 25% of iron by weight. Being naturally siderophilic, it tends to be concentrated as a native element in the core, and in this form it is the most important magnetic constituent of meteorites and all lunar rocks. However, the other common terrestrial elements are not sufficiently abundant within the Earth to use up all of the sulphur and oxygen, so that iron is forced to become chalcophilic and use up the sulphur, and ultimately to become lithophilic* and combine with the oxygen. It is present as sulphides, silicates and oxides in the mantle and crust; in the latter it averages about 5% by weight and 0.43% by volume. Native iron is unstable in the presence of water in natural Eh–pH conditions; only one possible occurrence has been reported in crustal rocks (Deutsch, Rao, Lauret & Seguin, 1972) and its properties (see Collinson, 1983) do not concern us here.

The iron minerals responsible for the magnetic remanence in most crustal rocks plot within the ternary system $FeO-TiO_2-Fe_2O_3$ (Fig. 2.1a). The oxides in this system have solidus temperatures of 1600–1400 °C, but because they comprise only a small percentage of most igneous magmas, they crystallise with the silicate phases at temperatures in the range 1200–800 °C. They usually crystallise as an early phase in basic magmas, but often as a late phase in acid magmas, especially in granites. Crystallisation is always complete by 750 °C, and well in excess of temperatures at which the minerals in this system can acquire magnetic remanence. The compositions of the natural oxide minerals in igneous rocks are shown in Fig. 2.2 and illustrate the tendency for natural titanomagnetites to be oxidised towards the titanohematite trend; they are more correctly referred to as titanomaghemites or cation-deficient spinels. The term 'titanomagnetite', however,

---

* Footnote: Chalcophilic and lithophilic ions have a natural affinity for combining with the large anions S and O respectively; siderophilic ions tend to remain in the native state.

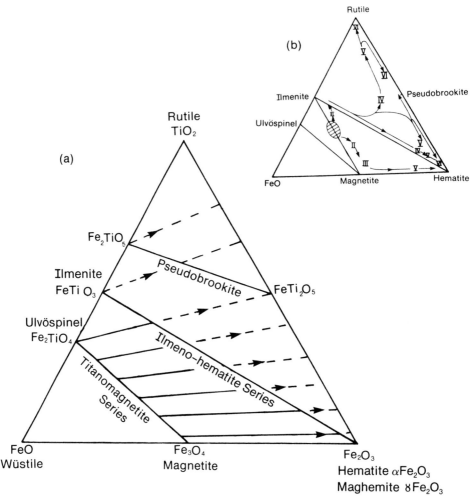

Figure 2.1 (a) The ternary system incorporating the iron-titanium oxides. The initial compositions may occur in the solid solutions magnetite-ulvöspinel (cubic structure) and ilmenite-hematite (rhombohedral structure) as shown in Figure 2.2. Commonly these grains exsolve into the end members of the solid solution series to produce a composite crystal. In addition, if the initial solution contains surplus oxygen, or existing titanomagnetite grains react with residual fluid phases containing oxygen, the composition changes towards the ilmeno-hematite series or beyond to the rutile series. The directions of increasing oxidation are shown by the arrowed lines. The actual oxidation may follow more complex paths and (b) shows the trends of oxidation of high temperature titanomagnetite in basalts defined by the oxidation indices I to VI of Wilson and Haggerty (1966) described in section 4.1. The bifurication of trends may indicate either separate minerals evolving as textures within the original titanomagnetite grain, or its complete replacement; after Watkins and Paster (1971).

is generally used to embrace both the true titanomagnetites and their oxidised equivalents (see also Table 3.1). As the temperature of crystallisation is reduced from basic to acid compositions, and from extrusive to intrusive environments, the compositions of these phases tends to become closer to pure magnetite in composition (Fig. 2.2). In practice, oxide phases only retain these homogeneous compositions if they are very rapidly cooled, and the high-temperature solid solutions normally exsolve to form intergrowths of a cubic phase close to magnetite in composition and a rhombohedral phase of the ilmenohematite series. This exsolution proceeds most vigorously at the post-crystallisation *deuteric* stages at temperatures of about 800–500° C; it is promoted by high oxygen fugacities in the residual fluids concentrated by magmatic crystallisation and circulated through the rock as it consolidates. However, it can probably also proceed at low temperatures and at a molecular level over geological time scales. The ultimate result of oxidation is to shift the composition of the oxide phases from the ulvöspinel–magnetite join in Fig. 2.1a in the direction of the arrows towards the hematite–rutile join, so that oxidation leads to the formation of hematite, rutile or anatase (TiO_2), and pseudobrookite (Fe_2TiO_5). It should not, however, be inferred that oxidation will follow this linear path. It generally proceeds through a series of intermediate phases, as in the sequence of deuteric alteration phases of Fig. 2.1b, and is described in more detail in the first part of Section 4.1.

Any of the oxide phases in igneous rocks may survive as detrital components in sedimentary rocks. In this environment, they are normally metastable and may

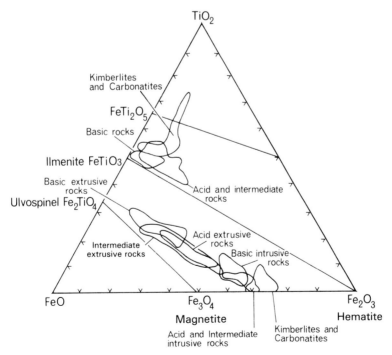

Figure 2.2 The compositions of natural titanomagnetites and ilmenites in igneous rocks. The compositional fields are calculated from approximately 130 analyses listed in Haggerty (1976).

alter to form new phases controlled by the redox potential (Eh) and the acidity (pH) of the environment. These changes are described in the second part of Section 4.2. Metamorphic conditions are characterised by elevated temperatures for long periods of time: they do not simply reverse the changes which take place in the deuteric and hydrothermal environments. Effects on the magnetic oxides differ in some important respects from the magmatic environment; they are described in Section 4.3.

2.2 The magnetite–ulvöspinel series

Magnetite has cubic symmetry and possesses an inverse spinel structure. The iron cations are located within two different lattices in the structure known as A and B. Of the 24 iron ions in one unit cell, sixteen Fe^{2+} and Fe^{3+} ions occur in the B lattice in six-fold (octahedral) coordination with the oxygen anions, while only eight Fe^{3+} ions occur in the A lattice in four-fold (tetrahedral) coordination with the oxygens. Since there are two cations in the B lattice for each one in the A lattice and their atomic moments are oppositely directed, there is a net magnetic moment giving the structure ferrimagnetic properties. Above a temperature known as the *Verwey transition* ($-155°$ C) the electrons can move freely between the octahedral sites and permit intermediate oxidation states; below this transition, electron exchange is no longer possible and the structure converts from cubic to orthorhombic symmetry. In addition, structural changes in the lattice cause the magnetic anisotropy (Section 3.1) to fall to zero at $-143°$ C, and below this temperature it has an increasing negative value; at this point the 'easy' axis of magnetisation of the magnetite structure switches from (100) to (111).

Ulvöspinel is known mainly as exsolution blebs in magnetite ores, but is probably quite widespread, especially in titanium-rich igneous rocks. It has an inverse structure, with Fe^{2+} in the A lattice and Fe^{2+} and Ti^{4+} in the B lattice. Since Ti^{4+} contributes no magnetic moment and Fe^{2+} cations are oppositely directed, this mineral is paramagnetic at room temperature, but it acquires a weak antiferromagnetism at temperatures below $-153°$ C. There is complete solid solution between magnetite and ulvöspinel at temperatures above $560°$ C, but at lower temperatures the solid solution is much more restricted and the two phases tend to exsolve (Fig. 2.3b). As the proportion of the ulvöspinel molecule increases, the cell dimensions increase linearly from a = 8.39 Å for magnetite to a = 8.53 Å for ulvöspinel. The Curie point decreases nearly linearly from a value of $578°$ C for pure magnetite. Although the titanomagnetites of many extrusive igneous rocks (Fig. 2.2) have predicted Curie temperatures (Fig. 2.5) of only $200°$ C or so, and would thus be expected to possess a TRM with a short relaxation time (Section 3.5), such low Curie temperatures are usually only observed in very recent and rapidly-chilled rocks (in the carapace of pillow lavas for example). Exsolution to ferrimagnetic magnetite and paramagnetic ulvöspinel probably results from a variety of processes over geological time so that the Curie points observed in most pre-Cenozoic rocks are close to pure magnetite in composition (second part of Section 5.6 and Fig. 5.10). The oxidation of titanomagnetite is accompanied by a rise

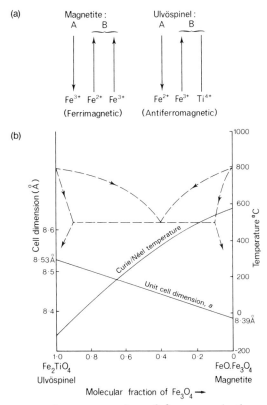

Figure 2.3 *(a) Diagrammatic representation of the magnetisation vectors in the A and B lattices of the inverse spinel structure of magnetite and ulvöspinel. (b) The ulvöspinel–magnetite series showing phase boundaries along which exsolution takes place into the end members during cooling, and the variation of cell size and the Curie/Néel point with composition (after Nagata 1961, and Tarling 1983).*

in Curie temperature; the effect is summarised for the compositional field between the two end-member series in Fig. 2.5.

With a saturation magnetisation of 90–93 $A.m^2 kg^{-1}$, magnetite has the strongest magnetic properties of any common crustal mineral. Saturation is achieved in fields of 50–150 mT. It is magnetically anisotropic with 'easy' and 'hard' directions of magnetisation along the (111) and (100) directions, respectively, but this effect is usually weak compared with the anisotropy associated with the shape of the magnetite grains. Coercivity increases with decreasing grain-size, from 3×10^3 Am^{-1} for 30 μm grains to 1.5×10^5 Am^{-1} near the SD:MD transition size (O'Reilly, 1983). Since the multidomain–single domain boundary in magnetite occurs close to the superparamagnetic boundary where particles are rendered magnetically unstable by thermal agitations, it seems probable that the stable NRM is carried by elongated or PSD grains (Section 3.3). Magnetic and optical properties are summarised in Table 2.1.

Table 2.1 Summary of the main properties of the opaque phases commonly encountered in the study of magnetic minerals.

	Density kg/m^3	Hardness	Saturation Magnetisation* (A.m^2 kg^{-1})	Reflectivity in air (%)	Characteristics in reflected light
Magnetite	5160–5220	$5\frac{1}{2}$–$6\frac{1}{2}$	90–93	21	Isotropic, weak grey to tan colours
Ulvöspinel	4800	$5\frac{1}{2}$–$6\frac{1}{2}$	paramagnetic	16	Isotropic, tan brown–reddish brown
Hematite	5300	5–6	0.2–0.4	25–28	Anisotropic, weak light grey colour, bluish tint, deep red internal reflections
Ilmenite	4800	5–6	paramagnetic	17–19	Anisotropic, white to brownish grey, pleochroic
Rutile	4250	$5\frac{1}{2}$–$6\frac{1}{2}$	paramagnetic	21	Anisotropic, medium grey colour
Pyrrhotite	4580–4650	$3\frac{1}{2}$–$4\frac{1}{2}$	1–20	37–42	Anisotropic, pale cream colour
Pyrite	5010	6–$6\frac{1}{2}$	paramagnetic	54	Isotropic, pale yellow colour
Maghemite	4700–4900	$5\frac{1}{2}$–$6\frac{1}{2}$	80–85	23	Isotropic, blue to grey
Goethite	~4300	$5\frac{1}{2}$–$6\frac{1}{2}$	≤ 1	13–18	Anisotropic, blue grey to yellow or brown grey, brown–yellow internal reflections.

* At room temperature. Reflectivity is the fraction of the incident light which is reflected from a polished grain surface and is measured by a prism photometer or a photoelectric cell; relative magnitudes are used qualitatively in routine work. Colour descriptions of the less distinctively coloured minerals (magnetite, ulvöspinel and hematite) tend to be subjective and other shades may be observed. The other phase commonly encountered in the study of magnetic minerals is pseudobrookite (density 4140 kgm^3, hardness $5\frac{1}{2}$–6, orthorhombic); it is distinguished from rutile by its dark grey colour. For comparative colour identification of these minerals see Battey (1967).

2.3 The hematite–ilmenite series

The ilmeno-hematites (αFe_2O_3–$FeO.TiO_2$) form solid solutions above 980° C, and at lower temperatures the intermediate compositions are represented by intergrowths of the end-members. Ionic replacement has the same effect as in the titanomagnetite series with Ti increasing the cell dimension and decreasing the temperature at which magnetic remanence is acquired (Fig. 2.4). The crystal structure is rhombohedral with a lattice composed of oxygen anion layers and Fe^{3+} and Ti^{4+} cation layers in six-fold coordination, each parallel to the axis of three-fold symmetry. Oppositely directed Fe^{3+} ions in A and B sub-lattices give a basic antiferromagnetic structure, but there is a small imbalance in the opposing spins. This canted antiferromagnetism is attributed to structural defects, but its origin is not completely clear. It results in the acquisition of a weak, very stable magnetism with a saturation remanence of 2–4 A. $m^2 kg^{-1}$ when hematite is cooled through its Néel temperature. It is observed between the Néel point of 680° C (although values of up to 710° C have been reported for this latter transition) and the *Morin transition* at −10° C at which point the ionic spins flip from the basal planes to the *c* axis; below this temperature the canting vanishes and hematite is purely antiferromagnetic.

A second source of remanence is also observed in natural hematite. It takes the form of a weak isotropic ferromagnetism, and is present both above and below the Morin temperature. It is believed to be due to interactions between the antiferromagnetism and either lattice defects or compositional impurities. This 'defect' ferromagnetism can be altered by stress (Dunlop, 1971). In fine-grained hematite the canted moment has higher coercivity than the defect ferromagnetism, but for

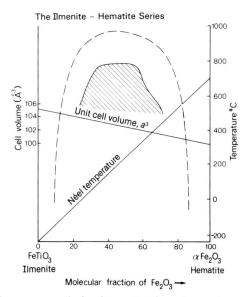

Figure 2.4 *The solvus curve of the hematite-ilmenite series and the variation of cell volume and Néel temperature with composition. Note that the miscibility gap in this series is not yet well understood (Merrill 1975) but is estimated by Lindsley (1973) to lie somewhere within the shaded area.*

large single-crystals the reverse is the case (Smith & Fuller, 1967). The stress-dependence of this remanence could be an important complication in the geological environment, but these large grains are probably not important remanence carriers. There have been no specific attempts to determine whether defect ferromagnetism is an important factor in palaeomagnetic studies, but the Morin transition is not often observed in sediments; this may be because the defect moment is more important, or because the transition is suppressed by small quantities of Ti^{4+} or other cations. Single-domain behaviour extends from 0.03 μm up to at least 100 μm and possibly 200 μm, but the highest coercivities are observed nearer the bottom of this size range. Coercivities are much higher than in the titanomagnetites; they are strongly dependent on grain-size, stress and thermal history and values observed on crushed natural hematites range from 10^4 Am^{-1}, for 300 μm grains increasing to 10^5 Am^{-1}. for 150μm grains, and then decreasing again to 4×10^4 Am^{-1} for 1μm grains (Chevallier & Mathieu, 1943). In polished section hematite is distinguished by a higher reflectivity than magnetite or ilmenite (Table 2.1).

Hematite is a common constituent of many igneous and sedimentary rocks. It occurs in sediments as black, polycrystalline (*specularite*) particles, often containing magnetite inclusions, and as a fine-grained coating on other matrix particles. This latter phase is a pigment ranging from brown through orange to red and purple and gives 'red' sediments their distinctive colour. It may be formed by the dehydration of goethite, by the weathering of Fe^{2+} in the lattices of clay minerals or other silicates, or by Fe^{3+} as an absorbed species on clay mineral surfaces. If hematite shows evidence of having pseudomorphed magnetite it is known as *martite*. It is stable as long as organic matter is absent; the latter encourages Fe^{3+} to reduce to Fe^{2+} and is accompanied by a colour change to drab greens and greys. Low Eh and low pH groundwaters have the same effect (first part of Section 4.2).

Pure hematite is already completely oxidised, but ilmenite, the other end-member of the series, undergoes oxidation at temperatures above *ca* 500° C so that the compositions move towards the rutile–hematite join on the ternary diagram (Fig. 2.1). At lower temperatures ilmenite is resistant to alteration and persists as a detrital mineral in sediments, although with prolonged oxidation it too, may be oxidised to hematite plus rutile. The solid-solution members in the range 0–50% ilmenite are canted antiferromagnetics; between 45 and 95% they become ferrimagnetics and then revert to an antiferromagnetic structure near 100% ilmenite.

Titanohematites with the general formula $xFeTiO_3.(1-x)Fe_2O_3$ occur as intergrowths for values of x between 0·1 and 0·9 as a common phase in metamorphic rocks (Section 4.3) but only rarely as a phase in igneous rocks. The compositions of interest in palaeomagnetism have x values between 0·45 and 0·9 (Fig. 2.4). The compositions between x = 0·45 and 0·6, although rare in nature, are of special interest because they show the phenomenon of *self-reversal*, in which a TRM is acquired antiparallel to the ambient magnetic field. It was first observed in a Japanese pumice of dacitic composition (Nagata, Uyeda & Akimoto, 1952) and is caused by a structure comprising two magnetic sub-lattices each with a different Néel temperature. The ions in the first lattice acquire their magnetism parallel to the ambient field at the higher Néel temperature. Due to their exchange coupling with the second lattice, they cause the moments in the second

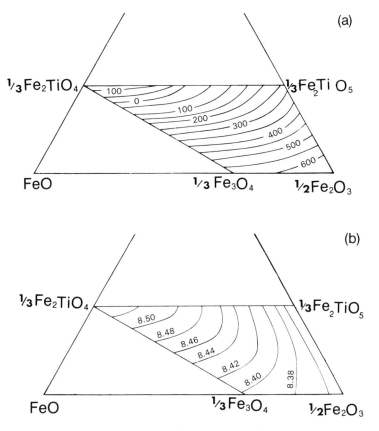

Figure 2.5 Contour diagrams showing (a) the variation of Curie temperature and (b) the cell edge parameter of titanomagnetites and their oxidised equivalents the titanomaghemites (after Readman and O'Reilly 1972).

lattice to become aligned in the opposite direction. The second lattice then becomes reversely magnetised at the lower Néel temperature; since it subsequently becomes more strongly magnetic than the first lattice at low temperatures, the structure has a net reversed magnetisation.

2.4 Maghemite

This mineral (γFe_2O_3) has the same chemical composition as hematite, but possesses a cubic inverse spinel structure with a vacancy at one in nine of the Fe^{3+} positions. The cell dimension is 8·35 Å. The similarity of the structure to magnetite gives it comparable magnetite properties and the saturation magnetisation (80–85 A. $m^2 kg^{-1}$) is nearly as high. The defect lattice is stabilised by small amounts of impurities which control the variable temperatures, (mostly in the range 300–350° C), at which maghemite converts to hematite; owing to this conversion, the true Curie point is not precisely known, but is believed to be close to 645° C (Ozdemir & Banerjee, 1984). Maghemite is formed by the low-temperature oxidation of magnetite (maghemitisation) in both subaerial and submarine environments. This conversion seems to be promoted by low hydrothermal temperatures and moderate Eh; these conditions are prevalent in the pillow-lava pile of crustal layer 2 in the ocean crust. Water is a catalyst, but is not an obvious prerequisite for maghemite formation.

The magnetic effects of maghemitisation are strongly dependent upon grain-size: maghemitisation of SD grains produces a decrease in coercivity and a large decrease in Q, while the maghemitisation of MD grains produces an increase in coercivity and no large change in Q, although it can produce a later CRM (Prévot, Le Caille & Mankinen, 1981). In practice, single-crystals of maghemite much larger than 1 μm are seldom observed because of their instability (Stacey & Banerjee, 1974) and the remanence of the ocean crust is dominated by SD and PSD grains. There is typically a reduction of remanent intensity by up to an order of magnitude producing the decrease in Q, but an increase in stability can result from grain-size subdivision by the cracking which takes place during volume reduction (Johnson & Hall, 1978). The decline in amplitude of the magnetic anomalies away from the mid-ocean ridges is believed to be due in part to maghemitisation, and the degradation of the palaeomagnetic record proceeds by the progressive acquisition of a secondary component (Kent & Lowrie, 1974). Maghemite can also form in soils by the action of burning: oxides and hydroxides of iron are first reduced to magnetite and may subsequently be oxidised to maghemite when air enters the soil again. Under conditions of slow oxidisation (pH ~ 7) and elevated temperatures maghemite is also produced in topsoils by the *in-situ* conversion of weakly magnetic oxide or hydroxide phases into a microcrystalline ($\ll$ 1 μm) phase in the clay fractions (Mullins, 1978). In this form it is widespread in laterites produced by the intense leaching of silica in igneous rocks in tropical environments; laterites are concentrated here as a residue rich in aluminium and ferric oxides. Maghemite is also responsible for the strong magnetic properties of lodestones (Nagata, 1961).

2.5 Goethite

The mineral group of iron hydroxides, which is loosely termed 'limonite', normally occurs as a complex of four crystal structures of which only one, goethite (α-FeOOH), carries a remanent magnetisation. It has an antiferromagnetic structure, but owing to impurities and lattice distortions, the cancellation of opposing moments is not quite complete and a small proportion of ions contribute to weak but stable ferrimagnetism with a Curie temperature of 110–120° C. The properties of this remanence are similar to hematite: it has a strong anisotropy and a weak maximum saturation remanence of *ca* 1 A.m^{2}kg^{-1}. Goethite and the other hydrated iron oxides (akagenite (βFeOOH), lepidocrocite (γFeOOH) and feroxyhyte (δFeOOH)) are formed in the weathering environment and occur as the yellow to brown coloured phases in weathered rocks and soils. Lepidocrocite may precipitate from natural waters and dehydrate to maghemite. The other hydroxides appear to converts to poorly-crystalline hematite (protohematite) at temperatures of 200–290° C and to true hematite at temperatures of ~ 350° C. In addition, goethite is metastable in the presence of water and oxygen in the diagenetic environment (Berner, 1969) and converts to hematite by the dehydration reaction:

$$2FeOOH \longrightarrow Fe_2O_3 + H_2O$$

Coarse-grained goethite is more stable than fine-grained goethite but, once formed, hematite will not rehydrate to goethite (Langmuir, 1971 and second part of Section 4.2).

The significance of goethite as a remanence carrier is not well understood. It may form as an alteration phase around pyrite and marcasite grains in some limestones and be the host of a low stability NRM removed at low temperatures (e.g. Turner, Vaughan & Tarling, 1978). It has been recognised by the combination of its high coercivity and low unblocking temperature as the remanence carrier in tan-coloured Jurassic limestones, where it has formed by the alteration of pyrite (Johnson, Van der Voo & Lowrie, 1984). Many other studies of sediments report a viscous PTRM exhibiting a progressive reduction in moment with thermal demagnetisation and the absence of a distinct high-blocking-temperature component for which goethite is a possible contender. However, because it is metastable in the subaerial environment and because hematite pigment can also have a relatively low blocking temperature, it is often not possible to attribute such remanence unambiguously to goethite.

2.6 Pyrrhotite

Between the rare iron sulphide, troilite (FeS), which is monoclinic and antiferromagnetic, and the common sulphide, pyrite (FeS$_2$), which is cubic and paramagnetic, there is a range of intermediate compositions with the general formula Fe$_{1-x}$S for $0 < x < 0.13$ containing ordered and disordered regions in the lattice. These include pyrrhotites, which range from monoclinic (pseudohexagonal) symmetry to true hexagonal symmetry. The majority are only slightly ferrimagnetic, but the compositional range $x = 0.09$ to $x = 0.13$ is strongly ferrimagnetic with a spontaneous magnetisation ranging up to 26 A.m^2 kg^{-1} and a Curie point of 320° C. The remanence of pyrrhotite is typically soft and comparable to low coercivity magnetite. Essentially it is the holes in the lattice created by the missing iron ions leading to an incomplete balancing of the ion spins, which produces this ferrimagnetic behaviour. In polished section it is difficult to distinguish pyrrhotite from troilite and pyrite (Table 2.1); a partially reversible thermomagnetic curve with a Curie point at *ca*. 320° C and converting to magnetite may identify this phase and distinguish it from maghemite.

Pyrrhotite occurs in similar environments to pyrite, namely in basic igneous rocks and low- to medium-grade metamorphic rocks. It may also form in anaerobic diagenetic environments where sulphates are reduced to form H$_2$S which reacts with iron to form troilite; continued production of H$_2$S encourages sulphur-oxidising bacteria to form elemental sulphur, which reacts with the FeS over a period of years to form microscopic aggregates of pyrite. The formation of pyrrhotite by this mechanism has not yet been demonstrated, but it seems to be present as a diagenetic phase in some marine cores (Kobayashi & Namura, 1974) and is generally to be expected in similar environments to pyrite; it has, for example, been reported as minute inclusions within pyrite grains, in limestones (Kligfield & Channell, 1981).

2.7 Other natural magnetic materials

Several minerals, such as cassiterite and zircon, have been reported to carry a remanence, but it is probable that this resides in the small amounts of ferrimagnetic impurities. The rare minerals: magnesioferrite ($MgFe_2O_4$, Curie temperature 440° C) and trevorite ($NiFe_2O_4$, Curie temperature 585° C) are ferrimagnetic in natural environments. Ferromanganese oxides belonging to the jacobsite solid-solution series $Mn_3O_4-Fe_3O_4$ are ferrimagnetic, with Curie temperatures of 300° C–580° C. In addition to minor occurrence as vein minerals, ferromanganese oxides grow authigenically in deep-water marine sediments. Their study is hampered by the microcrystalline or amorphous nature of manganese nodules, although there is now substantial evidence that their growth imparts a CRM which degrades the primary NRM in marine cores (Henshaw & Merrill, 1980).

Phases belonging to the Fe–Ni system may be important contributors to magnetic remanence in some serpentinised ultramafic rocks. Alloys of Fe-Ni-Co-Cu, with Curie temperatures in the range 620°–1100° C have been widely recognised in plutonic ultramafic rocks and their serpentinised equivalents (Haggerty, 1978): they may be expected to contribute a magnetic signature from the deep crust where the temperatures are above the Curie point of magnetite. At Burrow Mountain, California, serpentinisation has produced an Ni_3Fe phase characterised by Curie temperature of 590°–600° C and a saturation magnetisation 20% higher than magnetite (Lienert & Wasilewski, 1979). In addition, natural chromite spinels appear to carry a remanence with a blocking temperature of 350° C and a saturation IRM of ca. 70 mT (Kumar & Bhalla, 1984).

3
Outline theory of rock and mineral magnetism

3.1 Introduction: The properties of superparamagnetic, single-domain and multi-domain particles

Chapter 1 defined three size ranges for ferromagnetic particles, each characterised by distinctive magnetic behaviour, namely superparamagnetic (SP), single-domain (SD) and multi-domain (MD). These three fields are well substantiated by experimental studies such as the classic one illustrated in Fig. 3.1. It is generally believed that stable remanence resides in SD grains, while the less stable components are located in large MD grains or else in the hyperfine SP grains in which thermal activation plays a dominant role. However, SD behaviour encompasses only a very small part of the size spectrum in most magnetite-bearing rocks (Table 3.1), and a variety of experimental evidence, which will not be discussed here (see Dunlop, 1977), indicates that there is a broader size range above the SD–MD size transition, within which particles have a behaviour reminiscent of the SD range and are also able to retain a magnetic memory. This is the field of pseudo-single-domain (PSD) behaviour.

Néel (1949,1955) developed separate theories to describe the acquisition of a remanence by SD and MD grains. Although some of the assumpttions of these theories have still to be adequately tested, they provide a robust description of rock magnetic behaviour and are the only ones which have received wide acceptance. The constituent magnetic moments in a crystalline material have axes or planes of 'easy' magnetisation, and even magnetite with its cubic crystalline symmetry has anisotropic magnetic properties. The anisotropy is due to the shape and crystal structure, with subordinate contributions from stress and sometimes from an applied field while cooling through a critical temperature. The work done to magnetise a grain may be written in the form:

$$E_{AN} = E_o + K_1 f_1(\theta) + K_2 f_2(\theta) + K_3 f_3(\theta) + \ldots$$

where θ is the angle between the applied field and the direction of interest, and K_1, K_2, K_3, etc. are the anisotropy constants. E_{AN} is a minimum in the 'easy' and a maximum in the 'hard' direction or plane. Sometimes K_2 and higher terms are sufficiently small to be neglected, and the magnitude of K indicates the degree of anisotropy, with the sign of K_1 identifying the easy axis. SD behaviour is interpreted to be constrained entirely by crystal structure: to switch from one easy axis to another, or to an antiparallel sense, requires that an energy barrier be overcome.

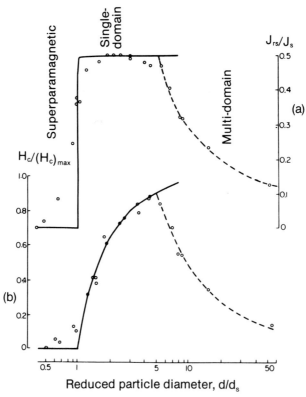

Figure 3.1 (a) *The saturation remanence and* (b) *the coercive force of spherical electro-deposited iron particles at room temperature. From Dunlop (1981) after Kneller and Luborsky (1983). The reduced particle diameter is the particle diameter divided by the size of the superparamagnetic-single domain transition. Note that the saturation magnetisation is constant and not size dependent in SD particles, but decreases according to d^{-n} above the SD–MD transition; this is the d^{-n} law where n is about 1 for titanomagnetite and about 0.65 for hematite. The coercive force (H_c) is below the theoretical microscopic coercive force $(H_c)_{max}$ for an appreciable size range above the SP–SD transition because the effects of thermal fluctuations are important in the smaller SD volumes.*

In the presence of an external field, SD grains acquire an extra energy related to the direction of the applied field. These grains cannot change their direction or magnetisation until the applied field exceeds the energy barrier; then the magnetisation can switch, or flip, into another easy direction. This critical field is the *switching field*. A uniaxial SD grain magnetised along its easy direction of magnetisation would have a square hysteresis loop (Fig. 3.2), with the coercive force equal to the switching field (section 3.2). If the field is applied perpendicular to the easy axis, the magnetisation reverts immediately the field is removed and there is no hysteresis. As the field is increased, the magnetisation increases linearly until all spins are aligned and saturation is achieved. In practice the grains within a rock tend to be randomly oriented, so that the behaviour of the assemblage of anisotropic grains is effectively isotropic. The hysteresis cycle then has an intermediate form shown in Fig. 3.2 which is similar, but not identical (see Merrill & McElhinny, 1983), to the hysteresis cycle of an assemblage of MD particles.

Table 3.1 A summary of domain-structure transition sizes deduced from theoretical and experimental studies

Mineral	Superparamagnetic/ Single-Domain (μm)	Single-Domain/ Multi-Domain (μm)
Magnetite Fe_3O_4	0·025–0·030	0·05–0·08
Maghemite Fe_2O_3	?	0·06
Titanomagnetite $(1-x)Fe_3O_4 xFeTiO_4$ ($x = 0·55$–$0·60$)	0·08	0·2–0·6+
Titanomaghemite $Fe_{(3-x)}Ti_xO_4$*		
$x = 0·6$, $z = 0·4$	0·05	0·75
$x = 0·6$, $z = 0·9$	0·9	2·4
Hematite Fe_2O_3	0·025–0·030	10–15
Pyrrhotite $Fe_{1-x}S$	?	1·6–3

After Dunlop (1981).

+ Radhakrishnamurty, Likhite, Deutsch & Murthy (1981) find that $\geq \sim 30\%$ substitution of Ti in titanomagnetite is sufficient to preclude MD structure, and they suggest that all natural titanomagnetites are in superparamagnetic or single-domain states.

* Titanomaghemite has the ideal formula $Fe_{(3-x)R}Ti_{xR}\square_{3(1-R)}O_4$, where R can take values between 1 and $8/(9+x)$ and $\square$ are vacancies in the cation sites of the structure. A formula unit contains $(1+x)Fe^{2+}$ and requires $\frac{1}{2}(1+x)O$ for complete oxidation. The oxidation parameter z is introduced to describe how far this process has gone to completion, so that an addition of $(z/2)(1+x)O$ will have oxidised a fraction z of the initial Fe^{2+} to produce an additional $z(1+x)Fe^{3+}$; when $z = 1$ all the iron is present as Fe^{3+}. A rigorous definition uses the term **titanomagnetite** for $z \leq 0·1$ and **titanomaghemite** for $z > 0·1$ (O'Reilly, 1983, and Section 2.1).

SD particles are always magnetised nearly to saturation, but once a domain structure is able to develop, lower remanent magnetisation states are probable, and are indeed, favoured by the demagnetisation effect. The contact between two domains is called the *Bloch Wall* and within this zone the atomic magnetic moments are progressively canted around from one domain to the next. The wall migrates laterally when a field is applied, so that the volume of domains with the magnetisation direction of the applied field grows at the expense of the other domains at a large angle to the applied field. Provided that the field is weak, the domain wall returns to its original position when the field is removed, but in stronger fields the wall may move past imperfections in the lattice. These form energy barriers, and not all of the barriers can be breached again when the field is removed. The grain then acquires an isothermal remanent remagnetisation (IRM). In very strong fields, the magnetisations of antiparallel domains can switch into alignment with the field, and the IRM reaches its saturated value. The magnetic moments within the domains will lie along the easy directions of magnetisation, but the canted spins within the Bloch Wall have components of magnetisation in the hard directions of magnetisation so that the wall zone has an intrinsic energy. It is the minimisation of the sum of the wall energy and the magnetostatic energy which determines the subdivision into domains.

The resistance of the magnetisation in some MD grains to changes in the applied field, and the consequent retention of a magnetic memory, is not well understood, but two possible models for this behaviour have been proposed. The first one suggests that the interiors of such grains contain imperfections or inclusions which nucleate SD-like regions (Verhoogen, 1959: Dickson, Everitt, Parry & Stacey, 1966). The second model proposes that these grains contain a central core of domains surrounded by a surface skin. This skin comprises fragments of domains at surface irregularities, closure domains at an angle to the 180° walls of the majority of domains, and additional moments associated with the edges of domain walls. These three features tend to be pinned and, collectively, they are only weakly coupled to the body domains. They are therefore resistant to migra-

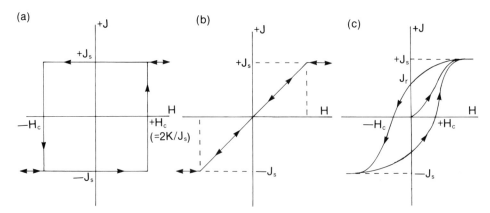

Figure 3.2 The hysteresis cycles of SD grains (a) subjected to an applied field parallel to their easy axes of magnetisation, (b) subjected to an applied field perpendicular to their easy axes of magnetisation, and (c) when the grains comprise a randomly-oriented assemblage.

tion (Stacey & Banerjee, 1974). Either model must explain the d^{-n} law (Fig. 3.1) and experimental results tend to support a variant of the second one (Parry, 1979).

3.2 The demagnetising factor, magnetic forces and relaxation times

The principal forces acting within a magnetised crystal structure are defined in Table 1.2. Although the exchange forces are strong, they only act between atoms in close proximity to one another. Three forces are effective on a macroscopic scale, and determine the domain structure of the crystal. These are the magnetostatic energy, the magnetocrystalline (anisotropic) energy and the magnetostrictive energy. The magnetostatic energy is dependent on the crystal shape: in an elongated grain, for example, it is easy to magnetise the grain along its length, and difficult to magnetise it at right angles to the length. This is because similar poles are then adjacent to one another and their mutual repulsion tends to make the dipoles rearrange themselves; there is then a large internal field opposing the magnetisation. This effect is known as *shape anisotropy*. The magnetostatic energy is a function of the *demagnetising factor*, N, which in turn depends on the shape of the body and its orientation with respect to the magnetising field; N links the internal field to the applied magnetic field. The coefficient N is usually evaluated in terms of a triaxial ellipsoid; it is $4/3\pi$ for a sphere. N tends to zero along the length of a very elongated ellipse and is 2π at right angles to the length of such an ellipse.

The crystalline structure imposes an intrinsic anisotropy on the alignment of domains, which is incorporated in the magnetocrystalline anisotropy energy. In the case of a cubic structure such as magnetite, this can be described in terms of the maximum and minimum anisotropy constants K_1 and K_2; these are functions of temperature and can be calculated from a knowledge of the combinations of Fe^{2+}, and Fe^{3+} and their symmetries in the structure. In this mineral the domains will be spontaneously magnetised along the 'easy' (111) axis; to switch a magnetisation through the 'hard' (100) direction requires a field given by:

$$H = 2\frac{K_1}{J_s} \quad (3.1)$$

Since the magnetisations of SD grains can only occur by spin rotations, equation (3.1) approximately defines the coercive forces of SD magnetite grains.

The magnetostrictive strain energy is caused by the mechanical deformation which accompanies the process of magnetisation. In a cubic grain it has a maximum value of:

$$E_{magnetostrictive} = \tfrac{3}{2}\lambda\sigma \quad (3.2)$$

where λ is the average coefficient of magnetostriction and σ is the internal stress amplitude. This energy is usually much smaller than the magnetostatic and magnetocrystalline energies.

Néel (1949) considered the SP to SD transition in terms of a relaxation time:

as particle size decreases, thermal fluctuations can randomise a magnetic moment within a grain over a sufficient time interval. In a uniaxial SD grain the energy of magnetisation is given by $E = Kv\sin^2\theta$ where v is the grain volume and θ is the angle between the magnetic moment and the axis. There are two positions of minimum energy in antiparallel ($\theta = 0$ and $180°$) directions. To switch a magnetisation from one position to the other requires that the potential barrier between those two positions, represented by the maximum energy at the $\theta + 90°$ positions, be overcome; this is given by $E_r = vK$. Since the energy of thermal fluctuation of the total moment of the whole domain is given by $E_t = kT$, where k is Boltzmann's constant and T is the absolute temperature, the magnetic moment can be rotated from one position to the other if $E_t > E_r$. Thus, an assemblage of SD grains with uniaxial anisotropy saturated at time t = 0 exhibits a remanent intensity, J_r which will decay according to the equation

$$J_r = J_s \exp(-t/\tau)$$

where J_s is the spontaneous magnetisation. The relaxation time, τ relates the energy of the potential barrier, E_r, to the energy of the thermal fluctuations, E_t, so that $1/\tau$ is proportional to the probability per unit time of the spontaneous magnetisation changing from one easy direction to another:

$$\tau = \tau_o \exp\left(\frac{E_r}{E_t}\right) = \tau_o \exp\left(\frac{vK}{kT}\right) \qquad (3.3)$$

τ_o is the time taken for the flip-over once it gets just over the potential barrier. It depends, amongst other things, on the elastic properties of the material and has a value of approximately 10^{-10} secs; it varies only slightly with temperature. K is a constant which depends on the energy barrier to be overcome; it describes the anisotropy energy per unit volume, and can be regarded as a measure of the shape, crystal and strain anisotropies which need to be overcome to reverse the magnetisation. In a single-domain grain it is simply related to the microscopic coercive force (or switching field), H_c, by the equation derived from (3.1):

$$K = H_c J_s / 2$$

For relaxation time, τ, of 1 second, $vK/2kT = 18\cdot4$, while for a relaxation time of 10^{17} seconds (i.e. older than the oldest rocks on Earth), $vK/2kT = 57\cdot6$. Thus, it requires only a small decrease in temperature or increase in particle volume to produce a dramatic change in the relaxation time. This leads to the concept of a *blocking temperature* (Stacey & Banerjee, 1974, and Section 3.5).

In a SD grain of magnetite simple values can be assigned to the three anisotropic energy barriers and their equivalent coercive forces. In summary, these are:

1. Magnetocrystalline: $E = 2K_1$ and $H_c = 2K_1/J_s$ \hfill (3.5)
2. Shape: $E = (N_b - N_a)J_s^2$ and $H_c = (N_b - N_a)J_s$ \hfill (3.6)
3. Stress: $E = 3/2\lambda\sigma$ and $H_c = 3\lambda\sigma/J_s$ \hfill (3.7)

N_a and N_b are the demagnetising factors along the minor and major axes of a prolate spheroid, and J_s is the spontaneous magnetisation.

Of these three contributions, it has already been noted that the stress effect is normally small and enormous stresses are required to explain coercivities >50 mT. (Strangway, Larson & Goldstein 1968). In addition, the magnetocrystalline anisotropy is unable to explain coercivities $H_c \geqslant 45$mT. (Evans & McElhinny, 1969). Thus, as is examined in the next section, it is necessary to appeal to shape anisotropy to explain coercivities >100 mT. which are observed in many magnetite-bearing rocks.

3.3 Single-domain behaviour: the theoretical approach

The grain-size represented by the SD-MD transition is estimated theoretically by using a method similar to that of Kittel (1949). The energy of a spherical grain in the SD state (e_1) is equated to the magnetostatic energy:

$$e_1 = \tfrac{1}{2} N J_s^2 (\tfrac{1}{6} \pi d^3) \tag{3.8}$$

N being the demagnetising factor ($4\pi/3$) for a sphere of diameter d. This is roughly twice the energy of the grain in a two-domain state (e_2), and the critical diameter, d_o, for the transition between the SD and MD states occurs when the decrease in energy resulting from partitioning the grain into two domains exactly balances the wall energy e_w:

$$e_1 = e_2 + e_w \tag{3.9}$$

and

$$e_w = \tfrac{1}{4} \pi d^2 w \tag{3.10}$$

where w is the energy per unit area of the wall (McElhinny, 1973):

$$d_o = \frac{9}{2\pi} \frac{w}{J_s^2} \tag{3.11}$$

Substituting the appropriate values for J_s and w yields estimates of $d_o \sim 0.03\ \mu$m for magnetite and $d_o \sim 0.15$ cm for hematite (McElhinny, 1973) in general agreement with the results of experimental work (Table 3.1). Thus, most hematite grains encountered in nature will be single domain. The estimate for the SD-MD transition in magnetite, however, is a size level at which thermal fluctuations would be expected to cause magnetic moments to change spontaneously; thus the SD and MD fields would appear to pass into one another, with no SD field between. This deduction is contrary to the observation of high coercivities, high Köenigsberger ratios and high stabilities observed in many magnetite-bearing rocks.

However, the simple treatment of the spherical grain yields only the smallest possible estimate for the SD–MD size transition. The critical size of this transition lies above the superparamagnetic threshold if the particles are elongated, and in theoretical terms (i.e. excluding the effects of grain imperfections, impurities and interactions) a shape anisotropy is required to explain both the high coercivities and a significant SD size range in magnetite grains. Thus, a grain with the shape of a prolate spheroid of axial ratio 10:1 has a demagnetising factor of 0.25 and different dimensional parameters that yield an estimate of $d_o \sim 160\ w/J_s^2 = 3\mu m$ for magnetite using the simple theory of equations (3.8) to (3.11) above (McElhinny, 1973). This is two orders of magnitude larger than the value for a sphere. The formal theory for elongate grains is given by Morrish and Yu (1955) and developed by Evans and McElhinny (1969). The SD:MD boundary is calculated in Figure 3.3 by equating the magnetostatic energy with the exchange energy involved in a circular spin configuration, and the superparamagnetic region is calculated from equation (3.3). The rapid transition from very short to very long relaxation times with increasing particle size is clear from this Figure. It shows that the region of SD behaviour in magnetite expands from zero for spherical grains up to the size range of 0.1–1 μm for very elongated grains; the theoretical microscopic coercivity increases correspondingly to a maximum of *ca.* 300 mT. Uncertainties in the calculations, effects of impurities in the magnetite lattice, and grain interactions when the grains are closely packed have the resultant effect of shifting the MD–SD boundary upwards towards the dashed boundaries in this figure. Comparable sets

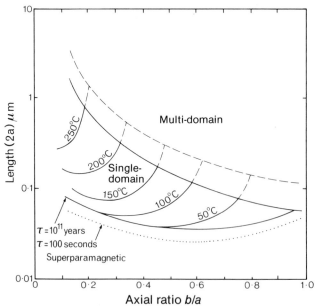

Figure 3.3 *The single domain, multidomain and superparamagnetic fields for magnetite grains with various lengths (2a) and axial ratios (b/a) at ambient temperatures (300°K). The upper dashed curve represents the possible error in determining the SD/MD boundary. The envelopes for the two relaxation times are shown to illustrate the rapid transition from superparamagnetic to SD behaviour. The solid curves show the peak a.f. field (in mT.) required to unblock the grains at room temperature. After Evans and McElhinny (1969).*

of curves are given by Strangway, Larson and Goldstein (1968) and Butler and Banerjee (1975). The approach of the former authors differs from Fig. 3.3 mainly by suggesting a temperature dependence of the SD–MD transition.

3.4 Coercivity

Coercivity is a measure of the resistance of the domain structure to changes. In MD grains it is related to grain diameter, d, by a relationship of the form:

$$H_c \propto \frac{1}{d^n} \qquad (3.12)$$

where n is an index in the range 0·25 and 1. Since changes in magnetisation result primarily from the motion of domain walls, the coercivity is determined by obstructions such as lattice dislocations and impurities. In SD grains the coercivity is dependent on the anisotropy, or directional control, of the magnetic energy of the particle and is given by equation (3.6). Although the magnetic field required to demagnetise a magnetisation is often referred to as the coercivity of that magnetisation (especially when referring to a.f. demagnetisation (first part of Section 5.5)), this is often less than the theoretical coercivity for two reasons:
1. The thermal agitation in small grains reduces their effective coercivity near the superparamagnetic field; for this reason the unblocking-field curves in Fig. 3.4, (which in the absence of this effect would be straight vertical lines) curve down and to the left towards parallelism with the superparamagnetic field.
2. The effective grain coercivities are affected by interactions between grains. Although this effect cannot be quantified, it probably acts to reduce the effective coercivity (Dunlop, 1968). However, Morrish & Watt (1958) conclude that interactions may increase the SD–MD transition size. Interactions are probably unimportant between the dispersed grains in rocks, although they may be responsible for the low coercivities in some organically-derived magnetite (Kirschvink & Lowenstam, 1979, and Section 4.2 *Limestones*).

3.5 Temperature dependence of unblocking of magnetic remanence

The temperature at which SD grains will unblock when they are heated in the laboratory can be derived from equation (3.3) by calculating the temperature at which the relaxation time becomes comparable with the experiment time. Unblocking curves for various temperatures are shown in Fig. 3.4 after Evans & McElhinny (1969). A general correspondence can be observed between the unblocking magnetic field (Fig. 3.3) and the unblocking temperatures; this results from the thermal agitation effect on coercivity noted above. It accounts for the value of the alternating field demagnetisation method (first part of Section 5.5) which preferentially demagnetises those grains with the lowest relaxation times and lowest blocking temperatures.

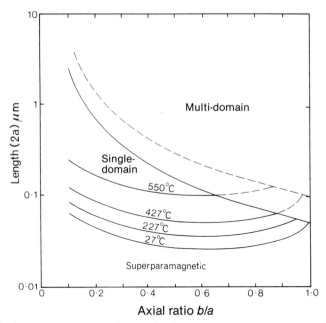

Figure 3.4 Blocking temperatures for single-domain magnetite grains. When heated the grains lying on a curve become unblocked at the temperature indicated. The dashed curve represents the possible error in determining the SD/MD boundary. After Evans and McElhinny (1969).

Equation (3.3) for the relaxation time incorporates the anisotropy constant. Since this is related to the coercivity, the equation can be rewritten as:

$$\tau = \tau_0 \exp\left(\frac{vH_cJ_s}{2kT}\right) \qquad (3.13)$$

which links relaxation time and coercivity. Two general consequences of equation (3.13) are noted here. Firstly, since H_c and J_s vary considerably between individual grains in an actual rock, there will be a range or *spectrum* of blocking temperatures. Secondly, for an assemblage of SD grains of various sizes and axial ratios, the relationship between coercivity and blocking temperature is only general: because of the sensitivity of relaxation time to grain volume, the grains with the highest coercivities do not necessarily possess the longest relaxation times. For a grain of given length, the coercivity increases, but the volume decreases as its width is reduced, so that the maximum relaxation time occurs for an axial ratio of 0·65. This probably explains why two component magnetisations have sometimes been observed, in which the component with the highest blocking temperature has the lowest coercivity (e.g. Buchan, 1978).

3.6 Thermoremanent magnetisation

The NRM in many igneous rocks is a strong magnetism acquired in a weak ambient magnetic field during cooling from the Curie point down to ambient temperatures. As a magmatic rock cools from high temperatures, a spontaneous magnetisation appears at the Curie point, and because its relaxation time is very short, it quickly reaches equilibrium with the applied field. As the temperature continues to fall, a point is reached where the relaxation time of the grain begins to increase very rapidly. At this *blocking temperature*, T_B, the magnetisation is no longer able to follow the ambient field, and it becomes frozen into the grain in such a way that subsequent changes in field direction are unable to affect it. For an assemblage of identical grains there would be a single blocking temperature, but because relaxation time is related to grain volume (for SD grains), as well as to temperature, a range of grains of different size will have a spectrum of blocking temperatures: at ambient temperatures the larger grains (provided that they exhibit SD or PSD behaviour) will have relaxation times of thousands of millions of years or more, and give the observed TRM, while the small grains will have relaxation times of only a few seconds or less, and thus behave superparamagnetically.

Thus, a TRM is generally acquired over the whole temperature range from the Curie point down to ambient temperatures, and the total TRM is the sum of the partial TRMs (PTRMs) acquired over each temperature interval (Fig. 3.5). This is the *Additivity Law of PTRM* due to Thellier (1951). For example, in Fig. 3.5:

$$\text{TRM} = \text{PTRM}_{T_c}^{550} + \text{PTRM}_{550}^{500} + \text{PTRM}_{500}^{450} + \ldots\ldots + \text{PTRM}_{100}^{0}$$

By the reverse process of heating up to a temperature $T < T_c$, the magnetisation of all grains with blocking temperature less than T is lost. The acquisitions of TRM, PTRM and VRM (Section 1.3) are not physically distinct processes; they merely represent gradations of the same thermal activation process acting over different

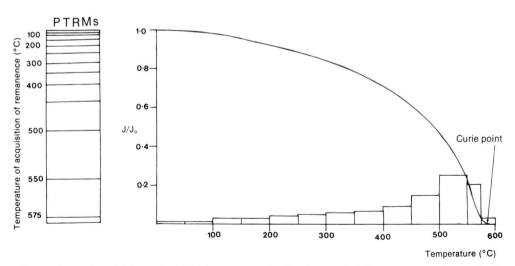

Figure 3.5 Acquisition of a TRM in a magnetite-bearing rock. The remanence is acquired over a temperature range below the Curie point (580°C in this example) so that the final TRM is the summation of the components acquired within each temperature interval (i.e. each PTRM) up to the Curie point.

combinations of time and temperature. Viscous effects are usually small at ambient temperatures, because only a small part of the blocking temperature spectrum will lie down at this low temperature, but rocks with a significant superparamagnetic size fraction are likely to acquire appreciable VRMs. VRM is predicted to be strongly size-dependent, and this has been demonstrated in hematite (Dunlop & Stirling, 1977). It accounts for the large secondary remanence acquired by some fine-grained red sediments.

Néel's theory of TRM in uniaxial SD grains relates the intensity of TRM (J_{TRM}) to the applied field H by the equation:

$$J_{TRM} = J_s \tanh\left(\frac{vJ_s(T_B)H}{kT_B}\right) \qquad (3.15)$$

where v is the volume of the particle, and J_s and $J_s(T_B)$ are the saturation magnetisations at room temperature and the blocking temperature, T_B, respectively. For the case of an assemblage of randomly oriented grains magnetised in a weak field ($\leqslant 0.2$ mT) this equation simplifies to:

$$J_{TRM} = \frac{J_s}{3}\left(\frac{vJ_s(T_B)H}{kT_B}\right) \qquad (3.16)$$

implying that the TRM is directly proportional to the applied field when this is weak.

The process of TRM acquisition in MD grains is more complex, as the whole domain pattern becomes frozen into the minimum energy configuration occupied at the blocking temperature. The TRM per unit volume of an assemblage of MD grains which occupy a volume fraction of $f \leqslant 1$ is related to the applied field H according to:

$$J_{TRM} = H \cdot \frac{J_s}{J_s(T_B)} \left(\frac{f}{N} - k\right) \qquad (3.17)$$

where N is the demagnetising factor and k is the volume susceptibility (Stacey & Banerjee, 1974). Values for these parameters in many rocks link the TRM to the applied field by the approximate equation $J_{TRM} \sim 0.1$ fH. This theory gives a general agreement with experimental results for large grains, but there is a rather broad range of sizes from *ca.* 20 μm down to SD sizes for which it is inapplicable. Stacey & Banerjee (1974) explain this transitional region in terms of a PSD grain-surface effect, whereby the domains are constrained within a grain volume comparable with their actual size. An expression for the TRM is then derived as:

$$J_{TRM} = \frac{16}{3\pi^3} \frac{(d_c^3 t J_s^2)}{kT_B}\left(\frac{J_s(T_B)}{J_s}\right)\frac{fH}{d} \qquad (3.18)$$

where d_c is the critical size of single domains and t is the average thickness of a domain wall. Thus TRM is predicted to be proportional to d^{-1} in the PSD range. Experimental results suggest that a particle-size dependence of weak-field TRM applies to magnetite over size ranges > 1 μm (i.e. the PSD and MD size ranges) and

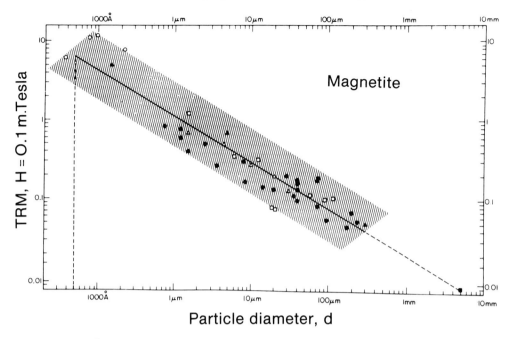

Figure 3.6 The dependence of a TRM acquired by magnetite in a field of 0.1T on the particle size. The PSD and MD size particles plot about the average $d^{-0.6}$ line shown but the SD data mostly plot well above this line. (After Dunlop (1981); the symbols refer to different experimental investigations listed in this reference).

is approximated by the relationship: TRM $\propto d^{-0.6}$ (Fig. 3.6). The prediction that the TRM intensity is proportional to the field magnitude in low fields has been generally supported by experimental observations (e.g. Dunlop & Waddington, 1975; Day, 1977). In strong fields this is no longer the case and the TRM reaches saturation in a field which is only a fraction of the isothermal saturation field.

In weak fields the intensity of the acquired TRM is however, also controlled by the duration of the magnetising field: a longer period of application will give more domains the opportunity to relax into the ambient direction. The TRM will therefore be controlled by the rate of cooling (Dodson & McClelland-Brown, 1980 and Section 12.5).

3.7 Chemical remanence

A CRM is acquired by chemical changes below the Curie or Néel temperature. It has properties comparable to TRM, and the Néel theory of TRM acquisition can be adapted to explain this. As the crystal grows, it is initially so small that it behaves as a superparamagnetic particle. The relaxation time increases logarithmically with increase in volume (equation 3.3) and the particle size may eventually pass a certain critical *blocking volume*, V_B, at which τ becomes large and the remanence is blocked. The CRM of a grain assemblage is then determined by the alignment of grain moments at the blocking volume. The value of V_B will corres-

pond with grain diameters of *ca.* 0.025 μm in magnetite and hematite (Table 3.1), although it depends on how fast the grains grow and thus the time available for blocking the remanence. Since τ is a function of v/T in equation (3.3), the essential difference between the processes of CRM and TRM acquisition is that the magnetism is acquired by increasing in volume in the former case, and decrease in temperature in the latter case; this would be consistent with the comparable coercivities observed for the two kinds of remanence (e.g. Nagata, 1961 and Fig. 5.6). If the particles grow further they become multi-domain and the ambient field will constrain the domain structure. Oppositely magnetised domains form and the total magnetisation decreases. This stage may readily be reached by authigenic magnetite (second part of section 4.2), but is probably not often achieved by hematite grains, which are the commonest carriers of CRM. In practice, the acquisition of a CRM by hematite may result from a range of processes including:
1. the precipitation and ageing of ferric oxyhydroxides
2. the pseudomorphing of silicates,
3. the *in-situ* oxidation of magnetite
4. the formation of authigenic overgrowths (second part of Section 4.2). The relative importance of direct precipitation versus the ageing of geothite to the formation of hematite CRM is presently unknown, and hence the validity of the simple CRM mechanism given above is unclear.

The intensity of a CRM produced in an assemblage of randomly oriented grains in a weak ambient field H is given by McElhinny (1973) as:

$$J_{CRM} = \frac{V_B J_s^2 H}{3kT} \qquad (3.19)$$

For hematite this yields an estimate of $J_{CRM} \sim 0.04H$, independent of the size to which the grains have grown. In theory, the remanent intensity of a hematite-bearing rock should depend not upon the actual amount of hematite present, but only on the number of grains which have reached the blocking volume.

The magnitude of the CRM is related to the TRM in the same grain by:

$$\frac{J_{CRM}}{J_{TRM}} \sim \frac{J_s}{J_{SB}} \cdot \frac{K_B}{K} \qquad (3.20)$$

Although J_s is larger at ambient temperatures than at the blocking temperature, T_B, the anisotropy constant (K_B) at this temperature exhibits a larger relative reduction, so that $J_{CRM} < J_{TRM}$ and the ratio in equation (3.20) has values of $0.1–0.4$.

3.8 Susceptibility

Two phenomena contribute to the susceptibility of magnetic minerals. The low-field susceptibility in magnetically soft MD materials is due largely to domain wall movements: if a field is applied parallel to one domain, it will expand relative to its oppositely-magnetised neighbour by movement of the wall between them. Secondly, domain rotation may occur when a field is applied at a high angle to a domain and causes the magnetisation to turn against the force of magneto-crystalline anisotropy.

When a grain is magnetised with an intensity J by an external field H, the effective internal field, H_I, is less than H by an amount NJ, the self-demagnetising field:

$$H_I = H - NJ$$

The *intrinsic volume susceptibility*, k_I, links J to H_I in small fields:

$$H_I = H - Nk_IH_I$$

so the observed volume susceptibility is:

$$k = \frac{J}{H} = \frac{k_IH_I}{H_I(1 + Nk_I)} = \frac{1}{1/k_I + N} \quad (3.21)$$

In an assemblage of grains only a fraction, f, of the total volume will be composed of magnetic material; hence:

$$k = \frac{f}{1/k_I + N}$$

For materials with a high susceptibility the demagnetising field produced by the whole sample will further reduce the observed susceptibility and the equation for k is modified to:

$$k = \frac{fk_I}{1 + (N + fN')k_I} = \frac{f}{1/k_I + (N + fN')}$$

where N' is the demagnetising factor appropriate to the shape of the sample. For the dispersed magnetite assemblages in most rocks this means that the susceptibility is reduced by up to a few per cent. For weakly-magnetic materials with k < 1 the measured volume susceptibility is a good measure of k_I, but for strongly magnetic materials like magnetite, k is large so that $k \rightarrow f/(N + fN')$. Thus for an assemblage of MD grains, susceptibility can be predicted from the magnetite volume and typical grain shape, because it is controlled only be the demagnetising factor and is practically independent of grain-size (Parry, 1965; Mullins, 1978).

The Köenigsberger (Q) ratio is given by:

$$Q = \frac{1}{Nk_I} \cdot \frac{J_s}{J_{SB}} \quad (3.22)$$

where the symbols have the same meanings as equations 3.20 and 3.21

For an assemblage of spherical particles ($N = 4/3\pi$), $Q \sim 0.5$ and is proportionately smaller for elongated grains.

Since coercivity measures resistance to the changes causing susceptibility, an inverse relationship between coercivity and susceptibility is to be expected in SD grains. An assemblage of randomly oriented grains has a susceptibility given by

$$k = \frac{\text{constant}}{H_c} \sim \frac{1}{3}\frac{J_s}{H_c} \quad (3.23)$$

and k ranges from 0·1 for infinite needles, to 1 for spheres. Although the susceptibilities for SD and PSD grains are similar, the TRM is typically higher than that of MD grains, and thus $Q > 1·0$. This difference in Q values for SD and MD particles provides a convenient test for the presence of stable remanence noted in Section 1.5 (Stacey, 1963). In the superparamagnetic range the susceptibility is proportional to grain volume.

3.9 Anisotropy of susceptibility

Although all magnetic minerals have magnetocrystalline anisotropy, the grains are randomly oriented in many rocks, and the resultant magnetism is isotropic. However, if there is an alignment of crystallographic axes within a grain assemblage, a *crystalline anisotropy* will be present. This effect is weak in magnetite, but it is strong in hematite because this mineral is much more readily magnetised in the basal plane than in any other crystallographic direction. Since this is the plane of the common platey habit of hematite, crystal structure is the common cause of anisotropy in hematite-bearing rocks. Rocks containing the minerals, geothite and pyrrhotite, may also have crystalline anisotropy because there is a systematic relationship between crystal axes and crystal shape. *Shape anisotropy* is present in any grain with a degree of elongation. It is the commonest cause of anisotropy in rocks, because elongated or platey grains are aligned by flow or crystal settling in consolidating igneous bodies, by the hydraulic action of currents during sedimentation, and by tectonic stress fields during regional metamorphism. Shape anisotropy is the only cause of macroscopic anistropy in rocks containing magnetite and maghemite, and is produced by the alignment of grain long-axes (which are also the axes of easy magnetistion).

Magnetic anisotropy can be determined by measuring the variation of any of several parameters over the surface of a sample, including susceptibility, saturation magnetisation, IRM and TRM, but low field susceptibility is almost universally employed, because it does not affect the magnetisation state of the sample, and it may also be sensitive to the alignment of paramagnetic phases (Section 5.3). Within an anisotropic sample, the magnetisation is generally non-parallel to the ambient field and the induced magnetisation has components along three orthogonal axes x, y and z defined by:

$$J_x = k_{xx}H_x + k_{xy}H_y + k_{xz}H_z$$

$$J_y = k_{yx}H_x + k_{yy}H_y + k_{yz}H_z \quad (3.24)$$

$$J_z = k_{zx}H_x + k_{zy}H_y + k_{zz}H_z$$

so that J is a tensor of the form:

$$J = k_{ij}H_{ij} \, (i, j = 1, 2, 3) \qquad (3.25)$$

of which only six components are independent because $k_{xy} = k_{yx}$, $k_{yz} = k_{zy}$ and $k_{zx} = k_{xz}$. In a manner analogous to the use of the stress and strain tensors in structural geology, this tensor is usually represented by a susceptibility ellipse, in which the three orthogonal principal axes define the maximum, minimum, and intermediate values of susceptibility. The anisotropy of susceptibility is defined in terms of these three *principal susceptibilities* (k_{max}, k_{int} and k_{min}); their directions are determined with respect to the reference frame of the rock and are conventionally plotted on a stereonet (Section 6.1).

If a rock comprises a fraction, f, of identical elllipsoidal grains, all aligned in the same direction, with semi-axes, a, b, and c (b = c) corresponding with the demagnetising factors N_a, N_b and N_b, the susceptibilities in the a and b directions from equation (3.21) are:

$$k_{max} = \frac{fk_I}{(1 + N_a k_I)} \text{ and } k_{min} = \frac{fk_I}{(1 + N_b k_I)} \qquad (3.26)$$

where k_I is the intrinsic volume susceptibility (Section 3.8). The difference between the maximum and minimum susceptibilities is:

$$k_{max} - k_{min} = \frac{fk_I^2 (N_b - N_a)}{(1 + N_a k_I)(1 + N_b k_I)} \qquad (3.27)$$

This dependence of $k_{max} - k_{min}$ on k_I^2 means that shape anisotropy is important in magnetite and maghemite because they have high susceptibilities, but it only makes a small contribution to the anisotropy in hematite, goethite and pyrrhotite which have low susceptibilities. In practice, a rock will contain a range of different sizes and shapes, although there will be a preferred alignment of long axes. The rock can then be treated in terms of a bulk equivalent susceptibility ellipse.

The degree of anisotropy, P, is given by:

$$P = \frac{k_{max}}{k_{min}} \qquad (3.28)$$

In addition to its use for resolving petrofabrics, anistropy is of importance in palaeomagnetism, because it is necessary to know what degree of anisotropy can be tolerated before the NRM direction is deflected through an unacceptably large angle from the true ambient field direction. This problem is discussed by Stacey (1960) and McElhinny (1973); the latter author shows that anisotropies of 10, 20 and 50% (p = 1·1, 1·2 and 1·5) are equivalent to a maximum deflection of the TRM direction by 2·7, 5·2 and 11·6°, respectively, although these deflections appear to be achieved by somewhat lower anisotropies if directional changes in the self-demagnetising factor are taken into account (Nagata, 1961). Thus, this effect is unlikely to lead to serious errors in the study of most igneous and sedimentary rocks, in which anisotropies are typically only a few per cent.

During sedimentation, magnetite grains will come to rest with their long axes horizontal, and platey hematite grains should come to rest with their basal planes parallel to the bedding plane. The long axes of magnetite grains are usually aligned with the direction of flow so that the k_{max} axis defines the direction of the hydraulic currents during sedimentation (Rees, 1961). A small degree of imbrication is often produced by hydrodynamic forces, and is shown by a small inclination between this axis and the bedding plane (e.g. Crimes & Oldershaw, 1967; Hamilton & Rees, 1970; Taira & Scholle, 1979). Hematite fabrics tend to show little net alignment in the bedding, but the k_{min} axes are normal to the bedding in most cases. These fabrics will tend to survive diagenesis, provided that the more elongated grains are the larger part of the magnetic fraction, and therefore are less liable to reorientation during dewatering than the smaller equidimensional grains. Depositional fabrics are unlikely to survive bioturbation (first part of Section 4.2). When this effect has been important it is probable that the fabric is of post-depositional origin and linked to gravitational compaction, for example. Diagenesis may preserve or enhance a depositional fabric if new mineral growth has an orientation controlled by pre-existing mineral cleavages and grain alignments; in other cases, however, the alteration of deposited magnetic minerals to hematite can destroy the depositional fabric (Addison, Turner & Tarling, 1985).

In igneous rocks, an alignment of grains may be produced by a combination of crystal settling and convective flow, especially in large layered intrusions and high-viscosity acid plutons. A weak flow fabric is also developed in hypabyssal and extrusive rocks, including dolerite dykes (King, 1967) and lava flows (Ellwood, 1978). Metamorphic rocks generally develop the highest degrees of anisotropy, and much interest centres on the quantitative relationship between the present susceptibility tensor (equation 3.25) and the ancient strain tensor (e.g. Rathore, 1979). Crystalline anisotropy may contribute to the total effect here as, for example, in the alignment of hemo-ilmenite grains in metamorphic rocks in the Adirondacks (Balsley & Buddington, 1958) and the basal planes of hematite grains in Welsh slates (Fuller, 1960). At low metamorphic grades the fabric may result from granular rotation, but at higher grades it is caused by new crystal growth preferentially along planes perpendicular to the maximum principal stress; it is promoted under pressure by ionic migration and stressed pressure points. Although a close correspondence is observed between the magnetic and strain fabrics in low- to medium-grade metamorphic rocks, there is much less alignment between the magnetic fabric and susceptibility in the higher grades (Hrouda & Janak, 1976; Hrouda, Janak & Rejl, 1978) suggesting that the regional stresses are only locally effective in orienting the small magnetic grains in a silicate matrix. Beckmann, Olsen & Sørensen (1978) propose a method to correct for the deflection of the magnetic vector resulting from the bulk anisotropy, although Irving & Park (1974) find that errors due to anisotropy are essentially random in some medium- to high-grade metamorphic rocks; they suggest that anisotropy has a small effect on the components of high coercive force, because the small percentage of grains carrying the stable remanence are not systematically oriented along the visible foliation in the silicates. Unfortunately, the effect of anisotropy in metamorphic rocks has not yet been widely investigated; although ordinary palaeomagnetic studies of medium- to high-grade metamorphic terranes have enjoyed a good measure of success this should not imply that their interpretation might not be affected in detail if more were known about the magnitude and variation of the anistropy effect. For a detailed discussion of the geological applications of magnetic fabric studies refer to Tarling (1983).

3.10 Magnetisation during slow heating and cooling

Below the blocking temperature, the relaxation time becomes so long compared with the time scale of cooling, that the magnetisation is no longer able to follow changes in the ambient field. However, if the cooling is slow, as it will be in plutonic terranes, it is to be anticipated that more time will be available for relaxation at lower temperatures, so that the blocking-temperature range for a distribution of grain-sizes will be broadened and lowered. Pullaiah, Irving, Buchan & Dunlop (1975) have developed a theory of thermal remagnetisation, originally examined by Chamalaun (1964), to compare the relaxation times at two temperatures T_1 and T_2 by using a formula derived from Néel's theory (equations 3.3 and 3.13):

$$\frac{T_2 \log (\tau_2/\tau_0)}{J_s(T_2)H_c(T_2)} = \frac{T_1 \log (\tau_1/\tau_0)}{J_s(T_1)H_c(T_1)} \qquad (3.29)$$

τ_1 and τ_2 are the relaxation times at temperatures T_1 and T_2 and τ_0 has the meaning assigned in equation (3.3). These authors accommodated the variation of J_s and H_c with temperature, and made the assumption that magnetocrystalline anisotropy could be neglected in magnetite, so that shape anisotropy is dominant and $H_c \propto J_s$. The resultant blocking curves for magnetite and hematite (Fig. 3.7) show that individual blocking temperatures are lowered and their collective range is broadened during slow cooling. These graphs suggest that grains with blocking temperatures of *ca.* 230°C would have a relaxation time of *ca.* 100 Ma at ambient temperatures, while grains with a slightly higher blocking temperature of 250°C would have a relaxation time of over 1000 Ma. Thus, rocks magnetised in Precambrian times, with laboratory blocking temperatures of less than *ca.* 250°C will by now have lost virtually all of their original magnetism, whilst grains with slightly higher blocking temperatures will have retained the bulk of their magnetism. In practice, it is found that a large fraction of the grains carrying the stable remanence in a rock have blocking temperatures within 100–200°C of the Curie temperature. There are two main fields of behaviour in Fig. 3.7, designated A and B. Field A covers magnetisations with narrow and high blocking temperatures, which are identified by square-shouldered thermal demagnetisation curves (Fig. 5.7); these remanences are only unblocked by prolonged, elevated temperatures at a little below their blocking-temperature range and therefore can remain stable over geological time. By contrast, assemblages in the B field have distributed blocking-temperature spectra, identified by sloping thermal decay curves; they are unblocked by lower temperatures, such as those achieved by burial, which need to operate for only relatively short periods of time. Reflecting these contrasts, the remanences with A and B type characteristics can be effectively separated by partial thermal demagnetisation (second part of Section 5.5). These graphs describe the ways in which TRMs are both lost during heating and acquired during cooling. The curves for hematite imply, for example, that a PTRM acquired in a red sediment at a temperature of 150°C for 10 Ma could be unblocked by heating to 400°C for 1 hour in the laboratory.

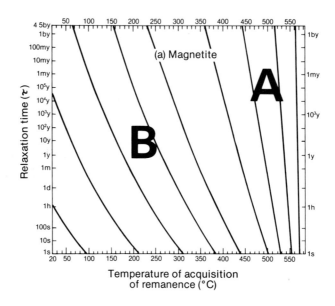

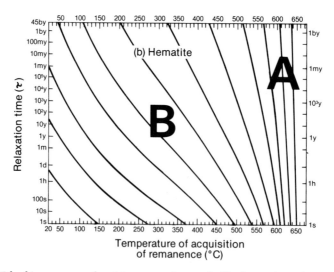

Figure 3.7 Blocking curves for (a) magnetite and (b) hematite after Pullaiah, Irving, Buchan and Dunlop (1975). The calculations are based on equation 3.29 assuming $J_s(T)$ functions for the two minerals and that $H_c(T)$ is proportional to $J_s(T)$ for magnetite and $J_s^3(T)$ for hematite.

However, although these curves may describe qualitatively the effects of prolonged heating and/or cooling on the magnetic history of a rock, their quantitative application appears to be very limited. Laboratory experiments by Pullaiah, Irving, Buchan & Dunlop (1975) suggested that the survival potential of magnetic remanence was greatly overestimated by these graphs. In part, this may be due to chemical changes promoted by prolonged laboratory heating, but also, it appears to be due to the change in the activation energy (H_cJ_s) with temperature, which decreases to zero at the Curie point. Dodson & McClelland-Brown (1980) have developed a theory incorporating this. The theory is described by curves showing the relationship between natural blocking temperature, T_B, and the laboratory-determined blocking temperature for magnetite and hematite, T_D (Fig. 3.8). The differences between these temperatures are larger at slower cooling rates and lower temperatures, but tend to zero as the Curie temperature is approached. Provided that a cooling rate can be estimated, the curves can be used to interpret the history of slowly-cooled igneous and metamorphic rocks from thermal demagnetisation studies.

Figs. 3.7 and 3.8 predict the result of a purely physical effect. In practice chemical changes occur, that will promote both the destruction of primary phases and the growth of new phases (second part of Section 4.2 and Section 4.3). Thus, the changes to be expected during regional and burial metamorphism are probably more profound than suggested by this theory. They may, however, be successfully predicted for relatively dry terranes, and some applications of Fig. 3.8 to a granulite terrane in West Greenland are described by Dodson & McClelland-Brown (1980) and Piper (1981a).

A number of distinctive characteristics are associated with the magnetisations acquired by plutonic terranes during slow uplift-related cooling. These include:
1. since the rocks have cooled through their blocking-temperature range during at least a number of cycles of the geomagnetic secular variation, the NRM in individual samples is a time-average of the magnetic field;
2. all rock facies formed before regional cooling will have similar directions of magnetisation, regardless of their different modes and ages of origin, provided that their blocking-temperature spectra are similar;
3. if appreciable long-term changes in the ambient geomagnetic field direction occurred during cooling, samples from different structural levels should have distributed directions defining this. In particular, samples from higher crustal levels should have been magnetised before those from lower crustal levels;
4. at any given crustal level, samples with higher blocking temperatures should record older directions of the geomagnetic field than those with lower blocking temperatures;
5. progressive demagnetisation should recover progressively higher blocking temperature, and therefore older, directions of the magnetic field.

These phenomena have been described from several slowly-cooled Precambrian plutonic terranes (Morgan, 1976; Piper, 1981a; Stearn & Piper, 1984). The analysis of Dodson & McClelland-Brown in Fig. 3.8 implies that the lowering of blocking temperatures is not dramatic, even when rocks have experienced some of the slow rates of uplift which characterised Proterozoic terranes (Watson, 1976). Thus, a remanence unblocked between 570°C and 550°C in the laboratory will have been blocked over a range of 565–527°C at a cooling rate of 6°C/Ma. It is

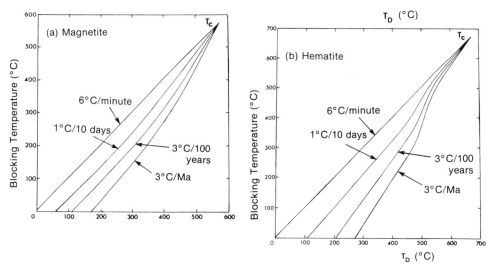

Figure 3.8 The natural blocking temperature, T_B, as a function of the laboratory demagnetisation temperature, T_D, for magnetite and hematite, calculated for four rates of cooling after Dodson and McClelland-Brown (1980).

possible that several millions of years of geomagnetic field behaviour may be recorded by some slowly-cooled rocks, but it is unlikely to be much more. A more realistic idea may be provided by the recorded field behaviour: some complete inversions of polarity have been described from the demagnetisation of plutonic rocks (e.g. Smith & Piper, 1982) but, currently, no double inversions — suggesting that cooling spanned more than one complete polarity event — appear to have been identified.

4
The magnetisations in rocks

The NRM in many rocks comprises at least two components, and it is the aim of partial demagnetisation techniques (Chapter 5) to recognise and separate these components. It is usually much more difficult, however, to identify their sources. Although optical (Table 2.1) and magnetic (Section 5.6) tests can identify potential magnetic phases, they are only of limited assistance in locating the origin of a remanence, for two reasons. Firstly, the source of the stable remanence surviving over millions of years will often be in SD and PSD magnetite grains which are not directly observable by routine optical investigations, and secondly, these tests recognise *all* of the potential magnetic phases although only a small fraction of them may actually be responsible for the stable remanence. It has been estimated that only 0·01% of the iron oxide minerals in a volcanic rock needs to be in SD form to account for typical observed NRMs (Strangway, Larson & Goldstein, 1968). Furthermore, the condition of the magnetic phases, and hence the volume contributing to the remanence, will be critically controlled by the crystallisation conditions. Plutonic rocks cool slowly to form large, discrete grains which generally have time to equilibrate with one another. Since they are kept at elevated temperatures for protracted periods of time, the bulk of the remanence will be continually reset as it follows the changing magnetic field. The resultant NRM is then essentially a viscous remanence acquired over a broad time interval, and represents a long time average of the ambient field. It seldom resides in the discrete homogeneous and annealed iron oxide grains, and is generally linked to exsolution phenomena in the silicate minerals. At the other extreme, rapidly-chilled materials such as the glassy carapace of pillow lavas and the pseudotachylyte veins resulting from frictional melting along fault planes, have a pure TRM acquired over periods of no more than a few seconds and residing in fine magnetite grains. These chilled grains often have a skeletal form, in which elongate grain shapes and large internal stresses resulting from rapid cooling are responsible for high coercivities. Since most lava flows cool to the Curie point of their magnetic phases within periods of a few hours to a few days, their TRMs also record essentially an instantaneous record of the magnetic field. However, extrusive volcanic regions are also low-pressure porous environments, particularly subject to the circulation of residual magmatic fluids. These effects may produce enhanced exsolution and oxidation phenomena, which can modify or replace the source of the initial TRM.

This chapter summarises what is presently known about the formation and modification of magnetic iron oxide and sulphide phases in igneous, metamorphic and sedimentary environments. Although this knowledge is linked to the acquisition of remanence, it will be apparent that the area of study linking rock magnetism and petrology is still imperfectly understood.

4.1 Igneous rocks

The volcanic and hypabyssal environments

If an igneous melt cools rapidly enough, homogeneous, but impure titanomagnetites ($Fe_{3-x}Ti_xO_4$) with compositions of ca. $0 \cdot 75 > x > 0 \cdot 45$ (see Fig. 2.2) are formed. The major cation impurities are Mg^{2+} and Al^{3+}, with smaller amounts of Mn^{2+}, Cr^{3+} and V^{5+}, and the corresponding Curie points are between $0°C$ and $250°C$. Such compositions are not stable (Fig. 2.3) and unmixing will normally occur. This is extremely slow at ambient temperatures, but it is readily promoted by the circulation of magmatic and hydrothermal fluids. It proceeds most actively at temperatures above $600°C$, and is therefore mainly a deuteric phenomenon. The ulvöspinel molecule becomes unstable and reacts with the oxygen and water-rich residual phase, according to the reaction:

$$6Fe_2TiO_4 + O_2 \rightarrow 6FeTiO_3 + 2Fe_3O_4$$

Since ilmenite does not form a solid-solution series with magnetite, exsolution of the two phases takes place as lamellae in the (111) planes of the titanomagnetite host. As the temperature falls, the titanomagnetite is progressively enriched in Fe and depleted in Ti, and the rhombohedral lamellae are progressively enriched in Ti. Several degrees of exsolution and oxidation of the ilmenite lamellae have been recognised; they are subdivided by Wilson & Haggerty (1966) and Wilson & Watkins (1967) into six classes:

1. Homogeneous titanomagnetite with no ilmenite exsolution.
2. and 3. Titanomagnetite with < (class 2) or > (class 3) 50% of the grain area covered by dense ilmenite lamellae.
4. The ilmenite lamellae are pseudomorphed by hematite and other decomposition products.
5. The original ilmenite lamellae are replaced by aggregates of hematite, rutile and pseudobrookite. The magnetite develops rods of spinel.
6. The grain consists of an aggregate of hematite, pseudobrookite and rutile, and the original lamellae have entirely disappeared, or are preserved only as a ghost structure.

From titanomagnetite oxidation classes 2 to 6, the accompanying discrete ilmenite grains show a progressive exsolution of ferrirutile phases as the titanomagnetite oxidation progresses.

Lava flows may exhibit all six classes of titanomagnetite exsolution and oxidation, often within a single flow unit (Ade-Hall, Khan, Dagley & Wilson, 1968a), while acidic rocks tend to be more oxidised than basic rocks, probably because they often have a higher water content (Haggerty, 1976). In hypabyssal intrusions, the cooling rate is sufficiently slow for the $Fe_{3-x}Ti_xO_4$ solid solution to exsolve into two separate cubic phases when the solvus curve is reached at ca. 600 °C (Figure 2.3). Hence, single-phase titanomagnetites are rarely observed in dykes and sills. Although the Ti and Fe oxides are often intergrowths, the magnetite belongs predominantly to class 1, and only rarely are grains of higher than class 2 observed (Ade-Hall & Wilson, 1969; Bird & Piper, 1981). Intrusives also contain much smaller amounts of discrete ilmenite and much larger amounts of primary iron sulphide than lavas (Ade-Hall & Lawley, 1970). These observations imply that a gaseous reducing phase rich in S_2 in the hypabyssal environment is converted into an oxidising gaseous phase in the extrusive environment. The reason for this is not presently well understood, but it appears to have an origin internal to the magma body, and to be unrelated to groundwater or rain-water contamination (Ade-Hall & Lawley, 1970).

The deuteric exsolution in classes 2 and 3 produces flat, plate-like and rod-shaped particles of magnetite within original titanomagnetite grains. The rod-shaped particles have uniform shape, and if their length:diameter ratios are in excess of ca. 15:1 they have SD structure because magnetocrystalline energy can overcome the tendency for the crystal to become magnetised along the easy (111) directions, and thus subdivide into a MD structure (Strangway, Larson & Goldstein, 1968). Furthermore, these rods are isolated from one another by intervening non-magnetic ilmenite, so that reduction of the coercive force by the proximity of other rods is slight. Some individual particles may have coercive forces approaching the theoretical maximum of 300 mT., and explain the high coercivities of magnetite phases in many igneous rocks. The higher degrees of oxidation further subdivide these grains and the effect is to produce a progressive increase in the stability of the magnetic remanence with oxidation class (Ade-Hall, Khan, Dagley & Wilson, 1968b). Since deuteric alteration is a post-magmatic phenomenon, it does not appreciably affect the time at which the primary remanence is acquired, but it acts close to the Curie points of the magnetic phases, and means that the remanence is a CRM rather than a TRM.

Hydrothermal phenomena represent the lowest grade of regional metamorphism (Section 4.3) in the zeolite facies. They achieve notable alteration in lava piles and operate in other permeable situations where the circulation of abundant groundwater has taken place; this circulation is motivated by the heat supplied by burial, exothermic reactions, later intrusions, and hot, residual, magmatic fluids. The resulting mineral assemblage is then a combination of deuteric and hydrothermal processes. In the less permeable hypabyssal environment, at greater depths, the effect of elevated pressure–temperature conditions may only be observed from subtle changes in the thermomagnetic spectra (third part of Section 5.6; Ade-Hall, Palmer & Hubbard, 1971). In lava piles the degree of metamorphism can be mapped (Walker 1960) by the zonation of diagnostic zeolite (hydrated Ca-Al silicate) minerals which are succeeded at depth by prehnite and then epidote. These zones and their temperature constraints are shown in Fig. 4.1.

The first effect on the magnetic phases appears to originate at depths > 900 m and T of 150°C. It is called *granulation* and takes the form of a subdivision of homogeneous class 1 titanomagnetite grains, with the development of a fine granular texture ($\leqslant 1$ μm in grain-size) which includes grains of rutile (Ade-Hall, Khan, Dagley & Wilson, 1968a). At higher temperatures, equivalent to the

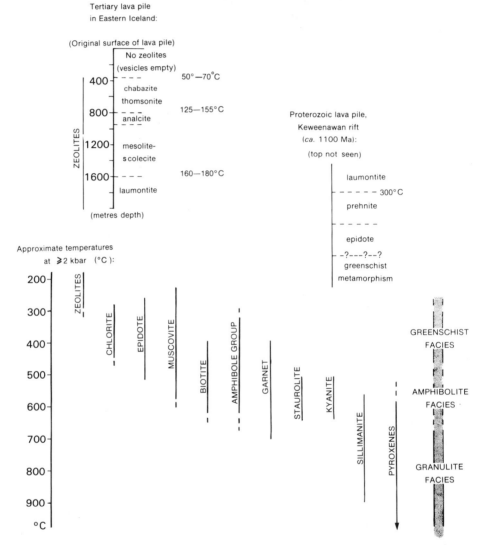

Figure 4.1 *The metamorphic facies with diagnostic minerals of the zeolite, greenschist, amphibolite and granulite facies. Compiled from Walker (1960), Ade-Hall, Palmer and Hubbard (1971) and Tarling (1983). The temperatures indicated for the hydrothermal transformations are existing data for $\leqslant$ 2kb or 5–6 km depth. The temperature ranges for the greenschist and higher grade minerals are approximate figures applicable to pressures of $\geqslant$ 2 kb.*

appearance of epidote, these grains show the development of flecks of titano-hematite, and then progressive replacement by a sphene-containing aggregate at $T > 250\,°C$. In class 2 and 3 grains the ilmenite lamellae break down to sphene and titanohematite in the laumontite zone. The effects of regional hydrothermal alteration in the class 4 to 6 grains are more difficult to distinguish from deuteric effects: in both cases the ilmenite is converted to an aggregate of titanomagnetite and ferrirutile; in classes 4 and 5 the host titanomagnetite appears to remain unchanged until it undergoes general alteration to titanohematite at epidote-grade conditions. Of the accompanying silicates, olivine is completely broken down by this stage (Ade-Hall, Palmer & Hubbard, 1971).

The net result of these mineralogical changes is to elevate the Curie point of the titanomagnetite as the influence of unsaturated hematite becomes prominent in the thermomagnetic determinations (Duff, 1979a and the third part of Section 5.6). The primary remanence in lava flows does not appear to be reset by hydrothermal metamorphism to the epidote grade, and a detailed record of the magnetic field behaviour during eruption has been recovered from lava piles at this grade of metamorphism (Robertson, 1973; Piper, 1977) probably because the original remanence was fixed as a stable component in grains with high degrees of deuteric oxidation; however, the titanohematite phase is probably responsible for a smaller, high-coercivity, secondary component acquired during burial (e.g. Palmer, Halls & Pesonen, 1981).

Ophiolites and serpentinisation

Although the extrusive lava pile comprising oceanic crustal layer 2 is often widely maghemitised to depths of $0\cdot5$ km, with the magnetic consequences noted in Section 2.4, this effect does not appear to be common in intrusive rocks of crustal layer 3. This may be because the grains are already deuterically oxidised, or because they are protected by silicate phases (Dunlop & Prévot, 1982). Furthermore, titanomaghemite will convert irreversibly to Ti-poor titanomagnetite if the lava pile is reheated to $300-350\,°C$ (Banerjee, 1980). The metamorphism of the complete crustal section to greenschist facies leads to a reduction in magnetisation, (section 4.3) although some meta-dolerites may preserve a fraction of their NRM, because large grains are more resistant to chloritisation. As a result, many ophiolite complexes have weak magnetic properties. The exceptions are those facies which have been serpentinised. Serpentinisation is caused by the action of hydrothermal systems on the ultrabasic facies in the ocean crust and upper mantle. Sea-water appears to provide the fluid phase and the equilibrium temperatures of the Mg-silicates imply formation temperatures ranging from $235\,°C$ for antigorite to $130\,°C$ for lizardite (Wenner & Taylor, 1971; Bonatti, Lawrence & Morandi, 1984). The iron is incorporated into disseminated pure magnetite grains, and is capable of producing an intense magnetisation. It commences with a fine dusting of magnetite grains within olivines; these grow and coalesce as serpentinisation proceeds. It progresses in the sequence: olivine before orthopyroxene, before clinopyroxene, so that it starts in dunites before harzbugites, and norites before gabbros. Saad (1969) recognised a linear relationship between the degree of serpentinisation and the NRM intensity and susceptibility, but Taylor (1984)

The plutonic environment

In the plutonic environment, the equilibration at lower temperatures over protracted periods of time (and hence lower TiO_2 content), the lower oxygen fugacities and the larger grain-sizes than in hypabyssal and volcanic environments all combine to make minute subdivision of titanomagnetite grains by exsolution lamellae a much less commonly observed phenomenon. Thus, grain subdivision is unlikely to produce single domains. Hence, attention has focused on the exsolution of Fe^{2+} and Fe^{3+} into oxides forming discrete rods and plates parallel to the cleavage planes in the silicate phases. In oceanic gabbros magnetite rods < 1 μm in diameter, with elongations up to 50:1, have been observed within plagioclase crystals, and appear to have exsolved at temperatures of $ca.$ 600°C (Davis, 1981). These extreme elongations would give the rods a SD structure (Section 3.3). Magnetite needles of the order of $10 \times 0 \cdot 5$ μm in size, in a plagioclase host, have also been identified as the carriers of the stable remanence in several large anorthosite intrusions, including the Michikamau (Murthy, Evans & Gough, 1971), the Morin (Irving, Part & Emslie, 1974), the Egersund (Hargraves & Fish, 1972), and the Cunene (Piper, 1974) bodies, and in dolerite (Hargraves & Young, 1969). Estimates of exsolution temperatures range up to 800°C (see Anderson, 1966), well in excess of the Curie point. Evans & McElhinny (1969) found similar grains within the pyroxenes of the Modipe gabbro, capable of explaining all of the observed NRM; the plagioclase in this example did not contain an opaque phase. Large titanomagnetite grains were also recognised in some of these studies, but were found to have MD structure and not to be the host of the stable remanence.

4.2 Sedimentary rocks

The depositional environment: depositional remanent magnetisations and post-depositional detrital remanent magnetisations

Clastic sedimentary rocks comprise the reconstituted detrital particles formed from weathering processes. They may have been deposited in air, water, or by gravity flow. Sedimentary particles generally include grains which have already acquired a remanence, and these grains are influenced by four forces during sedimentation: gravitational, hydraulic, thermal and magnetic. The magnetic force is dominant for ferromagnetic grains with diameters $< ca.$ 20 μm. These grains may become oriented in a fluid within periods of seconds as the grain rotates physically to align its fixed remanence with the ambient geomagnetic field (Collinson, 1965a); for very small particles ($< 0 \cdot 1$ μm) however, Brownian motions will tend to randomise this alignment. Thus, in theory, there is an intermediate grain-size range, over which the magnetic force predominates to produce a strong degree of alignment with the geomagnetic field; this prediction is confirmed by some laboratory analogues (Fig. 4.2). Above the size range controlled by Brownian (thermal) vibrations, weakly magnetised or non-magnetic particles will respond only to the gravitational and hydraulic forces. The former causes the particles to achieve a position of lowest gravitational energy at the sediment interface and the latter has an effect which is controlled by the particle shape; both effects counteract the acquisition of a DRM (Section 1.4).

Early attempts to reproduce DRMs, by re-depositing natural materials in the laboratory, showed that the observed magnetic inclinations were generally less than the inclination of the applied field, by an amount referred to as the *inclination error*. King (1955) found that the observed inclination (I_o) is related to the applied field inclination (I_A) by an equation of the form:

$$\text{Tan } I_o = f \text{ Tan } I_A \qquad (4.1)$$

King explained this by assuming that the disc-shaped particles in a mixture would have a strong shape anisotropy and would tend to have their magnetisations aligned in the plane of the disc; they would therefore, tend to be deposited in the plane of the sediment surface. Similar considerations also apply to elongated grains. If the mixture also contains spherical particles which accurately align with the magnetic field, f should be related to the fraction of platey or elongated particles. Griffiths, King, Rees & Wright (1960) subsequently showed however, that shape effects are inadequate to explain the magnitude of f; they also showed that the inclination error is independent of size for particles up to fine silt grade in size. They proposed that the second possible mechanism for the inclination error is the rolling of aligned particles into the nearest depression. Theoretical analysis of this mechanism shows that it increases from a minimum effect of zero, when the particle rotation axes are at right angles to the horizontal component of the magnetic

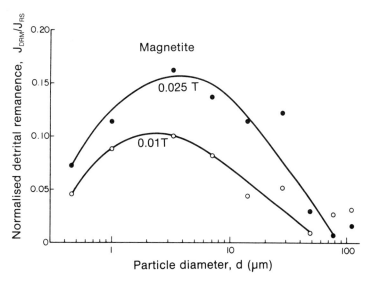

Figure 4.2 The dependence of DRM intensity on particle size for artificially-deposited magnetite grains in magnetic fields of 0.025 T and 0.01 T. (After Dunlop (1981) from data of C. Amerigian).

field, to a maximum effect when these axes are parallel to the field. This effect always operates to reduce the observed magnetic inclination, and is compatible with equation (4.1).

Two further effects may influence a DRM. Firstly, if deposition takes place on a sloping surface, a *bedding error* is present, which is approximately equal to the slope of the bedding plane (θ). It is present because particles rolling downslope roll through an angle ($\theta + \phi$) whilst particles rolling upslope roll through an angle of ($\theta - \phi$); the bias created by the difference between these two angles leads to the observed error (King, 1955; Griffiths, King, Rees & Wright, 1960). Secondly, as the fluid:sediment interface is approached, the fluid flow changes from turbulent to laminar, and the shear stress acting in the laminar layer can rotate the particles about horizontal axes, probably before they touch bottom (Granar, 1958). If the direction of flow is in a vertical plane containing the magnetic field direction, only an error in inclination is produced, but if the flow is oblique to the field, errors in both inclination and declination will result. Once rolling ceases, the position of minimum gravitational energy will leave the grains with an *imbricate* fabric, in which their long and intermediate axes are parallel to the sediment interface, and the hydraulic drag tends to align the long axes parallel to the current flow. This is the *shape fabric*, which can be conveniently determined from magnetic anisotropy studies (Section 3.9).

In practice, many field and laboratory studies have shown that these several sources of error, which may deflect a DRM from the ambient dipole field, are rarely significant in nature. The inclination error is only important when the magnetic grains have a similar size distribution to the other grains. (It may also be of little practical importance because sedimentation rates in nature are usually much less than in their laboratory analogues.) In addition, few remanences preserved in rocks are likely to be simple DRMs.

Initially, the water-lain sediment has a loose packing, with minimum grain contact, and is completely saturated with water. As sedimentation proceeds the grains are pressed into a tighter configuration and the intergranular water is progressively expelled. The smaller magnetic grains between the matrix of larger grains are rotated into the direction of the ambient field during this process. Irving (1957) first recognised that the processes imparting a detrital magnetisation could operate *after* deposition, and he introduced the term *post-depositional detrital remanent magnetisation* (hereafter referred to as a PDDRM). Irving invoked this mechanism to explain the uniform magnetic remanence in penecontemporaneous slumps within the Torridonian sandstones, where the remanence is carried predominantly by grains of hematite in a matrix of quartz grains. The reality of the mechanism was suggested by a laboratory analogue (Irving & Major, 1964) and has subsequently been produced by model sedimentation studies (Barton, McElhinny & Edwards, 1980; Hamano, 1980).

Clays and fine silts deposited from suspension in the laboratory are observed to faithfully record the applied magnetic field at shallow depths below the sediment:water interface, and within periods of a few days or less (Lovlie, 1974; Barton, McElhinny & Edwards, 1980). PDDRMs are acquired in intertidal (littoral) environments over periods of 1 (Ellwood, 1984) to 100 (Suttill, 1980) years. Since the sediments here are rapidly dried out by intermittent exposure, the burrowing activity of marine organisms (bioturbation) promotes the acquisition of PDDRM

(Ellwood, 1984). In sub-littoral environments, however, bioturbation precludes acquisition of PDDRM near the sediment surface, because it helps to create, or maintain, a high water content in the sediment. Bioturbation effects extend to ca. 8 cm in deep sea cores and to 50 cm or more in some shallow water deposits; only at greater burial depths where dewatering takes place can the magnetisation become fixed. The water content appears to be critical, because the PDDRM is set in the sediment when the water:particle ratio is reduced to ca. 1:1 (Hamano, 1980); only if sufficient intergranular water is present initially will the grains be reoriented into the ambient field (Verosub, 1977). Experimental and theoretical considerations suggest that the acquisition time of a PDDRM depends on sedimentation rate; it decreases as sedimentation rate increases because the lithostatic pressure acting within the sediment builds up more rapidly (Hamano, 1980). Over this acquisition time period, the PDDRM probably integrates the ambient field-changes to give a time-average resultant, which, however, is still short in comparison with changes in secular variation.

The influence of grain-size on the PDDRM is less well understood. Important contrasts have been observed within individual turbidite units, with the coarse-grained basal levels having lower inclinations and intensities than the slowly-deposited silty and laminated upper layers (Granar, 1958; Griffiths, King, Rees & Wright, 1960; Thompson & Kelts, 1974). Clearly, either grain-size or rate of deposition, or both, could be responsible for this effect. The collective field and laboratory observations imply that small ferromagnetic particles are realigned after deposition with the magnetic field within periods of a few days, but larger particles are incompletely reoriented, either because the geomagnetic couple is too weak, or because the grain packing does not permit sufficient grain rotation. PDDRM acquisition is evidently most efficient when the magnetic grains are appreciably smaller than the matrix grains (Tucker, 1980).

Unlike sands, clays are initially deposited with very high porosities (85–50%), well above the water:particle levels at which PDDRMs are acquired. Although this porosity may only be reduced below 50%, by burial to depths of 500–1000 m, compaction due to overburden pressure is not the only phenomenon acting here; both cementation and dehydration are observed close to the sediment surface (Selley, 1976) and probably act to fix the PDDRM within a few cm of the surface. At depths of > 2 km, diagenetic dewatering reactions, such as montmorillonite to illite, take place. These reactions increase the pore pressure and promote the alteration of existing phases, possibly to introduce a CRM.

Granular rotation into alignment with the ambient field during dewatering is not the only process acting near the surface of a sediment. Experimental studies employing the redeposition of Recent sediments in the laboratory are able to confirm the rapid acquisition of a PDDRM, but they produce a remanent intensity which is always larger than the NRM (Barton, McElhinny & Edwards, 1980). It appears that compaction of a dilute sediment into a more closely-packed structure partly counteracts the alignment into the ambient field. For equidimensional grains this scatter would be effectively random, while flattened grains would tend to move towards the plane normal to the axis of compaction. Provided that the magnetic grains comprise smaller particles in the matrix, the effect on the magnetic direction appears to be slight; however, these observations obviate the use of compacted sediments for palaeo-intensity determinations.

This discussion should not imply that the phenomenon of PDDRM is well understood. No satisfactory quantitative model presently exists and our understanding of the process is based largely on a few laboratory investigations and observations on Recent marine sediments. Deep-sea sediments are characterised by low deposition rates of about 0·1–5 cm/1000 yr. In general, when these sediments have a fossil content and an absence of authigenic nodule formation, they appear to preserve an accurate record of the geomagnetic field as a PDDRM below the level of bioturbation by burrowing organisms (Verosub, 1977). Conversely, the remanence appears to be only of secondary chemical origin where organic activity has been precluded and authigenic mineral formation is prominent. Lake sediments are characterised by much higher rates of deposition of *ca.* 0·1–1 m/yr, and this greatly enhances their potential as recorders of the secular variation of the geomagnetic field. Unfortunately, lake sediments possess a magnetic record showing larger changes in both declination and inclination than the ambient field. Regional trends can sometimes be recognised, but the degree to which records can be correlated from lake to lake is limited (Creer, 1978; Jacobs, 1984). Local and intermittent discrepancies may be due to hiatuses in sedimentation, to slumping* and to the influx of turbidity currents, but a more serious problem is likely to be the effect of hydraulic currents during deposition. In addition, coring procedures may often deform wet sediments. A systematic feature of lake sediment records is for the mean inclination to be shallower than the present field; this might be a record of the inclination error (King, 1955) or the possible compaction effect noted by Barton, McElhinny & Edwards (1980).

Because sediments are complex magnetic and textural systems, it is generally difficult to demonstrate that a remanence is a DRM or PDDRM. Clearly, an inclination error — if present — will be indicative of a DRM, but the reverse is not necessarily the case. In some special circumstances it may be possible to infer a DRM if there is evidence that the magnetisation stabilised soon after deposition. Verosub (1979) for example, studied deformed varves to demonstrate that remanence was imprinted on a varve within three years of deposition. Determination of the susceptibility ellipse (Section 3.9) is a technique of more general application. This is because sediments tend to contain a fabric which is the resultant of two effects, namely a lineation caused by the preferred orientation of the grains during deposition by currents, and a horizontal foliation imparted by sinking of non-spherical particles in the fluid under the force of gravity. As a result, sediments tend to yield an oblate susceptibility ellipsoid with a relative difference between k_{int} and k_{min} greater than that between k_{max} and k_{int}. The first of these two effects is important with magnetite, which is most readily magnetised along the length of the grains, while the second effect is more important in hematite (Section 3.9).

In general, a depositional fabric is evident when the maximum and intermediate axes of susceptibility lie in the bedding plane, and specifically, when the maximum susceptibilities lie in the palaeocurrent direction. Practical investigations suggest that anisotropy studies are most effectively applied to silt and fine-

---

* Some slump horizons which are observed to possess a coherent magnetisation (e.g. Addison, 1980) may in fact possess a *shear remanent magnetisation* (SRM). This is observed when unconsolidated sediments are subject to shear, and has been recognised as the process which magnetises wet clays when they are thrown into mud-brick moulds (Games, 1977).

sand grade material, but the demonstration of a depositional fabric does not necessarily mean that the remanent magnetism is of depositional origin. This can only be inferred from a comparison of the susceptibility axes and remanence directions: a coincidence of the axis of maximum susceptibility and the remanence direction indicates that the magnetisation is a DRM, while a spread of directions of maximum susceptibility between different samples suggests that the remanence is a PDDRM. Often, however, the imbricated or aligned grains responsible for the magnetic fabric are too coarse to be rotated into the ambient geomagnetic field, while the remanence is carried by a finer-grain-size fraction (typically $<20\ \mu m$), able to rotate and produce a PDDRM.

The diagenetic and weathering environments: chemical remanent magnetisations

As sediments accumulate, the weight of the overlying sediments progressively reduces the pore space beneath and expels the intergranular water. This purely physical process is accompanied by chemical changes which act to cement the sediment and form a consolidated rock. The chemical changes include solution and deposition by the water phase, the breakdown of unstable silicate minerals, and organic–inorganic reactions resulting from decay of incorporated plant and animal matter. These changes are promoted by the temperature increase due to both the geothermal gradient and exothermic organic reactions.

The resulting reactions will depend on the Eh (oxidation–reduction) and pH (acid–basic) conditions of the environment. They can be predicted from thermodynamic data relating to the major chemical components involved, although calculated relationships must be applied cautiously because only a limited range of components can be accommodated in one phase diagram, and possibly catalysts cannot be identified. Fig. 4.3 illustrates the $Fe^{3+}-Fe^{2+}-H_2O$ system showing solid–solid phase changes when hematite, rather than goethite, is the most stable fully-oxidised phase in the weathering environment. It is most appropriate to the subaerial and near-surface environment, with limited or intermittent access to water, and it predicts that hematite is the stable oxide phase formed by weathering within the field of most natural pH–Eh conditions, although magnetite may be expected to form as an authigenic or diagenetic mineral under mildly-reducing and alkaline conditions.

The groundwater in natural aquifers (mostly with pH = 3–8 and Eh = 0·3 to −0·1 volts) can carry Fe^{2+} ions in true solution, but it is mostly transported in the form of colloidal $Fe(OH)_3$. It is precipitated at various stages of deposition and diagenesis as amorphous $Fe(OH)_3$ or as a member of the ferric oxyhydroxide group (Section 2.5) which can convert to hematite by ageing, according to dehydration reactions of the form:

$$2Fe(OH)_3 \rightarrow Fe_2O_3 + 3H_2O$$

$$\text{and } 2Fe(OH)_3 \rightarrow 2FeOOH + 2H_2O \rightarrow Fe_2O_3 + 3H_2O$$

Hence goethite and ferric hydroxide are precursors of precipitated hematite, although it is uncertain whether this is always the case. The dehydration temperature is markedly dependent on pH, but it can be achieved under most geological conditions, and in the presence of water (although kinetic considerations suggest that it will take place slowly). Thus, red beds cannot necessarily be regarded as palaeoclimatic indicators of arid or seasonally dry climates. Furthermore, the oxidation of Fe^{2+} to Fe^{3+} is exothermic, and is used by some organisms as a source of energy.

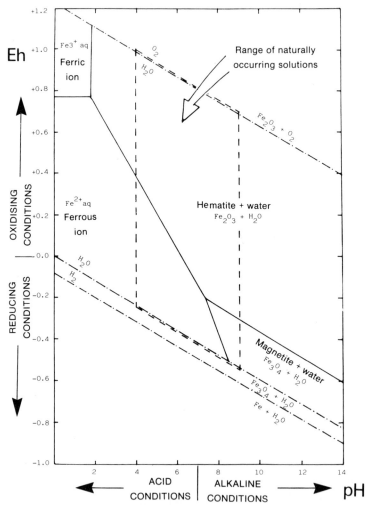

Figure 4.3 The stability fields of iron, magnetite and hematite at one atmosphere pressure in the presence of water after Garrels and Christ (1965). These conditions are representative of the subaerial and shallow subaqueous diagenetic environments except when the anions HS^- and HCO_3^- are abundant; these latter ions encourage the growth of pyrite and siderite respectively in the reducing field. Marine depositional waters are characterised by positive Eh and slightly alkaline conditions, whereas the interstitial waters tend towards negative Eh and neutral conditions in the diagenetic environment. Note that conditions may sometimes fall well outside of the parallelogram defining "normal" conditions, especially when abundant organic matter is present.

In the marine environment the dehydration reactions noted above cannot take place. Different equilibrium conditions occur and iron is intimately involved in the sulphur cycle. Fig. 4.4 shows the pH–Eh equilibrium diagram for the Fe–S–H$_2$O system, after Henshaw & Merrill (1980); it assumes that reactions with aqueous phases are important, so that the hydroxide:goethite, rather than the oxide: hematite, is the stable product of oxidation. This diagram is probably most applicable to marine conditions, where hydrous reactions are a permanent feature

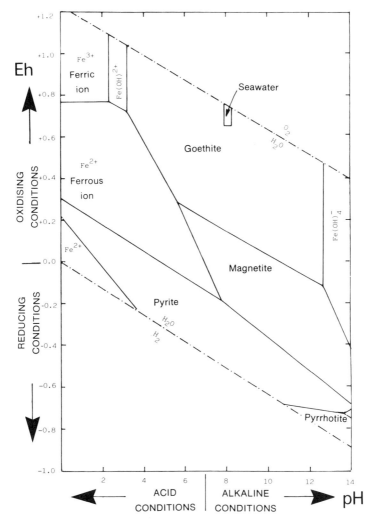

Figure 4.4 The stability fields of the iron sulphides, magnetite and geothite calculated by Henshaw and Merrill (1980) to accommodate aqueous species in the marine environment and using the assumption that geothite is more stable than hematite. This diagram is thought to be more applicable than 4.3 to intermediate and deep water marine environments and the parallelogram of normal conditions does not apply here. (Eh is measured in volts as the potential difference between an inert electrode immersed in the solution and a hydrogen electrode of known potential. pH is the negative logarithm of the concentration of free H$^-$ ions in grams per litre.)

of the environment. An important feature is the relatively large area of pyrite and magnetite stability fields incorporated within the range of natural pH–Eh conditions; both these minerals and goethite may be expected to form in aqueous conditions. SEM studies of magnetic extracts from deep-sea sediments have shown either rounded and abraided grains, which are probably the result of water transport, and angular grains, which are probably wind-transported volcanic dust (Lovlie, Lowrie & Jacobs, 1971). Authigenic mineral grains (i.e. formed *in situ*) are only formed at the surface in this environment under anoxic conditions but, as noted below, they may develop at shallow depths. They are also increasingly being recognised in shallow-marine sediments (Section 4.2, *Limestones*). Authigenic magnetite can be formed in sediments by the reduction of iron hydroxides or hematite and the concentration of magnetite in some anticlinal structures implies that the appropriate reducing conditions can be generated in sedimentary basins by the migration of hydrocarbons (Donovan, Forgey & Roberts, 1979).

The production of pyrite in acid, reducing environments is closely linked to the breakdown of organic matter. This usually takes place within the top few metres of an accumulating sedimentary sequence, so that it is largely complete within a few hundred years of deposition (Manning, Williams, Charlton, Ash & Birchall, 1979). Iron is a micronutrient and is mobilised by organic action, so that bacterial complexes ultimately fix iron sulphides, carbonates and phosphates. Although iron hydroxides may form as an intermediate phase, they react with the hydrogen sulphide released by the microbial decay of organic matter, to produce pyrite via intermediate metastable iron sulphides, such as mackinawite (tetragonal FeS) and orgreigite (cubic Fe_3S_4). Although this is the most common reaction in reducing environments, pyrrhotite may form and produce a CRM if conditions are more alkaline. Pyrrhotite is reported as the magnetic phase in some marine sediments of anoxic regions (Kobayashi & Namura, 1974), but it does not appear to be common in the marine environment.

Most marine sediments are oxidised when they are subjected to continuous current circulation, but they may become more anoxic with increasing depth below the surface, either because the oganic carbon content is high, or because burial has isolated the sediment from the source of oxidation. In deep waters, characterised by very slow deposition, the sedimentary column tends to be oxidised throughout, but diagenetic reactions continue to appreciable depths in the column, and include the formation of manganese nodules. Some of the oxyhydroxides in these nodules appear to be ferromagnetic (Henshaw & Merrill, 1980) and the acquisition of a CRM during diagenesis leads to a deterioration in the definition of the PDDRM record with increasing depth (Opdyke, Glass, Hays & Foster, 1966).

Red sediments

Sediments broadly grouped as 'red' owe their colour to finely-disseminated hematite and generally possess a fairly strong (10^{-4} to 10^{-6} $A.m^2kg^{-1}$) NRM. They occur most commonly in arid to semi-arid, fluviatile and aeolian sequences, and in certain shallow-marine and near-shore environments associated with moist climates. Their widespread development in the past has been favoured by low sea-levels and broad continental areas in intermediate to low latitudes, and they are well represented in late Precambrian to early Cambrian, Siluro-Devonian (Old Red

Sandstone), Permo-Triassic (New Red Sandstone) and Holocene–Pleistocene successions. In addition, red sandstone and arkose successions formed sporadically as graben infills during Proterozoic times after *ca.* 2250 Ma. The content of both Fe^{2+} and Fe^{3+} generally increases with decreasing grain-size, indicating a close association of hematite with silt and clay fractions (Turner, 1980). The alteration of silicate phases is an important cause of hematite formation in these sediments (Van Houten, 1973).

The hematite occurs as two principal phases, namely detrital and well-crystallised (specularite) grains in the size range 5–500 μm, and as a fine-grained diagenetic phase coating the detrital silicate grains, with a grain-size mostly <2 μm (Collinson, 1965a). The specularite grains are usually considered to carry a DRM or PDDRM, but they may also carry a CRM, because, generally, the grains are not entirely of detrital origin. Authigenic overgrowths occur commonly, and in some instances there is a suggestion that the entire grain may have grown *in situ* (Van Houten, 1968). The reported textures include single crystals, polycrystalline grains, crystals with ilmenite exsolution lamellae, and 'martite' or hematite grains with triangular exsolution lamellae after a cubic (magnetite) host (Walker 1976; Turner & Ixer, 1976). Fine-grained hematite forms a diagenetic phase, coating the detrital silicate grains; as an alteration product of detrital silicates such as pyroxenes and amphiboles; and as clay-oxide pellicles and pellets. Hematite crystallites have been reported to form along cleavage planes and to ultimately pseudomorph detrital biotites (Turner & Archer, 1977), and a common association of microcrystalline hematite with heavy-mineral bands suggests local derivation from a detrital phase. It is probable that the hematite coating on the silicates can originate both from direct precipitation of the breakdown products of unstable silicates and authigenic clays (Walker, 1976, Larson, Walker, Patterson, Hoblitt & Rosenbaum (1982)), and from conversion of an hydroxide phase (Berner, 1969). The only hematite phases which can be unequivocally interpreted as detrital from petrographic observation, are hematite-ilmenite intergrowths of high-temperature origin.

The magnetic components in red sediments are routinely distinguished by thermal demagnetisation and, in many cases, the pigment is found to have a lower blocking temperature spectrum than the specularite grains and the whole rock (e.g. Turner & Ixer, 1976). It may not be possible to recognise the components in this way if they have closely overlapping blocking temperature spectra; a combination of thermal and chemical demagnetisation techniques (third part of Section 5.5) is most appropriate in these cases (Roy and Lapointe, 1978).

As a result of diagenetic changes, red sediments may have a very complex magnetic history. The specularite grains have often been linked to primary DRM and PDDRM components, and the secondary pigment or overgrowths to secondary CRMs, but the distinction may not be as simple as this. In addition, it is possible that a PDDRM and CRM may be interactive for some time, because the chemical changes producing the CRM may occur in a medium that is fluid enough to permit re-alignment of a previously-formed DRM. The work of Walker (1976 see also Walker, Larson & Hoblitt 1981), has influenced some workers (e.g. Turner, 1980) who have argued that most, or all, hematite in red sediments is diagenetic in origin, and therefore that the magnetisations are unrelated to the time of deposition. This viewpoint is supported by the visible presence of authigenic overgrowths and ghost structures which suggest that many of the grains were originally detrital magnetites; it also questions the origin of large amounts of specularite in the sedimentary environment. However, this view conflicts with direct palaeomagnetic observations which suggest that many of these sediments contain prominent DRM or PDDRM components. These observations include:

1. Correlation of the dispersion of directions with grain-size. Irving (1957) for example, found that samples from fine-grained layers which contain a high percentage fraction of specularite grains, also exhibit good alignment of magnetic remanence; this is probably because winnowing action by currents allowed the particles to become well aligned with the magnetic field. In contrast, the directions in adjoining massive sandstone units with only a few percent of uniformly-dispersed specularite grains are more widely scattered, although they still exhibit the same mean direction. In addition, where the surface of a newly-deposited bed has been disturbed by ripple formation, the dispersion of magnetic observations has been found to decrease with depth below the ripple surface (Irving, 1964).

2. Slump horizons have sometimes been observed to have random directions of magnetisation which contrast with the uniformly-magnetised layers above and below, and it is clear that in these instances the remanence in the unslumped beds pre-dates the syn-depositional slumping (Irving, 1964; Smith, Stearn & Piper, 1983).

3. The large and non-systematic variations in magnetisation observed in some red beds are contrary to what would be expected from a later regional overprinting, in particular, the incidence of larger distributions in inclination as compared with declination suggest that the remanence is a DRM (Lovlie, Torsvik, Jelenska & Levandowski, 1984).

4. Although few studies of the susceptibility anisotropy in red sediments identify the preservation of a primary depositional fabric, there are some exceptions. In the Martin formation of Saskatchewan (1830–1650 Ma) for example, Aziz-ur-Rahman, Gough & Evans (1975) observed maxima of magnetic susceptibility (k_{max}) in the bedding plane, with similar directions to the palaeocurrents inferred from cross-bedding; the minima of susceptibility (k_{min}) grouped around the vertical (following correction for tilt). This does not, of course, necessarily mean that the remanence is of depositional origin. In the Martin formation the stable remanence appears to be carried by hematite of uncertain origin (Evans & Bingham, 1973), while the anisotropy results are probably dominated by the response of detrital magnetite grains (Turner, 1980).

This dichotomy of opinion on the origin of remanence in red sediments is largely resolved if the palaeomagnetic results are viewed in the context of the age of the sediment since most of the examples with a remanence attributed to a DRM or PDDRM are Precambrian in age, while most of those with a remanence linked to a diagenetic CRM are of Phanerozoic age. This, in turn, probably reflects the increasing redox potential of the atmosphere through geological time. The Earth's atmosphere had attained only *ca*. 3% of its present level of oxygen by late Precambrian times (Cloud, 1968). Oxygen levels began to rise markedly by 800–600 Ma and they seem to have reached at least 10% of the present level by the end of the Lower Palaeozoic times (e.g. Berkner & Marshall, 1967) to rapidly achieved the present levels by Carboniferous times. Thus, the diagenetic environment has been strongly oxidising only since late Precambrian times. For this reason (and possibly also because oscillating groundwater conditions did not persist for long periods during diagenesis) the magnetisation history of many Precambrian red sediments is remarkably simple. Some of the earliest red beds,* the 2250 Ma Firstbrook member of the Gowganda formation in the Huronian supergroup, have a remanence which is practically single-component and resides in a single specularite phase of probable PDDRM origin (Roy, Lapointe & Anderson, 1975). The 1100–1000 Ma Torridonian sediments have a composite secondary remanence of Caledonian and Recent origin, residing in no more than 10–15% of the total remanence; the remainder is a primary, specularite-carried PDDRM, defined in part by the evidence already noted (Smith, Stearn & Piper, 1983).

In contrast, many Phanerozoic red sediments have a very complex history of magnetisation. Some yield near-random directions of variable intensity and palaeomagnetic stability (e.g. Turner & Archer. 1975); larger specimens can often be subdivided into smaller samples with contrasting directions of stronger NRM, implying that bulk samples contain discrete, partially self-cancelling, zones of NRM acquired at different times during diagenesis (Larson & Walker, 1975). The St. Bees (Triassic) alluvial sandstone is an example of a formation with a complex history of magnetisation, in which a range of components has been demonstrated to reside in several secondary phases, acquired during a protracted diagenetic history (Turner, 1981). There are, however, exceptions: Baag & Helsley (1974) describe rapid and correlatable geomagnetic reversals in the Triassic Moenkopi formation, and perform Fourier analysis of magnetic directions to suggest that remanence in 2·2 cm high cores through the bedding was acquired as a PDDRM, or by rapid diagenesis within a secular variation period of $ca.$ 2000 years. This evidence is compelling, but contrasts with the larger number of other Phanerozoic studies showing dual polarities in the same sample and implying magnetisation over periods of 10^4–10^6 years (Roy & Morris, 1983).

When a DRM or PDDRM is preserved it implies that diagenesis was rapid and involved minimal alteration of detrital grains; this will be the case for mature sandstones, in which all phases except quartz and the stable heavy-mineral fraction have been removed prior to deposition, and when the circulation of groundwater is rapidly inhibited. The DRM or PDDRM components will be wholly or partially replaced by a complex range of CRMs when the sediment contains unstable detrital silicate phases, or if there is a long diagenetic history. Turner (1980) has generalised these considerations in terms of maturity versus diagenetic time fields (Fig. 4.5). The relative importance of diagenetic and primary magnetisations in red sediments have been central to a number of regional and global tectonic analyses of palaeomagnetic data and examples are given in Sections 9.4 and 10.4.

Limestones

Although weakly magnetised, limestones are attractive subjects for palaeomagnetic study, because they can usually be dated precisely from their fossil content, and they frequently occur in thick, continuous successions suited to

---

* The oldest red beds appear to be about 2700 Ma in age and are interleaved with strongly oxidised palagonitised basalts in Archaean greenstone sequences (Shegelski, 1979).

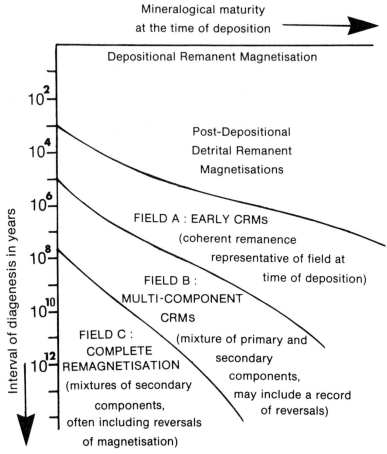

Figure 4.5 The types of magnetism in red sediments as a function of mineralogical maturity and the time period of diagenesis. Modified after Turner (1980) and Tarling (1983). Type A sediments contain a predominantly single component DRM or PDDRM acquired over a short time interval; they are capable of recording secular variation, and geomagnetic polarity transitions are stratigraphically defined. Type B sediments are multicomponent and antiparallel components, which can frequently be demonstrated by sample splitting or a combination of thermal and chemical demagnetisation, indicate acquisition over periods of $10^4 - 10^5$ years. Type C sediments have a remanence residing entirely in diagenetic phases which can often be shown to be syn- or post-folding and to contrast with the palaeomagnetic record of included volcanics; one single polarity is often present and directions are typically well-grouped. Examples of type A are found in Precambrian rocks; they are less common in Phanerozoic rocks but may include some Triassic formations. Examples of type B are found in Late Precambrian and Phanerozoic rocks and many good examples have been recognised in Devonian and Carboniferous rocks. Type C examples have been described from Permo-Triassic and Tertiary to Recent rocks.

magnetostratigraphic studies. Furthermore, limestones appear to accurately record the dipole field without inclination or declination errors, and their remanences often appear to be PDDRMs acquired below the levels of bioturbation, or diagenetic CRMs (Lowrie & Heller, 1982). Many limestones are reported to contain fine grains of magnetite, although magnetic tests (last part of Section 5.6) show that only a fraction of this is single-domain, except in fine-grained limestones which are dominated by SD carriers (Johnson, Lowrie & Kent, 1975). These grains are apparently of detrital origin, because they are associated with terrigenous input: impure bituminous limestones tend to be more strongly magnetic than pale-coloured pure limestones (Turner, Metcalfe & Tarling, 1979). The partial record of polarity transitions (Martin, 1976) and possible secular variation (Turner, 1975) in some limestones, support a PDDRM origin. Tests on some Carboniferous limestones have demonstrated a low net alignment of magnetic vectors which also indicates that the bulk of the magnetism was acquired during early diagenesis. In other cases, notably when the limestone has been subject to dolomitisation, the magnetite has grown as an authigenic phase, and takes the form of pure magnetite spheroids, sometimes, in excess of 100 μm in size (McCabe, Van der Voo, Peacor, Scotese & Freeman, 1983; Horton & Geissman, 1984). In principle, it may be possible to distinguish depositional and diagenetic fabrics by study of the magnetic anisotropy: a primary magnetite depositional fabric, with k_{max} in the bedding plane, can be partially or completely replaced as a result of dolomitisation and silicification because these effects are accompanied by oxidation of magnetite to hematite. This results in a secondary fabric caused by crystalline anisotropy, which is characterised by k_{max} at high angles to the bedding (Addison, Turner & Tarling, 1985). Buff-coloured limestones generally owe their colour to secondary oxidation of detrital magnetite and sulphides to limonite; goethite may then be responsible for a secondary remanence (Section 2.5). Goethite is sometimes observed as cubic pseudomorphs after pyrite grains, and it may convert to hematite. Hematite is not an important remanence carrier in limestones, and it is generally interpreted as a diagenetic phase from a goethite precursor.

Magnetite is also precipitated biologically in the teeth of chitons (a class of molluscs) where it takes the form of closely packed crystallites *ca.* 0·1 μm in size, falling largely into the SD size range of Fig. 3.3 (Kirschvink & Lowenstam, 1979). The bulk of this accumulates in sediments of intertidal and shallow subtidal origin, but some species live in deeper marine environments down to a maximum depth of 7000 m. Magnetic interactions between adjoining grains drastically reduce the resultant magnetic effect, but it is still capable of explaining the remanent intensities of some shelf limestones (Lowrie & Heller, 1982). In addition, certain aquatic mud bacteria have been found to contain linear chains of 5–20 magnetite grains, each *ca.* 0·1 μ in size (Blakemore, 1975; Blakemore & Frankel, 1981). These bacteria are magnetotactic and tend to swim along the lines of force of the Earth's magnetic field; most other types of bacteria are chemotactic or phototactic. The importance of magnetotactic bacteria through geological time is unknown, but they may possibly have contributed magnetite to a wide variety of sediments, ranging from Lower Proterozoic cherts through to coals, and recent peats and biogenic oozes, with no other obvious magnetic constituent (Kirschvink & Lowenstam, 1979).

Banded ironstone formations

Results from ironstone formations comprise an important part of the oldest palaeomagnetic data (2800–1800 Ma) described in Chapter 7 and also make up a small component of the sedimentary data from Phanerozoic times. The earliest examples are referred to as the Algoman-type iron formations (Gross, 1970) and occur in Archaean greenstone belts older than 2500 Ma. They comprise a reduced facies of ferrous sulphides, carbonates and silicates which alternate with silica-rich layers in thinly-bedded laminated deposits. They are generally interbedded with immature clastic deposits and quartzites and are relatively thin, with strike continuity of only a few tens of kilometres at the most. The close association with volcanics suggests a chemical origin for the iron and silicon, related to fumarolic activity under mildly-reducing and alkaline conditions (Fig. 4.3). The oxide facies may comprise up to 50% of the rock but it is mostly subordinate, with magnetite much more prominent than hematite; these rocks tend to have a maximum susceptibility axis in the plane of the bedding (Symons & Stupavsky, 1980). The oxide bands may grade laterally through carbonates (siderite-dolomite) into sulphides (pyrite-pyrrhotite) in a gradation which reflects the shallow- to deeper-water bathymetry of the primary basin of deposition (Goodwin, 1973).

The Superior-type ironstones are of early Proterozoic age and seem to occur in a remarkably narrow time-stratigraphic interval of *ca.* 2400–2200 Ma (Hutchinson, 1981). The ironstones are associated with shelf deposits, and tend to occur at the top of repeated sequences of dolomite, quartzite, and red through to black ferruginous shales; there is often no direct volcanic association, and individual laminae are traceable laterally for great distances, sometimes > 300 km. The oxide facies is much more abundant than the reduced facies, and hematite predominates over magnetite.

The development of banded ironstone formations in Archaean and early Proterozoic times reflects the widespread dispersal of ferrous iron in the sedimentary environment during these times (Goldich, 1973). With the increase in atmospheric oxygen, iron was throughly oxidised to the ferric state, so that it became less mobile in the aqueous environment. Phanerozoic ironstones are predominantly oolitic and pisolitic deposits, associated with shallow-water clastic deposits of littoral, estuarine-deltaic or lagoonal origin, in which hematite or siderite has diagenetically replaced calcareous oolites. They tend to form small lenticular deposits of no great lateral continuity.

4.3 Metamorphic rocks

The elevated temperature and pressure conditions imposed by regional metamorphism, promote the alteration of both the magnetic minerals which have stabilised in the sedimentary environment (Section 4.2) and the metastable minerals which have survived in igneous rocks (Section 4.1). The initial stages are covered by the advanced diagenesis and hydrothermal alteration processes noted in the preceding two sections. The lowest grade effects, associated with the formation of the zeolite minerals, are succeeded by chlorite-grade metamorphism (greenschist facies) at temperatures of 250–300 °C and pressures $\geqslant$ 2 kbar (Fig. 4.1). The Icelandic lava pile is progressively metamorphosed to zeolite grades with increasing depth of burial, and the decline of magnetic properties with depth predicts that zero susceptibility is achieved close to the beginning of greenschist-facies metamorphism (Bleil, Hall, Johnson, Levi & Schonharting, 1982); this depth is *ca.* 4 km beneath the top of the pile in this area, but is likely to be at little more than 1 km depth within most of the oceanic crust where the temperature gradients are higher (Ryall, Hall, Clark & Milligan, 1977). This decline in magnetic properties is a consequence of the incorporation of the iron into silicate phases, and greenschist-facies metamorphic rocks are widely reported to be essentially oxide-free and possess a negligible magnetism (e.g. Fox & Opdyke, 1973).

The iron oxides in metamorphic rocks reflect the conditions of metamorphism only when they have undergone complete recrystallisation. Complete recrystallisation can probably be achieved by protracted regional metamorphism to the upper greenschist–lower amphibolite-facies, since magnetite is widely reported in rock types falling within these facies, including schists and quartzites (Rumble, 1976). Also magnetite may form from the destruction of iron silicates such as biotite. Chemically this magnetite is nearly pure Fe_3O_4, and only at temperatures equivalent to the appearance of sillimanite (*ca.* 600 °C) can significant Ti be accommodated by the structure. Thus, intergrowths in magnetite grains are generally absent and are only reported to appear in the high amphibolite facies (Buddington & Lindsley, 1964). From the higher part of the greenschist facies (*ca.* 400 °C) into the higher grades of metamorphism, intergrowths in ilmenite and hematite become more common at the expense of magnetite, and according to the reaction:

$$\text{rutile} + \text{magnetite} \rightarrow \text{hemo-ilmenite} + \text{ilmeno-hematite}.^*$$

The compositions of the lamellae appear to reflect the metamorphic grade: the molecular fraction of hematite in the ilmeno-hematite increases from *ca.* 10% in low garnet and staurolite grades (*ca.* 500 °C) to more than 30% in the sillimanite grade (600–800 °C), while the ilmenite molecule in the ilmeno-hematite increases from *ca.* 30% to nearly 50% (Rumble, 1976). The progressive replacement of homogeneous titanomagnetite by hematite-ilmenite intergrowths is reflected in an increase in stability of magnetic remanence from the upper greenschist to the granulite facies (e.g. Piper 1981a). The extent to which this reaction proceeds will depend on the oxygen fugacity of the metamorphic environment. At the present time there are two contrasting views on the subject of redox potentials during regional metamorphism. One view is that the bulk chemical analyses and the overall geochemical cycle can be interpreted in terms of large-scale reduction during metamorphism, through the agency of organic matter buried with sediments, and/or by diffusion of H_2 from the mantle into the crust (Eugster, 1972). The alternative view is that the mineral reactions cited as evidence for reduction are in fact dehydration reactions, and that the mineral assemblages actually define widely varying degrees of oxidation (J. B. Thompson, 1972). The experimental evidence suggests that metamorphic rocks crystallise under com-

---

* Hemo-ilmenite is a two-phase grain, comprising lamellae of titanohematite (hematite with variable quantities of $FeTiO_3$ in solid solution) enclosed by ferro-ilmenite (ilmenite with variable amounts of Fe_2O_3 in solid solution) host. In ilmeno-hematite, titanohematite is the host.

parable oxygen fugacities to their plutonic igneous equivalents (Buddington & Lindsley, 1964). Magnetite is ubiquitous in the highest grades of the amphibolite facies, with estimated formation temperatures of 600–640°C but it is only sporadically reported in the granulite-facies rocks, with formation temperatures in the range 620–665°C (Buddington & Lindsley, 1964). No clear systematic change in the Fe_2O_3: FeO ratio with increasing grade has yet been demonstrated (Rumble, 1976).

Few studies have yet attempted to relate the stable magnetisms observed in metamorphic rocks to the oxide mineralogy but the remanence probably has a similar origin to the remanence in the plutonic igneous environment (Section 4.1). In a Proterozoic meta-dolerite that has experienced retrogressive metamorphism during uplift-related cooling, Morgan & Smith (1981) show that the large magnetite grains have MD structure with low coercivities and low blocking temperatures, while 0·01–0.5 μm size sheets and rods in palagioclase crystals are responsible for a high-coercivity remanence.

5

Field and laboratory methods

5.1 Objectives

A palaeomagnetic study is designed to determine the mean direction and/or intensity of the ancient geomagnetic field over the time interval, or intervals, during which a rock acquired its NRM. A number of factors need to be taken into consideration when designing such a study. Since the geomagnetic field undergoes a secular variation, with directional changes of *ca*. $10°-25°$ in 1000 years, it will normally be necessary to sample at a number of localities where the magnetisations are likely to have been acquired at different times. This is to gain an appreciation of the secular variation, and to derive a good estimate of the average field behaviour. It is particularly important when dealing with units such as lava flows, which cool through their blocking-temperature range over periods of time which are short compared with the secular variation cycles. At the other extreme, slowly-cooled metamorphic rocks may record a long time-average of the geomagnetic field, so that, in principle, a single sample may be sufficient to yield a palaeomagnetic direction. Long-term changes in the average geomagnetic field are caused by plate movements and polar wandering which have taken place at irregular rates of up to several degrees per million years (Section 12.8). These changes are evaluated by individual studies of successive geological formations. Inversions of magnetisation may be anticipated in a comprehensive survey and can prove useful as stratigraphic markers and also assist with the interpretation of NRM. Provided that the reversed field axis is precisely antiparallel to the normal axis (exceptions to this general rule are noted in Section 7.9), inversions will not normally affect the mean direction of magnetisation defined by the study. Since inversions take place over a time interval (10^3-10^4 years) which is short compared with the interval typically sampled by a palaeomagnetic survey, intermediate directions of the geomagnetic field recorded during an inversion are likely to comprise only a small fraction of the total collection; they are usually separated from the main body of data by employing an exclusion criterion.

5.2 Field sampling

A single outcrop which can be considered to have been magnetised over a short period of time (1–100 years) such as a dyke, a lava flow, or (in some instances) a single sedimentary bed, is regarded as a *site* for the purposes of palaeomagnetic sampling. It will normally be expected to yield a single spot record of the geomagnetic field direction, and hence a virtual geomagnetic pole (First part of Section 6.5); single outcrops in plutonic rocks are also traditionally regarded as sites, although they may then be expected to yield a time-averaged palaeomagnetic direction. Several pieces, or *samples*, distributed over a few square metres of the outcrop will normally be individually oriented at one site, in order to minimise orientation errors and test for variations in the magnetisation record. One or more *specimens*, which are usually drilled and sliced into cores, may then be cut from each of these samples for laboratory measurements. The deflection of a compass needle may be used to determine the approximate remanence direction in strongly magnetised rocks; this method has been widely applied to define the normal or reversed nature of the remanence in young Cenozoic volcanics (e.g. Einarsson, 1957) and it is an effective stratigraphic mapping tool in such environments, provided that the rocks have not acquired appreciable components of VRM or IRM. Although portable magnetometers and susceptibility meters may be used to test the NRM and Köenigsberger ratio at outcrops, it is, unfortunately, often true that only laboratory measurement and demagnetisation studies can decide the number of sites and samples necessary for adequate definition of the palaeomagnetic field. It may also be necessary to evaluate reasons for the scatter of magnetisations between samples and within samples — if these are appreciably different — and thus it is frequently necessary to follow up initial field studies with further sampling.

Five to eight specimens cut from three to six samples are usually considered adequate to define the NRM direction. Little advantage is gained by more detailed sampling at one site (Doell & Cox, 1963) unless more than one component of magnetisation is present. A minimum of five sites is necessary to obtain a reasonable average of the geomagnetic field direction, but it is preferable to have many more.

Block samples are obtained by first breaking off suitable pieces of outcrop, trimming as necessary, and then replacing in the original position for orientation. The orientation may be recorded either by drawing a strike and dip line on a single flat surface, using a compass and clinometer, or by marking a horizontal line around as much of the side of the block as possible and recording the azimuth of the magnetic field as arrowed lines across the top surface. If a flat surface is employed, the sample is reoriented in the laboratory with this surface horizontal and drilled with a vertical coring press. If the alternative method is used, the block can be mounted in a plaster mould in the field orientation for laboratory coring. In either case, the remanent magnetism of the outcrop may cause errors in the compass readings, and this should be tested for by sighting located topographic positions from the sample position. The problem can be overcome by employing a sun compass. The azimuth of the shadow cast by the sun on a horizontal plate graduated counter-clockwise in degrees is then recorded, together with the time of day to the nearest minute, and the coordinates of the site location. The method for calculating the azimuth from a sun compass measurement is given by Creer & Sanver (1967), and a simple sun compass suitable for the orientation of block samples is described by McElhinny (1973).

There are three main objections to block sampling. Firstly, the only samples which can be readily collected are associated with joints and projecting outcrops, which tend to be more weathered than the bulk of the exposure. Secondly, the precision of field orientation barely meets laboratory standards and further accuracy is lost during reorientation for laboratory drilling. Thirdly, a large weight and volume of material is accumulated from a small number of sampled units. The first problem may be locally overcome by utilising fresh road-cut and quarry exposures, but it is preferable, wherever logistically possible, to surmount all of these problems with the use of portable drilling apparatus. Palaeomagnetic drills are usually based on direct drive two-stroke chain-saw motors, modified to take a tubular steel drill-bit with a diamond-impregnated tip. They incorporate a pressurised water feed attachment to cool the drill-bit and wash out the rock dust from the drill-hole. An early example is described by Doell & Cox (1967) but several lightweight motors are now available which can be adapted for small-scale rock drilling operations. Drilled cores are oriented prior to extraction, by inserting a slotted brass or aluminium tube into the hole. The tube incorporates an inclinometer and a plate containing a sun compass and it can be adapted for magnetic compass readings if a sun-sight is not possible. The inclinometer is used to measure the inclination of the core axis above or below the horizontal, and the azimuth of the core is determined from the compass attachment. The core is then marked along a slot at the top or bottom of the tube by inserting a thin piece of brass wire. The rock core can be broken off at the base by tapping a blunt rod into the drill-cut; cores which do not break off conveniently can usually be reconstituted with adhesive. A diamond scribe and ink marker are used to make the orientation reference-line permanent, and to mark a symbol along it to indicate the top and bottom of the core.

A magnetisation referred to as *drilling-induced remanence* (DIR) is sometimes produced by the drilling operations involved in both types of sampling. It has been studied in some detail by Burmester (1977) and appears to be confined to rocks containing large MD magnetite grains; it is generally considered that a primary NRM of sufficient stability to yield useful palaeomagnetic information is unlikely to be affected by the drilling and slicing procedures. For further details of the collecting method, and the techniques for sampling soft sediments and archaeological materials, refer to Collinson (1983).

5.3 Magnetometers

All the standard techniques of palaeomagnetic measurement assume that the remanence of the sample corresponds with a simple dipole magnet at the centre of the specimen. This will only be precisely true if the specimen is a homogeneously magnetised sphere, and if the measuring instrument is far enough from the specimen. The most convenient practical approximation to this ideal shape is usually a cylinder cut from a drill-core, or sometimes a cube when cut from a soft sediment. Provided that the length:diameter ratio of the cylinder is in the range

0·8–0·9, the errors caused by the sample shape are less than the random errors of measurement (Collinson, 1983). This ratio is of more critical importance to magnetic anisotropy investigations, because of the self-demagnetising field caused by the shape effect (Stacey & Banerjee, 1974); an optimum ratio of *ca.* 0·87 is required for these studies. Rock samples are usually inhomogeneously magnetised, and the measurement procedure is designed to average this effect. To a convenient approximation, the inhomogeneity can be represented by a displacement of the sample dipole from the centre of the sample; it is then equivalent to a centred dipole plus a quadrupole. The components of the quadrupole can be cancelled by successive measurements, provided that sufficient orientations of the sample are introduced in turn to the measuring system.

Astatic and parastatic magnetometers

The astatic magnetometer is based on the behaviour of a system of two equal, antiparallel magnets suspended on a torsion fibre. It can be made insensitive to changes in the uniform ambient magnetic field, while responding to a field gradient produced by the remanence of a specimen positioned below the system. High sensitivity is achieved by carefully balancing the magnets, and making the controlling torque on the system very weak (Blackett, 1952). Performance can be further increased by using a number of small magnets in place of one larger one,

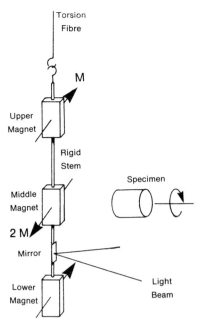

Figure 5.1   The Parastatic Magnetometer. This system is preferable to the astatic system in magnetically noisy environments because it is insensitive to temporal changes in the gradient of the field provided the gradient is uniform. Sensitivity can be improved by keeping the moment of inertia of the system to a minimum and equalising, or astaticising, the opposing magnets as completely as possible.

and by using a composite (parastatic) system such as that illustrated in Fig. 5.1. A perfect astatic system responds to the difference in horizontal field at the top and bottom magnets, and measures the vertical gradient dH/dZ of the horizontal field. A parastatic system is equivalent to two astatic systems, with the middle two magnets pointing in the same direction. It is unaffected by a horizontal field with a uniform vertical gradient, and measures the second vertical derivative d^2H/dZ^2 of the horizontal field. It therefore permits measurement of the sample in a side-on position (Fig. 5.1). Inhomogeneities can be cancelled in these systems by spinning the sample about the axis whose component is being measured, at a rate which is fast compared with the response time of the deflecting magnet system. Rotation of the magnet system was originally measured by the deflection of a light beam from an attached mirror, but it is nowadays routinely measured in terms of the back-off current required to produce a magnetic field nullifying the deflection. The recorded magnetisation is remanent if measured in a zero magnetic field or the resultant of remanent and induced magnetisations if measured in the Earth's field. The low-field susceptibility can therefore also be determined by comparison of the two measurements. The remanent component is isolated by measuring the anti-parallel deflections along three orthogonal axes and combining them to give the vector resultant, and/or the magnetometer can be oriented so that the Earth's field is perpendicular to the component being measured. The system is calibrated using a standard coil.

Spinner magnetometers

These magnetometers utilise the alternating voltage produced when a specimen is spun near a detecting coil, and about an axis in the plane of the coil. The amplitude of the e.m.f. is proportional to the component of the magnetic moment perpendicular to the rotation axis, and the phase can be used to relate the direction of the component to a reference direction in the sample. The induced e.m.f.s are measured about three orthogonal axes and compounded in the same way as the readings from astatic and parastatic magnetometers. Most spinners operate at frequencies of 5–100 Herz. Higher frequency air turbine spinners operate at 150–500 Herz to give an increased signal:noise ratio, but cannot be used for more friable samples, for which extra sensitivity is often required. The magnetic moment is normally measured with a flux-gate system, consisting of two flux-sensitive probes stationed in opposition near the rotating specimen (Fig. 5.2). As the sample rotates, the flux-gate probes are subjected to a fluctuating magnetic field, and the resulting signal is compared with a reference signal (which can be generated from the rotating system in a variety of different ways). The phase difference between the reference and signal voltages is proportional to the angle between the direction of the measured component and a fixed direction in the sample holder. In the Digico system (Molyneux, 1971) the signal is sampled by measurements taken 128 times each cycle and timed by the illumination of a photocell through radial slots in a rotating disc. The measurements are integrated over several cycles, and subjected to Fourier analysis by an on-line computer to produce phase and intensity information. The development of spinner magnetometers is described by Collinson (1983).

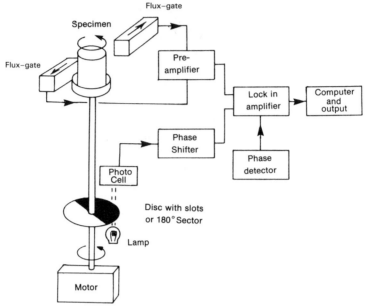

Figure 5.2 Block diagram of a Spinner Magnetometer. The sensing device can be either a pick up coil with a secondary coil in series designed to reduce the effects of strong externally produced alternating fields, or a system of one, two, or four fluxgate sensors; in the latter case it is necessary to employ magnetic shielding so that variations in direct and alternating ambient fields do not influence the detectors.

Cryogenic magnetometers

Parastatic and spinner magnetisations can be built to comparable sensitivities and are adequate for measuring NRMs of magnitude $>ca.\ 1 \times 10^{-6}\ A.m^2\,kg^{-1}$. The development of cryogenic technology has two important implications for palaeomagnetic investigations:

1. Remanence can be measured with a single insertion into the sense coils, because only one axis of the sample needs to be presented to the magnetometer (although repeat measurements of components are desirable). This permits magnetic and/or thermal cleaning to be automated with the measuring procedure without removing the sample from a field-free space, and it thus obviates the problem of viscous remanence acquisition between treatment steps.
2. The increase in sensitivity by one to three orders of magnitude, and the rapidity of the measurements permit study of a wider range of rock types than has been possible hitherto, and allow their magnetisations to be accurately resolved by repeated measurement.

Cryogenic magnetometers are based on the properties of a superconducting ring made of a metal with perfect diamagnetism, which causes the expulsion of all magnetic flux from the ring below a certain temperature; this is the *Meissner effect*. A ring has the property such that when an external field is applied to it while it is in the superconducting state, all flux is expelled from the interior of the metal, but not from the hole within the ring, so that induced currents flow in one direction on the outer face of the ring and in the opposite direction on the inner face. If the applied field is removed, the currents on the outside of the ring disappear, but the current on the inside face persists, to maintain the field within the ring at the same value. In this way, the magnetic field in which the ring was originally cooled is effectively trapped within the hole. Consideration of quantum mechanics shows that this trapped flux cannot take any arbitrary value, but is quantised into multiples of $h/2e$ (h = Planck's constant, e = the electronic charge); this is known as the flux quantum, ϕ and has the value $2 \cdot 07 \times 10^{-15}$ Weber. The ability of the superconductor to trap and maintain magnetic fields to better than one part in 10^9, can be utilised in magnetometry by trapping a zero field within the loop and then using this volume for measurements. The trapping ability is exploited by causing the ring to operate at the superconducting:resistive boundary and employing a sensing device known as the SQUID (superconducting quantum interference detector). This state can be achieved by raising the applied magnetic field to a point where the flux-cancelling current in the ring exceeds a certain critical value; the superconducting condition is then passed and the loop becomes resistive. This in turn causes the flux to enter the material of the ring to produce a back e.m.f., thus reducing the current circulating in the loop to below its critical level, and returning it to the superconducting state. Since the ring current is quantised, the flux entering the ring at this transition state is also quantised and the response of the loop to an increasing field has the sawtooth form shown in Fig. 5.3a. The magnitude of the critical current is a function of the cross-sectional area, and it can

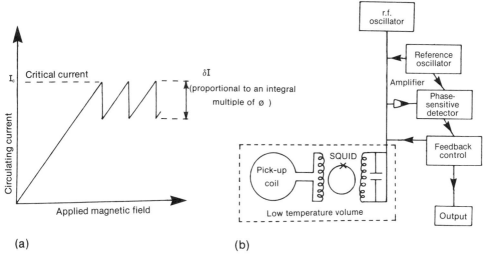

(a)   (b)

Figure 5.3 (a) *The circulating current in a superconducting ring as a function of applied magnetic field.* (b) *Block diagram of a cryogenic magnetometer. The ring is driven by an external field of just sufficient magnitude to cause the circulating current to exceed I_c. At each point in the cycle where I_c is exceeded a flux quantum enters the ring and the change in flux linkage within the ring can be detected by a voltage spike induced in the pick up coil. If a magnetic sample is introduced the resulting current in the ring causes the resistive switching to occur at a different point in each cycle and change the character of the output voltage. The changes detected by the pick-up coil are a measure of the change of magnetic field in the ring compared with the field present when it entered the superconducting state.*

be made very small if the current is concentrated through a small cross-sectional area (weak link) at one point.

In the magnetometer system (Fig. 5.3b) the loop carrying the controlled critical current I_c is inductively coupled to a tuned circuit externally excited at its resonant frequency in the r.f. range. The reversal of SQUID flux then takes place during a single r.f. cycle. When a d.c. flux is present it causes a d.c. shift of the function in Fig. 5.3a so that the peaks move to the left or right depending on the polarity of the flux. The effect is measured in terms of a feed-back current from a system known as the 'flux locked loop' required to cancel this d.c. shift.

The magnetometer is surrounded by an insulated evacuated space cooled by helium or a combination of nitrogen and helium. Pick-up coils connected to SQUID detectors are employed to measure the NRM along one axis or simultaneously along two or three mutually perpendicular axes. Mumetal shields are employed to blanket external field variations, and automated systems are presently under development to complete a.f. and thermal demagnetisation within this field-free environment (Shaw, Share & Rogers, 1984).

5.4 Susceptibility meters

Susceptibility measurements may be used to determine the Köenigsberger ratio, and hence to estimate the domain distribution and likely magnetic stability (Section 1.3) of rock samples. They are also employed in the interpretation of magnetic anomalies to compare the contributions of the remanent and induced components of magnetisation to the observed field, and they may be used to monitor any chemical changes which occur in the magnetic phases during progressive thermal demagnetisation (second part of Section 5.5).

The low-field susceptibility can be determined by astatic magnetometers, but it is more usual to employ an alternating field method, which is both independent of the sample NRM and rapid to use. Several systems are currently in use and are based either on transformer bridge circuits or balancing a change in the mutual inductance between two coils (Collinson, 1983). An example of the latter type is illustrated in Fig. 5.4. It comprises two ferrite rings, each containing an air gap. The primary windings carry on alternating current to produce a magnetic field of *ca*. 1 milliTesla across the air gaps. The ferrite chip is used initially to balance the circuit. In the balanced state there is no output when the primary coils are excited by a low-frequency oscillator, but when a susceptible rock sample is placed between the pole piece of one magnet, it alters the reluctance of the circuit and upsets the balance. The resultant signal is used as a measure of the susceptibility. The bridge can be calibrated over low susceptibility ranges by using pure paramagnetic chemicals of known susceptibility, in perspex containers equivalent to the rock shape. At higher values, the linearity of the bridge must usually be tested for, by using known amounts of dispersed magnetite powder, or standard rock samples measured by other systems.

A bridge system of this kind can be adapted to determine an anisotropy of susceptibility by rotating the sample between the pole pieces of the loop and measuring the susceptibility differences in three mutually perpendicular planes (Stone, 1967a). There are certain practical difficulties with this system, and it lacks

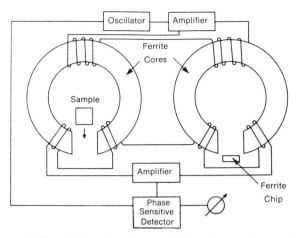

Figure 5.4 A susceptibility Bridge based on a balanced transformer circuit with ferrite-cored coils. After Collinson (1983).

sensitivity. The *torsion balance* or *torquemeter* is more commonly used to determine the anisotropy of susceptibility (King & Rees, 1962). In this instrument, the sample is suspended by a torsion fibre and subjected to a horizontal alternating field, applied through a pair of Helmholtz coils (Fig. 5.5a). During one half cycle, there is a couple between the applied field and the permanent magnetism, which tends to twist the sample in one direction, and during the other half cycle this couple acts in the opposite direction (Fig. 5.5b). Provided that the natural frequency of vibration of the suspended system is much lower than the frequency of the applied field, there will be no net deflection due to the remanent magnetism. If the sample has an anisotropy of susceptibility, the applied field and the induced magnetisation will not, in general, be parallel, and a resultant couple will be produced which tends to align the axis of easiest magnetisation or highest susceptibility with the applied field. This couple is in the same sense during both halves of the a.f. cycle. If the coil carrying the a.f. current is then rotated about the specimen, it will pass four points in each revolution when the induced moment and the applied field are parallel. At these points there is no resultant deflection and the applied field is crossing the planes containing the axes of the ellipse of susceptibility; they will be complemented by four maximum deflections when the applied field is at 45° to the susceptibility axes, so that the complete rotation produces a double sine torque curve (Fig. 5.5c). The voltage maxima are related to the difference $k_1 - k_2$ between the principal susceptibilities in the plane of measurement. Three sets of measurements in three planes mutually at right angles are required to determine the susceptibility tensor.

The torsion balance is sensitive, but it is time-consuming to use, and more rapid measurements are possible by adapting spinner magnetometers to determine anisotropy of susceptibility. When a sample is spun in a d.c. field, a signal is produced by the anisotropy with a frequency of twice the spinning frequency. This signal is compounded with a signal at the rotation frequency due to the NRM; the latter signal can be filtered out. The amplitude of the first signal is proportional to the applied field and to the anisotropy in the plane at right angles to the spinn-

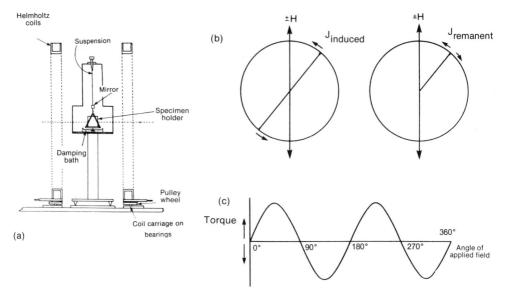

Figure 5.5 (a) *Section through a simple torsion balance.* (b) *Torque on a suspended sample as the direction of the alternating field is rotated through 360°.* (c) *Torque on a sample suspended in an alternating magnetic field due to the induced and remanent magnetisations. (After Stone (1967b)).*

ing axis. The *anisotropy delineator* is a modification of this system, in which the sample is located between two mutually perpendicular coils. A current passed through one of the coils produces an alternating magnetic field over the sample. If the specimen is anisotropic, the field is distorted and a sinusoidal voltage of twice the rotation frequency is induced in the second coil; this signal has an amplitude proportional to the difference in the principal susceptibilities in the plane perpendicular to the spin direction. The waveform is Fourier-analysed by computer, and the complete susceptibility tensor is recovered by spinning the sample about three mutually perpendicular axes (Collinson, 1983).

5.5 Demagnetisation techniques

Although a number of field tests (Section 6.4) can be applied to recognise the presence of primary and secondary magnetic components in a rock, partial demagnetisation treatment must be applied in the laboratory to properly isolate and determine these components. Demagnetisation or 'cleaning' methods aim to randomise or eliminate the magnetisations with lowest relaxation times and leave the relaxation times much longer than the age of the rock and most likely to contain the bulk of the primary remanence. If they are applied in a stepwise fashion, they will determine the spectrum of relaxation times in the sample. In practice, some of the primary remanence will normally be subtracted with the secondary remanence, but this will not be important, provided that the residual primary remanence is still sufficiently strong to be measured, or that it can still be recovered by a technique such as vector subtraction or remagnetisation circles (Section 6.3).

Alternating field (a.f.) demagnetisation

Since the magnetic minerals in a rock have a range of grain-sizes and shapes, they will also have a distribution of coercive forces known as the coercive force spectrum (Section 3.3). The alternating field (a.f.) method of demagnetisation subjects the sample to a peak alternating magnetic field, H, which is then smoothly reduced to zero. If θ is the angle between the remanence and the applied field, H, all SD grains with magnetisations of $<$ H cos θ follow the field as it alternates. As the field decreases, the sample is cycled through hysteresis loops of decreasing amplitude, so that domains with progressively lower coercive force are left stranded along their easy axes of magnetisation. Hence, all single domains with magnetisations $<$ H are effectively randomised, provided that the grains are randomly oriented in the rock. The multi-domain particles will be demagnetised by domain wall movement, or they will be left with randomly-directed residual moments. The peak field, H, may be incremented in stages, such as 5 or 10 milliTesla, so that the lower part of the coercive force spectrum is selectively removed to leave a progressively narrower portion of the higher part of the spectrum.

Multi-domain magnetite grains are demagnetised in fields of a few tens of milliTesla or less, and it is primarily the SD grains of importance in palaeomagnetism which can survive appreciable a.f. treatment. The coercive force of these particles is a function of their shape anisotropy (Section 3.3 and Fig. 3.3) and elongate grains may require fields of >250 mT. for demagnetisation. The a.f. cleaning method is very effective when applied to titanomagnetites, due to the general relationship between unblocking alternating fields and temperatures (Figs 3.3 & 3.4). The unblocking field curves cut across the unblocking-temperature curves, and therefore a given a.f. field can demagnetise grains with a wide spectrum of blocking temperatures. In addition, because of the angle the unblocking-field curves make with the blocking temperature curves, any given a.f. field will probably (depending on the distribution of lengths and axial ratios of the grains present), demagnetise more grains with lower blocking temperatures than with higher blocking temperatures. The effect is to isolate those remanences residing in grains with the longest relaxation times.

A TRM and CRM acquired in a given magnetic field have similar demagnetisation characteristics (Kobayashi, 1959) and have a much higher coercivity than an IRM which is destroyed by a field comparable with the inducing field (Fig. 5.5). A VRM, however, requires a peak demagnetising field larger than the field in which it was acquired; the actual value will be a function of the time period over which the magnetising field has operated, but since it is generally less than the coercivity of the TRM or CRM, the a.f. method is an effective technique for removing the IRM and VRM and isolating the primary remanence. Due to its very strong crystalline anisotropy, hematite requires demagnetising fields well in excess of 500 mT. for effective demagnetisation. Coils developing fields of this magnitude cannot be constructed without introducing significant magnetising fields into the cleaning procedure, and a.f. cleaning is not usually a practical proposition for demagnetising the remanence carried by this mineral. Fig. 5.6 illustrates the way in which the normalised intensity of magnetisation diminishes with progressive a.f. treatment in different mineral phases. Many examples of directional behaviour with a.f. treatment are illustrated in the literature.

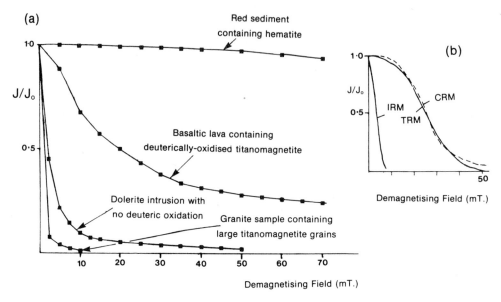

Figure 5.6 (a) A.f. demagnetisation behaviour of some contrasting rock types (the dolerite and granite begin to exhibit erratic behaviour in demagnetising fields higher than those illustrated). (b) The demagnetisation behaviour of TRM, CRM and IRM in titanomagnetite grains after Kobayashi (1959).

If the demagnetisation cycle takes place in the presence of a constant field an anhysteritic remanent magnetisation (ARM) is acquired. ARMs are unknown in nature, but they have properties comparable with a TRM, and may complete dominate the remaining NRM. Hence, it is important to reduce the steady magnetic field around the sample to a low value during the magnetic cleaning procedure. It is also important for the a.f. signal to be as near to a pure sinusoidal form as possible, because random transient signals or harmonics of the a.f. waveform may induce IRMs into the sample. In addition, a rotational remanent magnetisation (RRM) is sometimes acquired when a sample is rotated in an alternating magnetic field (Wilson & Lomax, 1972; Edwards, 1980); it is believed to be related to the phenomenon of gyromagnetic remanence (Stephenson, 1980). Collectively these effects limit the usefulness of magnetic cleaning in high fields; they are reflected in the tendency for directional and intensity information to become erratic at advanced stages of treatment, and it is generally not possible to recover the highest part of the coercivity spectrum by this method.

The a.f. demagnetisation equipment comprises a coil tuned by a bank of capacitors and connected to the power supply, which will in turn be controlled by some form of alternating current reduction mechanism. The specimen holder is located within the part of the coil with the most uniform flux distribution. This system is contained within a low-field coil system or magnetic shielding, designed to reduce the ambient steady field to a minimum. Rotating the sample about two or three axes during demagnetisation is known as *tumbling*; it allows more efficient cleaning by exposing many sample directions to the applied field, and helps to randomise any components of ARM. The relative rates of rotation about these axes

Field and laboratory methods

are designed to optimise these two objectives (McElhinny, 1966; Hutchins, 1967). RRM effects can be largely eliminated by reversing the sense of rotation of the tumbler at short intervals.

Thermal demagnetisation

The relaxation time of a magnetisation is strongly dependent on the absolute temperature (equation 3.13). Thus, if the temperature of a sample is raised until the relaxation time of the lower part of its blocking temperature spectrum is reduced to a period of a few minutes or so, this lower part will be unblocked in the time period of the experiment, and cease to contribute to the total NRM. Provided that the cooling takes place in zero field, the grains carrying this unblocked fraction will not acquire a new PTRM. In the same way that progressive a.f. cleaning aims to isolate narrower fractions of the high part of the coercivity spectrum, so progressive thermal demagnetisation to increasing temperatures can succeed in isolating narrower and higher portions of the blocking-temperature spectrum (Fig. 5.7). The resulting spectrum may be distributed or discrete (Section 3.10 and Fig. 5.7). Sometimes, important differences are present between the results from a.f. and thermal cleaning, because of the complex relationship between the coercive force spectrum and the blocking temperature spectrum. Also, components can sometimes be isolated by thermal cleaning, but not by a.f. cleaning. Thermal cleaning is most effectively applied to hematite-bearing rocks, both because they cannot generally be treated by the a.f. method, and because they cannot undergo

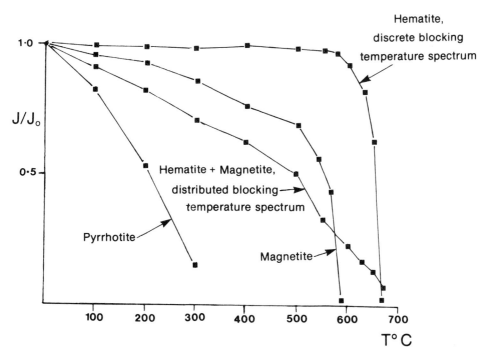

Figure 5.7 Some typical thermal demagnetisation (J/J_o versus temperature) graphs of rock samples possessing pyrrhotite, magnetite and hematite remanences.

further oxidation during heating; it is also effective for removing PTRMs in both hematite and magnetite acquired by thermal processes, such as the burial of sedimentary rocks and the slow cooling of plutonic rocks. It is not, however, always possible to separate the components of magnetisation residing in red sediments; this probably reflects a complex mixture of depositional and diagenetic magnetisations which have overlapping grain-size and compositional spectra (third part of Section 4.2).

The stepwise heating is conducted in a furnace made of non-ferrous materials and designed to produce a uniform temperature over the sample volume. The heating coils are wound in a non-inductive way in order to reduce stray magnetic fields to a minimum. The samples, which may range from 4 to 50 in number, are separated in the furnace, so that their magnetic fields do not interfere with one another. The ambient field needs to be cancelled to a degree which depends on the stage of treatment: the cleaning of strongly magnetised rocks to a few hundred degrees may not be significantly affected by a spurious PTRM acquired in a field of 100 nT. but the residual moments remaining just below the Curie point can be dominated by PTRMs acquired in a field as low as 10 nT. Thus, a cancellation of the magnetic field to 1–2 nT. is generally aimed for. This can be achieved by suitable magnetic shielding, or by surrounding the furnace with a large coil system, and a control circuit based on three orthogonal flux-gate sensors which detect variations in the North, East and vertical components of the magnetic field adjacent to the demagnetising unit; the signals from the flux-gates are used to modify the currents in the coils and cancel the field changes. The problem of PTRM acquisition can be overcome by measuring a sample continuously as it is heated. The apparatus incorporates a magnetometer adjacent to, or combined with, the furnace, but it involves severe design problems and can only cope with one sample at a time. The main practical limitation of thermal treatment is encountered when chemical changes (such as magnetite to hematite) and phase changes (such as maghemite to hematite) are promoted by the heating procedure. These effects can be monitored by susceptibility measurements.

It is also possible to thermally demagnetise rocks containing minerals of the magnetite–ilmenite series by cooling them down to low temperatures in a field-free space. It relies on a magnetic transition which takes place at $-143\,°C$ in magnetite, but at progressively lower temperatures with increasing titanium substitution, when the magnetic anisotropy constant, K_1, passes through zero as it changes sign. Multi-domain remanence is destroyed by cooling to this temperature, while SD and PSD magnetisations are largely unaffected. This response has been studied by a number of workers, notably Ozima, Ozima & Nagata (1964) and Merrill (1970) and is usually achieved by repeatedly cycling the sample down to the liquid nitrogen temperature ($-196\,°C$); it has not yet been generally applied as a routine cleaning technique.

Goethite-bearing rocks are difficult to treat with standard methods, although some form of treatment is essential, because the low Néel temperature means that this mineral is very susceptible to VRM acquisition. Owing to its high coercivity, it is virtually unaffected by a.f. cleaning, while the conversion to hematite can obscure the demagnetisation record at relatively low temperatures during thermal demagnetisation. The most effective treatment is to boil goethite-bearing samples in water and cool them in a zero field (Lowrie & Heller, 1982).

Chemical demagnetisation

This technique is now widely employed to separate the magnetic components in red sediments. It was first described by Collinson (1965b), who demonstrated that immersion of rock samples in concentrated hydrochloric acid selectively dissolved the fine grains with high surface area:volume ratios, while leaving the large grains with low surface area:volume ratios intact. Since the former typically include the diagenetic hematite pigment carrying a CRM, and the latter include the detrital specularite grains carrying a PDDRM, or at least a CRM of different origin, it is possible to separate the two components by this method. The samples are periodically removed from the acid, washed in distilled water and remeasured to monitor the progressive effect of the acid; the iron lost from the specimens can be determined by analysis of the acid solution. Holes may be drilled through the samples to increase the surface area and speed the process (Henry, 1979), and high-pressure cells have been developed to improve acid penetration in low-porosity sediments. An example of a pressure cell that does not require continuous pumping is described by Smith (1979).

If the remanence is dominated by specularite grains it tends to decline with progressive treatment as a concave J/J_0 versus time graph (Fig. 5.8) indicating an increased resistance to acid leaching with time. This probably corresponds with the removal of authigenic overgrowths of relatively high surface area:volume ratios to expose rounded and more resistant detrital grains with lower ratios (Turner, 1981). If the remanence is dominated by pigment, the J/J_0 versus time curves tend to be convex (but over a shorter period of treatment), probably because the pigment grains have a small size distribution and are removed within a narrow time interval. If grains other than hematite are present, they are dissolved in the order: magnetite before goethite, before hematite, before rutile, provided that these

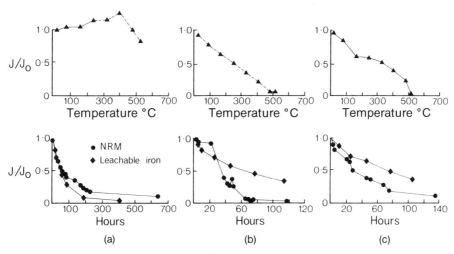

Figure 5.8 Examples of chemical and thermal demagnetisation of samples from the St. Bees red sandstone (Triassic) in which the magnetisation resides in (a) predominantly specularite grains, (b) predominantly pigment, and (c) a combination of both. (After Turner 1981).

phases have similar sizes and crystal developments (Henshaw & Merrill, 1980); since these parameters are usually very variable, however, chemical leaching cannot be reliably used as a criterion for identification.

The correspondence between chemical and thermal demagnetisation may be rather limited, both because the acid leaching dissolves pigment and specularite grains at the same time (although at different rates) and because the pigment may be protected by overgrowths of quartz and feldspar cement, or be enclosed within clay and silicate phases. Although an inverse relationship might be expected, because coercivity in hematite is generally related to particle size, it is usually observed that the pigment has a lower blocking-temperature spectrum than the specularite grains, so that its remanence is removed at an earlier phase of both progressive thermal and chemical treatment. Collinson (1974) for example, studied the contributions of specularite and pigment separates to the bulk NRM in red sediments, and demonstrated that the specularite fraction has a narrower blocking-temperature spectrum and coercivity spectrum than the pigment in most examples. This is probably because the remanence of hematite resides in both the fundamental spin-canted component and the defect component. The latter component is dominant in the fine grain-sizes of the hematite pigment, but is subordinate in the larger specularite grains (Dunlop, 1971) and it has a lower average blocking temperature than the spin-canted moment; the proximity of the pigment grain-sizes to the superparamagnetic transition size (*ca.* $0 \cdot 03$ μm) may also be an important factor here. A complete analysis of the remanence in some red sediments may require the application of both thermal and chemical techniques. In the Carboniferous Maringoan formation for example, Roy & Park (1974) recognised three components removed successfully by acid leaching and residing in the pigment, authigenic overgrowths and specularite grains, respectively; thermal treatment was unable to separate the last two components. An integrated approach applies chemical and thermal cleaning to paired specimens of the same sample (Roy & Lapointe, 1978).

5.6 Laboratory tests for magnetic minerals and domain structure

Domain structure

The complete analysis of a palaeomagnetic remanence requires the identification of the mineral phase carrying the remanence, and an interpretation of its domain state. Optical investigations (Chapter 2 and Table 2.1) are of limited value, because stable remanence is dominated by SD grains with sizes below the limits of resolution by optical analysis. The best that can usually be achieved is to measure the observed grain-sizes and assess whether the tail of the observed distribution extends into the SD range (e.g. Evans & Wayman, 1970). Obviously, the routine a.f. and thermal demagnetisation procedures yield powerful clues for the identification of magnetic remanence carriers: hematite-held remanence normally has high coercivities, while magnetite, maghemite and pyrrhotite-held remanence normally has moderate to low coercivities (Fig. 5.6); hematite-held remanence may be expected to survive thermal treatment to temperatures of *ca.* 680°C, some magnetite-held remanence may be expected to survive temperatures of ≤580°C, while pyrrhotite and maghemite-held remanences usually survive treatment to no more than 300°C or so (Fig. 5.7) although with some impurities present maghemite may survive to 700°C. In addition, a variety of magnetic tests outlined in this section are also routinely applied to investigate composition and domain size of the remanence carriers. They are often simple and effective, but they detect all of the components with a susceptibility, and not just those carrying the natural remanence.

The form of the hysteresis loop can be expressed in terms of the saturation moment and the coercivity; the area enclosed by the loop is a measure of the work done in taking a unit measure of the magnetic material around the cycle once. Its actual shape depends upon both composition and domain structure. However, since the coercivity spectrum of hematite extends above 5 T. while that of a magnetite is <0·1 T., the presence of a high-coercivity component will be a clear indication of hematite (second part of this Section). The absence of a high-coercivity component does not, however, always exclude the presence of hematite, both because only low-coercivity hematite grains may be present, and because magnetite has a much higher saturation magnetisation (Table 2.1) that may be reflected in a higher initial remanence, masking the contribution of small amounts of hematite. Curie point determinations (third part of this Section) are the most effective method for distinguishing between magnetite, maghemite and pyrrhotite. Two tests (concluding part of this Section) are available for the identification of domain structure, but the extent to which they can be applied to all rock types is not yet fully understood.

Isothermal remanence tests

A sample will acquire an IRM when it is exposed to a DC magnetic field. This controlled field is progressively increased until the IRM reaches a saturation value, and the behaviour is expressed as an IRM acquisition curve (Fig. 5.9). Following saturation, an increasing antiparallel DC field is usually applied, and the 'coercivity of remanence' is determined as the back field required to reduce the saturation remanence to zero; this is called the *destructive field*.

IRM tests can be used to detect very small amounts of magnetic material, and they are mainly employed to distinguish members of the magnetite–ulvöspinel series from the hematite–ilmenite series. Magnetite usually saturates in fields of <150 mT. while the fields required to saturate hematite range from 200 mT. for coarse specular hematite to 2 Tesla or more for fine pigmentary hematite; the higher part of this range is seldom reached by standard electromagnets. Maghemite and pyrrhotite exhibit similar IRM behaviour to magnetite. Since the saturation magnetisation of hematite is much less than that of magnetite, a small amount of magnetite can dominate the IRM acquisition curve when both minerals are present; the hematite may, however, be recognised by a gradual rise in the IRM curve and a coercivity spectrum extending up to a maximum of *ca.* 5 T. A continuing curvature into these high fields may help to distinguish hematite from goethite (which can have even higher coercivities) although this effect is not diagnostic. Sometimes the IRM curve is stepped and this effect may be caused by the resultant contribution of two components with differing coercivities; an

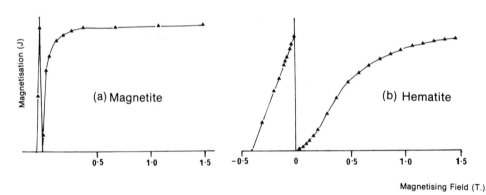

Figure 5.9 *The acquisition of IRM and the reverse field demagnetisation of examples of sediments containing (a) magnetite and (b) hematite. No scale is indicated on the magnetisation axis but weight for weight hematite will have a saturation moment only about 2 percent of that of magnetite.*

example from an anorthosite, in which hemo-ilmenite and ilmeno-hematite grains produce this effect is described by Carmichael (1961). For a given magnetic component, the saturation remanence is indicative of the concentration of the component, but since it also depends on domain structure, and is sensitive to grain variations, saturation remanence can only be used in a semi-quantitative way.

Curie point (Thermomagnetic) determinations

To determine the Curie temperature, small pieces of the rock sample in a moveable container are heated in a strong magnetic field. The saturation remanence is monitored, as the temperature rises, by movements of the container which result from changes in the induced magnetisation. The recording system can be used to trace out a *thermomagnetic curve* of the change in saturation remanence during a heating and cooling cycle (Fig. 5.10). When more than one magnetic material is present, the curve will be the summation of the magnetisations of the two minerals, and the lower Curie point corresponds with the inflection point on the curve (and not to the extrapolation of the curve to the temperature axis). The applied magnetic field is seldom adequate to saturate hematite, and a peak is produced as the temperature rises, because the non-saturated part of the magnetisation is unblocked and aligns itself with the applied field (Duff, 1979a); this 'hump' is suppressed as the saturating field is increased. A similar effect can be produced in titanomagnetite (Day, 1975), although this mineral is normally saturated by the applied field. In both cases these peaks could also indicate the growth of new minerals during the heating.

Magnetic materials which are metastable, or which crystallised at low temperatures, frequently undergo chemical changes as they are heated in air. These changes may produce new minerals with higher or lower saturation remanences so that the heating and cooling curves are dissimilar. The pattern of the changes is often diagnostic of the magnetic mineralogy although the interpretation may be ambiguous. Some examples are illustrated in Fig. 5.10.

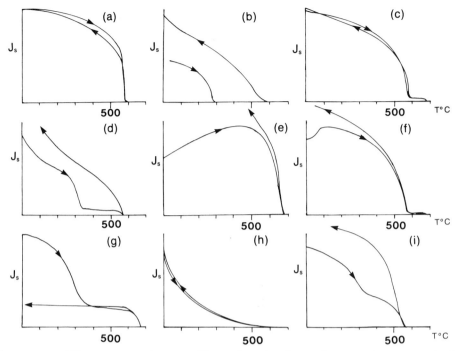

Figure 5.10 Thermomagnetic curves illustrating the change in saturation magnetisation with temperature during heating and cooling cycles in a variety of magnetic mineral assemblages. The examples shown are: (a) pure magnetite stable during heating and cooling; (b) Ti-rich titanomagnetite from a Recent lava flow exsolving during the cycle to low-Ti magnetite; (c) Precambrian lava flow containing both magnetite and hematite (note that the low J_s of hematite produces a small tail which contrasts with the large magnetite deflection); (d) igneous intrusion containing pyrrhotite which alters to magnetite during heating; (e) hematite (note that the peak on the heating curve becomes lower with increasing saturating fields); (f) sample containing magnetite and unsaturated hematite; (g) maghemite converting to hematite on heating; (h) chlorite-grade metamorphic rock dominated by paramagnetic phases which obey the Curie Law (section 1.1); and (i) unmixing of titanomagnetite-maghemite intergrowth during heating. The J_s axis is arbitrary; values will generally be much higher for magnetite-bearing rocks than for hematite-bearing rocks.

Low temperature susceptibility behaviour

The measurement of susceptibility down to the temperature of liquid nitrogen has been used by Senanayake and McElhinny (1981, 1982) to classify titanomagnetite-bearing rocks on the basis of their domain-structure and titanium content; it follows earlier investigations by Radhakrishnamurty, Likhite & Sahastabude (1977). The variation of susceptibility with temperature of the basaltic rocks studied by the former authors has a form falling within one of three broad groups of behaviour (Fig. 5.11).

Group 1 samples exhibit a continuous decrease in susceptibility. These are homogeneous titanomagnetites characterised by low Curie points and low coercivies; they have compositions of $Fe_{3-x}Ti_xO_4$ with $x > 0.3$. These samples are dominated by MD grains and the effect is interpreted as the strong temperature-dependence of the anisotropy constants K_1 and K_2 on temperature in these grains. Group 2 samples exhibit little change in susceptibility. They are titanomagnetites extensively subdivided by ilmenite lamellae, have high coercivities and high Curie points, and are believed to contain predominantly single domains resulting from grain elongations so that the shape anisotropy is dominant (Section 3.3). Group 3 samples have a peak in their susceptibility curves due to the change in the sign of K_1 at about 140°C. Curie points are comparable to Group 2 but the coercivities fall between groups 1 and 2. These titanomagnetites tend to be magnetite-rich with $x \leqslant 0.15$ and their magnetic structure is interpreted to be mainly MD.

This method of investigation is not yet routinely applied in rock magnetic studies because the experimental work is currently based largely on basaltic rocks. However, it may eventually prove suitable for selecting rock samples for further palaeomagnetic (Groups 2 and 3 but not Group 1) and palaeointensity (Group 2) investigations.

The pure magnetite structure changes from cubic to orthorhombic at a (Morin) temperature (-155°C) slightly lower than that at which the crystalline anisotropy disappears. The magnetic susceptibility drops suddenly at this point and an isothermal remanence acquired at a lower temperature will almost disappear when the sample is heated above -155°C. Hematite becomes perfectly antiferro-

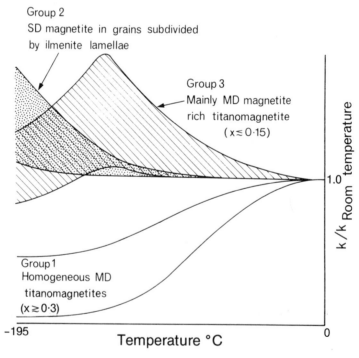

Figure 5.11 The behaviour of susceptibility at low temperatures in basaltic rocks according to Senanayake and McElhinny (1981). Sample behaviour tends to fall into three fields designated Groups 1 to 3 and related to composition or petrology as indicated.

magnetic at $-10°C$, as the configuration of the magnetic spins switches from an alignment in the basal plane of the crystal to a configuration perpendicular to this plane and parallel to the crystallographic c axis. The disappearance of saturation remanence at these temperatures can be used as a further identification criterion for these minerals.

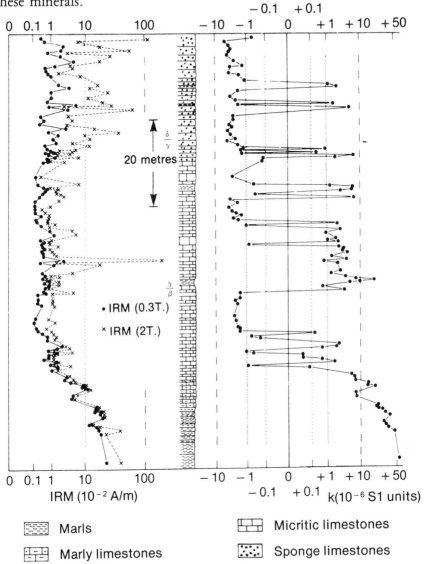

Figure 5.12 Medium and high field IRM and susceptibility through a shallow water marine marl-limestone succession from the Jurassic of southern Germany. The limestones are diamagnetic and the marls are paramagnetic. The change from similar IRM values at the base to dissimilar values at the top is the signature of a change from low coercivity magnetite to high coercivity hematite or goethite. The importance of magnetite at the base is confirmed by the correlation of IRM and susceptibility while the lack of correlation at high levels indicates separate causes for the IRM (high coercivity ferromagnetic phase) and susceptibility (paramagnetic clay mineral fraction). After Lowrie and Heller (1982).

Since the magnetic susceptibility depends on a range of factors (section 3.8), it cannot generally quantify the amount of magnetic mineral present in a rock; to be a useful parameter in this context requires (i) only one strongly susceptible phase should be present, (ii) it should be in the MD range above the size ranges which show a prominent susceptibility variation with grain size and (iii) the grains should have a similar shape (Mullins, 1978). Measurement of the frequency dependence of susceptibility (for example, at 1 and 10 KHz) can be used to indicate the nature of a magnetite size fraction because susceptibility is approximately inversely proportional to the logarithm of frequency in the SD size range but independent in the superparamagnetic range (Mullins and Tite, 1973). The application of a laboratory ARM has been combined with susceptibility measurements to give an ARM:susceptibility index which can be used to indicate variations in magnetite grain sizes.

In magnetostratigraphic studies it is important to separate changes due to geomagnetic field behaviour from changes due to fluctuations in the magnetic mineral content and composition. This is best achieved by comparing the concentration-dependent parameters of NRM intensity, saturation IRM and ARM (either before or after treatment at a specific level of demagnetisation) with susceptibility. The susceptibility can be dominated by paramagnetic and diamagnetic phases if the ferromagnetic concentration is low. In the example illustrated in Fig. 5.12 from a section of Jurassic marls and limestones the susceptibility changes from positive to negative reflect changes from paramagnetic clay-rich facies to diamagnetic carbonate-rich facies. The coincidence of medium field (0.3 T) and high field (2 T) IRM in the lower part of the succession shows that magnetite is the dominant magnetic phase here while the difference between these values in the higher part of the succession indicates that high coercivity hematite or goethite are the main carriers; they must be present in small amounts because they make a negligible contribution to the susceptibility.

The Lowrie–Fuller test

Lowrie and Fuller (1971) proposed a test for titanomagnetite domain structure based on a comparison of NRM or a weak field TRM with a laboratory induced IRM. Subsequently Johnson, Lowrie and Kent (1975) showed that a weak field ARM could be substituted for the TRM in this test thus obviating the problem of chemical changes during heating experiments. These workers conclude that NRM or ARM with coercivities larger than an IRM acquired in a saturating field are indicative of SD grains of magnetite-titanomagnetite (i.e. grain sizes $< 0·5$ μm) while the converse result is indicative of MD grains (Figure 5.13); grains in the size range $0·5-0·15$ μm are PSD and give results identical to SD grains. More recent studies by Bailey and Dunlop (1983) and Dunlop (1983) have attempted to provide a firmer experimental and theoretical basis for the Lowrie–Fuller test; these workers conclude that the NRM and ARM behaviours are a manifestation of the coercivity spectrum which is only indirectly related to domain size but still provides an indication of it. Dunlop (1983) recognises four types of Lowrie–Fuller characteristics in magnetite-bearing rocks, namely: (i) pure SD behaviour due to grains $\leqslant 4$ μm in size which are the remanence carriers in many volcanic and hypabyssal rocks; (ii) pure MD behaviour due to grains $\geqslant 15$ μm in size respon-

Field and laboratory methods 97

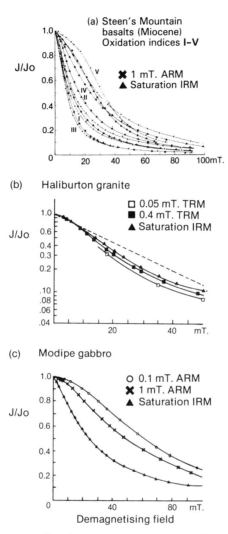

Figure 5.13 The Lowrie–Fuller Test. (a) Progressive change in the Lowrie–Fuller characteristics of rocks from a single palaeomagnetic collection illustrating MD-type behaviour to SD-type behaviour with increasing deuteric oxidation class. (The class III sample in this case has anomalously soft MD-type characteristics.) (b) MD behaviour showing the approximation to an exponential decay on a semi-logarithmic plot. (c) SD-behaviour in a rock with a bimodal magnetite size distribution but dominated by fine magnetite inclusions in pyroxenes. After Dunlop (1983).

sible for the unstable NRM in many plutonic rocks; (iii) mixed SD and MD characteristics in rocks with a grain size distribution in the range 4–15 μm; and (iv) pronounced SD behaviour but exhibiting low coercivity demagnetisation curves for a strong-field induced ARM and caused by a bimodal magnetite distribution of large discrete grains and fine inclusions in silicates.

6
Palaeomagnetic directions, palaeomagnetic poles and apparent polar wander

6.1 Palaeomagnetic directions and their presentations

Spherical projections

The direction of the palaeomagnetic field is referred to the local horizontal plane, in terms of a *declination*, D, which is the angle measured clockwise from true geographic north, and an *inclination*, I, which is the angle measured up (regarded as negative) or down (regarded as positive) from the horizontal plane. Directions of magnetisation can be visualised as lines radiating from the centre of a sphere which intercept the surface of the sphere. If the sphere is projected onto a horizontal or vertical plane through its centre, the points representing the direction of magnetisation can be projected in the same way. The two projections commonly used are the *stereographic* and the *equal area* (Fig. 6.1). The stereographic projection has the property that small circles drawn on the sphere project as circles on the plane; this is useful in palaeomagnetism because most of the statistics are circular functions, and a circle of confidence around a mean direction of NRM can be plotted directly as a circle. The equal-area projection preserves the relative areas of the divisions on the sphere, and is useful for providing a correct impression of distribution of points over the surface of the sphere; on this projection a circle pro-

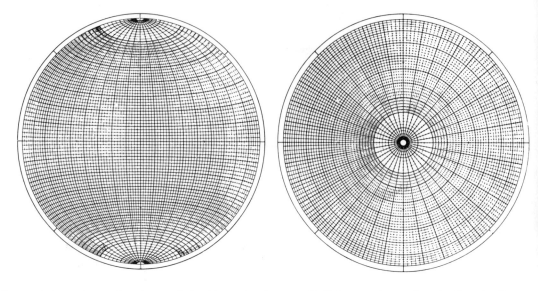

(a) Equatorial equal-area (Lambert or Schmidt) projection   (b) Polar equal-area (Lambert) projection

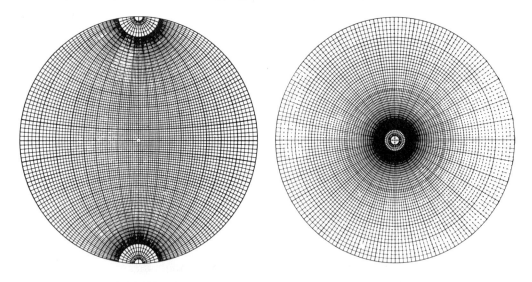

(c) Equatorial stereographic (Wulff) projection   (d) Polar stereographic projection

Figure 6.1 The standard spherical projections employed for plotting palaeomagnetic directions and poles. After Collinson (1983).

jects as an ellipse with its minor axis along a radius and equal to the diameter of the circle. Projections onto the vertical plane are described as *equatorial*, and projections onto the horizontal plane are described as *polar*. In polar projections it is normal practice to project the lower hemisphere only. Points on the upper hemisphere (negative inclinations) are plotted as open symbols, and on the lower hemisphere (positive inclinations) as closed symbols.

Orthogonal projections

Spherical projections allow presentation of the direction, but not the magnitude, of the magnetisation vector; to illustrate the changes in magnitude, it is necessary to employ a separate J/J_o versus a.f. or temperature graph, as shown in Figs. 5.6–5.8. The magnitude and direction can, however, be illustrated on the same diagram as an orthogonal plot, in terms of the components of the vector along three Cartesian coordinate axes, x, y and z, where x is conventionally referred to the North, y to the East and z to the vertically downward direction. A vector of magnitude J is then the resultant of three orthogonal components (Fig. 6.2a):

$$x = J \cos D \cos I$$
$$y = J \sin D \cos I$$
$$z = J \sin I$$

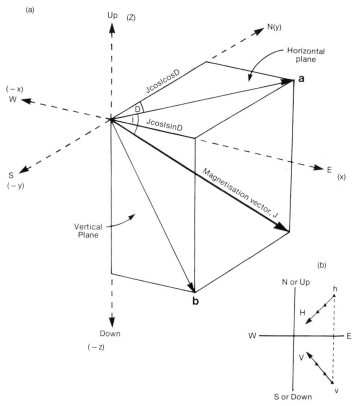

Figure 6.2 The Orthogonal projection. (a) The representation of a magnetisation vector as orthogonal components and the projections on a horizontal plane H and vertical plane V. (b) The projection of the end of the vector on to the horizontal (h) and vertical (v) planes. The arrow shows the movement of these points with progressive demagnetisation if the magnetisation is single component.

In the method described originally by As (1960), Wilson (1961) and Zijderveld (1967), the end of the NRM vector is projected onto two orthogonal planes, one horizontal (h in Fig. 6.2) and one vertical (v in Fig. 6.2). By convention, solid symbols denote projections onto the horizontal plane and open symbols denote projections onto the vertical plane. The axes are scaled in units of magnetisation. As the rock sample is progressively demagnetised, these projections will trace out paths defined by the collective changes in D, I and J. It is advantageous to take the intersection of the two planes (either E–W or N–S) as the abscissa, so that any change in the direction of the traces must lie on a common perpendicular to this axis.

If the NRM is a vector comprising only a single component, its magnitude will be progressively reduced by demagnetisation, but its direction will remain unchanged and the projections (h) and (v) in Fig. 6.2b will migrate along two straight lines towards the origin of the coordinates. If the magnetisation is multi-component, then the demagnetisation path will trace out a straight line, provided that only one component is being subtracted at a time (last part of this Section), but these straight-line segments will not pass through the origin. The declination D_n, and inclination I_n of the nth component which defines such a straight-line segment can be calculated from the orthogonal plot by measuring D_n in the horizontal plane (N versus E or E versus N), in an eastward sense from the N axis. If I_{app} is the angle in the vertical plane (Up versus N or Up versus E) between the abscissa and the other line incorporating the vertical component (open symbols), then I_n is given by

$$\tan I_n = \tan I_{app} |\sin D_n| \text{ for the Up versus E projection}$$

and

$$\tan I_n = \tan I_{app} |\cos D_n| \text{ for the Up versus N projection}$$

The advantage of orthogonal plots is that they give an immediate impression of the number of components constituting an NRM, and the degree to which their blocking-temperature or coercivity spectra overlap (Section 6.3). They are capable of showing both single-component and multi-component behaviour, which may not be evident from plotting the data on a spherical projection, where only the magnetisation direction can be illustrated. Their disadvantage is that two lines are required to describe each component, and they are not suitable for plotting the behaviour of more than one or two samples. In addition, the axes need to be scaled with care, because a single large decrease in intensity can produce a large movement towards the origin of the projection; this would suggest that a single component has been subtracted, when, in fact an expanded projection could show that this had not been achieved. To remove this illusion it may be necessary to illustrate the vector plot in two stages, with the advanced stages of demagnetisation illustrated on an expanded scale. Orthogonal projections are being used increasingly in palaeomagnetic studies, to illustrate the behaviour of representative samples with progressive demagnetisation, while conventional spherical projections are employed to illustrate groups of directions at various stages of treatment.

6.2 Combining directions of magnetisation

Fisher statistics

If each direction of magnetisation is regarded as a vector of unit magnitude, it can be specified in terms of three vector components, or direction cosines which are conventionally taken as the North, East and vertical (down) components. These are:

North Component: $\cos D \cos I = x_i$

East Component: $\sin D \cos I = y_i$

Down Component: $\sin I = z_i$

The resultant of N directions of magnetisation has a length R given by:

$$R^2 = \left(\sum_{i=1}^{N} x_i\right)^2 + \left(\sum_{i=1}^{N} y_i\right)^2 + \left(\sum_{i=1}^{N} z_i\right)^2$$

Provided that the N directions are non identical, R is less than N. The directions of magnetisation which are being combined will obviously be well-grouped if R approaches N, and the significance of the grouping can be tested by comparing N and R. For any given value of N, there is a certain probability that the vectors come from a random population if R is less than a critical value, R_o. The 95% significance level is normally used in palaeomagnetic studies, and Table 6.1 summarises these values for populations ranging from N = 3 to N = 20 after Watson (1956) and Irving (1964).

The direction cosines of the resultant are:

$$x = \frac{1}{R}\sum_{i=1}^{N} x_i, \quad y = \frac{1}{R}\sum_{i=1}^{N} y_i, \quad z = \frac{1}{R}\sum_{i=1}^{N} z_i \qquad (6.1)$$

Table 6.1. The 95% significance level, R_o, for a population of N vectors. (If the calculated value of R is < R_o, the population of directions can be considered to be random at this probability.)

N	R_o	N	R_o	N	R_o
3	2·62	9	4·76	15	6·19
4	3·10	10	5·03	16	6·40
5	3·50	11	5·28	17	6·60
6	3·85	12	5·52	18	6·79
7	4·18	13	5·75	19	6·98
8	4·48	14	5·98	20	7·17

(After Watson (1956), for values of **N** between 21 and 100 see Irving (1964)).

The mean declination $\bar{D}$, and mean inclination $\bar{I}$ of these N directions are given by:

$$\bar{D} = \tan^{-1}\left(\frac{y}{x}\right), \text{ and } \bar{I} = \sin^{-1}\left(\frac{z}{R}\right) \qquad (6.2)$$

Most of the subsequent palaeomagnetic calculations are based on the direction cosines of equation (6.1). Although mean directions are usually calculated in palaeomagnetic studies by assigning an equal weight to each of the magnetic directions, this is merely a convenience in the absence of a method for weighting the magnitude of the individual vectors. It means that vectors with similar directions are combined, while their intensities of magnetisation may vary by up to several orders of magnitude. This method will be valid if the range of intensities is caused by a variation in the magnetic mineral content between the samples, but it could also mean that the directions are of quite different origin; further analysis may be necessary to clarify this possibility. Thus, the magnetic vectors which are being combined should come from the same population, i.e. they should represent magnetisations acquired by the same mechanism at about the same time. Although tests are available to determine whether or not groups of vectors can be legitimately combined on statistical grounds, a subjective evaluation must usually be applied to isolate the sets of vectors for group analysis.

Fisher (1953) developed a method for the statistical analysis of a population of directions, which is routinely employed in palaeomagnetism. It is based on a circularly-symmetrical density function, in which the directions are regarded in terms of their projections as points onto a sphere. In a manner analogous to the two-dimensional Gaussian distribution, it is proposed that they have a probability density, $P(\psi)$, given by:

$$P(\psi) = \frac{\varkappa}{4\pi \sinh \varkappa} \exp(\varkappa \cos \psi) \qquad (6.3)$$

where ψ is the angular distance between any point on the sphere and the mean of the population; when $\psi = 0$, the density is a maximum. The precision parameter, $\varkappa$, varies from infinity when all the directions are identical, down to zero when the directions are uniformly distributed over the surface of the sphere. The value of $\varkappa$ is estimated in terms of an approximation k (k $\geqslant$ 3) where

$$k = \frac{N-1}{N-R} \qquad (6.4)$$

Owing to the infinite possible variation in the value of k, actual values should be treated with some caution: a value greater than *ca.* 10, implies that the calculated mean direction is close to the true mean direction of the whole population and k increases rapidly for low values of N up to *ca.* 8 (Tarling, 1983). Differences between high values of k ($>$1000) do not indicate significant improvements in the estimate of the mean direction, especially for large N values. In view of this limitation, it is more useful to consider the angular distance between the true mean direction and the calculated mean for a particular probability. Fisher (see also McFadden, 1980b) showed that the angular radius of the circular cone of confidence, α, about the resultant vector is given by:

$$\cos \alpha_{(1-P)} = 1 - \frac{N-R}{R}\left[\left(\frac{1}{P}\right)^{1/N-1} - 1\right] \quad (6.5)$$

At the 95% confidence level for $N \geq 5$ (McFadden, 1980a),

$$\alpha_{95} \simeq \frac{140}{\sqrt{kN}} \quad (6.6)$$

Small values of α_{95} imply that the mean direction has been precisely estimated, while large values imply that it has been poorly estimated. Equation (6.3) incorporates a measure, ψ, of the angular distance of a point from the mean, so circles can be defined enclosing percentages of directions within specified distances of the true mean. The 95% probability that a single direction makes an angle α with the true mean direction is given by:

$$\psi_{95} \simeq \frac{140}{\sqrt{k}} \quad (6.7)$$

Thus, a cone with an apical half angle of ψ_{95} contains 95% of observations and is a measure of the *scatter* of directions. It is independent of N, whereas α_{95} is an estimate of the *precision*, because it decreases as the number of observations (N) increases.

Circular probabilities which are equivalent to the quartile distance and the standard deviation of the two-dimensional normal (Gaussian) distribution are:

$$\psi_{50} = \frac{67.5}{\sqrt{k}}$$

and

$$\psi_{63} = \frac{81}{\sqrt{k}}$$

Fisher's statistics apply only to single-peak distributions, and when both normal and reversed polarities occur in a palaeomagnetic study it is usual practice to calculate a mean direction by inverting one set of directions and then applying the statistics to the combined group.

Comparison of mean directions

It is often necessary to test whether or not two directions are significantly different; for example, to find out whether a measured direction is significantly different from the present Earth's field direction or whether the palaeomagnetic directions from two limbs of a fold are identical. If two populations of samples N_1 and N_2

yield resultant vectors R_1 and R_2, respectively, the F ratio can be calculated from:

$$F = (N-2) \frac{(R_1 + R_2 - R)}{(N - R_1 - R_2)} \quad (6.8)$$

where R is the length of the vector sum of the resultants R_1, R_2 of the separate populations and $N = N_1 + N_2$. Large values of F suggest that the directions cannot be considered to be the same, because $R_1 + R_2$ will be much greater than their vector sum R. For a rigid comparison, F is referred to F ratio tables for 2 and $2(N-2)$ degrees of freedom. This test is only valid when the two populations being compared have similar precisions. Their similarity is tested by a second F test with $F = k_1/k_2$ for two separate groups and with variances of $2(N_1 - 1)$ and $2(N_2 - 1)$ degrees of freedom. There are two possibilities (McFadden & Lowes, 1981). If the two populations pass this null test, the hypothesis that they share a common true mean direction can be rejected at the level of significance, P, if:

$$\frac{(R_1 + R_2 - R)^2 / (R_1 + R_2)}{2(N - R_1 - R_2)} > \left(\frac{1}{P}\right)^{1/(N-2)} - 1$$

If the two populations do not have a common precision parameter ($k_1 \neq k_2$), the hypothesis of a common mean direction is rejected if:

$$\frac{r[(R_1 + R_2)^2 - R^2]}{2[(N_1 - R_1) + r(N_2 - R_2)](R_1 + rR_2)} > \left(\frac{1}{P}\right)^{1/(N-2)} - 1$$

where $r = \varkappa_2/\varkappa_1$ and is estimated from the ratio k_2/k_1; since this test involves an approximation, it should not be used to derive a conclusion when its results are marginal. Table 6.2 lists values of $(1/P)^{1/(N-2)} - 1$ for the 95% probability level, after McFadden & Jones (1981). F ratio tables can be consulted in standard texts on statistics.

Sometimes, palaeomagnetic samples are derived from bore cores, of which the azimuth, and hence the reference declination, of the samples is unknown. If appreciable VRM components are present, these cores can be oriented by assuming that the viscous remanence has been acquired along the ambient field direction

Table 6.2 Values of $(1/p)^{1/(N-2)} - 1$ for $p = 0.05$ — the 95% probability level

N	$(1/p)^{1/(N-2)} - 1$	N	$(1/p)^{1/(N-2)} - 1$	N	$(1/p)^{1/(N-2)} - 1$
3	19·00	10	0·4542	17	0·2211
4	3·472	11	0·3950	18	0·2059
5	1·714	12	0·3493	19	0·1927
6	1·115	13	0·3130	20	0·1811
7	0·8206	14	0·2836	25	0·1391
8	0·6475	15	0·2592	30	0·1129
9	0·5341	16	0·2386		

(After McFadden & Jones, 1981).

(e.g. Kodama, 1984), provided that:
1. a drilling-induced remanence has not been acquired.
2. The samples are measured before thay have opportunity to acquire a new VRM. When this is not possible, the end-result of the study is a distribution of inclinations. These form part of a three-dimensional distribution, and can be analysed to derive a mean inclination and the Fisher precision estimate, k, according to the methods described by Briden & Ward (1966) and Kono (1980).

Site and group mean directions of magnetisation

The palaeomagnetic results from any single sampling site can be influenced by a number of factors which may cause the measured remanence to deviate from the true palaeomagnetic field direction at the sample locality. the factors include:
1. The influence of local magnetic anomalies during initial magnetisation;
2. Variable degrees of departure from magnetic isotropy;
3. Variable inclination and compaction errors (in the case of sediments).
4. Orientation errors during collection.
5. Incomplete removal of secondary components in the laboratory.

All of these factors contribute to the scatter of magnetisations between the samples collected at the site; this is referred to as the *within-site scatter*. Provided that sufficient samples have been collected and analysed to average these effects, a single site should record the magnetic field at a single point in time, and the dispersion of magnetic directions between sites — known as the *between-site scatter*, will represent a record of the secular variation of the geomagnetic field (Section 12.6). Thus, sample directions are first averaged at the site level, and then the site directions are averaged to determine the mean magnetisation direction for the rock formation, from which the palaeomagnetic pole can be calculated (Section 6.5). The magnetometer measurements will tend to average out the inhomogeneities of magnetisation within individual samples, although a criterion will be required to separate instrumental noise from meaningful remanence measurements in the most weakly magnetised samples; usually these samples are measured several times, and the agreement between successive measurements is tested. Once a result has passed the acceptance test, there is usually no case for preferring it to any other determination, so that it will be assigned a unit weight in the analysis. However, since one objective for collecting several samples at each site is to average out any random orientation errors, it is important that each field sample should be given unit weight, regardless of the number of cores cut from it for measurement. It may then be appropriate to average at a sample level before combining the sample data to produce the site mean.

Most palaeomagnetic collections are heterogeneous statistical samples, particularly after laboratory treatment and analysis. As a result, both the numbers of cores and the dispersions of directions generally differ at two or three levels of observation, namely within individual oriented blocks, within individual sampling sites, and within the rock formation. There is no general way of dealing rigorously with these observations: unit weight can be assigned to observations from each specimen within an oriented block, to each oriented sample from a single site, or to each site within a rock formation. The most realistic procedure is to apply Fisher's statistics at two levels, firstly giving unit weight to each sample to derive a site mean, and secondly by assigning unit weight to each of these means to obtain an overall mean for the rock formation. For the reasons 1. to 5. given above, this will yield a maximum estimate of the geomagnetic field variation during the period of magnetisation of the formation, and the error estimates will be too large. It is also a physically sound procedure because any significant differences of scatter identified in this way will almost certainly have a geological significance. The disadvantage of this procedure is that not all the data are fully utilised, because it does not take into account errors in the site means or variations in the number of observations at each site; as a result too much weight is given to sites with low precisions, and too little to those with high precisions. Also, although it is a realistic procedure for many igneous bodies, its application to sediments and metamorphic rocks is less clear, both because long time periods may be represented by the magnetisations at individual sites, and because the age of these magnetisations is often not clear.

The alternative procedure of assigning a unit weight to each *specimen* is sometimes applied when large numbers of sample observations are available from only a few sites. It assumes that each specimen records a spot reading of the geomagnetic field, and it is only strictly applicable when the dispersions between specimens from individual samples can be attributed to secular variation. As this is not generally true, the method is a misuse of Fisher statistics and, in principle, any error estimate can be obtained by multiple sampling at a single collecting site or by cutting any number of cores from an individual block. It can only be justified statistically if the dispersions observed at each level of analysis can be considered to be the same. A method for making fuller use of the data in this instance is known as *two-tier analysis*, and is described by Watson & Irving (1957) and McElhinny (1973). It requires that the dispersions are small and approximately equal from site to site. It cannot yield significantly different estimates of ancient field directions at the sample and site levels of analysis, but it can define tighter error limits than those derived from the site level alone.

Although site means are usually combined to produce a formation mean, this procedure is only strictly valid if the sites have the same within-site dispersion, k_w. An F ratio test may be carried out to evaluate this possibility and it may be appropriate to exclude sites failing the test.

If k_w varies significantly from site to site, the site mean directions can still be combined, provided that the within-site and between-site dispersions obey the Fisherian distribution, and the product $k_w N_i$ (where N_i is the number of samples at the ith site) is statistically constant at each site (McElhinny, 1967). If k_w is constant, the total angular dispersion ψ_T is related to the combined between-site dispersion from N sites, ψ_B, and the within-site dispersion, ψ_w, by the relationship:

$$\psi_T^2 = \psi_B^2 + \frac{\psi_w^2}{N}$$

where N is the total number of samples. The precision parameter for the resultant from all N samples (k_T) is related to the group mean precision (k_B) and the site mean precision (k_w) by:

$$\frac{1}{k_T} = \frac{1}{k_B} + \frac{1}{k_w N}$$

This formula is a specific example of the general formula for determining a precision k if there are n sources of error with precisions k_1 to k_n so that:

$$\frac{1}{k} = \sum_{i=1}^{n} \frac{1}{k_n}$$

Fisher and Bingham distributions

It is necessary to carry out a rigorous test on a large number of sample or site directions to ensure that they obey the Fisherian distribution (equation 6.3). It is usually merely assumed that sets of palaeomagnetic directions have this distribution for the application of the foregoing analysis. However, this test can be applied by first determining the distribution of angles between each direction and the estimated mean, and then by comparing the distribution of azimuthal angles with the distribution expected from equation (6.3) (Irving, 1964). Most tests have shown (see summary in Onstott, 1980) that directions and poles are seldom distributed precisely according to the Fisherian formula. They are usually distributed rather more tightly about the mean that this formula predicts, and provided that the distribution is approximately circular, the application of Fisher statistics gives a reasonable but conservative estimate of the precision (Tarling, 1983). Onstott (1980) has advocated the use of the Bingham (1964) density function to palaeomagnetic analysis, because it can generate distributions which are circular, elliptical, or distributed along a great circle, according to the parameters selected. It would appear to be more appropriate to the palaeomagnetic problem because:

1. Almost all actual distributions of palaeomagnetic directions have maxima which are more peaked and closely distributed about the mean than the corresponding Fisher distribution.
2. A circular distribution of directions converts into an oval-shaped distribution of poles on application of the axial dipole formula (Section 6.5), and elongate distributions of poles are generated by departures from dipole symmetry and polar wander (section 6.6).
3. A streaking of directions along a great circle occurs when the magnetic vectors are mixtures of two remanence components.
4. Reversed directions do not have to be switched to their antiparallel positions to calculate an overall mean.
5. The Bingham distribution can accommodate other bipolar data, such as axes of magnetic susceptibility.
6. Intermediate directions can be accommodated by the function.

This last point is probably not important because intermediate directions are usually a small component of the total sample; since they are not considered to be applicable to the main dipolar field axis a criterion is normally adopted to exclude them (Tarling, 1983). Unfortunately, because the calculations required to evaluate the Bingham distribution parameters are tedious and conceptually more complex than the parameters defined by the Fisher distribution, they have not yet been widely employed. An example of the use of Bingham statistics is given by Kirschvink (1980).

6.3 Identification and separation of magnetic components

The NRM is normally the vector sum of the primary magnetisation and one or more secondary magnetisations. One of the latter components is likely to be a viscous remanence, acquired in the present magnetic field, but ancient secondary components of remanence may be as interesting as the primary remanence because they often record a response to regional uplift, folding, thermal metamorphism or metasomatism (Van der Voo, Henry & Pollack, 1978). The aim of laboratory demagnetisation studies is, in part, to reverse the history of the remanence acquisition in nature and to recover the directions of these components.

The magnetic vector can be described by a line in three-dimensional space. The tip of this vector is a point which traces out a path with progressive demagnetisation. Depending on the composition of the vector, this path may trace out a straight line or a curved line. A straight line usually results from the progressive removal of a single magnetisation; if this is the only component present the point migrates towards the origin (Section 6.1). A curved path lying in a plane occurs when two discrete magnetic components (not colinear) are being simultaneously removed at different rates. A path which does not lie in a plane is produced by the simultaneous removal of three or more components. The single magnetisation in the linear example is simply demonstrated by the clustering of directions on a spherical projection and the linear convergent paths on an orthogonal vector plot. The two remanences indicated by a co-planar path can often be distinguished by vector subtraction, convergent remagnetisation circles and possibly, linear segments on an orthogonal plot (concluding part of this section). Each of these methods works best when the coercivity or blocking-temperature spectra are either discrete or show rather limited overlap. The isolation of more than two magnetisations represents a more difficult problem and usually requires that the blocking-temperature or coercivity spectra of the components are discrete.

Vector subtraction

The general co-planar situation comprising two vectors is illustrated in Fig. 6.3 at three arbitrary stages of demagnetisation 1, 2 and 3. Over this range of demagnetisation treatment, both of the components J_A and J_B are reduced in intensity when a fraction of each component is subtracted. The resultant magnetisation changes magnitude and direction, as shown by the arrowed lines. The solid bars are the *subtracted vectors* over each interval of treatment.

Since both components are being eliminated simultaneously, it will not be possible to recover either component A or B over the interval of treatment 1 to 3. However, if the spectra of the two components only partially overlap, as shown in Fig. 6.4a, stepwise demagnetisation will yield the direction of the A component as a stable end-point. At initial stages of treatment the B component is un-demagnetised, so that the successive subtracted vectors remain parallel to one another. The constancy of these successive directions defines the range of treatment over which the A component can be determined. In the rather special case shown in (b) with the spectrum of the A component completely overlapped by the B component, successively parallel subtracted vectors are also present; however, because

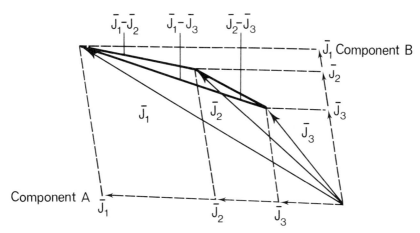

Figure 6.3 Vector Subtraction. The general case when two components are reduced simultaneously by cleaning treatment at three arbitrary stages of cleaning 1, 2 and 3. The dashed lines are the components, the solid arrowed lines are resultant vectors, and the thick lines are the subtracted vectors.

the intensity of B is changing, the vector differences do not have the same direction as the A component. In this instance it would be necessary to know the special relationship and the contribution of B to the total spectrum, in order to determine the correct direction of the A component. In the other two examples shown here, the spectra of A and B overlap at their lower (d) or both (c), extremities. In case (d) it will not be possible to isolate the A component by vector subtraction, and in case (c) neither component can be isolated by this method.

Hoffman & Day (1978) describe a modification of the vector subtraction method which is applicable to more than two components. They plot the successive subtracted vectors at each stage of treatment on a stereographic projection (Fig. 6.5), and fit great circles to the data. The point of intersection determined from any two successive great circles defines the direction of the component between them (see also Section 6.3, *use of vector plots*).

Remagnetisation circles

The technique of converging remagnetisation circles, employed originally by Khramov (1958) and Creer (1962b), has been developed by Halls (1976, 1978) to determine components of magnetisation when the demagnetisation procedure has not isolated a stable end-point. It may be applicable to cases in which the vector subtraction method cannot be used, because it is relatively insensitive to the spectral composition of the remanence. Like the Hoffman-Day method, this technique utilises only the directions of magnetisation. It can be applied to a two-component system when the earlier component has been disturbed by folding or by some non-structural mechanism such as secular variation, prior to the acquisition of the later component. In the general situation, the directions of magnetisation migrate during demagnetisation along great circular paths which pass through the primary and secondary remanence directions. However, if the primary component has been dispersed in some way, the demagnetisation will not define a single great circle,

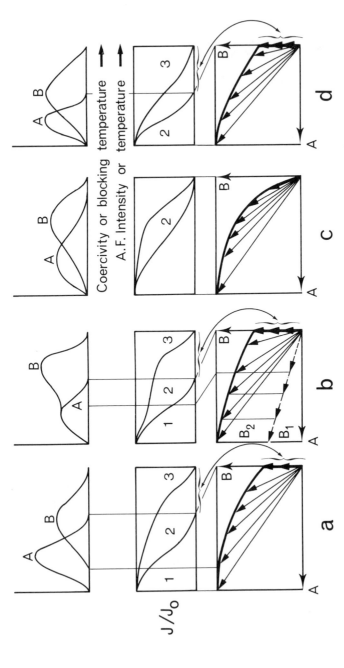

Figure 6.4 The demagnetisation behaviour of two superimposed remanences with a variety of relationships between a lower (A) and higher (B) distributed coercivity or blocking temperature spectra. The regions of the demagnetising curves are specified as: 1, resultant vectors progressively change direction but successive vector differences remain parallel to one another; 2, resultant vectors progressively change direction but successive vector differences are not parallel to one another; and 3, resultant vectors do not change direction implying that the A component has been completely removed; a stable end point is then achieved and gives the direction of the B component. The lower figures show the change in direction and intensity of the resultant magnetisation vectors with stepwise cleaning. The arrowed lines are the vectors at each stage of treatment and the thick lines are the subtracted vectors. In the case (a) when the spectra do not entirely overlap it is possible to recover the direction of A and B uniquely from the behaviour of subtracted vectors. In case (b) it is not generally possible to recover the A remanence uniquely, while in case (c) it is not possible to recover either remanence direction. (Modified in part after Halls (1976)).

Palaeomagnetic directions, poles and apparent polar wander 113

but will trace out a number of different great-circle paths, which tend to converge towards an intersection point defining the direction of the secondary component.

Figure 6.6a illustrates an example where the primary magnetisation has been dispersed by folding. The directions of magnetisation in samples 1, 2 and 3 migrate away from the lower coercivity or blocking-temperature component, but

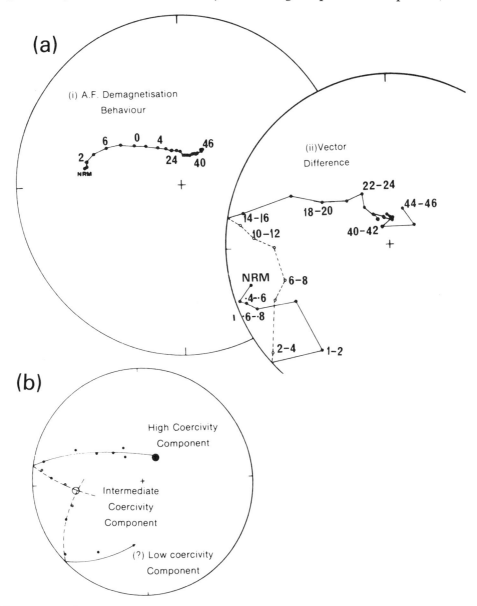

Figure 6.5 The use of subtracted vectors to determine directions of magnetisation. The example is from a.f. demagnetisation of a pillow lava of Jurassic age. (a) demagnetisation data (fields are in mT.) illustrated on a stereographic plot and vector differences calculated from the data. (b) Great circles fitted to the vector differences and inferred components. After Hoffman and Day (1978).

114　*Palaeomagnetism and the Continental Crust*

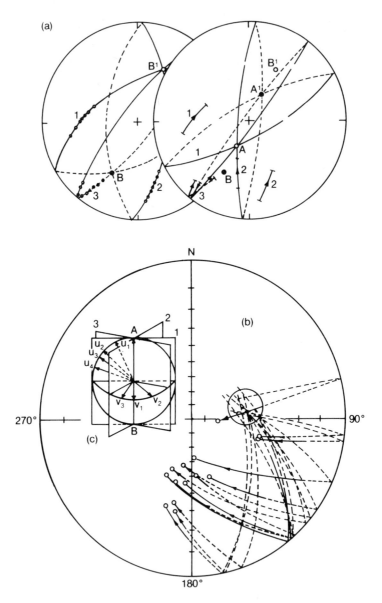

Figure 6.6 *The application of convergent remagnetisation circles. (a) an example where the circles defined by three samples converge at B and A before and after structural unfolding. (b) An example of remagnetisation circles obtained on samples of Precambrian igneous rocks from Slate Island, northern Lake Superior. The solid portions of the circles are defined by data and the dashed parts are extrapolated with the arrows indicating the direction of movement with progressive cleaning. The open circle is the area of great circle convergence. (c) determination of the axis of convergence of the remagnetistion circles from the intersection of remagnetisation planes. After Halls (1976).*

no stable end-point is reached. The points B and B' are the common intersection points of these remagnetisation circles before structural correction, and either B_1 or B_2 is the direction of the B component. The points A and A' are the intersection points of the same circles after structural correction, and one or other of these points is the direction of the A component. The ambiguity is removed because the data points defining these circles must lie between the true primary and secondary directions; hence A and B are the primary and secondary components, respectively.

In practice, the intersections of the remagnetisation circles will usually be imprecise because of dispersion. Halls (1976) describes a least-squares method for calculating the best fit to the intersection of the remagnetisation planes. The principle of the method is illustrated in the inset Fig. 6.6c: three remagnetisation planes are defined by progressive demagnetisation data such as the directions $U_1 \rightarrow U_4$, and they intersect along the line AB. The normals to the planes 1, 2 and 3 are V_1, V_2 and V_3, respectively. Thus the normal to this plane, fitted to V_1, V_2 and V_3 by least squares, is the direction AB.

Halls (1978) describes three situations which may produce converging remagnetisation circles:
1. A conglomerate carrying clasts with dispersed directions of primary NRM may be partially remagnetised, so that the NRM directions are biased towards the secondary magnetisation direction. With cleaning, remagnetisation circles will be produced (provided that there is some difference between the component spectra) and the circles will converge on the direction of the secondary component.
2. If a folding episode takes place between the imposition of the primary and secondary remanences, the directions of both of them can be recovered by the method outlined above. To obtain a range of widely differing circles it is clearly important to determine the remagnetisation circles from samples covering a wide range of structural attitudes.
3. If the primary remanence was acquired over periods of time much longer or shorter than the secondary remanence the required scatter could be produced by the influence of secular variation or apparent polar wandering on the more slowly acquired remanence.

Unqualified use of this technique can lead to a bias in the estimation of the end point directions and Schmidt (1985) discusses the conditions which are appropriate and inappropriate to its general application.

Use of vector plots

Orthogonal vector plots have the property that the projections on the two orthogonal planes follow linear paths when a single component is being subtracted by cleaning treatment. Provided that appreciable segments of the blocking temperature or coercivity spectra are discrete, the individual components of a multicomponent magnetisation will be defined by these linear segments. Three possible situations involving two components J_A and J_B are illustrated in Fig. 6.7 after Dunlop (1979a). In the first case the coercivity or blocking temperature spectra of the two components are discrete. With cleaning the direction migrates along a great circle path between the directions of J_A and J_B until J_A is completely removed and J_B is recovered. On the orthogonal plot the points migrate along a straight line to a position corresponding to the components of the J_B vector. Sub-

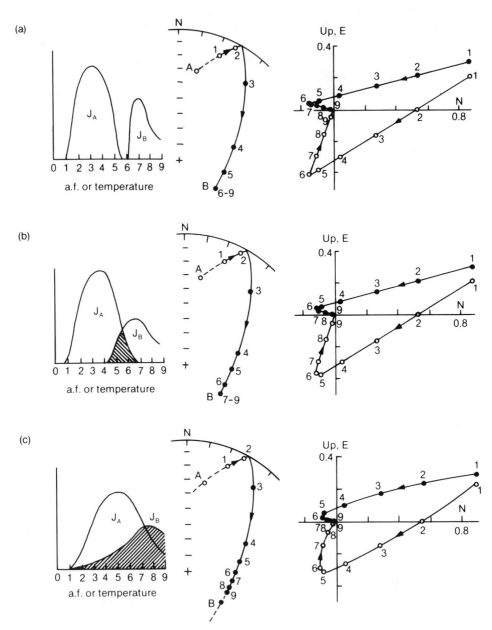

Figure 6.7 Stereoplots and orthogonal vector plots produced by a two component magnetisation when there is no overlap (a), partial overlap (b), and complete overlap (c) of the coercivity or blocking temperature spectra. After Dunlop (1979a).

sequent treatment progressively subtracts this vector and the points migrate along a straight line towards the origin; the direction and intensity of J_A can also be determined from the angles which the initial straight line segments make with the axes of the diagram.

In the second and more common situation when J_1 and J_2 have spectra which overlap, the two components are still represented by two discrete straight-line segments on the Cartesian plot, but the changeover from the pair of lines corresponding with J_A to the pair of lines corresponding with J_B becomes rounded. The degree of rounding becomes more pronounced as the overlap in the spectra increases, but the separate directions of J_A and J_B can still be determined, provided that the reference point is taken as the point of intersection of the extended straight-line segments. Finally, in the third situation there is complete overlap of the spectra. The traces on the orthogonal plot are no longer straight lines but become broad arcs. J_A cannot be defined, and a static end direction is not achieved (and hence J_B recovered).

These principles can be extended to the case in which three or more magnetisations are superimposed, provided that linear segments are present. Each component can then be defined by the process of linear least-squares fitting. The method is applicable to individual samples; unlike the method of remagnetisation circles, it does not require multiple samples or a scattering mechanism. Furthermore, components may still be resolved when the data are 'noisy' and vector subtraction is not applicable.

Multicomponent analysis by the Hoffman-Day method and graphical techniques

Although it is generally considered that components can only be resolved if there are non-overlapping regions in their coercivity or blocking-temperature spectra, it may sometimes be possible to isolate a component by the Hoffman-Day method, even though it may never by itself form a straight-line segment (Kirschvink, 1980). Fig. 6.8 illustrates a comparison between the orthogonal plots and the sequence of subtracted vectors for situations involving increasing overlap between three spectra A, B and C. The vector plot changes progressively from three discrete straight line segments to an entirely curved path. The adjacent component pairs which form the two planes incorporating components A and B, and B and C continue to form great circles intersecting in component B; thus, B alone can be found even when it is not defined by a linear segment on the vector plot or a discrete grouping on the stereonet (Fig. 6.8). Demagnetisation planes must be used here in place of the subtracted vector plots because they have the same intersection defining component B.

Foss (1981) describes a simple graphical method for determining the magnitude and direction of the vectors in a multicomponent remanence which can sometimes achieve an accuracy of up to about one degree, and is therefore suitable for most purposes. The angle (θ) between the vectors at any two stages of demagnetisation treatment and the intensity ratio are entered on graphs to derive the magnitude of the separated vector and the angle between this vector and the vector at the initial stage of treatment. Provided that only one component is being

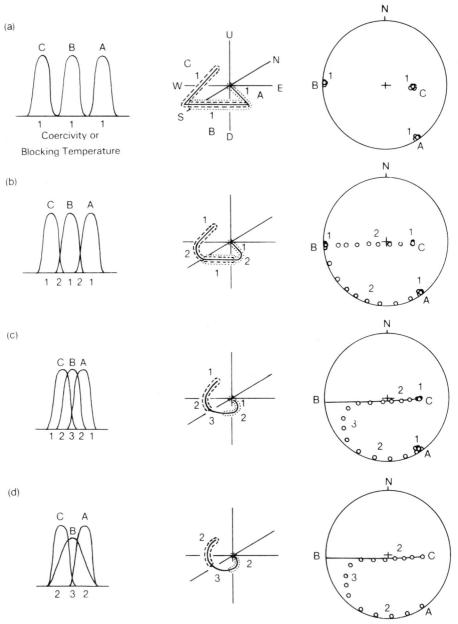

Figure 6.8 *The demagnetisation of three components showing increasing overlap of their coercivity or blocking temperature spectra and illustrated as orthogonal plots and stereographic plots. The small numbers give the number of different magnetic components being removed simultaneously due to overlap of the spectra. In the orthogonal plots the AB plane is surrounded by a dotted envelope; thick lines show the linear regions. The stereonet shows the direction of the subtracted vectors removed sequentially by demagnetisation. After Kirschvinck (1980).*

subtracted at a time by the demagnetisation treatment, two constituent components A and B can be resolved by using a second graph.

Other methods

The earliest statistical methods for distinguishing magnetic vectors were a range of *stability indices* based on successive changes in direction (Tarling and Symons, 1967), on successive changes in direction and intensity (Wilson, Haggerty and Watkins 1968; Briden, 1972), or on the rate of change of the total subtracted vector (Symons and Stupavsky, 1974). They are applicable to both a.f. and thermal demagnetisation and have been routinely used to define the optimum field for cleaning treatment. They are most effective when applied to rocks containing a recent VRM component added to a discrete ancient remanence, and they have been successfully applied to Cenozoic volcanics and a considerable number of older rock suites. However, they can only be used to separate components of magnetisation if the spectra have little or no overlap. The application of stability indices is analysed by Giddings and McElhinny (1976) who advocate a distinction between indices used to define optimum cleaning fields and indices used to define the stable range of demagnetisation treatment. They are also fully described by Collinson (1983) and Tarling (1983) and will not be described here. Furthermore, recent technological developments have substantially reduced the effort required to progressively demagnetise, measure and analyse the remanence in single specimens, and the blanket treatment of palaeomagnetic collections based on the stability indices of pilot samples is no longer justified or necessary.

A method for accommodating a large overlap in the coercivity spectra of successive components is described by Stupavsky and Symons (1978). It assumes the exponential decay of NRM predicted from the physical theory of a.f. demagnetisation and fits successive exponential curves to the J/J_o versus a.f. demagnetising field curves. The least squares fit of two or more curves is obtained to derive best values for the successive component vectors represented by the data. The method does not assume the number of magnetic components present but instead runs a sequence of models and isolates the best one in terms of standard deviations and goodness of fit parameters.

Kirschvink (1980) has applied the classic multivariant technique of principal component analysis to palaeomagnetism. This method applies a linear transformation of the orthogonal coordinate axes to a new orthogonal reference frame corresponding to the geometry of the data set and determined by a least-squares fit to the data. Each of the three axes in the new reference frame has a measure of variance (σ^2) associated with it; these are known as *eigenvalues* and are designated λ_{max}, λ_{int} and λ_{min}. Two of these eigenvalues will be zero if the data points fall along a line while λ_{min} will be zero if the data lie in a plane; if they are all finite the data define a curve. (The normals to the planes defined in the second case can also be useful because they are directions in which the magnetisation cannot lie; in large collections they may be employed to define zones which are inapplicable to the data analysis). Multivariant analysis may be employed at two levels. Firstly, if criteria for linearity and planarity are defined, computer analysis can be employed to locate data sets corresponding to linear and planar groups and hence to determine the vector, or vectors, to which they correspond by the methods previously described. Secondly, it may be used as an alternative to the Fisherian method to determine the mean of a group of specimen or site directions. In this instance, the vector corresponding to λ_{max} defines the mean direction of the data set and since the technique is axial rather than directional, dual polarities can be conveniently accommodated. In addition, this analysis can be used to determine whether or not the data set is circularly distributed about the mean direction (as is required by the Fisherian analysis): the ratio of the major to the minor semi axes of an ellipse drawn about the mean direction is given by $\sqrt{\lambda_{int}/\lambda_{min}}$ and should have a value close to unity; larger values indicate an elliptical data set and hence the presence of composite magnetisations. In an alternative approach to the analysis of demagnetisation paths Kent, Briden and Mardia (1983) have developed an algorithm* to determine the linear and planar segments in demagnetisation data; the computer looks for all possible segments in the data set, tests whether or not they pass the linearity or planarity criteria, and then ranks them in order of precision; unlike the method of Kirschvink (1980), this algorithm takes into account the errors of the individual measurements.

6.4 Dating components of magnetisation

The components of magnetisation isolated by palaeomagnetic study are dated by reference to the geological history of the rock formation. Absolute ages may be assigned either indirectly by reference to the regional history, or directly by geochronologic information. Alternatively, the magnetisation may be dated by reference to existing knowledge of the temporal variation of the palaeomagnetic vector in the sample area. The primary magnetisation will then generally refer to the age of sedimentation, igneous cooling, or metamorphism, and the secondary components will refer to later regional or local events. Unfortunately, it is not always possible to decide which are the primary and secondary components. It has often been assumed that the relative ages of superimposed magnetisations are defined by their blocking temperature spectra, the component with the highest blocking temperature being the oldest and the successively lower blocking temperature components being progressively younger. This simplistic view presupposes that all the components are PTRMs. It will not be correct, for example, when a CRM overprint carried by hematite is present. In sedimentary rocks positive evidence of the DRM, PDDRM, CRM or PTRM nature of the remanence will be required to link it to primary sedimentation, to diagenesis, or to later thermal-tectonic events. It is often possible to design the palaeomagnetic collection to incorporate fold, conglomerate, contact and unconformity tests described in the latter part of this section; subsequent laboratory analysis may then be able to assign a relative date to the magnetic components with respect to a geologic event.

Geochronological dating

Whilst the primary components of remanence in a rapidly-cooled igneous rock may be directly related to radiometric ages derived from methods such as K–Ar,

---

* An algorithm is any step by step method for deriving the solution to a problem.

Ar^{40}–Ar^{39}, Rb–Sr, U–Pb or Sm–Nd, it is generally true that many rocks, especially those in orogenic environments, have experienced later heating events. These events may be of sufficient temperature and/or duration to reset the K–Ar and Rb–Sr systems. In the same way that a primary magnetic remanence may be partially or completely overprinted by secondary magnetisations, so these isotopic systems can also be reset at the times of burial-heating or metamorphism. In metamorphic terranes the radiometric record may be very complex if the radiogenic nucleides have been incompletely dispersed from their minerals of origin by later events. A range of apparent ages may then be observed between the times at which the minerals became closed systems after each event. In the highest grade metamorphic terranes, only the zircon U–Pb, Sm–Nd and whole-rock Rb–Sr isochron ages are likely to record the primary emplacement and initial crystallisation ages, but in lower-grade terranes it is usual to find that the most retentive and coarser-grained minerals yield the oldest apparent ages, and the least retentive and finest-grained minerals record the youngest apparent ages. In terranes affected by a thermal event, the sequence of apparent mineral ages observed is generally in the order U–Pb (zircon) > Rb–Sr (K-feldspar) > K–Ar (hornblende) > Rb–Sr (muscovite) > Rb–Sr (biotite) $\geq$ K–Ar (biotite) > K–Ar (K-feldspar) (Hart, 1964; Moorbath, 1967).

The temperatures at which minerals begin to lose significant amounts of argon on heating have been estimated from laboratory experiments and the outgassing effects around small igneous intrusions, but they are not known with any great accuracy. Observations indicate a minimum outgassing temperature of 150–250 °C for biotite and 200–300 °C for muscovite. Consistent differences can be recognised between the ages recorded by these minerals in several orogens (Harper, 1967). Hornblende generally has a greater argon retentivity than the micas, although outgassing temperatures decrease from ca. 550–600 °C for Mg^{2+}-rich amphiboles down to low temperatures comparable with medium-grained biotites (275°–325 °C), with increasing substitution of the larger Fe^{2+} ion (O'Nions, Smith, Baadsgard & Morton, 1969). The closure temperature range for sanidine is 200–300 °C (Moorbath, 1967). Although little is known about the diffusion behaviour of radiogenic strontium, indirect geological evidence suggests that it is rather similar to that of radiogenic argon.

It will be apparent from the foregoing discussion that the assignment of ages to magnetic components in most rocks, and especially those from a plutonic environment, requires a careful assessment of the whole range of radiometric evidence. TRM components with laboratory blocking temperatures of 500–680 °C will have been acquired before the closure ages defined by most metamorphic minerals (Section 3.10 and Dodson & McClelland-Brown, 1980) and it is probable that Rb–Sr mineral ages on feldspars, with an estimated closure temperature range of 450–500 °C, will provide the closest estimate of magnetisation age. The closure ages indicated by less retentive minerals may, however, be closer to the age of PTRM components or even CRM components if the metamorphic event has promoted late-stage chemical changes. The regional interpretation of palaeomagnetic data, which is often required to evaluate apparent polar wander paths (Section 6.6), has sometimes been considered in the context of the regional distribution of K–Ar mineral ages (e.g. McWilliams & Dunlop, 1978). This variation is plotted in the form of 'chrontours' which illustrate the regional isotopic closure, and hence the uplift-related cooling, of the terrane. In these cases, the ages of the TRM components may be expected to be appreciably older than the 300–150°C temperatures defined by the mica ages, although the spatial variation of muscovite and biotite mineral ages may allow a range of magnetisations to be placed in their correct age sequence. In contrast, examples are described in Section 10.3 where the uplift related magnetisations are apparently *younger* than the muscovite and biotite chrontour ages and are therefore likely to be TCRMs (Section 1.4).

A mineral becomes a closed isotopic system by a mechanism which has analogies with the blocking of magnetic remanence. Both processes are dependent on a thermal activation mechanism, and have associated relaxation times, which are exponential functions of temperature. In the isotopic example, diffusion proceeds rapidly at high temperatures, so that no accumulation of daughter isotopes can occur; as the temperature falls, diffusion is inhibited and each mineral grain becomes a closed system.

Dodson (1973) derived the following formula for the radiometric closure temperature of a mineral grain (T_C):

$$(T_C)_{\text{radiometric}} = \frac{E}{|R \ln(A\tau D_0/a^2)|} \quad (6.9)$$

where E is the activation energy of the diffusion process, a^2/D_0 is a characteristic time incorporating a dimension of the grain (a) and a diffusion coefficient (D_0); A is a geometrical parameter, R is the gas constant and τ is the relaxation time which is related to cooling rate (dT/dt) by the formula:

$$\tau = -\frac{RT^2}{E(dT/dt)}$$

This formula predicts that closure temperature is relatively insensitive to cooling rate for most geological examples; thus a change in dT/dt by a factor of ten from 5°/Ma to 50°/Ma changes T_C by only *ca.* 10%. Equation (6.9) allows the isotopic closure temperature of a metamorphic mineral to be estimated, but magnetic and isotopic closure temperatures cannot be directly linked to each other under conditions of slow cooling, because their activation mechanisms depend on temperature in a different way (Dodson & McClelland-Brown, 1980, and c.f. York, 1978).

The limitations of conventional K–Ar dates in polymetamorphic terranes have led to increasing interest in the step-heating version of the ^{40}Ar–^{39}Ar dating method. This method is potentially able to isolate valid from invalid determinations (i.e. those which have experienced argon loss or gain), and identify metamorphic events as discrete plateaus on 'apparent age' versus temperature graphs; it also benefits from the insensitivity of argon loss to pressure (Berger & York, 1981). Although these studies are still in their infancy, they have been successfully applied to the dating of multicomponent remanences in a number of Precambrian metamorphic terranes (e.g. Berger, Dunlop & York, 1979; Dallmeyer & Sutter, 1980). It appears that K-feldspars are sensitive monitors of low temperature (250–150°C) and short-term (*ca.* 1 Ma) heating events (Berger & York, 1979). These events can be distinguished from isotopic closure in biotites, defined by plateaus above experimental temperatures of 600°C, which relate to geological events at temperatures of *ca.* 300°C. Studies with hornblendes may yield plateaus indicative of events at temperatures of up to 600–700°C (and therefore comparable with magnetisation ages), but some investigations have proved less successful because this mineral does not release all of its argon until it is completely dehydrated, and it yields false plateaus below experimental temperatures of 900°C (Berger & York, 1979, 1981).

The radiometric dating of Precambrian sediments is another geochronological problem related to palaeomagnetic interpretation. Owing to the combined factors of an inherited mineralogy and an isotopic closure delayed until complete dewatering, the Rb–Sr isochron ages may be 10–20% higher or lower than the depositional age (Perry & Turekian, 1974; Chaudhuri, 1976). This problem is gradually being surmounted by a better palaeontological understanding of Riphean and Vendian times, which obviates a sole reliance on Rb–Sr age determinations.

The fold test

This test compares the directions of magnetisation from the limbs of a folded structure, before and after they are restored to their pre-folding configuration. It was first applied by Graham (1949) to the (Silurian) Rose Hill formation. A comparison of the dispersions in each case can indicate whether a magnetisation was acquired before or after the time of folding. The scatters about the two mean directions are compared in each case, using a test similar to that described in the third part of Section 6.2. The original statistical test of McElhinny (1964) calculated the precision of the sample or site mean directions before (k_a) and after (k_b) tilt correction, and tested the ratio k_a/k_b. This test may have some particular applications (e.g. McClelland-Brown, 1983) but is not generally applicable because the overall population cannot have a Fisherian distribution in both the pre- and post-folding orientations, and because the precisions of the means from each limb are not independent (McFadden & Jones, 1981). The test appears to require a greater degree of distortion than would be needed for statistical significance, so that some studies which have failed McElhinny's test may nevertheless indicate pre-folding magnetisations. McFadden & Jones (1981) develop a test in which the group mean direction from one fold limb is compared with the group mean direction from the other fold limb. To apply the test, it is first necessary to confirm that the precisions from the two limbs are comparable. The subsequent test is independent of precision and calculates the value of:

$$\frac{[(R_a + R_b - R)^2/(R_a + R_b)]}{2(N - R_a - R_b)} \qquad (6.10)$$

where R_a and R_b are the lengths of the resultant vectors from each limb, R is the resultant vector of all site mean directions, and N is the number of sites. If this is less than:

$$\left(\frac{1}{0\cdot 05}\right)^{1/(N-2)} - 1$$

(see section 6.2 and Table 6.2) the two limbs can be regarded as having a common mean direction after unfolding at the 95% confidence level. The theory can be simply extended to the simultaneous consideration of more than two limbs.

There are several problems which may arise from the application of the fold test. If the fold is a plunging structure it will not be sufficient to rotate the palaeomagnetic vector about the local strike direction, and there must also be a correction for the component of plunge (Fig. 6.9). The effect of failing to accommodate a plunge component is to produce an error in declination but not in inclination; this declination anomaly could be confused with tectonic rotation about a vertical axis (Macdonald, 1980, and Section 6.6).

It may also be necessary to consider the style of folding, because simply returning a bed to its pre-folding orientation does not accommodate the internal strain experienced by the rock (Facer, 1983). When the fold style is concentric (parallel), internal movement has taken place along planes parallel to the bedding, and the resultant shear could rotate grains carrying a detrital remanence; the remanence carried by a pigment might not be affected in the same way. In similar folds, however, the fold shape is achieved by shear parallel to the axial plane (Fig. 6.10). In each case the angle between the bedding and the palaeomagnetic inclination changes and becomes asymmetrical on opposite limbs as the dip of the beds increases. If the fold has been compressed, the error is more severe, although the cases shown in Fig. 6.10 are limiting ones, because actual folds usually involve the operation of both mechanisms. The important observation is that unfolding a shear fold by flexure fold techniques may leave an appreciable difference in the vectors from either limb, and the correction is thus an imperfect one.

A fold test will only define the age of a remanence as pre- or post-folding. Refined palaeomagnetic analyses are discovering an increasing number of examples of remanences which pre-date a folding episode, but are still secondary (see Roy & Morris, 1983). Indeed, it is highly likely that the processes associated with the folding episode, including such factors as the internal strain already noted, will partially or completely remagnetise the rock. An interesting example is described by McClelland-Brown (1983) from a Permo-Carboniferous fold in Devonian

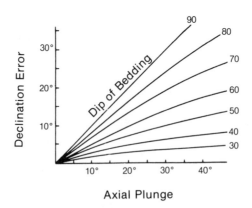

Figure 6.9 The correction to magnetic declination for the effect of axial plunge of a fold axis. After Norman (1960) and Tarling (1983).

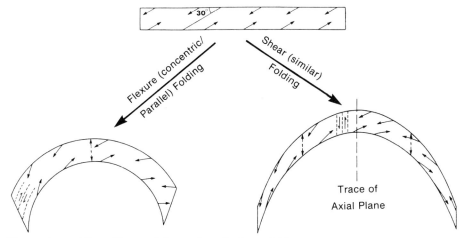

Figure 6.10 The effects of flexural and shear folding on a primary DRM or PDDRM inclined at 30° to the bedding plane. In each case the layer is modified by internal movements along shear planes (dashed) which may produce rotation of the grains responsible for a DRM or PDDRM; the sense of shear along these planes is indicated: the fold axes are assumed to trend E–W so that the beds dip N and S. Modified after Facer (1983).

sediments of South Wales; a discrete high blocking-temperature (hematite) component was acquired here when folding was *ca.* 25% complete, and a low blocking-temperature (magnetite and hematite) component was acquired when folding was *ca.* 75% complete.

The conglomerate test

If the conglomerate pebbles in a sediment, or agglomerate boulders in a volcaniclastic rock can be petrographically linked to the formations from which they were derived, it is possible to determine whether or not the magnetisation in the parent formation pre-dates the formation of the conglomerate or agglomerate. It was first suggested by Graham (1949) and can be applied by comparing the vector sum, R, of the magnetic directions from a number of clasts, calculated by the Fisher methods, with a second vector sum, R_o, expected from the same number of directions with a uniform (random) distribution at the 95% probability level (Table 6.1). If the magnetisations of the clasts are random, it suggests that the remanence of the formation pre-dates their erosion (Fig. 6.11). The test has been applied, for example, to show that the magnetisations in Torridonian sandstones (Section 7.8). pre-date Devonian erosion and sedimentation (Irving & Runcorn, 1957) and include magnetisations pre-dating penecontemporaneous conglomerates (Stewart & Irving, 1974), while clasts of Lewisian metamorphic rocks in basal Torridonian sediments retain a remanence which has survived erosion and deposition (Smith & Piper, 1982). An application to agglomerate clasts is described by Briden & Mullan (1984).

There are two reservations which should be noted. Firstly, if a significant grouping of directions is observed in a collection of clasts, it does not automatically

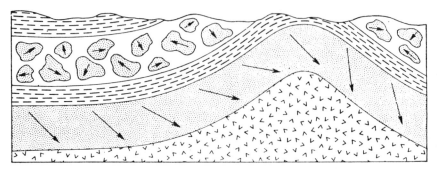

Figure 6.11 The principle of the fold test and the conglomerate test where the stippled horizon preserves a magnetic remanence indicated by the arrows and predating the fold and conglomerate formation. After McElhinny (1973) from Cox and Doell (1960).

mean that the conglomerate has been overprinted by a later magnetisation; if the clasts have a shape anisotropy inherited from the parent rock, this may be preserved as a fabric in the conglomerate. Secondly, an unambiguous test for randomness of directions may require a large collection of samples. By testing model assemblages, Starkey & Palmer (1971) found that a secondary remanence equal to 50% of the primary remanence could go undetected in a sample of 25 clasts, and a secondary remanence of 25% could go undetected in a sample of 100 clasts; a history of remagnetisation in a conglomerate or agglomerate may therefore be difficult or impossible to demonstrate in small samples. A noteworthy example of the analysis of secondary remanence is reported from the Keweenawan Copper Harbour conglomerate by Palmer, Halls & Pesonen (1981).

A variant of this test can be applied to slump horizons in sediments which have experienced penecontemporaneous deformation. In this instance it may not be possible to define the original bedding, but directions of magnetisation can be compared with undeformed layers above and below. Application of this test has shown, for example, that the remanence in many, but not all, of the slump horizons in Torridonian sediments post-dates the slumping (Irving & Runcorn, 1957; Smith, Stearn & Piper, 1983 and the third part of Section 4.2).

The contact test

When an igneous intrusion is emplaced into country rocks, it acts as a heat source, with the resultant increase in temperature diminishing away from the contact. The supply of heat may be prolonged if the intrusion is a magma conduit. It can have the additional effect of mobilising fluids in the contact zone to promote chemical changes. Near the contact, temperatures may be expected to exceed the Curie temperature, and the country rock will become totally remagnetised: similar directions of magnetisation in the igneous rock and the country rocks are then a record of the magnetic field at the time of cooling. Since the magnetic mineralogies of the intrusion and country rocks are unlikely to be identical, a similarity of remanence direction is a strong indication that this remanence is primary and dates from the time of intrusion. The remagnetised zone may be subdivided into a *metamorphic zone*, in which extensive changes in mineralogy have taken place, and a *heated zone*, with little change in the mineral phases. If the metamorphism is associated with growth of new magnetic mineral phases, there may be a contrast in magnetic properties between these zones (Fig. 6.12) which is often reflected in high magnetic intensities at the contact.

The value of the contact test is enhanced if a systematic change in magnetic properties can be demonstrated away from the contact (Everitt & Clegg, 1962). In the heated zone the Curie point will probably not have been reached, but a PTRM may be present, and can be removed by thermal cleaning to recover the pre-intrusion remanence. Beyond this, the country rock shows no influence of the intrusion, and a remanence pre-dating the intrusion can be recognised, provided that the country rock was magnetised in a direction contrasting with the remanence

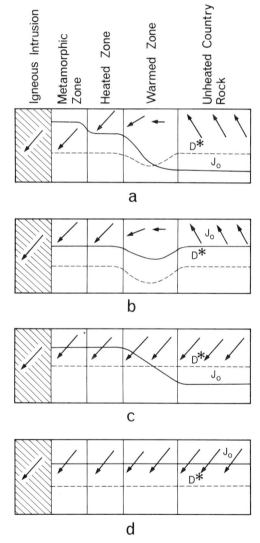

Figure 6.12 The Igneous contact test illustrated by four possible situations. The arrows indicate the directions of magnetisation. Note that example (c) can be distinguished from regional remagnetisation by the increase in intensity (J_o) towards the contact. Modified after Irving (1964) and McElhinny (1973).

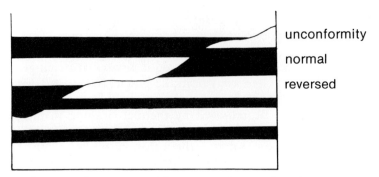

Figure 6.13 The unconformity test. The zones of normal and reversed magnetisation (magnetozones) are truncated by an unconformity in the sequence and imply that the magnetisation defining the lower zones is older than the unconformity. After Kirschvinck (1978).

in the intrusion. The dispersions of magnetisations may also change systematically away from the contact and reflect the magnetic mineralogy of the rock types. Several possible situations can be envisaged (Irving, 1964; McElhinny, 1973, and Fig. 6.12). A uniform direction of magnetisation in igneous intrusion and country rock may indicate either that the country rock is little older than the intrusion, or that they were collectively reheated at some later period of regional metamorphism or deep burial. The presence or otherwise of a systematic change in magnetic intensity and susceptibility away from the margin of the intrusion may sometimes allow discrimination between these two possibilities.

The unconformity test

Kirschvink (1978) describes how unconformities can be used to determine the relative age of remanence when reversals of magnetisation are present; he has applied this to rocks of early Cambrian age. Sediments which record a DRM or PDDRM acquired during, or shortly after, the deposition of the sediment, will record a polarity zonation similar to that shown by the black and white zones in the lower part of Fig. 6.13. In the example illustrated here, sedimentation was interrupted and erosion took place to truncate the section and produce the unconformity; renewed sedimentation then resulted in deposition of the upper sequence. This will contain a younger polarity zonation which does not match the zonation below the time break (provided that the remanence is a DRM or a PDDRM). This situation implies that magnetisation in the beds below the unconformity is older than the episode of erosion which created the unconformity. If, however, zones of normal and reversed polarity exhibit continuity across the unconformity, it implies that the sediments carry a CRM post-dating the younger sedimentation.

6.5 The geomagnetic field and palaeomagnetic poles

The dipole field and axial geocentric dipole model

In common with a vector of remanent magnetism (Section 6.1), the geomagnetic field at any point on the Earth's surface is described in terms of *magnetic elements*, comprising a declination, D, an inclination, I, and a total intensity, F. The horizontal (H) and vertical (Z) components of the field are therefore given by H = F cos I and Z = F sin I, and F = $(H^2 + Z^2)^{1/2}$. At the present time some 90% of the geomagnetic field can be explained in terms of a magnetic dipole at the centre of the Earth, inclined at $11\frac{1}{2}°$ with respct to the axis of rotation; this is the *dipole field*.

The dipole axis intersects the surface at the *geomagnetic poles*. These are not quite coincident with the *magnetic poles*, where the magnetic field is observed to be directed vertically downwards, because *ca.* 10% of the total field remains after the best-fitting geocentric dipole has been subtracted; this residual field is the *non-dipole field*. The magnetic field is constantly changing in a way which is known from geomagnetic observations extending back over several hundreds of years (Merrill & McElhinny, 1983), and archaeomagnetic observations extending back over several thousands of years (Tarling, 1983). These changes are described as the *secular* (ie slow) *variation*, and they appear to be roughly cyclic, with a periodicity of several hundreds of years. Together with palaeomagnetic studies covering the last few millions of years, the collective observations conform to a model in which the time-averaged magnetic field is an *axial geocentric dipole* corresponding with a dipole at the centre of the Earth, and directed along the rotational axis so that the geographical and geomagnetic axes coincide. The average magnetic inclination is therefore +90° at the geographic north pole and −90° at the geographical south pole, and the geographical and geomagnetic equators and lines of latitude are coincident.

This time-averaged model for the geomagnetic field can relate the ancient directions of magnetisation to the magnetic field at the time these magnetisations were acquired. It assumes that the palaeomagnetic vector records a time-averaged field and, to a first approximation, this corresponds not with a dipole inclined to the axis of rotation — like the present instantaneous field — but to a dipole aligned along the axis of rotation. This, of course, requires that the palaeomagnetic study has sampled the ancient field sufficiently thoroughly to give a good time-average of the field. The validity of the geocentric axial dipole assumption is examined in the next section.

A dipole of strength M (A.m²), sited at the centre of the Earth (radius R metres) produces horizontal and vertical components of the field at a latitude, λ, on the Earth's surface given by:

$$H = \frac{\mu_0 M \cos \lambda}{4\pi R^3} \tag{6.11}$$

and

$$Z = \frac{2\mu_0 M \sin \lambda}{4\pi R^3} \tag{6.12}$$

where μ_0 is the permeability of free space, and H and Z are measured in Teslas.
The total field intensity is given by:

$$F = (H^2 + Z^2)^{1/2} = \frac{\mu_0 M}{4\pi R^3}(1 + 3\sin^2\lambda)^{1/2} \tag{6.13}$$

The inclination of the magnetic field is given by $\tan^{-1}(Z/H)$, hence:

$$\tan I = 2 \tan \lambda \qquad (6.14)$$

Ancient palaeolatitudes are calculated from palaeomagnetic inclinations using this equation (see Fig. 6.14). In palaeomagnetism the co-latitude, θ, is also used; this is the angle $(90 - \lambda)$ between the sample point on the Earth's surface and the pole. Then:

$$\tan I = 2 \cot \theta \qquad (6.15)$$

The coordinates of the magnetic pole in degrees north and degrees east (N'E') are determined from the site mean coordinates (N,E) and the magnetic field direction (D,I) from the equations (see Merrill & McElhinny, 1983):

$$N' = \sin^{-1}(\sin N \cos \theta + \cos N \sin \theta \cos D)$$

(for $-90° \leq N' \leq +90°$) \hfill (6.16)

and: $\qquad E' = E + \beta$ (for $\cos \theta \geq \sin N \sin N'$)

or: $\qquad E' = E + 180 - \beta$ (for $\cos \theta \leq \sin N \sin N'$)

where (for $-90° \leq \beta \leq 90°$):

$$\beta = \sin^{-1}(\sin \theta \sin D / \cos N') \qquad (6.17)$$

Provided that a single palaeomagnetic site yields a mean direction representing a spot reading of the geomagnetic field, these equations will give an instantaneous position of the magnetic pole, which (like the present magnetic pole) will be displaced from the geographical poles. This spot record is a *virtual geomagnetic*

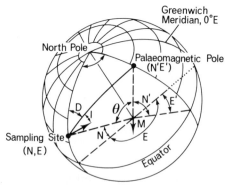

Figure 6.14 *The calculation of the palaeomagnetic or virtual geomagnetic pole with coordinates (N', E') from the direction of magnetisation with declination D and inclination I at a sampling site with coordinates (N, E) assuming that the field source is a geocentric dipole M.*

pole (VGP). The average of a number of site mean directions will, in general, have averaged secular variation, and the pole calculated from the overall mean direction is then a *palaeomagnetic pole*.

If a set of directions of magnetisation is circularly-distributed according to the Fisher model, their corresponding VGPs will have an elliptical distribution; similarly a circular distribution of poles converts into an elliptical distribution of directions. Thus, a circle of confidence about the mean direction will convert into an ellipse of confidence about the palaeomagnetic pole. This ellipse tends towards a circle as the mean inclination increases towards the vertical. Since the secular changes of geomagnetic field are likely to approximate to a circular distribution in the sense that they may be regarded in terms of random fluctuations about the mean (or a continuous track with a bias towards the rotation axis), it would be more statistically correct to apply Fisher's statistics to the distribution of VGPs. In practice, however, most workers determine the statistical parameters from the directions of magnetisation, and then extrapolate the results to determine error parameters for the derived palaeomagnetic pole. These are given by the semi-angles of the 95% confidence oval about the palaeomagnetic pole, in the co-latitude direction (δp) and at right angles to it (δm). These parameters are given by:

$$\delta p = \tfrac{1}{2} \alpha_{95}(1 + 3\cos^2\lambda)$$

and

$$\delta m = \alpha_{95} \frac{\cos \lambda}{\cos I} \tag{6.18}$$

If $I = 90°$, $\delta p = \delta m$; otherwise $\delta p < \delta m$.

A calculation of the pole position from the VGPs automatically corrects for differences in latitude and longitude of the site positions. Normally these differences are small and will not appreciably affect the palaeomagnetic pole derived from directions, but if the spread of sampling sites is large, it may be necessary to reduce all directions of magnetisation to the equivalent direction at the centre of the sample area, before averaging them.

The axial geocentric dipole — a valid assumption?

It is possible to test whether or not the geocentric dipole assumption provides a sound model for interpreting the palaeomagnetic field in several ways:
1. palaeomagnetic poles from all continents and ocean basins covering the last 10 Ma plot around the geographical poles rather than the present geomagnetic poles (Tarling, 1983), although few poles actually plot precisely on the geographical poles. The average pole determined from these data is statistically identical to the geographical pole. This observation establishes that the palaeomagnetic field has been axial (but not necessarily geocentric) during this last part of geological time.
2. Palaeomagnetic poles from magnetisations of the same age, from different

locations within a rigid continental block (with preferably, as wide a latitudinal variation as possible) should be in close agreement with one another. This has been demonstrated by the close correspondence of Phanerozoic palaeomagnetic data distributed over large continental areas such as Africa, North America and Eurasia (McElhinny, 1973; Irving & Irving, 1982; Chapters 10 & 11) and is widely applicable to Precambrian times (Piper, 1982a, and Chapters 7 & 9).

3. There should be a definite probability function for the magnetic inclination, I, if palaeomagnetic studies have been randomly distributed over the globe. This distribution can be readily obtained by estimating the surface area of the globe corresponding with any given limits of I. This test has been carried out by Evans (1976) for data covering the last 600 Ma; he shows that the observed distribution of I corresponds much more favourably with the distribution predicted from a simple dipole field source than with more complex quadrupole and octopole sources.

4. The intensity of the geomagnetic field should increase from the palaeo-equator to the palaeopole according to equation (6.13). Unfortunately, this test is not practical, both because the intensity of the field undergoes large short-term changes and because palaeo-intensity determinations do not permit the dipole contribution to be separated from the non-dipole contribution (McFadden & McElhinny, 1982).

5. The angular dispersion of the palaeomagnetic observations can be related to the secular variation of the geomagnetic field in which they were magnetised, and this parameter is critically dependent on the sources of secular variation of the geomagnetic field. The angular dispersion is considered either in terms of the dispersion of palaeomagnetic directions, or in terms of the dispersion of the corresponding VGPs. The dispersion of the former decreases towards high palaeolatitudes, whereas the dispersion of the latter increases. The way in which these parameters change with latitude (Creer, 1962a) has been predicted by several models for the dipole and non-dipole fields, which are critically assessed by Merrill & McElhinny (1983). The observed dispersion of palaeomagnetic data covering the last 195 Ma changes with latitude in a way which corresponds closely with the predicted variations, although this agreement accommodates the non-dipole, rather than the dipole field (Merrill & McElhinny, 1982; McFadden & McElhinny, 1982 and, Section 12.6).

6. The planetary magnetic fields lie within 15° of the axes of rotation. This implies that planetary magnetic fields are generated in a way which requires a close association between the geographic and geomagnetic axes and it thereby supports an axial location for the sources (Busse, 1976). The recently discovered field of Uranus appears to be an exception to this rule.

While these tests support the assumption that the mean palaeomagnetic field was *axial*, they tell us little or nothing about the degree to which it was *geocentric*. The analysis of large numbers of late Cenozoic and Quaternary palaeomagnetic studies on a regional basis has shown that there are small but significant departures from geocentricity. Wilson (1970, 1971) divided the globe into a number of regions and showed that the mean palaeomagnetic directions from each region tended to yield pole positions which do not coincide precisely with the present geographic pole; instead they tend to plot on the far side of the pole and to the right hand side with respect to an observer at the sample location. This observation is consistent with a model in which the mean dipole source is an axial dipole displaced by a small distance to the north of the geocentre. Subsequently Wilson and McElhinny (1974) have subdivided the younger Cenozoic data into three groups: a Quaternary group including data younger than 2 Ma in age, a Pliocene–Pleistocene group incorporating data between 7 and 2 Ma in age, and a Miocene–Pliocene group incorporating data between 25–7 Ma in age. While the mean pole positions are statistically coincident with the geographic pole and establish the axial nature of the source over these time intervals, the average directions of magnetisation from fifteen regions distribute over the globe yield palaeomagnetic poles all but one of which is far sided with respect to the sample locality (ten of them are also right handed). Wilson and McElhinny argue that post-25 Ma plate movements are unable to explain these pole distributions and that the most likely cause is a geomagnetic field configuration averaged by an axial dipole displaced by 555 (25–7 Ma), 316 (7–2 Ma) and 143 (2–0 Ma) kilometers to the north of the equatorial plane. According to this model the inclination is shallower than an axial geocentric dipole in the northern hemisphere and steeper than it in the southern hemisphere; the departure from the inclination predicted from an axial geocentric dipole is greater in equatorial latitudes and again in intermediate latitudes and is least at the poles (McElhinny, 1973). This is not of course, a real description of the source but is merely a convenient model for the palaeomagnetic directions observed at the surface; it does not accommodate the right handed positions of the poles and so far no model seems to have been proposed to account for this. While these complexities defined by the geomagnetic field over the last 25 Ma are second order effects, they do imply that the true position of the geographic pole cannot be precisely determined from the palaeomagnetic vector, and even a comprehensive survey may yield a pole position differing by up to about 5° from the true pole of rotation when the source has been assumed to be a geocentric one.

To test the longer term behaviour of the field we must have recourse to palaeoclimatic evidence because only this can provide an independent measure of past geographic latitudes. Since climatic belts are subject to seasonal variations and are controlled by the oceanic and continental distributions, they are not generally parallel to the lines of palaeolatitude; hence this is necessarily a coarse test, but the evidence from Phanerozoic times strongly supports an average axial location for the dipole.

The palaeoclimatic indicators which have been used for this purpose include reef-building corals, evaporites and glacial deposits (Blackett 1961; Runcorn, 1961; Opdyke, 1962; Irving and Briden, 1964; Drewry, Ramsay and Smith, 1974). coal deposits (Briden and Irving, 1964), palaeowind directions (Opdyke and Runcorn, 1960; Opdyke, 1962), and phosphate deposits (Cook and McElhinny, 1979). The latitudinal distribution of hydrocarbon deposits has been examined by Irving and Gaskell (1962) and Irving, North and Couillard (1974), and the distributions of a number of fossils have been examined in the context of palaeolatitudes determined from the axial dipole assumption (Runcorn, 1961; Irving and Brown, 1964; Middlemiss, Rawson and Newall, 1971; Hallam, 1973; Ross, 1974; Turner and Tarling, 1982). This evidence is discussed in some detail in Section 10.7. With the reservation that some of these parameters are controlled by atmospheric circulations which are strongly influenced by the prevailing continental sizes and shapes, they all have distributions according in a general way with the axial dipole assumption: individually none of them have the precision to confirm the assumption, but the observation that they are all collectively consistent with the model is a very strong indication that is a valid one.

Reversals and excursions of the geomagnetic field

Although many theories have been proposed to explain the geomagnetic field (see Rikitake, 1966), the great majority of scientists now accept that it is caused by a dynamo action resulting from motions in the liquid outer core of the Earth. This view has been supported by Elsasser (1946) and Bullard (1949), and placed on a firm theoretical basis by the work of Backus (1958) and Herzenberg (1958).

The most outstanding property of the magnetic field is its ability to change its polarity. The identification in the 1950s of an increasing number of rocks with magnetisations roughly opposed to those of the present field, accompanied studies of materials which possess the property of acquiring a TRM in a direction antiparallel to the applied field (Nagata, 1961). In natural materials, this property of *self reversal* is largely restricted to a limited compositional range in the ilmenite–hematite solid solution series (Uyeda, 1958 and section 2.3) and is unlikely to be an important mechanism in most rocks. The great number of field observations which confirm the reality of field reversal (inversion)* include studies of:
1. Baked contact zones adjacent to igneous intrusions which show agreement in polarity between the intrusion and the baked rock (Wilson, 1962).
2. The records in sedimentary sequences, lava piles and some igneous intrusions, which show a continuous change in polarity via intermediate directions.
3. The identification of worldwide simultaneous zones of one polarity (see Cox, 1969, for a summary).

There remains an enigmatic correlation between polarity and deuteric oxidation state: reversely magnetised lavas (Ade-Hall & Wilson, 1963; Wilson, 1964), dykes (Ade-Hall & Wilson, 1969; Piper 1975a; Bird & Piper, 1981) and plutonic metamorphic rocks (Balsey & Buddington, 1958) are often statistically more oxidised than normal ones and the same correlation has been found in Precambrian lavas (Piper, 1977). The opposite correlation has not been observed, but it is not everywhere present (Ade-Hall & Watkins, 1970), and no explanation can be offered at the present time.

Detailed records of magnetic field inversions in lava sequences (Van Zijl, Graham & Hales, 1962, Watkins 1969; Lawley, 1970), in igneous intrusions (Dunn, Fuller, Ito & Schmidt, 1975), in red sediments (Herrero-Bervera & Helsley, 1983) and in deep-marine sediments (Opdyke, Kent & Lowrie, 1973) suggest that the geomagnetic field first begins to fall in intensity for perhaps several thousands of years before the inversion. The field vector then begins to move, so that the VGP follows an irregular path across the equator and towards the position of opposite polarity. Over the course of this passage the intensity drops to a value of ≤ 20% of the typical normal or reversed values. Estimates for the time taken

---

* The word *reversal* has been used to refer to both (i) in the reversed state of the field, and (ii) in the act of inversion either from N to R or R to N. To avoid this ambiguity it is best to use *reversal* for (i) and *inversion* for (ii).

by the whole cycle are mostly in the range 4000—10,000 years, although the field direction may take no more than 1000–2000 years to invert (McElhinny, 1973).

The behaviour of VGP and field intensity during an inversion define the balance of field sources in the outer core: if inversions are due to a collapse in the dipole field only, the non-dipole components would remain, and these would not produce a unique VGP path, either at the same time when observed from different points over the Earth, or during different inversions. However, if the dipole field simply flips over from one position to the other, essentially the same VGP path should be observed over the whole Earth. Studies of recent inversions suggest that there is no common path, and indicate that the transition field is controlled by non-dipole moments. However, it is unlikely that each inversion will have the same characteristics, and this may not always be true; in particular, strong intermediate fields lasting for appreciable periods of time have been identified during a few reversals (Shaw 1977 and section 12.6).

In addition to polarity inversions, the VGP has sometimes moved by large angular distances (>40°) away from its typical position near the geographical pole, and then returned to this pole without attaining the opposite polarity. These movements are described as *magnetic excursions*. They are recorded as TRMs in aboriginal fireplaces and some lava flows, and as DRMs or PDDRMs in deep-marine and shallow-lake sediments. Most excursions have so far been reported from within the Bruhnes normal epoch (<700,000 years) and the evidence for some of these is contentious. In particular, some aberrations of the magnetic field vector in soft sediments may have a sedimentological explanation, or be due to deformation during coring. Other excursions are locally well established, but are not observed in contemporaneous rocks a few hundreds or thousands of kilometres away; they are, therefore, regional and not global effects. Dipole sources at the core–mantle boundary, of opposite polarity to the main dipole field, could apparently produce anomalous fields observable over only a few 1000s of km at the surface and are possible cause of magnetic excursions (Barbetti & McElhinny, 1976).

For further discussion of inversions and excursions, refer to Merrill & McElhinny (1983) and Jacobs (1984). The history of the magnetic field through time is described in sections 12.2 to 12.6). Here it is noted that palaeomagnetic poles which are anomalous when compared with roughly contemporaneous data from the same crustal plate may record excursions of the geomagnetic field. This explanation could be applicable if the poles are calculated from results which have averaged a short time period ($10^2 - 10^3$ years), but this is less likely if the magnetisation time is significantly longer.

6.6 Apparent polar wander (APW)

Definition and presentation

In general, palaeomagnetic poles of the same age, from the same general region may be expected to agree with one another, and this agreement will show that the axial dipole assumption is a valid one (although it tells us nothing about the geocentricity of the source). If they fail to agree, there are two probable explanations: either non-dipole components were significant at the time of magnetisation, so that the palaeomagnetic field did not approximate closely to the axial dipole, or relative movements have taken place between the sample locations, which invalidates the assumption that they originate from the same crustal plate. Since the axial dipole assumption appears to be a good one (second part of Section 6.5) the latter explanation will be generally applicable, and it is therefore necessary to consider as a single data set, only those palaeomagnetic poles derived from the same rigid crustal plate. The time sequence of palaeomagnetic poles from a single plate defines an *apparent polar wander path* (APWP). The divergence of these paths from the present geographical pole provides the essential palaeomagnetic evidence that continents have changed their positions on the globe, while the comparison of such paths between different continental blocks provides the quantitative evidence for continental drift, and provides a method for determining the past relative positions of the continents.

APWPs are plotted on standard map projections, with the continental block retained in present-day coordinates. These plots describe the way in which the palaeomagnetic north or south pole has moved with respect to the crustal plate.* (The equivalent alternative procedure for presentation retains the average palaeomagnetic pole for a single time period at the position of the present pole, and illustrates the inferred movement of the crustal plate over the globe. This is less manageable and is also deceptive because it wrongly suggests that palaeolongitude is known; it is not commonly employed nowadays).

APWPs incorporate two sources of error: the first is the error on the pole determinations, which is roughly reflected in terms of the confidence ovals about each pole determination, and the second is the error in the age assignments to the poles. These errors mean that a subjective evaluation of the paths is still needed, especially for Precambrian data where complementary stratigraphic age assignments are not usually available. However, an average APWP can be derived from the 'moving window' technique (Irving & Irving, 1982 and Chapter 11) which averages overlapping groups of poles to derive successive means. This procedure usually gives unit weight to each pole; it thus averages out the errors in individual observations, if these errors are random. It is best applied when the data are of fairly uniform quality and evenly spaced in time. Parker & Denham (1979) proposed a method for interpolating the time sequence of poles based on cubic splines. A more sophisticated method has been developed by Thompson & Clark (1981) to accommodate the variable precisions of individual poles; it defines the APW as a running mean to a least-squares fit of cubic splines. This method has provided smoothed APWPs and 95% confidence bands for post-400 Ma palaeomagnetic data, but cannot yet be efficiently applied to the more complex body of pre-400 Ma data.

Interpretation of APW: the rate of continental movement

The time average of the present geomagnetic field is axially symmetrical, and has been so during most of the Earth's history (see second part of Section 6.5, with possible exceptions noted in Section 7.9). Hence, changes in the palaeomagnetic inclination at any single point on the globe reflect changes in palaeolatitude (equa-

---

* In most of the subsequent text it is convenient to use the term *plate* for single rigid slabs of continent, although in plate tectonic terms it should be recognised that they may form only component parts of larger plates incorporating oceanic crust as well.

tion 6.14) and rotation is revealed by changes in the declination; they indicate nothing about the translation of the point through longitude because lines of longitude are loci of constant magnetic inclination. For this reason the rate of movement represented by an APWP (expressed in degrees/Ma or cm/year) cannot be translated into the absolute rate of movement of the continent. High rates of APW do not necessarily imply high rates of continental movement, and neither do low rates of APW imply little or no movement: any combination of fast or slow APW and continental movement can be produced by appropriate geometries of the pole and continent (Fig. 6.15). Obviously the minimum rate of movement over a given time interval is constrained by the change in palaeolatitude, but the APW information indicates more than this. The geometry of motion of a sizeable continental area between two positions on the globe is described in Fig. 6.16. The resultant motion between two positions on the globe, as illustrated in Fig. 6.16a, can be completely expressed in terms of the rotation of the continent through an angle θ about a unique pole of rotation (*Euler pole*) on the Earth's surface. The line connecting this pole of rotation to the centre of the Earth is the axis of that

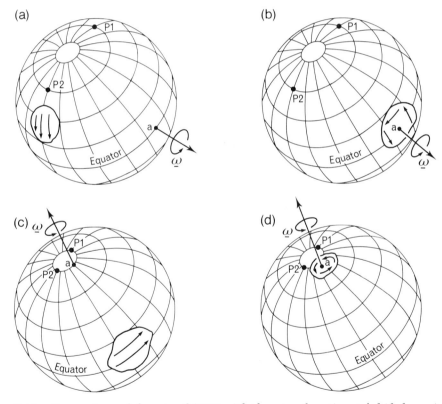

Figure 6.15 Comparison of the rate of APW with the rate of continental drift for a given angular rotation, ω, about a Euler pole a. APW (or continental drift) is fast when the palaeomagnetic pole (or continent) is about 90° away from the pole of rotation and slow when the palaeomagnetic pole (or continent) is near the pole of rotation. Note that the original palaeomagnetic pole P_1 is moved to P_2 by the rotation of the continent and the resultant motion is visualised by moving P_1 rigidly connected to the continent to the new palaeomagnetic pole P_2. After Gordon, McWilliams and Cox (1979).

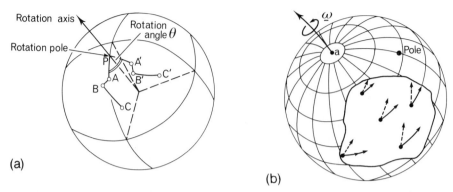

Figure 6.16 (a) *Rotation about a Euler pole. ABC is a line on the Earth's surface and rotation of ABC through an angle θ about the axis running from the centre of the Earth to the point P moves it to A'B'C'. This is the rotation axis; P is the Euler pole and θ is the rotation, or Euler, angle.* (b) *The linear velocity (solid lines) at points over a continental area resulting from an angular rotation about a Euler pole at a. The northward component of this linear velocity towards the pole is indicated by the dashed lines. The palaeolatitude method of computing continental velocity from APW determines the minimum value of this northward component.*

rotation. The linear velocity vectors of each point on the continent are indicated by solid lines in Fig. 6.16b and are parallel to the small circles about the axis of rotation. It is clear that linear velocity varies over the continental area: not all points can be moving towards the pole of rotation everywhere, and the northward component of linear velocity varies over the continental area, as shown by the dashed lines. One approach is to calculate the value of the northward component of motion at all points on a grid covering the continental area; this will, in general, yield a higher average value of velocity than a calculation of the palaeolatitude change at a single point on the continent. However, it is also possible to calculate a total velocity vector commensurate with the rotation of the rigid plate and the constraint provided by the APWP.

For two closely spaced portions on an APW path of AGE 1 and AGE 2, the mean palaeomagnetic pole can be estimated from the midpoint of the great circle connecting P1 and P2 (Fig. 6.17). It is convenient to use this point as an axis ($\hat{Z}$) of an orthogonal coordinate system, so that a second axis ($\hat{X}$) is defined by the intersection of the palaeo-equator with the meridian perpendicular to the great circle connecting P_1 and P_2. The third ($\hat{Y}$) axis lies along the meridian 90° from $\hat{X}$. The point P_1 can be rotated to P_2, along any number of small circles, by rotation about axes lying in the XZ plane. These axes intercept the great circle through $\hat{X}$ and $\hat{Z}$ at Eulerian poles such as a, b and c which correspond with widely different different velocities of the continental body: angular velocities ω_1 and ω_b correspond with large velocities V_a and V_c, while the angular velocity ω_b corresponds with a lower velocity V_b much closer to the minimum root mean square velocity (V_{RMS}) which can be determined from the APW path. Two geometrical relationships follow from Fig. 6.15. Firstly, the angular velocity, ω, for a given rate of APW decreases as the axis of rotation moves from the palaeopole towards the palaeo-

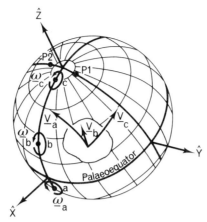

Figure 6.17 The minimum velocity of a continent as defined by its APW path. The reference Z axis is the bisectrix of the short segment of APW path linking poles P1 and P2. Angular rotations about any pole on the meridian containing the X and Z axes will move P1 and P2. The linear velocities V_a, V_b and V_c correspond to rotations of ω_a, ω_b and ω_c about points a, b and c on this great circle. The minimum velocity of the continent as defined by the APW path is closest to the minimum velocity V_b. After Gordon, McWilliams and Cox (1979).

equator, and secondly, for a given ω, V_{rms} decreases as the axis of rotation moves towards the centre of the continent. The relative importance of these two effects depends on the palaeolatitude of the continent, the direction of APW relative to the palaeomeridian of the continent, and the size of the continent.

Thus this method of calculating the velocity of continental movement from APW, which is due to Gordon, McWilliams & Cox (1979), calculates the position of the Eulerian pole giving the minimum rate of continental drift, and V_{rms} is defined as:

$$\sqrt{\frac{\int(\bar{\omega}\bar{r})^2 dS}{area}} \qquad (6.19)$$

where $\bar{r}$ is the position vector on the surface of the Earth and dS is the element of surface area. The minimum value of this function is calculated according to the method described by Gordon, McWilliams and Cox (1979) by employing two constraints, namely:

$$\omega_y = 0$$

(i.e. the Euler pole of rotation must lie along the great circle containing $\hat{X}$ and $\hat{Z}$), and

$$\omega_x = \frac{P_2 - P_1}{AGE\ 2 - AGE\ 1} \qquad (6.20)$$

(i.e. ω_x must satisfy the rate of APW)

This method will, in general, yield higher estimates of the rate of continental movement from APW than the estimates derived from palaeolatitude changes alone, although they will normally be less than the absolute rates of movement. This method is routinely employed in subsequent chapters to calculate the rates of continental movement from the APW through geological time.

The temporal change in palaeomagnetic poles described by APW can be explained in one of two ways:
1. The continents may move with respect to the spin axis (and hence the average geomagnetic dipole axis) of the Earth.
2. It is also theoretically possible for the whole outer shell of the Earth to move with respect to the rotation axis (Goldreich & Toomre, 1969); this would produce *true polar wander* (TPW). Either explanation would be commensurate with the wealth of geological evidence for lateral continental movements, but the second explanation requires that the same movements are common to all plates, while the first explanation predicts that different movements will, in general, be recorded by each plate. Hence, the term *continental drift* is restricted to relative motions of the lithospheric plates. Obviously, if the APWP of a single plate is treated in isolation there is no *a priori* reason for favouring either explanation, but comparison of the collective APWPs of the major plates strongly suggests that most or all of the observed APW in Cenozoic times is explicable in terms of independent movements of the lithospheric plates, with respect to a stationary spin axis (Jurdy & Van der Voo, 1974). There are, however, strong indications that a large component of APW in pre-Cenozoic times is explicable in terms of true polar wandering (Creer, Irving & Runcorn, 1957; Jones & Gartner, 1979; Piper, 1983b), and discussion of this subject is deferred until the APWPs have been treated in detail (Section 12.7).

Interpretation of APW: tectonic models

The bulk of the continental crust is known either by direct outcrop or by a combination of borehole and geophysical evidence to be underlain by Precambrian crust, probably of great antiquity (>2500 Ma). The components of this ancient crust are shown in Fig. 6.18 as the (blank) continental shields. They comprise the building blocks of the Precambrian crust, and their shapes are probably determined by fragmentation along primitive long-lived zones of crustal weakness. With the exception of limited areas which have been underthrust beneath other continental crust in Phanerozoic orgenic belts — such as the contemporary underthrusting of Tibet by the Indian subcontinent (Klootwijk & Bingham, 1980 and Section 11.3) — the known areas and shapes of these blocks must be accommodated by models for the growth and movement of the Precambrian continental crust, with the proviso that these areas may have changed somewhat, as a result of compression, attenuation and rifting. The contemporary continents are quite different in shape from their Precambrian and Palaeozoic forebears because they have been fragmented or have collided with other continents since these times. The collision zones are *sutures*, and they lie along extinct orogenic belts such as the Appalachians, the Caledonides and the Urals (Chapter 10), or active orogenic belts such as the Alps, the Himalayas and the Circum-Pacific belts (Chapter 11). These sutures are identified by the presence of ophiolites, accretionary flysch sequences

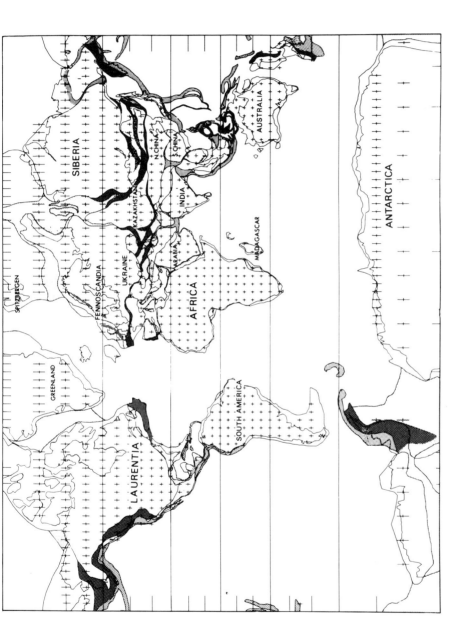

Figure 6.18 World suture map showing the distribution of the continental shields, comprising outcrop or basement of Precambrian crust (blank), bordered by areas of Palaeozoic (dense shading) and Mesozoic–Cenozoic (intermediate shading) accretion; these latter areas generally do not have a basement which is older than Later Precambrian in age. The light shading indicates areas of oceanic crust infilled with sediments and/or rift type volcanic rocks and includes deltas, extensional zones, and gaps left by incomplete closure in areas of continental collision. Modified after Scotese, Bambach, Barton, Van der Voo and Zeigler (1979).

and sediments indicative of a deep-marine environment (Dewey and Bird, 1970). In addition biogeographical patterns are disrupted across these lines (Wilson, 1966 and Sections 9.6 and 10.7) and the regions bounding them have contrasting geological histories.

The suture zones are the relics of ancient plate convergence, and record to varying degrees the histories of closure of ancient ocean basins. They can be interpreted in the context of processes and events which are taking place in the modern ocean basins (Dewey & Bird 1970). The destructive margins of the plates are defined by ocean trenches along which ocean lithosphere is being consumed. They are typically the sites of island arcs built up from the calc-alkaline igneous products from partial melting of the largely-amphibolitised basaltic crustal section. Where these trenches develop near to continental margins they may be characterised by paired metamorphic belts with high P–low-T blueschist-facies metamorphism on the ocean side and high-T–low-P greenschist-facies metamorphism on the landward side. The back arc side may be the site of incipient ocean-floor spreading and the accumulation of synorogenic sediments derived from the volcanic arc (second part of section 11.4). When a continental arc is part of a crustal plate it will eventually arrive at a destructive margin. Owing to its buoyancy it cannot be carried down into the subduction zone, and so the arc shifts to consume the ocean crust behind it, and converts the margin of the continental crust into a Cordilleran margin (e.g. the Andean belt of South America (sixth part of section 11.4)) which then becomes the site of crustal accretion. The continental crust then thickens by the addition of igneous melts derived from subducting oceanic lithosphere: Cordilleran margins may show few signs of compression, but are characterised by a complex volcanic history of partial melting and fractionation (Brown & Hennessey, 1978). Eventually, another continent will arrive at the trench and collide with the opposite Cordilleran margin to produce a continent–continent collision (such as the Alpine-Himalayan belt), characterised by a complex compressive zone of overthrusting and underthrusting; this can result in an orogenic belt of the grandest scale (section 11.3 and third part of section 11.4). The belts then become the sites of molasse-facies sedimentation directed away from the trench and over the underthrust plate. When collision and suturing are complete, consumption of the sinking slab of lithosphere ceases and the plate boundaries are reorganised to permit the consumption of oceanic lithosphere elsewhere.

It is possible to interpret many Phanerozoic sutures in the context of these general observations of contemporary plate tectonic processes. The Caledonian-Appalachian belt, for example, was the result of island-arc- and Cordilleran-type mechanisms in early Ordovician times, followed by continent–continent collision in Siluro-Devonian times, and late strike-slip motions between the sutured plates in Devonian and Carboniferous times (Chapter 10). Similarly, the Himalayan orogenic belt is the end result of island arcs, and subsequently of continental blocks, colliding with the Eurasian landmass after the beginning of Mesozoic times (Sections 11.3 and third part of Section 11.4). However, further back in time it becomes increasingly difficult to apply the tenets derived from contemporary observations: the blueschist facies of the paired metamorphic belts (high P–low T in proximity to low P–high T) of accretionary margins for example, are with one or two exceptions, not observed before Cambrian times. Therefore, palaeomagnetic evidence must be used to estimate the past plate distributions and motions. As described in Chapters 8 and 9, this evidence also provides a possible solution to the temporally-changing characteristics of plate tectonics.

Since the breakup of the last supercontinent — Pangaea — commencing *ca.* 150 Ma ago, the continental shapes and areas, as shown in Fig. 6.18, have probably been little altered, except along the Alpine-Himalayan front zone. The movements, collisions and disruptions of these plates have traced out APWPs with a form unique to the individual rigid plates, except for transient periods when enlarged plates were formed by the suturing of two or more smaller plates. Conversely, the APW signature can be used to interpret the manner in which these events have taken place. Four tectonic models for the interpretation of APWPs are illustrated in Fig. 6.19, together with APW signatures which might develop in each case. In the multi-continent model A, a larger continent is formed by the welding of smaller continents. The APWPs of the constituent continents are unrelated until the time of suturing, after which a common path will apply. This situation is clearly applicable, for example, to the Eurasian landmass in Phanerozoic times, where the Siberian plate has been welded onto the Ukrainian plate along the Uralian orogenic belt; it is anticipated that these two divisions will have distinct APWPs prior to welding in Permo-Triassic times (Chapter 10), after which a single APWP is applicable to the entire Eurasian landmass (Sections 10.5, 11.2 and 11.3). If the larger continent in Model A is subsequently disrupted, each fragment will record a memory of the interval of continuity, in terms of a common segment of APW.

Model B is the situation for a single integral continent in which each tectonic division yields an APWP compatible with every other one; tectonic deformation is not precluded, but it must be below the limits of palaeomagnetic detection (i.e. ⩽ 1000 km). This situation is, of course, only applicable to Phanerozoic times, insofar as continental blocks such as Africa and South America (with the exception of their peripheral orogenic belts) can be considered to have behaved as rigid blocks. It is, however, a relevant model for the Precambrian development of these plates. The large segments of Precambrian shield comprising the stable blocks of Fig. 6.18 are themselves the end-result of crustal fractionation, growth and stabilisation in the earlier part of the Earth's history. Much discussion on the temporal development of plate tectonics prior to breakup of Pangaea has centred on the possible modes of development of this crust. One view considers it to be the end result of development as multiple cratonic nuclei, either by lateral accretion or by continuous breakup and collision in a manner analogous to contemporary plate tectonics (Model A). The alternative view is that this crust developed as a primary single unit which has subsequently been disrupted to form the shield fragments of Fig. 6.18. As will be shown in Chapters 7 and 9, palaeomagnetism has been instrumental in resolving this question. The implication of model B is that each part of the crust comprising a single shield yields palaeomagnetic poles, compatible with only a unique APWP. Since the geological record is inevitably incomplete, only limited segments of APW can be defined from any one region, such as a cratonic nucleus; but they should overlap and collectively conform to the single APWP to satisfy this model (Fig. 6.19).

It is instructive to examine the significance of the shape of the APWP. During the interval 1–4 the continent rotates about an external Eulerian pole; the APW then traces out a track which is a small circle centred on this pole. During the

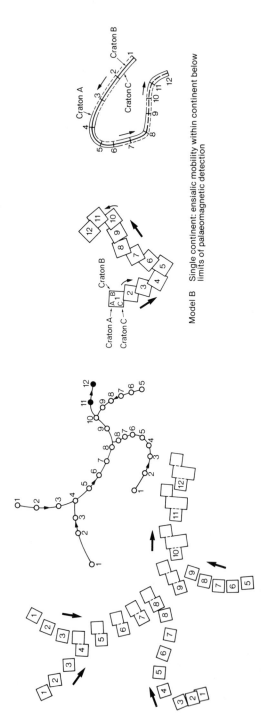
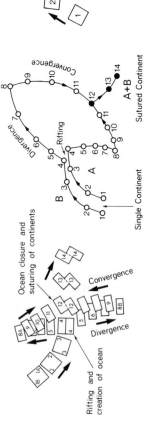

Figure 6.19 Some tectonic models for the evolution of the continental crust and their possible APW signatures. Adapted from Piper, Briden and Lomax (1973) and Irving, Emslie and Ueno (1974).

interval 5–8 it rotates about a different Eulerian pole, so that the bend between 4 and 5 corresponds to a change in position of the pole of rotation. Changes in the motion of the continent are defined by bends in the track (Section 7.10). The APWP is thus a sequence of small circles defined by the sequence of the poles of rotation and by the rates and durations of rotations about them (Irving, Emslie & Ueno, 1974). Since bends correspond with the changes in poles of rotation, they should correspond with continental collisions, or with a change in the driving mantle convection system (Sections 7.10, 7.11 and 12.11).

The *Wilson cycle* of basin opening and closure may define an APWP of the form of Model C. Going backward in time from 14 to 1 there is a common polar track after final suturing, preceded by complementary polar loops, consisting of two or more small circles, and corresponding with convergent and divergent phases. Prior to initial rifting, A and B will have similar tracks which are unlikely to overlap because the initial rift and final suture will not generally coincide. Unfortunately, the Wilson cycle has been widely promoted as a general explanation for lithospheric tectonics, without adequate definition or justification by observation. Thus, the cycle of Model C requires subdivision of a coherent block above a rising and bifurcating convection cell, followed by ocean basin formation, and then a reversal of the convective mechanism to bring the continental blocks back into juxtaposition. Wilson's original (1966) description of orogenic processes described *closure* of an ocean basin, culminating in formation of the Caledonian-Taconic orogenic belt, to be followed by *opening* of a new ocean basin (the North Atlantic) in Mesozoic times on, or close to, the suture by the Palaeozoic collision. In 1968 he extended this analysis to a more general proposal for lithospheric tectonics, to be generally considered in the context of ocean basin opening and closure. However, examination of the present tectonic régime shows that the sequential opening and closure of a single ocean basin is an exceptional occurrence. The opening of one basin is complemented by the closure of another, with a time scale (more than 10^7 years) which makes the possibility of the same margins again coming into continuity, very remote. For this reason contemporary tectonics should be envisaged in terms of Model A, followed by subdivision and the creation of two separate APWPs. The opening and closing of the Iapetus Ocean between Laurentia and Baltica remains the classic example of the Wilson cycle (Chapters 9 & 10); it may also be applied to certain transitional orogenic belts in Gondwanaland (Section 8.5), although this has not yet been demonstrated palaeomagnetically.

Model D has been proposed as an explanation for Proterozoic tectonics by Onstott & Hargraves (1981). It regards mobile belts within Precambrian shields as ensialic, but the sites of strike-slip motion on a scale large enough to be palaeomagnetically detectable. A common APWP applicable to the contiguous plate is divided at the time of strike-slip motion by relative movement of the cratonic forelands on either side of the mobile belt. Thus the APW signature is superficially similar to Model A, but the older separated portions of the APWP have segments of common length and shape. Although proposed as a model for *ca.* 2000 Ma-old data from Africa and South America, it now appears that those data are explicable in terms of APW, because older data from the same area conform to a single path and obey Model B (Sections 7.2 and 7.3). However, this model may be relevant to a number of Phanerozoic situations (Chapter 10 and section 11.4).

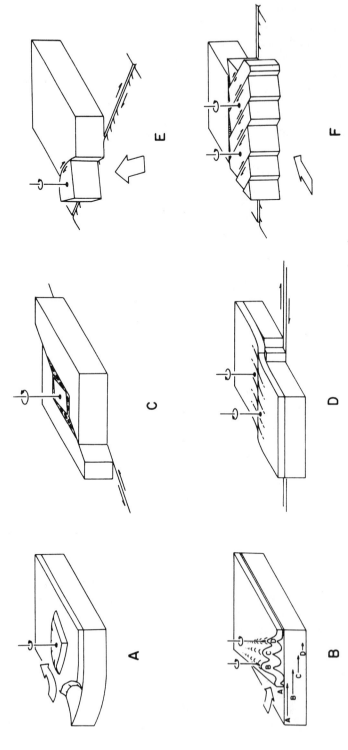

Figure 6.20 A variety of small scale structures which cause rotation about a vertical axis and which may in turn cause the palaeomagnetic vector from a nearby site to be rotated: (a) A horizontal sheet of rock is detached along a plane of weakness and slides over adjacent terrane as a thrust slice or décollement; (b) Another variety of décollement which results primarily in folding; the differential folding shown results in greater rotation about a vertical axis at fold B than at fold D after correction for bedding tilt. (c) Fault blocks in strike slip fault zones rotate about vertical axes; rigid or quasi-rigid blocks are surrounded by a deforming matrix of mylonite, cataclastic rocks or a tectonic mélange. (d) Plastic drag near strike slip fault zones produces rotation about vertical axes which is related to the sense of shear; the example shows right lateral (dextral) shear. (e) Oblique subduction or transpression (Harland 1971) rotates crustal fragments along strike slip faults subparallel to the trend of the subduction zone. (f) Regional compression or shear stress rotates the blocks about vertical axes; in this example the boundary faults rotate with the blocks. After Macdonald (1980).

More variants of models A to D could be generated, but little purpose would be served by them. Owing to the indeterminacy of palaeolongitude, it becomes difficult or impossible to distinguish between the ramifications of Models A, C and D, even with a high APW resolution. A good example is presented by the tectonic evolution of the Western Cordillera of North America (Beck, 1980) where microplates have been moved northwards by a combination of strike-slip and rotational movements against cratonic North America (concluding part of Section 11.4). Both the declination and inclination of the remanence vector have changed with respect to the craton, and although a model analogous to D is favoured on geological grounds, the recovery of APW from individual microplates does not permit this model to be preferred to the exclusion of Models A or C. The most testable model is B, because it demands the conformity of palaeomagnetic poles to one path only. The applicability of this model to Precambrian times (Chapters 7 & 9) permits a confidence in the palaeomagnetic model for these times which cannot be attained during the Palaeozoic (Chapter 10).

Interpretations of APWPs are more complicated when the data come from small tectonic blocks and accreted (*suspect*) terranes.[1] In this instance, the interpretation is best undertaken by utilising the raw directional data rather than the derived APWP. Most commonly rotational movements take place about vertical axes sited at the centres of the crustal blocks as they touch, and are then accreted onto, the adjacent plates and subsequently, when movement takes place within strike-slip and transform zones (Fig. 6.20). These movements may involve rigid rotation in a matrix of cataclastic rocks or a tectonic mélange of plastically-deforming rocks. The sense of motion reflects the sense of movement, with anticlockwise rotation occurring in zones of right-lateral (dextral) slip, and clockwise movement in zones of left-lateral (sinistral) slip. In areas of combined folding and faulting, rotation may take place on inclined fault surfaces about non-vertical and non-horizontal axes. This response is most important within small tectonic blocks, and the scale of the scatter in the palaeomagnetic directions will depend on the size of the rotated block: it will only be significant if the sample site spacing is greater than the size of the blocks. It has a resultant effect of producing an expanded distribution of declinations. Hence, the challenge to the palaeomagnetic worker is to distinguish between terrane rotations, local structural rotation about vertical axes, and apparent tectonic rotation about inclined axes (Macdonald, 1980). If Fisher statistics are inappropriately applied to such a distribution, they estimate a mean inclination slightly steeper than the true inclination, and lead to a pole position which is 'near sided' with respect to the sampling area. Palaeomagnetic poles from an orogenic terrane affected by tectonic rotations generally have a distribution along a small circle centred on the sampling area (Macdonald 1980); such a distribution is an expression of local tectonics and should not be interpreted as APW relative to the whole region.

The palaeomagnetic data base

Prior to the age of the oldest presently-surviving ocean crust (~ 150 Ma),*

---

[1] The spelling 'terrane' is used here for a 3-D block of crust to distinguish it from the word "terrain" which should be used to refer to the ground surface only.

*The oldest fragments of ocean crust are actually *ca.* 190 Ma in age, but magnetic lineations are not readily recognised for crust older than *ca.* 150 Ma in age. Possible reasons for this are discussed in Section 12.2.

palaeomagnetic information is the only source of quantitative data on the past orientations of the continents. The main limitation of the data is that they cannot determine palaeolongitude. The older ocean crust has either been subducted or is so deformed against the continental margins that it cannot be used to determine the size of ancient oceans in the ways which have been possible for late Mesozoic and Cenozoic times. This limitation should be borne in mind when examining Phanerozoic reconstructions presented here and elsewhere; it is not applicable to Model B in Fig. 6.19 and the Precambrian interpretation of Chapters 7 and 9.

With the exception of a few isolated results from early Archaean rocks (3500–3300 Ma), the data base begins at 2850 Ma. The coverage is inevitably very uneven, both in areal terms and in age terms. Precambrian results, for example, are still heavily biased towards the Laurentian Shield of North America, Greenland and NW Scotland, while some major terranes including the Antarctican, North China (Sino-Korean) and South China (Yangste) Shields, contribute practically no data. Furthermore, each crustal terrane has been built up by a limited number of igneous and thermal-tectonic events of short duration. The Laurentian Shield, for example includes major (largely basic) igneous events represented by the Matachewan (*ca.* 2650 Ma), the Priessac-Nipissing (*ca.* 2150 Ma), the Mackenzie (1220–1190 Ma), Gardar (*ca.* 1280, 1220 and 1170 Ma), Keweenawan (*ca.* 1150 Ma) and Franklin (*ca.* 650 Ma) episodes. These are prominent and widespread events, and because they have been extensively studied, the data base is heavily weighted towards them. These igneous provinces are intruded into metamorphic basement consolidated during the Kenoran (*ca.* 2700 Ma), 'Hudsonian' (*ca.* 1800 Ma) and Grenville (*ca.* 1100 Ma) mobile episodes and, with a few exceptions, these events collectively represent the possibilities for reconstructing the Precambrian APWP of the Laurentian Shield.

The palaeomagnetic data base is rendered more heterogeneous by the variable quality of the data. Modern studies have made more thorough investigations of magnetic stabilty, and have separated the magnetic components in larger collections. Also, they have tended to displace earlier studies of the same rocks, and often the remanences identified by the early studies have proved to be secondary or composite in nature. The present analysis is based in part on sixteen compilations published by the Geophysical Journal of the Royal Astronomical Society between 1960 and 1980. From 1960 to 1965 these were compiled by Irving (1960a,b, 1961, 1962a,b 1965) and Irving & Stott (1963). Only a minority of these studies were subjected to magnetic cleaning, and although most are largely of historical interest, some have been substantiated by later work. The remaining compilations, covering results from 1964 were collected by McElhinny (1968a,b, 1969, 1970, 1972a,b) or McElhinny & Cowley (1977, 1978 & 1980). Additional listings covering specific time intervals or terranes are due to Irving & Hastie (1975) and Irving, Tanczyk & Hastie (1976a,b,c). The extensive body of data from the USSR has only received partial inclusion in these listings, although it has since been selectively reviewed by A. N. Khramov and published by McElhinny, Cowley, Brown & Wirubov (1977) and McElhinny, Cowley & Brown (1979). Data published since 1980 are compiled with the older data on computer files at the University of Liverpool, and form the basis of the abbreviated listings in a companion publication (Piper, 1987). The computer storage has the advantage that the results can be arranged in age sequence and continuously updated as new results become available. Since the tectonic analysis is based on the comparison of data from single rigid plates, the palaeomagnetic results are compiled plate by plate in these listings. The minimum criteria for inclusion in the following analysis are similar to those adopted by McElhinny (1973); in general they require the use of samples which have been subjected to field and laboratory stability tests or, in a few cases, to field stability tests (Section 6.4) only.

Phanerozoic magnetisations can often be precisely dated by the stratigraphic constraints, and sometimes by radiometric dates as well. This is not the situation with Precambrian studies, where age control provided by the Riphean and Vendian time divisions (Appendix A) is quite coarse, or is not available. The end-result of a Precambrian study is generally a sequence of remanence directions, of which even the primary one cannot usually be dated to better than 50 Ma. The problem of dating Precambrian remanences is extensively discussed in the literature describing individual studies, and will be illustrated here by two examples. The Big Spruce igneous complex in the Slave Province of northern Canada is dated as 2171 ± 40 Ma by Rb–Sr whole-rock methods, and yields four components of magnetisation (Irving & McGlynn, 1976a): a predominant D remanence is attributed to initial cooling, while subordinate X, Y and Z remanences, with generally lower blocking temperatures, fall on the APWP between 2000 and 1200 Ma. The latter are provisionally linked to later tectono-thermal and/or magmatic episodes at 2100–1900 Ma, *ca.* 1800 (Hudsonian, section 7.5) and *ca.* 1220 Ma (Mackenzie, section 7.7). These assignments are plausible but cannot be confirmed and their conformity with the APWP at these respective points does not mean that these are their only possible ages.

In the Grenville belt, a zone along the SE margin of the Canadian Shield, subjected to high temperature and pressure metamorphism at *ca.* 1100 Ma., Buchan & Dunlop (1976) have made a definitive study of the Haliburton gabbro-anorthosite intrusions of Ontario. They recognise three components with distinctive coercivity and blocking-temperature spectra. A high-coercivity A remanence is carried by fine rods of magnetite or a hemo-ilmenite, both of which are probably primary magmatic phases. A secondary B component is a strong magnetite-held remanence with medium coercivities; it can be removed with moderate alternating field treatment, often to progressively recover the A component in samples where both vectors are present. A third C component has variable coercivities, and is distinguished by its contrasting direction from the A and B components. Buchan & Dunlop interpret these three components as recording the mineralogy, temperature and duration of heating at different stages of the slow uplift-related cooling between *ca.* 1000 and 800 Ma, following the high-grade regional metamorphism. The A and C components have subsequently been linked to ^{40}Ar:^{39}Ar plateau ages, by Berger, York & Dunlop (1979) and are dated at *ca.* 960 and 820 Ma respectively.

The selection criteria employed here may be regarded as too lenient by some workers, but it will become apparent that little is to be gained by application of more rigid criteria. The bulk of the data conform to remarkably simple and testable models with the secondary poles falling along the APWP at some point after the primary poles; many of the early studies also have a simple explanation because — unless they are the resultant of two approximately equal components

— their poles frequently fall on, or close to, younger parts of the APW path. However, the definitive pole positions established by thorough demagnetisation studies and with a remanence age defined by field tests and/or high-quality radiometric dating must form the basis of any APWP. Clearly, in the discussion and analysis these poles must take precedence over any number of second-rank poles which are either poorly dated or based on inadequate laboratory and field analysis.

7
Archaean and Proterozoic palaeomagnetism

7.1 Introduction

Palaeomagnetic investigations in several continents have found that data from Precambrian shields are best interpreted in the context of little or no detectable motion between the constituent cratons,* since Lower or Middle Proterozoic times. This case has been made separately for Africa by Piper, Briden & Lomax (1973), Piper (1976a), McElhinny & McWilliams (1977) and McWilliams & Kröner (1981), for Australia by McElhinny & Embleton (1976), for North America by Irving & McGlynn (1976b) and Roy & Lapointe (1976) and for the Fennoscandian Shield by Piper (1980, 1982a) and Pesonen & Neuvonen (1981). These analyses support the tectonic view which favours a dominance of intracrustal mechanisms in Proterozoic times (e.g. Engel, Itson, Engel, Stickney & Cray, 1973; Shackleton, 1973; Kröner, 1977) and they imply that deformation of coherent continental crust generally took place below the limits of palaeomagnetic detectability. They argue against the opposing view which advocates a continuity of the Phanerozoic régime of diverse movements between multiple continental fragments, according to Models A or C of the third part of Section 6.6. There are dissenting views from this majority opinion: Burke, Dewey & Kidd (1977) for example, have interpreted the palaeomagnetic record in terms of Model C (although their treatment of the data has been criticised by palaeomagnetic workers (McElhinny & McWilliams, 1977), while Onstott & Hargraves (1981) have claimed to recognise limited differences between the palaeomagnetic records of adjacent cratonic nuclei, which

---

* In this book the term 'craton' is restricted to the ancient nuclei which had stabilised by late Archaean times (>2700 Ma), while the term 'shield' is applied to the broad tracts of Precambrian crust which are usually covered in part by younger Phanerozoic rocks and comprise several cratons separated by later mobile belts (see Chapter 8).

they explain in terms of large-scale strike-slip motions along ensialic mobile belts (see Section 8.5) between them; this is Model D of the third part of Section 6.6.

The present continents are the dismembered fragments of the supercontinent Pangaea (Chapter 10) and there is no *a-priori* reason why the APWPs from the Precambrian shields should be single coherent paths, but unique in themselves. Unfortunately, the definition of Precambrian APWPs is constrained by the preserved rock record in each shield (Section 6.6); they can never be defined in their entirety, and this limits the degree of correlation that is possible from shield to shield. However, Piper (1975b) showed that the *ca.* 1250–1000 Ma data from the African and Laurentian Shields defined comparable APW loops which brought the Precambrian continental areas into direct continuity when superimposed. This operation also aligned a number of structural lineaments of Precambrian age. That this correlation was unlikely to be fortuitous was subsequently demonstrated by the conformity of the 2150–1950 Ma data from the two shields on the same reconstruction (Piper, 1976b); in addition the 1000–800 Ma path from Africa was also shown to match the APWP defined from the Grenville Province of North America (Piper, 1976a) although it was not then established that this latter path was an integral extension of the APWP for the whole of the Laurentian Shield (Irving & McGlynn, 1976b). Since the early 1970s the continuous expansion of the data base has reinforced and extended the correlation between the African and Laurentian data, whilst demonstrating or suggesting that the records from the Fennoscandian, Indian, Siberian and Australian Shields also satisfy a single reconstruction which is rigid on the scale resolved by the paleomagnetic data (Piper, 1982a, 1983a). The configuration of the Siberian Shield in this reconstruction has already been defined by geological evidence (Sears & Price, 1978) although it cannot be effectively tested by the available palaeomagnetic data until Upper Proterozoic times (Fig. 7.7). The continuity of the Ukrainian Precambrian inlier with the Fennoscandian Shield in their present-day configuration is suggested by the continuity of structural trends and age provinces (e.g. Semenenko, Scherbak, Vinogradov, Tougarinov, Eliseeva, Cotlovskay & Demidenko, 1968) and is borne out by the palaeomagnetic data from *ca.* 2000 Ma (Fig. 7.3 onwards). The collective assembly is known as the 'Proterozoic Supercontinent', and only this model for the Precambrian palaeomagnetic data will be examined in detail here (Table 7.1), although some counter-arguments will be noted and assessed.

The current data set is patchy, both in areal terms and in age coverage, and there are practically no data from some shields. However, since continuity can be demonstrated between such large segments of the Precambrian crust as Africa, Australia, North America and Fennoscandia, it is likely that the remainder of the Precambrian crust obeyed the same model. The remaining continental blocks with large areas of known or probable Precambrian crust (North and South China and Khazakhstania) have been inserted in the reconstruction by utilising the two criteria of:
1. geological evidence such as the alignment and continuity of lineaments and
2. areal and shape considerations, so that fragments are inserted to preserve crustal continuity, like the residual pieces of a jigsaw. East Antarctica, the remaining major segment of Precambrian crust, is inserted in its Pangaean configuration with

respect to India (Fig. 7.1(b) and see, for example Hurley & Rand, 1969; Petrascheck, 1973, and Kratz, 1974).

Two general points need to be made at this stage. Firstly, for any given Precambrian data point, there are possible errors in both the pole position and the age assignment: a well-defined pole is of little value if the age is poorly known. To give an elementary measure of this problem the data can be considered at three levels. Results from igneous units with ages constrained by stratigraphic evidence or linked to radiometric ages are regarded as *first-rank data*. Metamorphic rocks with ages linked to cooling during uplift, and sediments constrained within narrow age limits may yield pole positions which approach the quality of the igneous data, although their age control is less good and qualifies them as *second-rank data*. First- and second-rank data are plotted as large symbols in the diagrams which follow. Data from sediments with only broad age constraints, from secondary magnetisations (overprints) in igneous, metamorphic and sedimentary rocks, together with magnetisations from low-grade metamorphic rocks with poorly-constrained ages assigned from the regional geological history, are regarded as *third-rank data*; they are plotted here as small symbols. (Note that overprints are sometimes difficult to define in tilted rocks, because the attitude at the time the overprint was acquired is often difficult or impossible to determine). Small symbols are also used where there is an ambiguity in the assignment of the pole to the APWP; this can happen when the path passes over the same position at two times within the age limits assigned to the pole.

Table 7.1 The Euler rotational operations applicable to the Precambrian Shields to reconstruct the Proterozoic Supercontinent with the Laurentian Shield retained in present-day coordinates. Anticlockwise rotations are positive and clockwise rotations are negative.

	Euler pole		Rotation
	°N	°E	
Australia	41·0	270·0	102·0
East Antarctica*	77·7	103·5	117·7
India	50·5	238·0	137·0
Africa	73·0	138·0	−146·0
Arabia*	75·5	142·1	−148·8
South America	−51·0	274·5	122·5
Siberia	66·0	309·0	86·0
North China*	−19·2	18·0	−111·0
South China*	−17·4	26·3	−103·0
Khazakhstania	77·0	208·0	141·5
Fennoscandia–Ukraine	21·0	8·0	41·0
NW Scotland	27·7	88·5	−38·0
Greenland	70·5	265·6	−18·0

* Based on geological/geometrical relationships only. After Piper (1982a) with adjustments.

Secondly, model B can be justified by its testability, which in this instance, counterbalances the heterogeneous and incomplete nature of the data base. Models A, C and D permit either tectonic or magnetic explanations to be invoked for an apparent difference in pole positions between two cratons, but model B requires that the data conform to a single APWP, either at the time of rock formation if they are primary magnetisations, or at later times if they are of secondary origin. However, we need to know the degree to which the definition of APW is constrained by the assumption of a single, coherent body of crust in Proterozoic times. Data from the Laurentian shield (comprising the Precambrian crust of North America, Greenland and NW Scotland) can be used to gauge this assumption, because they are more abundant than results from the remaining shields, and the concensus of geological opinion believes that this shield has behaved as a coherent unit since Archaean times (see papers in Kröner, 1981). Accordingly, the APWP identified from this shield alone is defined by solid lines in Figs. 7.1 to 7.8, while extensions to the path suggested by data from elsewhere are indicated by dashed lines. It is noteworthy that only over short periods at *ca.* 2450, 2000 and 1550 Ma are complexities introduced into the APWP by data from elsewhere, and even then some third-rank Laurentian data (small symbols) of uncertain significance generally plot in these positions. The Laurentian APWP can therefore be used as a 'master' curve for analysis of the Precambrian data. The collective APWP is represented as a broad track (or swathe) *ca.* 20° in width; this has become standard practice (Spall, 1971, 1972; Piper, Briden & Lomax, 1973; Irving & McGlynn, 1976b) and gives a reasonable representation of the individual polar errors. When making these comparisons, it is critical to know that paths of the same normal (N) or reversed (R) polarity are being correlated. The identification of the polarity of Precambrian APWPs is not a simple matter, because some rapid APW movements took place in late Precambrian (Section 9.3) and early Palaeozoic (Section 10.3) times, which are currently difficult to interpret. However, the palaeomagnetic study defines the relative proportions of normal and reversed (antiparallel) field directions in any rock collection; this is indicated in the figures by the percentage of sites or samples defining the poles (closed fraction of the symbol) and antipoles (open fraction of the symbol). Clearly, if poles are correlated with poles, and antipoles with antipoles, then the validity of the overall interpretation is reinforced. This proves to be a rather efficient test because the Precambrian palaeomagnetic record is strongly biased to one (probably N) polarity (Irving & McGlynn, 1976b, and Section 12.4).

7.2 Archaean–early Proterozoic APW (2900–2200 Ma)

The continental reconstruction shown in Fig. 7.1(b) is derived by the rotational operations listed in Table 7.1. The same operations are applicable to palaeomagnetic data sets from each shield and the collective data for they interval 2800–2200 Ma are plotted in Fig. 7.1(a). In this and subsequent figures, the poles are referenced in the caption and derived from new listings (Piper, 1987) which give their source in older listings (concluding part of Section 6.6) or the original publication. The age assignments are based on a variety of U–Pb, Rb–Sr, Sm–Nd, K–Ar, ^{39}Ar:^{40}Ar and fission-track age determinations, or by reference to the local geological history (Section 6.4); the absolute ages are adjusted to the new decay constants wherever posssible (Steiger & Jäger 1977; Dalrymple, 1979). Space does not permit proper evaluation of age and stability evidence for each pole, and it is necessary to refer to the original research paper for this. The palaeomagnetic analysis is addressed in general terms here and the interpretation is not based on individual pole determinations. Since the analysis has, of necessity, to work with a data base heavily biased towards the Laurentian Shield, it is convenient to retain this shield in its present coordinates (adjusting Greenland for the Mesozoic separation across the Davis Strait and NW Scotland for the opening of the Atlantic (Bullard, Everitt and Smith, 1965), and rotate the data from all other shields towards this nucleus.

Only small areas of the present continental crust had stabilised in early Archaean times, and only where unmetamorphosed supracrustal successions of this age are preserved can it be certain that the present surface levels had cooled by these times. By late Archaean times, however, large areas of granite-greenstone terrane had been added to the present shields, and these are the source of most of the data covering this interval. In the Laurentian shield there are five Archaean blocks which, in order of decreasing size, are the Superior, the South Greenland (including parts of coastal Labrador and NW Scotland), the Slave, the Bear Tooth terrane of Wyoming and Montana, and the Minnesota River Valley Inlier (Figure 7.1c). The magnetisations within the Archaean craton of South Greenland appear to have been acquired during uplift and cooling following a *ca.* 1800 Ma mobile episode (Section 7.5) and the Superior, Slave and Bear Tooth terranes at 2900–2700 Ma provide the oldest recoverable remanences from this shield. The greenstone belts in the Superior craton comprise volcanic complexes of ultrabasic to acid lava piles which also incorporate sediments and are extensively cut by intrusions of comparable age (Goodwin & Ridler, 1970). All three rock groups have now been studied palaeomagnetically and they define the form of an APWP (referred to as 'Track 6' by Irving & Naldrett, 1977) in some detail. These studies include contact tests, and they have identified remanence components pre-dating intrusion of the Matachewan dykes at 2693–2634 Ma (Irving & Naldrett, 1977; Schutts & Dunlop, 1981). They are reinforced by investigations of the banded ironstone formations, in which magnetic components demonstrably pre-dating folding and Kenoran metamorphism at *ca.* 2650 Ma are linked to a range of U–Pb ages on interleaved volcanics (Symons & Stupavsky, 1979, 1980; Symons, Quick & Stupavsky, 1982); in addition, the Griffith mine GM1 component (**2**) at the beginning of this track may pre-date formation of the main Abitibi belt and record one of the earliest discernable events in this shield. The paths joining pre- and post-folding remanences in these formations define a consistent APW sequence (Figure 7.1a) which incorporates the range of poles from igneous units, as well as partial overprints in igneous and metamorphic units (Dunlop, 1979b; Schutts & Dunlop, 1981). Kenoran magnetisations are ubiquitous in this terrane, while magnetisations defining the earlier part of Track 6 are only sporadically represented in the igneous units, although they are widely preserved in the ironstone formations. Igneous results (**17–27**) from the Superior Geotraverse (Dunlop, 1979b) lie collectively along this track, although since it is not conclusively established that they all belong to this sector of the APWP, they are treated as third-rank data here; the balance of the evidence however, shows them to be magnetite components which

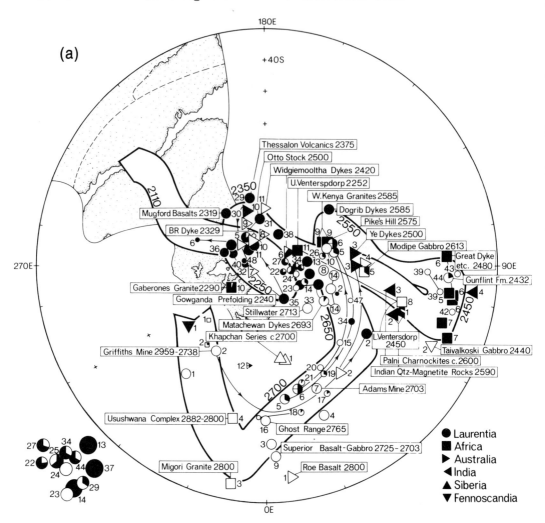

Figure 7.1 (a) APWP for the continental crust from 2800 to 2200 Ma. This path is derived by retaining the Laurentian Shield in present day coordinates and rotating the remaining shields towards it (Table 7.1) on a Lambert Equal Area Projection centred at the North Pole according to the reconstruction discussed in the text (Table 7.1) and illustrated in (b). The palaeomagnetic poles from each shield are indicated by the symbol and are shown as closed symbols when they are poles and open symbols when they are antipoles. When the palaeomagnetic study has yielded mixed polarities, the symbol is divided according to the ratio of poles to antipoles; where this is not known the symbols are crossed or divided 50:50 if reversals are present. The poles are referenced in new listings (Piper 1987) and the assigned ages are given in brackets as inferred from geological relationships or radiometric dating (UP = Uranium-Lead, RS = Rubidium-Strontium, KA = Potassium-Argon, AA = $^{39}Ar:^{40}Ar$, FT = Fission track; mineral age studies are indicated as H = Hornblende, B = Biotite, M = Muscovite; KAI = Potassium Argon isochron age).

The poles are numbered: LAURENTIAN SHIELD: 1, *Algoman iron formation, prefolding remanence (2703 UP on associated Keewatin volcanics)*; 2, *Griffiths Mine, GM1 and GM2 prefolding remanences (2959–2738 UP) and postfolding (? Kenoran) remanence*; 3, *Kamiskotia Complex (>2693 RS)*; 4, *Dundonald Sill*; 5, *Sherman Mountain iron formation prefolding SM1 remanence (>2703 UP) and postfolding SM2 (ca. 2600 Ma) remanence*; 6, *Moose Mountain iron formation prefolding MM1 remanence, prefolding MM2 remanence (ca. 2750–2700) and postfolding C remanence*; 7, *Adams Mine prefolding A remanence (2703 UP) and postfolding B remanence*; 8, *Archaean gabbro*; 9, *Archaean basalt-gabbro (2755–2703 UP)*; 10, *Pike's Hill (Munro) lavas (Kenoran metamorphism, 2525 AA)*; 11, *Archaean gneiss*; 12, *Dogrib dykes (Slave craton, 2585–2521 UP, 2310–2165 KA)*; 13, *Chibougamau greenstone sills (2665, 2563 RS)*; 14, *Matachewan dykes, north east and south west magnetised groups from six studies (2693, 2634 RS)*; 15, *Skead Group (>2755 UP)*; 16, *Ghost Range Complex (2765 UP). (Results 17–27 come from a geotraverse through the Superior Craton between 267.5 and 270°E and 48.5 and 50°N. Not all of these poles can with certainty be assigned to this track and poles for which the assignment is uncertain are plotted here as small symbols; see Dunlop (1979b))*; 17, *Shelley Lake granite (2595 AA)*; 18, *Burchell Lake granite and Shebandowan igneous rocks*; 19, *Gulliver greenstone*; 20, *Ignace Martin gneiss*; 21, *Wabigoon gabbro*; 22, *Shelley Lake granite, Group 2 (Kenoran overprint)*; 23, *Huronian granite paragneiss*; 24, *McKenzie Lake diorite*; 25, *Burchell Lake granite (2600–2550 on nearby granites)*; 26, *Sapure Iron formation*; 27, *Watcomb granite*; 28, *Poobah alkaline complex (2706–2556 KA)*; 29, *Thessalon Volcanics (2375 RS, 2240 KA)*; 30, *Mugford basalt (2319 RS)*; 31, *Otto syenite stock (2500 UP, 2470 RS, 2520 KA)*; 32, *Soudan andesites*; 33, *Stillwater Complex (2900 SN, <2730, 2692, 2491, 2450 RS, pole position corrected for Laramide deformation)*; 34, *Unmetamorphosed Kaminak dykes (2370 UP)* 35, *Gowganda prefolding A remanence (≥2240 RS)*; 36, *Gowganda formation (≥2240 RS)*; 37, *Coleman member (>2100 RS)*; 38, *Huronian argillites, Firstbrook Formation (≥2240 RS)* 39, *Cobalt Formation, two poles (<2650 SN, >2100 RS)*; 40, *Big Spruce Complex, D remanence (2171 RS)*; 41, *Gowganda Formation, Postfolding B remanence (<2240 RS, >2116 RS)*; 42, *Nagaunee Magnetite ores*; 43, *Soudan hematite ores*; 44, *Gunflint Formation (2432 AA, >1620 RS)*; 45, *Pensive Lake Sill*; 46, *Bighorn dykes (2020–1390 KA)*; 47, *Owl Creek Mountains dykes (2750 UP).* FENNOSCANDIAN SHIELD: 1, *Varpaisjarvi diorites (2684 UP)*; 2, *Taivalkoski gabbro (2440 UP).* SIBERIAN SHIELD: 1,2, *Verkneanabar, Daldy and Khapchan Series (2900–2700 Ma).* AFRICAN SHIELD: 1, *Jeppestown amygdaloidal lava (3100–2800 UP,RS)*; 2, *Zoetlief Series (2700–2500 UP)*; 3, *Migori Granite (2800 RS)*; 4, *Usushwana Complex (2882, 2874 RS, 2800 SN)*; 5, *Modipe Gabbro (2613 RS)*; 6 *Great Dyke of Zimbabwe, three poles (2480 RS)*; 7, *Satellite dykes of the Great Dyke (2545 RS)*; 8, *Lower Ventersdorp lavas, two results (2450 RS)*; 9, *West Kenya Granites (2470 RS, 2580 KA) and Mumias granite (2530 RS)*; 10, *Gaberones granite (2290 RS)*; 11, *Upper Ventersdorp lavas (2252 RS).* INDIAN SHIELD: 1, *Kistaropalli quartz-magnetite rocks (2590 RS)*; 2, *Makdampur quartz-magnetite rocks (2590 RS)*; 3, *Palni Charnockites (2600 UP)*; 4, *Sukinda chromite formation (early Proterozoic).* AUSTRALIAN SHIELD: 1, *Roe Basalts (ca 2800)*; 2, *Duffer Formation B remanence (3000 Ma?)*; 3, *Duffer Formation C (≥2600 Ma)* 4, *YE dykes (2500 RS)*; 5, *Ravensthorpe dykes (2500 RS)*; 6, *Widgiemooltha dykes (2420 RS)*; 7, *YA dykes (ca 2500 or 1700 Ma)*; 8, *Black Range gabbro dyke (2329 RS)*; 9, *Cajaput giant dyke (ca 2400 Ma)*; 10, *Mt. Goldsworthy lode ore MG3 (3000–2000 Ma)*; 11, *Koolyanobbing-Dowd's Hill ore*; 12, *Koolyanobbing 'A' deposit.*

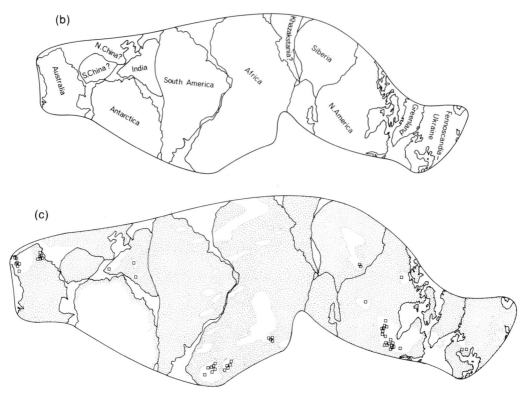

(b) The configuration of the major continental blocks in the Proterozoic Supercontinent; Mollweide projection.

(c) The locations of the 2800–2200 Ma palaeomagnetic studies within the continental reconstruction. The stippled area is the approximate extent of the crust reworked by later thermal and tectonic events.

have been largely displaced by higher blocking temperature hematite components defining a Kenoran overprinting effect (see also Berger & York, 1979). In the greenstone divisions at Munro township, the basalt lavas (9) carry a post-folding remanence identical to that in gabbro intrusions (8) with which they are believed to be essentially consanguinous (Goodwin & Ridler, 1970). This latter remanence is referred to U–Pb zircon ages in the range 2725–2703 Ma (Nunes & Jensen, 1980) and it pre-dates intrusion of the Matachewan dykes. The Matachewan results (14) include both polarities; the remanence in these dykes appears to post-date the Kenoran metamorphism and define the younger extremity of this arcuate track.

Poles from the Usushwana complex (2880–2800 Ma), in the Kaapvaal craton of southern Africa, and from the Migori granite (2800 Ma) in Kenya, lie at the older end of this track, while poles from the Modipe gabbro (2613 Ma) and the West Kenya granites (2580–2420 Ma) from the same two cratonic nucleii, lie at the younger end. Collectively, the African and Laurentian data confirm that the integral reconstruction is applicable to these two shields back to *ca.* 2850 Ma, and reinforce an assignment of this track to the interval 2850–2600 Ma (Morris, 1977a). Data from the Stillwater complex (33) are linked by U–Pb and Sm–Nd studies to an age of 2710 Ma (Nunes, 1981) and extend the Laurentian shield record to the Bear Tooth uplift; a single pole (**12**) comes from the Slave craton, while poles from the Khapchan series of the Siberian Shield (*ca.* 2900–2700 Ma) may extend the record from this shield back along this track. The Koolyanobbing poles from the Yilgarn craton, Australia, are derived from hydrothermal hematite, which is probably linked to greenstone magmatism and granite intrusion here at 2750–2500 Ma, although in common with results from other hematite ores, they cannot be directly dated. In the Fennoscandian Shield (which provides the most voluminous body of Precambrian palaeomagnetic data after the Laurentian shield) a Saamo-Karelian terrane in eastern Finland and the NW USSR was stabilised at *ca.* 3000–2700 Ma. Although locally affected by later events, this terrane was uplifted with deposition of sediments after 2000 Ma. This nucleus is bounded by progressively younger terranes to the west in Finland, Sweden and Norway. Two results come from the oldest terrane: a pole from the Vasparaisjarvi diorites (**1**) plots on the oldest part of the path, although its assigned age is too young to confirm this assignment, while a second pole (**2**) plots on the APWP at a position which accords closely with its assigned age of 2440 Ma.

The first loop in the Precambrian APWP is identified at 2600–2450 Ma. This is not well-defined by Laurentian data because few rock divisions are represented here between the intrusion of the Matachewan dykes and the Huronian sedimentation. It is, however, collectively defined by results from the Great Dyke, and its satellite dyke swarms cutting the Zimbabwe craton (except at their southern extremity where they were influenced by the Limpopo mobile belt and subjected to uplift-related cooling at *ca.* 2000 Ma; Section 7.4). Other poles falling on the loop include four from the Archaean–Lower Proterozoic nucleus of southern India, and one from the Fennoscandian Shield (**2**); Laurentian poles which may belong here include the Gunflint formation and four other ironstone formation results of unspecified Lower Proterozoic age.

An important component of the 2350–2150 Ma Laurentian record comes from the Huronian supergroup. Morris (1977a) interprets the collective data in terms of an anticlockwise loop, beginning with *ca.* 2350 Ma igneous units, moving to the pre-folding remanences in the Huronian sediments, and then back to a remagnetisation event in these rocks. This event is recorded in authigenic magnetite and hematite and is widespread as a pre-Nipissing (2110 Ma, Section 7.3) folding and exhumation episode in the Southern Province of the Superior terrane. This path (Loop 2) incorporates poles from granites in southern Africa, and six poles from Australian dyke swarms and giant dykes in the Yilgarn and Pilbara cratons (**4–9**); the rapid movement between the *ca.* 2450 Ma poles and this loop is identified between the Lower and Upper Ventersdorp lava successions in South Africa (**11**; Henthorn, 1972).

The conformity of this oldest part of the Precambrian palaeomagnetic record with a single APWP is remarkable, and belies the reservations which have frequently been made about the use of such data. Furthermore, the polarities of data points from the different shields also correspond well, and give added confidence that poles of the same polarity are being correlated. The earliest part of the track (2850–2700 Ma) is defined mainly by antipoles (open symbols in this and subsequent diagrams) and the later path is defined mainly by poles (closed symbols). The 2450–2280 Ma Australian dyke swarms are the main exception, but since they

160   *Palaeomagnetism and the Continental Crust*

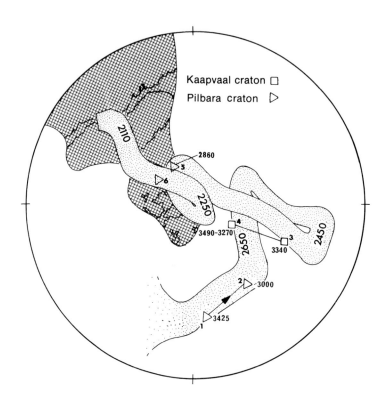

Figure 7.2 Summary APWP for the interval 2800–2200 Ma from Figure 7.1 with pre-2800 Ma palaeomagnetic poles from the early-formed cratonic nuclei, the Pilbara Craton in NW Australia; (1, Duffer Formation A component; 2, Duffer Formation B component) and the Kaapvaal Craton of Southern Africa (3, Komati Formation; 4, Kaap Valley Intrusion). Also plotted are two poles from the Pilbara Craton which do not conform to the continental reconstruction and imply that this craton was not part of the reconstruction at the time these rocks were magnetised; these poles are 5, Millindinna Complex (2860 Ma); 6 Mt. Tope Volcanics; pole 6 is believed to be about 2800 Ma in age (Schmidt and Embleton 1985). Note that these early Archaean data from the Australian and African Shields plot in contrasting regions on the primitive reconstruction of Figure 7.1 in a way which is not again evident until Lower Cambrian times (Figure 9.4).

plot within a loop covering some 200 Ma, this is not unexpected. Since all the data points conform to the single path, the loose exclusion criteria have not impaired the test, and more rigid criteria would merely operate to reduce the overall number of data points. The data base for this interval of geological time proves to be particularly robust; this is probably because the results are mostly derived from cratonic interiors which have remained stable since the time of magnetisation; the smaller number of sedimentary results come from units deposited in already stabilised environments which possess a simple and resolvable diagenetic structure (Roy & Lapointe, 1976, Morris, 1977a) acquired in a low redox-potential environment.

Most of the early Archaean nuclei pre-dating the time interval described in Fig. 7.1 have experienced profound thermal overprinting in subsequent times, and no remanences dating from the last formative or tectonic events are likely to have survived. Two possible exceptions are the Kaapvaal craton of southern Africa, which has an exposed history of supracrustal volcanism and sedimentation going back to *ca*. 3500 Ma, and the Pilbara craton of western Australia. Two data points from the former terrane come from the Komati Formation lavas, part of a greenstone sequence dated 3340 Ma (Hale and Dunlop 1983), and the Kaap Valley rocks dated 3490–3270 Ma (Kröner, McWilliams and Layer 1983). Results from the Pilbara craton comprise the Duffer Formation primary and B components (McElhinny and Senanayake 1980); although the latter has been assigned an age of 3000 Ma it could equally well be some 400 Ma younger (see Fig. 7.1). The Komati, Kaap Valley and Duffer A poles show no agreement with one another on the post-2800 Ma reconstruction (Fig. 7.2). This observation is compatible with relative movement between these two early-formed nucleii before formation and consolidation of the bulk of the continental crust in late Archaean times (section 8.2). These results also plot along the APWP of Fig. 7.1 at younger ages, and although this might be taken to indicate that they have been remagnetised in later times, the field and stability tests make this supposition unlikely. While the results from the Yilgarn craton conform closely with the data from elsewhere on the reconstruction in Fig. 7.1, the significance of the data from the smaller Pilbara craton to the NW is less clear: the poles from the iron ores of the Hammersley Range appear to conform to the later APWP of Fig. 7.3 (although their ages are poorly known) as do younger results. The Archaean data from the Mildunna Complex (2860 Ma) and the slightly-younger Fortesque group volcanics (Schmidt & Embleton, 1985) are divergent (Fig. 7.2) and, like the older Duffer A pole, imply relative movements between the cratonic nucleii prior to 2750 Ma. However, the youngest pole in this study from the Roe basalts, approaches the contemporaneous poles from the African and Laurentian shields (Fig. 7.1), and is a further suggestion that the nucleii were accreting into the single, coherent body in late Archaean times.

7.3 The 2200–2000 Ma APW path (Fig. 7.3)

The continuity between the APWPs of Figs. 7.1 & 7.3 is defined in the Laurentian Shield by the migration between the primary and secondary magnetisation in the *ca*. 2250 Ma Huronian sediments and the suite of poles from the Nipissing diabases which intruded them after folding (Roy & Lapointe, 1976). In the 1970s,

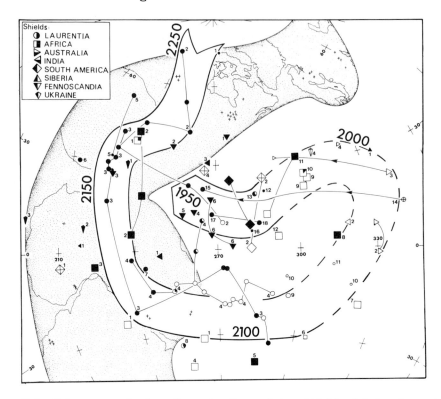

Figure 7.3 Palaeomagnetic data from the Precambrian Shields defining Loop 2 (ca. 2150–2000 Ma). LAURENTIAN SHIELD: 1, La Grand Riviére Proterozoic dykes; 2, Nipissing diabases, C group (2116, 2111 RS); 3, Priessac (formerly Abitibi) dykes (2102 RS, 2150 AA); 4, Nipissing diabases, D group (2116, 2110 RS); 5, Pioneer iron ore and host greenstone; 6, Marathon dykes (2172 RS, >1810 KA); 7, Indian harbour dykes (2080 KA); 8, Sokoman iron formation, prefolding remanence (2200–2100); 9, Metamorphosed Mackay ('X') dykes (2300 UP, 2165 KA); 10, Beartooth dykes 4 + 5 (2550 RS, 1690 KA) and 6 (2550 RS, 2000 KA); 11, Duck Lake sill (2090–1490 KA); 12, Post-Sokoman dolerite dykes; 13, Indin dykes (2044 RS, 2000–1800 KA); 14, Western River Formation; 15, Hornby Bay Group sandstone; 16, Slave Craton overprint, SE directions; 17, Tholeiite dykes, granite falls (2080 KAH, 1950, 1650 AA, 1800 KA); 18, Hornblende andesite dykes (<17, 1930–1690 KA); 19, Marquette Range Supergroup; 20, Rove Slate. AFRICAN SHIELD: 1, Cunene anorthosite complex, three poles (2112 RS); 2, Bushveld gabbro (2051 RS, 2096 AA, 2050 UP) three poles from critical zone and east and west main zone; 3, NW Sahara tonalites (2115 RS); 4, Transvaal System lavas (2203 RS); 5, Tarkwaian intrusions (2100 KA, UP); 6, Post-Obuasi dyke (<2100); 7, Obuasi greenstone body (2100 KA, UP); 8, Nimba granulite (2000 AAH); 9, Aftout plutons (1985–1950 RS); 10, Cunene anorthosite, younger facies (2030 RS); 11, Harper amphibolite (1970, 1830 KA, 1964 AAH, 1894 AAB); 12, Orange River lavas (2100 RS). FENNOSCANDIAN SHIELD: 1, Hypia metadolerites (2200–2100); 2, Eno metadolerites; 3, Nilsia area dolerite dykes (2160 UP); 4, Kuusamo dykes and greenstones (2160 UP); 5, Nilsia area dolerite dykes (2160); 6, Akujarvi granulites (two poles) and quartz diorites (1925 UP). UKRAINIAN SECTOR: 1, Oktyabr'sk alkaline complex (2060 KA); 2, Bug-Podolia Seriya gneisses (2000 KA): 3, Amphibolites (2240–2140 KA); 4, Ingulets Complex (2000–1700 KA).

all of the 2100–1700 Ma poles from this shield were popularly constrained with a single APW loop (the 'Coronation Loop' of Irving & McGlynn, 1979). Several lines of evidence, including the evaluation of the complex remanence record in the Nipissing rocks, have shown that such a view is no longer tenable. The collective data (**2,4** Fig. 7.3) from these rocks and the contemporaneous Priessac dyke swarm (**3**) define a nearly-continuous swathe of poles. The C group (**2**) have N polarity only,* but the contact tests of Roy & Lapointe (1976) have been able to demonstrate that at least some of these record a primary remanence, dating from the time of intrusion referred to 2150 Ma by ^{39}Ar:^{40}Ar dating. A similar conclusion applies to comparable pole positions from the Priessac dykes (Irving & Naldrett, 1977). The southerly D poles are more common and exhibit both polarities; contact (Morris, 1979) and fold (Morris, 1981a) tests, show that these directions too, record a primary remanence. The time sequence of the igneous episodes recorded by the C and D magnetisations is presently disputed. The evidence of Roy & Lapointe (1976) supports a C → D sequence, while Symons (1971) found that sills carrying a D remanence are cut by dykes carrying a C remanence, suggesting that the movement D → C was applicable here. The latter view was given qualified support by Morris (1981b) who argued that the E–W smear of D poles (Fig. 7.3) is due to differential rotation of the sampled units within small fault blocks; the more compact grouping of D poles does not appear to be influenced by this effect. Furthermore, a negative fold test has been recorded in the D group (Symons, 1971, and pole **1** in Fig. 7.4), and some of these poles may record later overprints acquired during low-grade Penokean metamorphism (Morris, 1979). Accordingly, there seem to be two populations in the D group. One group was magnetised during initial intrusion and defines the earlier part of loop 2 as shown in Fig. 7.3, in common with other 2150 Ma data points, while the other group appears to be a later overprint and predictably correlates with other 2000–1900 Ma poles at the younger extremity of loop 3 or the beginning of loop 4 (Fig. 7.4). Unfortunately, in the absence of complete field tests, the selection between most of these data points is arbitrary and the D poles plotted in Fig.

---

* The ensuing discussion will show that the APWPs plotted in Fig. 7.1 to 7.8 are probably the path of the North pole with respect to the continental crust; some rapid APW motions in early Palaeozoic times mean that it is not yet possible to be certain of this (see Sections 9.5 & 10.3).

SIBERIAN SHIELD: 1, *Olenek region, gabbro diorites (intruded by 2085–1850 Ma granite)*. SOUTH AMERICAN SHIELD: 1, *Supamo-Pastora group (2227 UP, 2290 RS)*; 2, *La encrucijada rocks (2064, 1958 RS, 1972 AAH, 1882 AAB)*; 3, *Imataca granulite complex (2055–2000 RS)*; 4, *Guri dam rocks (2020, 1952 RS, 1968 AAH)*. INDIAN SHIELD: 1,2, *Visakhapatnam charnockites (2600 UP, ca. 2000 Ma magnetisation)*; 3, *Kondapalli charnockites (2600 UP, ca. 2000 Ma magnetisation)*; 4, *Group V dykes, Dharwar craton*. AUSTRALIAN SHIELD: 1,2, *Mt. Tom Price and Mt. Newman ores (2000, 2020 RS on associated rocks)*; 3, *Mt. Goldsworthy lode ore MG1 (ca. 2250–2200, <MG3 pole plotted in Figure 7.1)*; 4, *Mt. Goldsworthy BIF (jaspilites)*; 5, *Mt. Goldsworthy, MG2 crust ore (<MG1). Lambert equal area projection centred at 270°E, 0°N; the format symbols and abbreviations are the same as for Figure 7.1. The crosses refer to sample locations of the palaeomagnetic studies and the stippled area is the continental crust.*

7.3 may in fact be applicable to Fig. 7.4. In any case, it should be observed that the distribution of C poles comprises an arc centred on the sampling locality and therefore defines a tectonic effect rather than real APW (Macdonald, 1980; Morris, 1981b); the same effect must also influence the Priessac dykes to some extent. Other data for this interval are too few to define the true APW very pecisely, but it is evident that the Nipissing C and D and the Priessac distributions are coincident with *ca.* 2150 Ma poles from dyke swarms (1–5, **Fig. 7.3**) intruding the already stabilised Saamo-Karelian terrane in the Fennoscandian Shield, plus the granulites along its western margin (**6**, this latter point is an uplift remanence which could be *ca.* 1925 Ma in age; see Fig. 7.4 and Pesonen & Neuvonen, 1981). Palaeomagnetic data from the Ukrainian Precambrian inlier (mainly due to Kruglyakova, 1961) are plotted here and in subsequent figures on the same reconstruction as the Fennoscandian Shield. Unfortunately, knowledge of the Ukrainian results and their age constraints are inadequate to properly evaluate this procedure, although in results **1** to **3** and in subsequent figures there is general conformity of these data with those from elsewhere.

In Africa, poles from the older facies of the Cunene anorthosoite complex, bordering the Kasai (Congo) craton (**1**), and early igneous phases in the Dorsal Eglabs mobile terrane along the northern margin of the West Africa craton (**3**) fall on this track; more recent age data for the Bushveld gabbro (**2**, Onstott, Hargraves, York & Hall, 1984) requires an assignment to the earlier part of Loop 3. Other dyke swarms from the Superior (**6**), the Slave (**9,13**), and possibly the Bear Tooth (**10**) cratons of the Laurentian Shield contribute to the definition of this loop. The south-east extremity of the Nipissing C and Priessac groups is recognised in poles from the 2100 Ma Obuasi and Tarkwaian greenstone belts at the southern margin of the West African craton (**5–7**) but the immediate continuation of the path is otherwise unrepresented by the present record. The return limb of the loop appears to be defined by the *ca.* 2000 Ma sediments at the base of the succession in the Athapuscow aulacogen* bordering the Slave craton (**14,15**, see also Section 7.5) and the Hammersley Range ores of Australia bracketed in the age range 2200–2000 Ma; Porath & Chamalaun (1968) describe poles from several stages of hematite genesis in the latter area, with the outer part of the lode representing a primary deposit (Fig. 7.1) and the inner parts developing by later hydrothermal and metamorphic processes (**4,5**). The predominant reversed polarity of the Australian and African poles lends some credence to their correlation with this part of the APWP, but the Indian and Siberian poles plotted here cannot be assigned to this loop with any degree of certainty. The termination of the loop however, is fairly well constrained by the assembly of *ca.* 2000 Ma poles from West (**8,11**), North (**9**) and South (**10,12**) Africa and the Guyana Shield of South America (**2–4**).

By adopting a zero APW option Onstott, Hargraves, York & Hall (1984) regard the Guyana Shield poles **2** and **3** and West African pole **11** contemporaneous with pole **2** (and poles **7** and **8** in Fig. 7.3) from southern Africa; if this is the case

---

* An aulacogen is a major graben structure with a long history of rifting and sediment infilling. The Coronation geosyncline and Athapuscow aulacogen developed around the north-west and southern margins, respectively, of the Slave craton, and were the sites of thick sedimentation between *ca.* 2000 Ma and 1700 Ma; most of the present palaeomagnetic data from this area come from the Great Slave supergroup in the Athapuscow aulacogen (see also Sections 7.5 & 8.6).

they can be brought into agreement by a clockwise rotation of the Kaapvaal-Kalahari craton by 60° with respect to the Guyana–West Africa Shield. This possibility cannot be directly refuted by the conformity of the older data in Fig. 7.1, because there are no pre-2000 Ma data from the Guyana–West African region to establish whether that region was then part of the reconstruction. However the interpretation is unsound for several reasons:

1. The ages of the magnetisations cannot be linked directly to the ^{39}Ar:^{40}Ar ages within the analytical errors of the latter, for reasons discussed in Sections 3.10 & 6.4, and no magnetisation of this age can be interpreted to better than ±25 Ma.
2. This interpretation does not permit APW, although the differences interpreted in a tectonic context are actually smaller than the differences between contrasting blocking-temperature/coercivity components observed in the same rock unit in many Precambrian studies (see for example, concluding part of Section 6.6),
3. Acceptable poles from West Africa (9), which are contemporaneous with the southern African poles within the uncertainties of the age data, are also coincident with them.

Although major rotational movements between western and southern Africa are not supported by the palaeomagnetic data, they cannot preclude a *ca.* 1000 km dextral shear between the West Africa and Guyana Shields as originally suggested by Onstott & Hargraves (1981); such a displacement is within the scatter of data points and could for example, be invoked to explain the difference between African poles 11 and Venezuelan poles 2 in Fig. 7.3, although an explanation in terms of APW is equally likely.

7.4 The 2000–1750 Ma APW path (Fig. 7.4)

A well-defined group of poles from post-tectonic intrusions in the Dorsal Eglabs zone of the NW Sahara (1,6) and the Kaapvaal craton (2,3,7) coincide with less well-dated but probably *ca.* 1900 Ma poles from the Laurentian Shield (1–3), and Fennoscandian results (2,4) to define the pole position at *ca.* 1900 Ma. A slightly younger group of poles assigned to the interval 1900–1830 Ma implies APW movement of some 40° to the south west and is well-represented by Laurentian data from the Minnesota River Valley Inlier (11,12) and sedimentary formations at intermediate levels of the Athapuscow aulacogen. It is even better defined by a large and coincident group of Fennoscandian poles from rock units which range from dyke swarms in the Saamo-Karelian terrane (5,8,12,16) to supracrustal sediments and volcanics deposited on the Archaean terrane (9–11) and numerous poles from gabbro-diorite intrusions linked to 1890–1850 Ma magmatism during the Svecokarelian mobile episode. This position is represented in the Indian Shield by the 1830 Ma Gwalior Traps (5), and some of the Indian charnockites (1,6) are believed to record a remanence of this age (McElhinny, 1969). A number of poles from dyke swarms intruding the Indian Shield also plot here or along the immediate extension of the path; some are probably Gwalior feeders, but since their ages are only very generally constrained within the mid-Proterozoic interval 2200–1400 Ma, they may belong to another part of the path (see also Kumar & Bhalla, 1983).

Although the Sudbury lopolith in the Superior terrane has been the subject

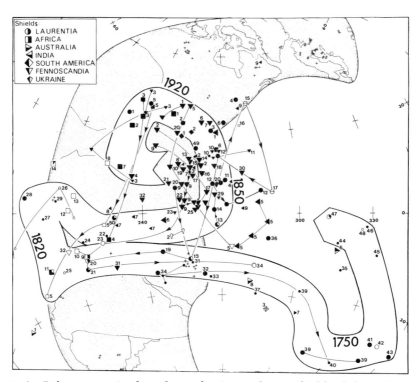

Figure 7.4 Palaeomagnetic data from the Precambrian Shields defining Loop 3 (ca. 1950–1750 Ma). The poles are numbered: LAURENTIAN SHIELD: 1, Nipissing diabase post-folding remanence (⩽1995 KA); 2, Otish gabbro (1935 KAB, 1925, 1810 KA); 3, Metamorphosed Kazan dykes (2000–1900 KA); 4,5, Sudbury Complex N1 and N2 components (1915–1645 RS, 1944 UP, 1860, 1785 RS, 1849,1849 UP); 6, Spanish River carbonatite (1852 RS); 7, Sokoman iron formation, post-folding remanence (1900–1800?); 8, Kahocella Group, primary component (1873 RS); 9, Sakami Formation, primary component; 10, Hornby Bay Group sandstone; 11, Minnesota River Valley adamellite (1850–1825); 12, Minnesota Hornblende andesite dykes (1930–1690 KA); 13, Akaitcho River Formation (1907–1897 UP); 14, Seton Formation (1333 RS); 15, Gibraltar Formation; 16, Charlton Bay Formation overprint; 17, McLeod Bay Formation; 18, Douglas Peninsula Formation (2000–1700); 19, Takiyvak Formation (ca. 1850); 20, Stark Formation (1845–1630); 21, Tochatwi Formation (1873, 1845–1630); 22, Chibougamau greenstones, CV overprint (?Hudsonian); 23, Spanish River Complex, dykes and fenites (<1852, Pole 6); 24, Big Spruce secondary X component; 25, Nonacho sediments, site1; 26, Nipissing E directions (?Hudsonian overprint); 27, Murdoch, Thompson Lake and Menihek metavolcanics and intrusives (2105–1816 RS); 28, Meta-igneous rocks, Labrador Trough (1873, 1855 KA); 29, Douglas Peninsula Formation overprint; 30, Sudbury Complex N3 directions; 31, Sakami Formation, post-folding remanence; 32, Peninsula Sill (1800 RS); 33, Seward Formation, A component (?1800); 34, Pearson Formation, A component (1809 RS, 1870–1800 KA); 35, Menihek Formation, A component; 36, Martin Formation (1930–1780 UP, 1835–1650 RS,KAM, 1635 KA, <1970); 37, Retty Peridolite, Labrador, A component (?1800); 38, French River anorthosite, relict magnetisation C; 39, Persillon volcanics, two poles; 40, Sudbury Complex, N4 component; 41, Qingaaluk sediments (<39); 42, Nastapoka basalt, (<39 and 41); 43, Eskimo Volcanics (1750 RS,

of palaeomagnetic study over a long period of time (see Morris, 1980, for summary), it has proved difficult to integrate the data with results from elsewhere in the Superior craton, both because they appeared to be discordant with other Laurentian data, and because the age of the body was poorly constrained by the radiometric investigations (Irving & McGlynn, 1976b). New U–Pb studies appear

---

1750–1625 KA); 44, *Big Spruce Complex Z component*; 45, *Grenville Province relict A remanence*; 46, *La Grande Riviére Proterozoic dykes, secondary component*; 47, *Et-Then group primary (1835–1630 KA, 1250 KAB)*; 48, *Pearson Formation, B component.* LAURENTIAN SHIELD, GREENLAND SECTOR: 49, *Angmagssalik norite-charnockite (1800 UP)*; 50, *Nordfjord gneisses.* FENNOSCANDIAN SHIELD: 1, *Seili gabbro anorthosite and amphibolites (ca. 1850 UP)*; 2, *Svappavaara gabbros, two poles (1880–1870 UP)*; 3, *Åva intrusives five poles from migmatites, amphibolites, granites, monzonites and lamprophyres (1815–1803 UP)*; 4, *Svecokarelian gabbros, Allivuotso and Tsokkdaivi, (1900 Ma)*; 5, *Keuruu dykes, three poles (1900–1800 UP)*; 6, *Northern Karelia sandstones (1870–1610 Ma)*; 7, *Nordingra-Ornskjoldsvik dykes (1850 Ma)*; 8, *Nilsia tonalitic dykes (1860–1830 UP)*; 9, *Pukhta-Pedaselsk (Iotniy) Sandstones (1950–1850 UP)*; 10, *Jatulian redbeds and lava (1870–1610 KA)*; 11, *Central Karelia Sandstones (1870–1610 Ma)*; 12, *Iisalmi dolerite dykes*; 13, *Haukivesi quartz diorite (1880 Ma)*; 14, *Mikkeli gabbro-diorite (1880 UP)*; 15, *Haukivesi quartz diorite (1880 Ma)*; 16, *Central Karelia dolerite dykes (1870–1610)*; 17, *Vlivieska gabbro, three poles (1885 UP)*; 18, *Skoksha Group sediments (1950–1850 KA, UP)*; 19, *Haukivesi area dykes (1840 Ma)*; 20, *Vittangi gabbros (1880–1707 Ma)*; 21, *Pap gabbro-norite (1880–1820 KA)*; 22, *Onega gabbro-dolerites (1880–1820 Ma)*; 23, *Ryboretskii Sill, Lake Onega (1800–1600 Ma)*; 24, *Pohjanmaa synorogenic intrusions (1880 Ma)*; 25, *Pielavesi gabbro (1892 UP)*; 26, *SW Finland gabbro diorites (ca. 1880 Ma)*; 27, *Hyvinkaa gabbro (1875 UP)*; 28, *Forstrom gabbro (ca. 1900 Ma)*; 29, *Iisalmi dolerite dykes*; 30, *Tammela gabbro-diorite (ca. 1880 Ma)*; 31, *Laanila dykes*; 32, *Foglo intrusions (1890–1800 UP).* UKRAINIAN SECTOR: 1, *Korosten Complex (1750 KA, RS, UP)*; 2, *Korosten Granite (1750 KA, RS).* AFRICAN SHIELD: 1, *Aftout gabbro (1985–1950 RS)*; 2, *Losberg gabbro (1910 RS)*; 3, *Phalabowra Complex (2060 UP)*; 4, *Vera Hill syenite*; 5, *Waterberg Formation, lower sites (1950–1740 RS)*; 6, *Diorite intrusions, NW Sahara (<1910)*; 7, *Vredefort ring, granophyre dykes (2006, 1950, 1928 RS)*; 8, *Limpopo gneisses, A component (2020–1935 RS)*; 9, *Vera Hill syenite*; 10, *post-Phalabowra dykes (1825 KA)*; 11, *Mashonoland dolerites (1810, 1910 RS)*; 12, *Bubi swarm, SW extension of Great Dyke, overprint (1930–1810 RS)*; 13, *Sabanga Poort dyke (2030–1935 RS)*; 14, *Great Dyke extension, Ruri and Crystal Springs swarm (1900–1810 Ma overprint).* SOUTH AMERICAN SHIELD: 1, *La encruciajada rocks (1850 AAB)*; 2, *La ceiba rocks (1820 RS, 1804 AAH)*; 3, *Telles Pires rocks, Brazil (>1610).* INDIAN SHIELD: 1, *Kondapalli charnockites (2600 UP, ca. 2000 Ma magnetisation)*; 2, *Hyderabad dyke, two poles*; 3, *Group IV dykes, Dharwar craton*; 4, *Anatapur dykes, group 2*; 5, *Upper Gwalior traps, four poles (1830 RS, >1700)*; 6, *Visakhapatnam charnockites (2600 UP, ca. 2000 Ma magnetisation)*; 7, *metabasic dykes at Hirapur Dhaiya and Gomoh, three Poles*; 8, *Anatapur dykes, group 3*; 9, *Karimnagar area dykes, four units (<2000 Ma, probably contemporary with unit 5*; 10, *Kolar dyke.* AUSTRALIAN SHIELD: 1, *YF dykes (1700 RS)*; 2, *Edith River Volcanics (1760)*; 3, *Hart dolerite (1800 RS).* Lambert Equal area projection centred at 270°E, 0°N; the format, symbols and abbreviations the same as for Figure 7.1.

to fix the ages of this body at *ca.* 1850 Ma (Krogh, McNutt & Davis, 1982) and the oldest N1 remanence here differs only marginally from the well-defined cluster of 1850 Ma poles from elsewhere; the N2 component (5) is a metamorphic magnetisation post-dating initial deformation and could belong here or be applicable to the path at 1700 Ma in Fig. 7.5. The Sudbury N1 pole is coincident with one result from the Athapuscow aulacogen. The remaining stratigraphic succession in this latter area appears to record most of the APW during the time interval under discussion here (Evans & Bingham, 1973; McGlynn & Irving, 1978; Irving & McGlynn, 1979). These poles are connected in stratigraphic sequence in Fig. 7.4 (13 → 14 → 15 → 17 → 18 → 20 → 21 → 34 → 47) although this is not, of course, necessarily the age sequence of the magnetisations; a distinct record of secondary overprinting is also present in these rocks (e.g. Bingham & Evans, 1976). The excursion to the westerly extremity of the loop at 1800 Ma is defined by two poles in the upper part of the succession (20,21). It has been argued that a 60° clockwise rotation could explain the difference between these results, and the other data from the Athapuscow (Irving & McGlynn 1976b, 1981). While the sampled region is one of greater tectonic complexity than the remainder of the aulacogen, the susceptibility ellipsoids observed in these sediments are broadly comparable with the orientations of ellipsoids elsewhere in this area, and compatible with palaeocurrent indicators (Gough, Rahman & Evans, 1977); they seem to provide no indication that the region has been tectonically rotated. Although this dispute is not yet resolved, it is no longer necessary to argue that these poles are aberrant on the basis of the APWP, because this westerly migration of the path is now recognised in other sediments from the circum-Slave area (18 → 29). Elsewhere, it is identified from the sequence of poles from igneous events in the Kaapvaal craton (3 → 4,10), the Superior craton (6 → 23) and in the uplift and overprint remanences (12,13,14) in the northern margin of the Limpopo mobile belt, between the Kaapvaal and Zimbabwe cratons (Jones, Robertson & McFadden, 1978, Morgan & Briden 1981). It is also probably recorded in southern Africa by the stratigraphic sequence of poles (5) in the lower part of the Waterberg red beds (McElhinny 1968b) and by the Mashonaland dolerites (11). The hairpin at *ca.* 1800 Ma is identified by the high blocking-temperature components in the Nagssugtoqidian belt in east Greenland (50) and an overprint (26) of probable 'Hudsonian' age (Section 7.5) in the Superior craton (Roy & Lapointe, 1976).

The long southerly extension of this loop after *ca.* 1825 Ma is essentially identical to the latter part of the Coronation Loop of McGlynn & Irving (1978) and is recognised from characteristic and overprint remanences from the highest part of the Athapuscow aulacogen. The extension to 1750 Ma is poorly-defined by Laurentian data only, and is represented predominantly by results from igneous rocks and sediments of the Labrador Trough (Schmidt, 1980; Schwartz & Fujiwara, 1981). A fold test implies that at least one of these poles (43) is primary and linked to the Rb–Sr age of 1750 Ma, while another (39) pre-dates faulting, which in turn pre-dates an overlying formation yielding an identical pole (42). Although these pole positions may be aberrant and interpreted in terms of relative movements between terranes on either side of the Labrador Trough, there are three points which argue against this possibility:

1. This is the only point at which the Sudbury N3 component coincides with other Laurentian data.

2. The geological evidence favours an ensialic development of the Labrador Trough within a zone of stretched and attenuated crust which forms a sinuous border to the already-stabilised Superior craton (e.g. Baragar & Scroates, 1980).
3. Large relative movements are precluded by correlations of the older Laurentian and Fennoscandian data (this section and Section 7.2).

7.5 The 1750–1600 Ma APW path (Fig. 7.5)

The palaeomagnetic record for this interval is intimately linked to the widespread thermal-tectonic mobile events which affected the continental crust between *ca.* 1900 and 1700 Ma (Section 8.5). In the Laurentian Shield they include the Churchill (Canada), Nagssugtoqidian (Greenland) and Laxfordian (NW Scotland) terranes and are broadly grouped as the 'Hudsonian' episode. In the Fennoscandian Shield they are represented by the Svecokarelian and Svecofennian episodes and were widely accompanied by syn- to post-tectonic magmatism. Much of the palaeomagnetic record from these terranes is an uplift and cooling remanence, and the bordering cratons and their supracrustal successions have been variably overprinted by the thermal effects.

The return track of the APWP from the 1750 Ma group is recognised in the time sequence of uplift remanences in the Nagssugtoqidian terrane of West Greenland, and linked to mineral ages in the range 1790–1650 Ma. The direction of APW here is defined by progressive demagnetisation studies on single samples, by movement of low to high blocking-temperature components back along the track, and by the lateral and vertical changes of remanence within the terrane (Morgan, 1976; Piper 1981a, 1985b). The complete data from the Nagssugtoqidian define a long sequence of poles (**40**) from magnetite-held components defining the SE → NW trend of this APWP, and a sequence of hematite-held components (**52**) from a later local event plotting on the return limb of Loop 5 at *ca.* 1630 Ma (Piper, 1985a). Hudsonian uplift remanences or overprints are also widely represented in the Churchill and Nain terranes bordering the northern margin of the Superior craton (**8,9,17,23,24,26**); an overlapping, or slightly older, part of the same trend has been identified from high to low blocking-temperature magnetite and secondary hematite components in the Flin Flon–Snow Lake Belt of Saskatchewan (**19 → 20**, Park, 1975). Furthermore, it is prominently present as an overprint in the Athapuscow aulacogen (**14,22,25**), and as a relict remanence within the *ca.* 1100 Ma Grenville terrane along the south-east margin of the Shield (**6,7** and Section 7.8). Although the ages of most of these poles are inferred within the general geological history of the area concerned, their assignment to this part of the path is supported by better-constrained results from dykes and supracrustal rocks (**15,18,19,28**).

Elsewhere, this APWP is identified from the Roraima sandstones (**2**) and dolerite dyke suites intruding these rocks (**3–6**) in the Guyana craton of South America. Few African data points can be assigned to this interval with certainty, although all rock units of this general age conform to the path as defined from the Laurentian Shield. One set of 1700 Ma Australian dykes (**2**), accords with the path at this age, although a second group with the same age (**3**) is somewhat discordant, for no clear reason. The well-studied Lunch Creek lopolith (**7**) conforms to the

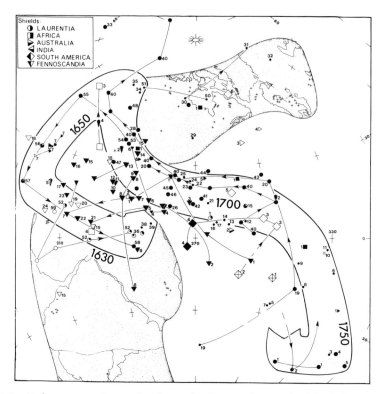

Figure 7.5 Palaeomagnetic results from the Precambrian Shields defining Loop 4 (ca. 1750–1600 Ma). The poles are numbered: LAURENTIAN SHIELD: 1, Persillon Volcanics, two poles (ca. 1750 Ma); 2, Sudbury Complex N4 directions; 3, Qingaluuk sediments (ca. pole 1); 4, Nastapoka basalt (ca. pole 1); 5, Eskimo Volcanics (1750 RS, 1750–1625 RS, KA); 6, French River Anorthosite, relict Hudsonian magnetisation, two poles; 7, Grenville Province, relict C magnetisation; 8, Menihek Formation, A component; 9, Big Spruce Complex, Z component; 10, Pearson Formation, B component (<1870–1800 Ma); 11, La Grand Rivière Proterozoic dykes; 12, Post-Priessac overprint, Superior craton (?Hudsonian); 13 Pearson Formation C component; 14, Slave Craton overprint (SE directions); 15, Sparrow dykes (1770 AA, 1550–1390 KA); 16, Retty Peridotite, B component; 17, Menihek Formation, B component; 18, Dubawnt Group sediments and lavas (1787 RA, 1735, 1716 KA); 19, Flin Flon, C (high blocking temperature) and B remanences (1816 UP, 1805–1775 RS, 1770–1735 KA); 20, Flin Flon A2, A1 and A3 remanences in order of probable age (1770–1735 KA); 21, Nonacho Sandstones, post-folding remanence; 22, Coronaton Geosyncline, overprint; 23, Metamorphosed Kaminak dykes (1892–1615 KA); 24, Castignon Complex, B component; 25, Mcleod Bay Formation, overprint; 26, Daly Bay Metamorphics, A component (1690–1580 KA, mean 1622 KAB); 27, Eskimo Volcanics, secondary component; 28, Wind River Dykes (1880–1680 KAP); 29, Kaminak lamprophyre dykes; 30,31, Sudbury Complex, N5 and N6 components; 32, Daly Bay Metamorphics, B component (1690–1580 KA); 33, Mt. Nelson Formation, Purcell Supergroup; 34, Menihek Formation, C remanence; 35, Wakuak gabbro-Willbob basalt secondary (<2105–1816 Ma); 36, Flaherty Volcanics, (1790, 1755, 1640 RS, 1693 KA); 37, Haig intrusions (1790 RS, 1620 KA); 38, Sutton Lake inlier diabase sill; 39, Seward Formation, B component. LAURENTIAN SHIELD, GREENLAND SECTOR; 40,

path at *ca.* 1650 Ma, although age data from this body are conflicting; instead, the pole might be applicable to the path illustrated in Fig. 7.6. In the same region, the Middleback Range ores (**4–6**) are assigned to this general interval, in accordance with two of the three poles.

Sequence of uplift remanences, Nagssugtoqidian mobile belt, twelve poles (1790–1650 KA, RS); 41, Itivdleq dykes north of Itivdleq-shear belt (1750 KAI); 42, Itivdleq dykes and gneisses, shear belt (1790–1650 KA); Kaellinghaetten gneisses; 44, Nordre Stromfjord gneisses; 45, Godhåb and Amitsôq gneisses; 46, Kangamiut dykes south of Nagssugtoqidian Front (< 1930 RS, 1790–1650 KA); 47, Post-Nagssugtoqidian pseudotachylyte; 48, Kangamiut dykes north of Nagssugtoqidian Front; 49, Sagdlerssuq dykes and gneisses (1620 KAI); 50,51, Ketilidian metavolcanics (< 1840–1780 UP, 1700–1600 KA); 52, sequence of B magnetisations, Nagssugtoqidian mobile belt (< 1690 RS). LAURENTIAN SHIELD, NW SCOTLAND (LEWISIAN) SECTOR: 53, Scourie dykes, uplift remanence (1650–1625 KAH, > 1630–1390 KA); 54, Lewisian A2 Central Zone remanence (as for 53); 55, Lewisian A3 Southern Zone remanence (< 1700–1670 RS, 1690–1678 KAH); 56, A4 remanence (1685–1472 KAH); 57, A5 remanence (1504–1468 KAH); 58, Lewisian B1 directions (? late uplift cooling, ca. 1600 Ma); 59, Badcall dykes (1639 RS). FENNOSCANDIAN SHIELD: 1, Kallax gabbro (1775–1530); 2, Tarendo gabbro (< 1775 UP, 1690–1530 RS); 3, Tarendo silicic rocks (age data as for 2); 4, Niemisel gabbro and postgabbro dyke, two poles (1690–1530 RS); 5, Storalulevatten gabbro, high blocking temp. component (1690–1530 RS); 6, Nottrask gabbro, two poles (1840 RS, magnetisation age 1690–1530); 7, Jokkmokk basic rocks, two poles (1690–1530); 8, Sangis gabbro, three poles (1690–1530 Ma); 9, Vuollerim gabbro, two poles (< 1775 UP, 1690–1530 RS); 10, Radmansö gabbro-diorite (1760–1475 Ma); 11 Harads amphibolite (1690–1530 Ma); 12, Nordingrå area basic dykes (1770–1700 Ma); 13, Uppsala district metabasite dykes (> 1570, Svecofennian (ca. 1700 Ma) remanence); 14, Korstrask gabbro (< 1775 UP, 1690–1530 RS); 15, Gallivare basic rocks, four poles and Svecofennian diorite; 16, Makaala Piikkio quartz diorite (ca. 1650); 17, Loftahammar gabbro (1690–1530 Ma); 18, Maarta quartz diorite (1700–1600 Ma); 19, Dundret gabbro, three poles (< 1890–1875, 1775–1530 Ma magnetisation); 20, Kuissari dolerite (1670–1630 UP); 21, Stora-Lulevatten gabbro, low blocking temperature component; 22, Aland rapakivi granite (1659–1589 UP); 23, Dala Porphyries (1634 RS, > 1570); 24, Kumlinge Brando dykes (1602–1556 RS). AFRICAN SHIELD: 1, Dykes intruding Gueld-el-hadid and Yetti series, NW Sahara, four poles for swarms with contrasting orientations (1760–1350 KA); 2, Sub-Rust der Winter sediment (1790 RS); 3, Waterberg Formation, higher sites, two poles (< 1760 UP); 4, Van dyke mine, dolerite dyke (1615 RS). SOUTH AMERICAN SHIELD: 1, Cuchivero-caicara acid metavolcanics (1701 RS, 1700 KA); 2, Roraima Sandstone (> 1660); 3, Roraima diabases I, two poles (1700, 1670 KAH, 1660 RS); 4, Roraima diabases II, two poles (as for 3); 5, Kabaledo Sill (1750 KA); 6, Blackwatra E–W dykes, (1540 KA, 1610 RS). AUSTRALIAN SHIELD: 1, Fortesque Group, Opthalmian orogeny overprint (ca. 1700 Ma); 2, YF dykes (1700 RS); 3, GB dykes (1700 RS); 4, Iron Monarch ores, positive, Middleback Range (1800–1500); 5, Iron Prince ores, Middleback range (1800–1500); 6, Iron Prince ores, negative, Middleback Range (1800–1500); 7, Lunch Creek Lopolith five poles (< 1740, 1730 UP, 1498 RS, 1431 KA). INDIAN SHIELD: 1, Kerala dyke (1670 KA). Lambert equal area projection centred at 270°E 0°N; the format, symbols and abbreviations are the same as for Figure 7.1.

In the Fennnoscandian Shield the SE → NW track is well-recorded in the Svecofennian terrane of Sweden. The bulk of the magmatism here occurred somewhat later than that in the bordering *ca.* 1900 Ma Svecokarelian terrane in Finland which yields much of the Fennoscandian data shown in Fig. 7.4. The magnetisations of the numerous gabbro-diorite bodies in the Svecofennian are constrained between a late-orogenic granite episode at 1780 Ma, and the deposition of post-orogenic supracrustal rocks and the last major granite magmatism, at 1600 and 1530 Ma respectively. The main mobile episode was therefore contemporaneous with the post-Hudsonian uplift and the poles are predictably distributed along the same track in Fig. 7.4; they confirm the path tentatively suggested by Pesonen & Neuvonen (1981). Unfortunately, the information is currently inadequate to confirm the trend of the APW movement, as has been possible in Greenland, with the exception of two studies (5 and 7). The youngest Svecofennian poles coincide with the *ca.* 1630 Ma extension of the Laurentian path, and the other Svecofennian poles here include representatives of the rapakivi granite suite (22, Fig. 7.5) and all have age estimates of *ca.* 1650 Ma. The mixture of polarities at this part of the path contrasts with the predominantly normal polarity along the preceding SE → NW swathe.

The younger part of the suite of uplift poles, derived from the Hudsonian terranes of Greenland and Canada, overlaps with the post-Laxfordian (<1800 Ma) remanences observed in the Lewisian terrane of NW Scotland. The latter include results from the Scourie dyke swarm (53,54) in a relict Central Zone in this terrane, and the time sequence of uplift remanences observed in a Southern zone (55–57; Smith & Piper, 1982) terminating in a later overprint (58). At present, the age of this APW sequence is only generally constrained within a 1690–1490 Ma spectrum of K-Ar and Rb-Sr mineral ages, but the continuity between the *ca.* 1700 Ma SE-NW swathe and the younger group of *ca.* 1630 Ma poles suggests that it is an integral part of the loop (referred to here as 5) defined by these remaining data. A number of results, including the extension of the sequence of uplift poles from West Greenland, plot close to the present pole; the APWP has not been extended to include these data in Fig. 7.5 because they include the lowest coercivity/blocking-temperature components, and they may incorporate Recent overprints.

7.6 The 1600–1350 Ma APW path (Fig. 7.6)

Following stabilisation, uplift and erosion of the Hudsonian terranes, the Precambrian shields had largely attained the structures and fabrics which they preserve at the present day (Section 8.5); subsequent activity was confined to narrower zones, and took place mostly at peripheral locations in the continental crust. Temperature gradients in the crust were however, still high, and permitted widespread emplacement of large (anorogenic) plutons, and these rocks provide much of the palaeomagnetic evidence relating to these times. Rock units from the interval 1600–1450 Ma are poorly represented in the Laurentian Shield (see Irving & McGlynn, 1981), and the APWP during these times is only currently described by results from the Fennoscandian Shield (Piper, 1980; Pesonen & Neuvonen, 1981; Bylund, 1985). The data here are derived from numerous *ca.* 1530–1500 Ma dyke swarms and anorogenic intrusions, and they define a flattened loop, probably executed in little more than 100 Ma. The loop commences near the *ca.* 1600 Ma position, well represented by data points from several shields and plotted in Fig. 7.5. As defined in Fig. 7.6 it begins with results from the Kumlinge dyke swarms (1,2) and extends rapidly to the *ca.* 1500 Ma Nordingrå poles (6) near the apex of the loop. The *ca.* 1530 Ma E–W dyke swarms and giant dykes of Sweden yield a number of poles (9,10,13,15 and 20) of both polarities, which lie along the full length of one limb (Bylund, 1985). This return path is also represented by the Jotnian sediments and lavas (16,17). Following the convention, this APWP is outlined here by dashed lines, because it is not defined by data from the Laurentian Shield, although preceding (Fig. 7.5) and succeeding (Fig. 7.7) data give no indication that it is not also applicable to the Laurentian Shield. The oldest data from the latter shield plotting on Loop 6, lie near the lower part of the return path, and comprise the sequence of poles from the Harp complex and associated rocks of Labrador (1–3). The excursion of the path to the east is included here to incorporate single but well-defined poles from the African (3), Australian (2) and Laurentian (10) shields; it is only a tentative interpretation because age estimates do not completely agree, and the Laurentian pole might by interpreted in the context of a local rotation (Irving & McGlynn, 1981). Two Fennoscandian poles (7,11) are also included here, because they plot only on this loop, although the existing age estimates are too young to endorse this interpretation.

Anorogenic intrusions assigned to the interval 1450–1350 Ma are more widely represented in the Laurentian Shield, notably along the SE sector (6,11,13,24) bordering the younger Grenville terrane (Section 7.8), and in the Elsonian terrane (Stockwell, 1973, and Section 8.6) which is dominated by igneous intrusion without major deformation, and occupies the largely-concealed southern margin of this shield. The limit to the southerly extension of the return limb of Loop 6 appears to be defined by the palaeomagnetic record in the Belt supergroup. Sedimentation in growing aulacogens accompanied rifting in many parts of the Precambrian crust in Middle and Upper Proterozoic times (Burke & Dewey, 1973); the Belt succession is a product of this sedimentation along the western margin of the Shield, and locally exceeds 20,000 m in thickness. Both early and later investigations (e.g. Collinson & Runcorn, 1960; Evans, Bingham & McMurry, 1975) have identified a similar axis of magnetisation throughout much of the thickness of this succession (14–23). Although the early studies did not incorporate demagnetisation studies, they are included here because they agree very well with later work. The Beltian rocks have yielded ages ranging from ca 1400 Ma near the base to 1300–1100 Ma at the top. However, the lower part of the succession is intruded by rocks dated at 1335 Ma, and the conformity of all of the palaeomagnetic poles with 1400–1300 Ma data from elsewhere suggests that deposition may have taken place during little more than 50 Ma, near the older spectrum of the radiometric ages. There is no evidence here for the large amount of APW observed in the slightly-younger Grand Canyon supergroup (Section 7.7), and since these poles fail to conform to the later APWP there is little scope for invoking large scale remagnetisation, unless it took place soon after deposition; indeed, the existing results show a simple N → R → N pattern of polarity through the succession, with the lower inversion occurring in the Siyeh and Purcell divisions (17,18) and the upper inversion occurring in the Klintla (23; Evans, Bingham & McMurry, 1975).

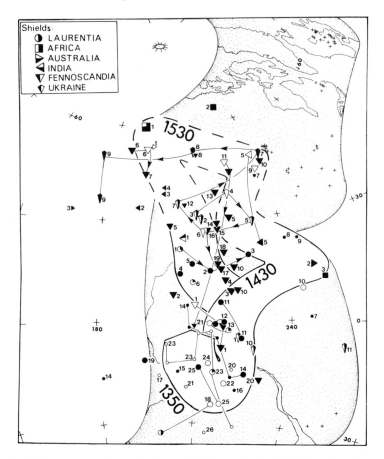

Figure 7.6 Palaeomagnetic results from the Precambrian Shields assigned to the interval ca. 1600 to 1350 Ma. The poles are numbered: LAURENTIAN SHIELD: 1, Harp Lake complex (1450 UP); 2, Harp country rocks, thermal aureole (1450–1350); 3, Harp dykes (1450–1350); 4, Laramie Range anorthosite; 5, Arbuckle granites (1400–1320 RS); 6, Nain anorthosite (1480, 1418 RS); 7,8,9, NRM results from mid Proterozoic sediments: Barron Quartzite (Wisconsin, two poles), quartzite (Minnesota), red beds (Wisconsin); 10, Western Channel diabase (1392 RS, 1400 KAB); 11, Croker Island complex, two poles (1440 RS); 12, St. Francois tuffs and igneous units (1514–1408 UP, 1378–1246 RS); 13, Michikamau anorthosite (1499, 1479 RS, 1450–1400 KA). Lower Belt Supergroup Formations (>1335 Ma): 14, Spokane (Grinnell) Formation argillites (six poles); 15, Appekunny Formation; 16, Waterton Formation; 17, Upper Siyeh Formation. Upper Belt Supergroup Formations: 18, Purcell Lava and red sediments (two poles); 19, Miller Peak Formation; 20, McNamara Formation; 21, Missoula Group, Montana; 22, Shepard Formation; 23, Klintla A, B and C Formations (seven poles); 24, Shabogamo gabbro north of the Grenville Front (1375 RS, SN); 25, Sibley Group, normal and reversed groups (1340 RS); 26, Bonito Canyon Quartzite. FENNOSCANDIAN SHIELD: 1, Kumlinge dykes, three poles (1602–1556 RS); 2, Kumlinge-Brando dykes (1602–1556 RS); 3, Bangsfjallet Complex (1520 Ma); 4, Bunkris dyke (1546–1516 Ma); 5, Halleförsnas giant dyke, two poles (1518 RS); 6, Nordingrå gabbro-anorthosite normal and reversed groups, (1552 RS); 7, Nordingrå granite (<6, 1415 RS); 8, Eckero dykes (1650–1250 Ma); 9, East Breven dyke

The Avzyan supergroup in the south Urals yields a set of poles which plot on this loop but correlate with no later part of the path. It therefore seems likely that the characteristic magnetisations are some 100–200 Ma older than the 1350–1100 Ma distribution of K–Ar ages from these rocks. This interpretation would accord with the observation that poles from the overlying Zil Merdak group, assigned to the Riphean, yield a sequence of poles following the 1250–1200 Ma E–W swathe described in the next section.

7.7 The 1350–1150 Ma APW path (Fig. 7.7)

The poles from higher formations in the Belt supergroup probably define the limit to the southward motion of the return limb of Loop 6, because *ca.* 1250 Ma poles lie some 20°–30° north of these data; indeed, there is some suggestion that this northward movement of the path is recorded in the Beltian results by the difference between the pre-Klintla and Klintla poles in the upper part of the succession. The E → W latitudinal movement is identified by both the Fennoscandian and Laurentian shields. In the Greenland sector, basaltic activity near the northern (**29,30**) and southern (**31,32**) margins of the shield defines the earliest part of the movement at 1280–1250 Ma (Piper, 1977), while the collective motion is defined from the Gardar Igneous Province of South Greenland by early lava sequences (**31 → 32**), successive later dyke swarms (**33 → 34 → 35**) and intrusive complexes (**→ 36**, Piper & Stearn, 1977) with the sequence confirmed by detailed Rb–Sr studies (Patchett, Bylund & Upton, 1978). This magmatism was paralleled in the Canadian sector (see Fig. 9.1) by shield-wide dyke intrusion represented by the NNW-trending Mackenzie dyke swarm in central and northern regions (**22**) and

---

(1548–1392 RS); 10, *East-West dyke swarm, southern Sweden, two poles (1548–1532 RS);* 11, *Ragunda complex (1320 RS);* 12, *Gavle Granite;* 13, *Eskilstuna dykes (1560–1510 Ma);* 14, *Åva dolerite dykes (1550 Ma);* 15, *Glysjon dala dykes (1530 Ma);* 16, *Dala (Jotnian) sediments;* 17, *Dala (Oje) basalts (1650–1550 KA, < 16);* 18, *Foglo dykes (1523–1518 RS, UP);* 19, *Foglo-Sottunga dykes (1523–1518 RS, UP);* 20, *West Breven dyke (1548–1392 Ma);* 21, *Post-Ragunda dykes (<11). UKRAINIAN SECTOR: Avzyan Supergroup, South Urals (1350–1100 KA):* 1, *Revet sequence;* 2, *Green and Ushakova sequences (1250–1150 KA);* 3, *Maloinzer and Kataska sequences (1350–1263 KA);* 4, *Kataska sequence, Limestones and sandstones;* 5, *Maloinzer sequence;* 6, *Ushakovka sequence;* 7, *Green sequence, two poles;* 8, *Revet-Green sequence;* 9, *Revet sequence, Limestones and sandstones;* 10, *Turchinki gabbro (1400–1200);* 11, *Uman granite (1400–1200). AFRICAN SHIELD:* 1, *Barby Formation lavas (>1290 RS);* 2, *Ezelsfontein Formation (>1178 RS);* 3, *Pilanesberg dykes (1282 RS). AUSTRALIAN SHIELD:* 1, *Nullagine lavas;* 2, *Gawler Range volcanics (1525 RS);* 3, *Morowa Lavas (1390 RS). INDIAN SHIELD:* 1, *Kaimur Sandstone (>1145 RS);* 2, *Newer dolerites, Singbhum, main group (1960, 1540 KA, 1600–900 Ma);* 3, *Tirupati dykes, main group;* 4, *Group 3 dykes, Dharwar craton;* 5, *Sukinda orthopyroxenites (magnetised during E. Ghats episode, 1625–1570 Ma). Lambert equal area projection centred at 200°E, 0°N; the format, symbols and abbreviations are the same as for Figure 7.1.*

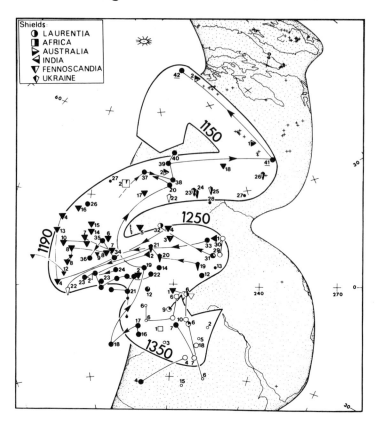

Figure 7.7 Palaeomagnetic results from the Precambrian Shields assigned to the interval ca. 1350–1150 Ma. The poles are numbered: LAURENTIAN SHIELD: Upper Belt Supergroup Formations: 1, *Miller Peak Formation*; 2, *McNamara Formation*; 3, *Missoula Group, Montana*; 4, *Purcell Lavas and red sediments*; 5, *Shepard Formation*; 6, *Klintla Formation, seven poles*; 7, *Sibley Group, normal and reversed groups*; 8, *Osawattamie granite bore core (1361 UP)*; 9, *Sherman granite and andesite dyke (1377 RS)*; 10, *Shabogamo gabbro north of Grenville Front (1375 RS, SN)*; 11, *Mistatin pluton (1310 RS)*; 12, *Big Springs granite bore core (1339 UP)*; 13, *Cheneaux metababbro, probable emplacement remanence (1310 RS)*; 14, *Seal Lake Group red beds (<1563 Ma)*; 15, *Bonito Canyon Quartzite (>16)*. Poles from Grand Canyon Supergroup, Middle formations, in stratigraphic order; 16, *Bass Limestone*; 17, *Bass Limestone (middle-upper)*; 18, *Hakatai Shales (lower middle)*; 19, *Hakatai Shales (middle-upper)*; 20, *Shinumo Quartzite*; 21, *Sudbury dykes and contacts, four poles (1205, 1203, 1194, 1189 RS)*; 22, *Mackenzie dyke swarm, five poles (1251, 1232 RS)*; 23, *Coppermine group lavas and sediments (1240 RS, 1360–865 KA)*; 24, *Muskox intrusion (>23, 1214 RS, 1150 KA)*; 25, *Metalamprophyre dyke, Labrador (>958 KA)*; 26, *Arizona (Gila County) sills (1150 UP, 1140 KA)*; 27, *Keweenawan diorites, three poles*; 28, *Aillik dykes, Labrador (995–685 KA)*. LAURENTIAN SHIELD, GREENLAND SECTOR: 29, *Zig-zag dal basalts (<1380 RS)*; 30, *Midsommerso dolerites (1250 RS, probably same igneous event as 1)*; 31, 32, *Gardar lavas, Lower and Upper groups (1300–1250)*; 33, *BDO (WNW) dolerite dykes (1243, 1238 RS)*; 34, *NW Lamprophyre dykes (1249–1228 RS, <33)*; 35, *NNW dolerite dykes (1238, 1236 RS, <34)*; 36, *Gabbro ring dyke, Kûngnat Complex (1219 RS, <35)*; 37,

the Sudbury dyke swarm in the south (21, see also Fig. 9.1); both episodes are linked to Rb–Sr ages in the range 1230–1190 Ma (Patchett, Bylund & Upton, 1978). The Mackenzie dykes radiate southwards from an intrusive centre comprising the Muskox (24) which is overlain unconformably by 1000 m of dolomites, prior to deposition of the Coppermine group lavas and sediments (23). Contrasting data from the lower and upper Coppermine group are explicable in terms of rotation about a fault zone (Baragar & Robertson, 1973); following corrrection for this effect, these results show the westward APW movement implied by contemporaneous data from elsewhere.

In the Fennoscandian Shield, the widespread post-Jotnian basaltic igneous episode in Sweden and Finland defines a large group of poles, coincident with the younger part of the path defined by the Greenlandic and Canadian data. The magnetisations from both shields have uniform normal polarity, and imply that these rocks were magnetised during a protracted period of constant polarity. The post-Jotnian episode has been extensively dated, with the bulk of the determinations falling in the range 1270–1215 Ma (Patchett, 1978). There is no significant difference between the range of age estimates from the Mackenzie and mid-Gardar events and the post-Jotnian event, although the distributions of palaeomagnetic poles suggest that the post-Jotnian activity is slightly younger: only in three intrusive suites (4,7,8) does the latter part of the E → W swathe appear to be recorded.

Although these igneous episodes may record rather short events in geological terms, that they give good information on APW prior to 1150 Ma is shown by the

Hviddal syenite giant dyke (1150 RS); 38, Gabbro giant dykes, Tugtutôq (1150–1144 Ma); 39, Narssaq gabbro and ultramafics (1150–1144 RS); 40, NE dyke swarm (>1144 RS); 41, Ilímaussaq intrusion, marginal syenites (1144 RS); 42, Ilímaussaq intrusion, interior fractionated rocks (1126 RS). FENNOSCANDIAN SHIELD: 1, Post-Ragunda dolerites, normal and reversed groups (<1320 Ma); 2, Satakunta Sandstone (1370–1300 Ma); 3, St. Tuna dolerites (1371 RS); 4, Sundsjö-giman dolerite dykes (1229–1156 RS); 5, Alvho dolerite sills (1215 RS); 6, Vaasa dolerite dykes (1225 RS, 1270 UP); 7, Satakunta intrusions, two poles for sills and dykes respectively; 8, Nordingrå dolerites and overprint on Nordingrå Complex (1245 KAI); 9, Lybersgnup dyke; 10, Ulvo dolerite; 11, Market dolerite dykes (1270 RS, UP); 12, Gnarp dolerites (1245 RS); 13, Alvdalsan dolerite sills (1231 RS); 14, Dala (post-Jotnian) dolerites (1290–1215 RS); 15, Vast-norrland dolerites (1245 KAI, 1270–1213 RS); 16, Gavle dolerites; 17, Emadalen sill (1223 RS); 18, Bamble sector B1 remanence. UKRAINIAN SECTOR: Zil-Merdak Supergroup (1100–1000); 19, Nugush and Biryan sequences; 20, Biryan sequence; 21, Nugush sequence; 22, Bedershinsk sequence, two poles, (1000 KA); 23, Lemerzinsk sequence sandstones; 24, Nugush sequence; 25, Biryan sequence; 26, Nugush and Biryan sequences. AFRICAN SHIELD: 1, Kisii Series lavas, Kenya (1213–974 AA, 1105 mean KA); 2, Koras Group, Kalkpunt Formation (1200–1000). AUSTRALIAN SHIELD: 1, IB intrusives (1149 RS); 2, Giles Complex, two poles (<1600, 1060 RS, adjacent volcanics = 1250–1140 Ma). INDIAN SHIELD: 1, Cuddapah Traps (Lower Vindhyan, ca. 1400–1200 Ma); 2, Bhima Series ferrous shales and limestones, two poles (1300–1200 or 950–750 Ma). Lambert equal area projection centred at 200°E, 0°N; the format, symbols and abbreviations are the same as for Figure 7.1.

palaeomagnetic record in the lower part of the Grand Canyon supergroup. This 3000–4500 m succession has been studied in Arizona (Elston & Grommé, 1974), where the lower part of the succession pre-dating the Cardenas lavas (1065 Ma) defines appreciable APW (in contrast to the somewhat older Beltian sequence), and includes a loop (Loop 7) which accords closely with data from elsewhere in the shield. The first part of this loop begins near the Mackenzie-Gardar-Jotnian poles, and continues via five poles comprising a NE path (16 → 20). This movement is also recorded by the youngest events in the Gardar Igneous Province and identified by the poles 36 → 37 → (38 + 40) → 41 which conform to the sequence of geological events and Rb–Sr isochron ages; collectively they imply that this earlier limb of Loop 7 was rapidly executed in 50 Ma. Unfortunately, there are few data points from the other shields covering this time interval, but the accordance of African (1,2), Indian (1,2) and Australian (1,2) poles with the APWP defined from the Laurentian and Fennoscandian data is to be noted; only in the latter instance however, does the radiometric evidence give a confident assignment to this path.

7.8 The 1150–900 Ma APW path (Fig. 7.8)

The results from this time interval commence with the data from the Keweenawan succession; this is exposed in the Lake Superior region (see Fig. 9.1) and represents part of the infill of a complex NE–SW trending *en-echelon* trough (the Mid-Continent Rift system). With over 60 reported palaeomagnetic poles, it is the most intensively studied rock sequence in the world. Lavas and sediments were deposited within the aulacogen at intervals during its development, although the total time span of the activity may have been little more than 40 Ma. The U–Pb studies of Silver & Green (1975) showed that deposition of the Lower and Middle Keweenawan may have been restricted to the interval 1140–1120 Ma, while the age of the Bear Lake rhyolite, intruding the Upper Keweenawan (986 Ma, data of Chaudhuri quoted by Henry, Mauk & Vander Voo, 1977) would appear to define a minimum age for the bulk of the succession. The early work of DuBois (1962), defined in outline the distribution of poles shown in Fig. 7.8, while Robertson & Fahrig (1971, see also Spall, 1971) incorporated these data and others into a loop which they referred to as the Great Logan Loop. This remains substantially the same as the path shown in Fig. 7.7 & 7.8 and referred to here as Loop 7, although some of the key reference points have changed, and there is some evidence that the depth of this loop is an artifact of a multipole geomagnetic field source (Section 7.9).

Although it was originally believed that the lowest part of the Keweenawan lava succession recorded a segment of the older limb of Loop 7 (see discussion in Irving & McGlynn, 1976b, and Halls & Pesonen, 1983), this has not been substantiated by later work (Palmer & Halls, 1985) and the collective data appear to plot along the return limb of this loop as a single, long NE → SW arcuate swathe directed away from the sampling area. Following standard practice, this is how the data are plotted in Fig. 7.8, but it should be noted that ages as old as 1170–1150 Ma have been obtained on the Logan diabases (see poles **20,24**) by $^{39}Ar:^{40}Ar$ dating (Hanson, 1975). Thus the age, position and polarity of these poles are virtually identical to Greenland poles 37–40 on the outward limb of the loop in Fig. 7.7, and it is probable that some of the Keweenawan data from intrusive rocks do record this limb. However, because the outward and return limbs are so close in time and space, it is impossible to be sure of the assignment of those data which are not part of a well-defined sequence of events. A most significant feature of the return limb of Loop 7 is the consistent R → N polarity change recorded in several lava sequences (Osler, Mamainse, North Shore and Gargantua). Dual polarities are also recognised in several of the dyke swarms (which may have been feeders of the lavas) and intrusive complexes. The quality of the collective data set is reinforced by contact tests and positive fold tests; in addition, the penecontemporaneous nature of the remanence in the lavas is emphasised by the tight groupings of directions within individual flows, and the similarity of directions between strongly and weakly deuterically-altered zones (Robertson, 1973). The stratigraphic relationship of the earlier R poles is not well-constrained, but the higher units, namely the Lower North Shore (**8**) and Higher North Shore (**22**) volcanics, the Portage Lake volcanics (**29**), Copper Harbour conglomerate (**37**), Nonesuch shale (**53**), Freda sandstone (**54**) and Jacobsville sandstone (**55**), plot in a sequence in accordance with their stratigraphic position, which defines the latter part of Loop 7. This appears to have been an interval of essentially constant polarity: no reversals have been identified in 900 m of Nonesuch shale and Freda sandstone, while one reversal has been noted in the overlying Jacobsville sandstone at the top of the Keweenawan succession (Roy & Robertson, 1978). In the latter formation, the highest blocking temperature and most chemically-resistant components correlate with their stratigraphic locations to identify a progressive south-easterly motion of the APWP. In general, the lavas and sediments in the Keweenawan yield the best-constrained data because their attitudes are well-known, while the intrusive suites have been subjected to warping during the rift development, which is less well-understood (Halls & Pesonen, 1983).

Secondary magnetisations comprise a small but significant part of the Keweenawan palaeomagnetic record. The secondary component in the Copper Harbour conglomerate is demonstrably pre-folding, as is a similar component in the Freda and Nonesuch sediments, while components which post-date folding are prominent in the higher stratigraphic levels. These letter fall roughly into two groups: one, which is considered to be a burial-induced CRM (Watts, 1981), may be relevant to the continuation of Loop 7 in Fig. 7.8, or to the subsequent path (Fig. 9.3), while the second group lies close to the present field direction, and is either of early Cambrian or Recent origin (Section 9.4). The latter (normally-magnetised) segment of Loop 7 is also represented in the Laurentian Shield by eight poles from the upper part of the Grand Canyon succession (**41** → **49**).

Although the Lower and Upper Torridonian sediments of NW Scotland have been dated at 968 and 777 Ma, respectively, by Rb–Sr isochrons on shale horizons (Moorbath, 1969), the palaeomagnetic poles from these rocks do not correlate with other Laurentian data at these ages (Stewart & Irving, 1974; Smith, Stearn & Piper, 1983). Most of the remanence in these rocks can be shown to date from the time of deposition, or soon after (third part of Section 4.2) and since the stratigraphic sequences of poles in the Lower and Upper Torridonian conform to the APWP at *ca*. 1100 and 1000 Ma in Fig. 7.8, it appears likely that these are true ages of

180   *Palaeomagnetism and the Continental Crust*

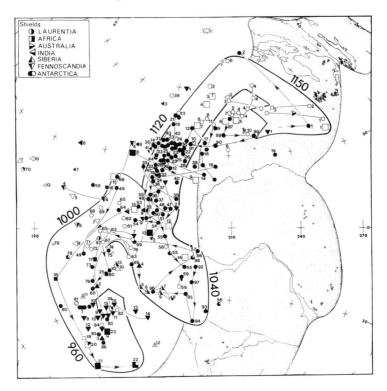

Figure 7.8 Palaeomagnetic results from the Precambrian Shields assigned to the interval 1150–900 Ma. The poles are numbered: LAURENTIAN SHIELD: 1, *Alona Bay lavas*; 2, *Baraga County dykes*; 3, *Mamainse lavas, reversed, (1040 RS, 1055–800 KA)*; 4, *Logan Sills, reversed group, Canada (1115 RS, 1050 KA)*; 5, *Marquette County dykes*; 6, *Lower Cape Gargantua volcanics (approximately contemporaneous with 3)*; 7, *Thunder Bay dykes and baked rocks, reversed*; 8, *North Shore volcanics, reversed*; 9, *Boulter intrusion*; 10, *South Trap Range (Powdermill) lavas*; 11, *Duluth Complex, reversed (1115 RS, 1200–1000 KA)*; 12, *Duluth Complex, normal, six poles*; 13, *Cook County gabbro*; 14, *Thunder Bay dykes and baked rocks, normal*; 15, *Nemagosenda carbonatite (1035 KA)*; 16, *Isle Royale lavas, lower and upper flows, two poles*; 17, *Lester River Sill*; 18, *El Paso (Franklin Mountains) rocks, three poles*; 19, *(Keweenawan) Ironwood-Portage lavas, six poles (<10)*; 20, *Logan dykes, normal, Canada (1133 RS, 1020 KA)*; 21, *Keweenawan gabbros, Mellen, Minnesota and Clam Lake bore core, three poles*; 22, *North Shore Volcanics normal, two poles*; 23, *Mamainse Lavas, four poles (1026 RS)*; 24, *Logan diabase, normal, U.S.A.*; 25, *Upper Cape Gargantua Volcanics (1090 KA)*; 26, *Endion Sill*; 27, *Osler lavas, normal*; 28, *Beaver Bay Complex, two poles (1090 KA)*; 29, *Portage Lake lavas, two poles*; 30, *Gogebic quartz porphyry*; 31, *Copper harbour lavas, primary*; 32, *Michipicoten Island lavas, two poles (938 RS)*; 33, *Basalt dykes, Minnesota*; 34, *Gogebic rhyolites (<19, grade into unit 36)*; 35, *Kearsage rhyolite (<19, 1007 RS)*; 36, *Copper Harbour Lavas, secondary components 1 and 2, two poles*; 37, *Copper Harbour conglomerates, secondary components*; 38, *Logan diabase reversed, USA*; 39, *Keweenawan Volcanics, Slate Island, A remanence*; 40, *Seal and Croteau rocks, secondary component. Grand Canyon Supergroup, higher middle formations (c.f. Figure 7.7)*: 41, *Dox Sandstone, upper lower*; 42, *Dox Sandstone, lower middle*; 43, *Dox Sandstone, middle*; 44,

Dox Sandstone, upper middle; 45, Dox Sandstone, upper; 46, Cardenas lavas (Uppermost Unkar Group, 1065 RS); 47, Sandstones associated with Cardenas lavas; 48, Nankoweap Formation (<Unkar Group); 49, Nankoweap Formation, upper member; 50, Nonesuch Shale, two poles (1065–1025 RS) and Fond du Lac, two poles; 51, Pikes Peak Granite (1040–1020, mean RS and KA); 52, Michigan basin borehole red beds; 53, Freda and Nonesuch Sandstones, two poles (1046 RS); 54, Freda and Nonesuch secondary component; 55, Jacobsville sediments, JIC, JIB, JIA and JISA components; 56, Grenville gneisses and pyroxenite, Ontario and Quebec, two Poles: 56, Shabogamo gabbro near Grenville Front; 57, Rapitan Formation X component (900–825 KA); 58, Cheneaux metagabbro, Grenville components, two poles (<1310); 59, Mealy dykes, Labrador; normal and reversed groups (1380 RS, 1222–1078 KA, 964–955 KAB); 60, Whitestone anorthosite WY, WX, WW and Z components (1100–1025 UP, 1000 KA, RS, 980 AAH, 945 AAB); 61, Sudbury Complex N7 Grenville uplift remanence; 62, Sudbury dyke swarm remagnetised at Grenville Front; 63, Umfraville gabbro, two poles (1180 UP); 64, River Valley anorthosite-Fall Lake Complex; 65, Michael gabbro (1488 RS, 2080–974 KA, Grenville magnetisation?); 66, Thanet Gabbro, high blocking temperature component; 67, Croteau Group (1563 RS, Grenville remagnetisation?); 68, Seal Group Volcanics, A magnetisation; 69, Seal Group diabases (Grenville remagnetisation?); 70, Town Mountain Granite (1050–1000 RS, KA); 71, St. Urbain anorthosite; 72, Grenville Front anorthosite; 73, Chibougamau greenstone, ch (Grenville?) overprint; 74, Indian Head anorthosite (900, 830 KAB); 75, Michigan Basin borehole, igneous unit; 76, Lake St. Jean anorthosite, normal and reversed groups, two poles; 77, Wilberforce Pyroxenite; 78, Montagnais dykes; 79, Frontenac NE dykes; 80, Magnetawan metasediments; 81, French River anorthosite, A remanence; 82, Larrimac and Bryson diorites; 83, Mealy dykes, secondary remanence (<59): 84, Haliburton intrusion, A component (960 AA); 85, Allard Lake anorthosite (1000 KA); 86, Morin anorthosite, high coercivity component. LAURENTIAN SHIELD, NW SCOTLAND SECTOR: Lower Torridonian, Stoer Group, poles listed in approximate stratigraphic order: 87, Main outcrop; 88, Rubha Reidh; 89, Enard Bay; 90, Achiltiebuie; 91, Stoer Bay. Upper Torridonian, Diabaig, Applecross and Aultbea Formations, listed in approximate stratigraphic order: 92, Diabaig Formation; 93, Cape Wrath; 94, Applecross Formation, lower sites; 95,96, Applecross Formation, Coigach, lower and upper sites; 97, Applecross Formation, Loch Maree; 98, Aultbea Formation, Coigach; 99, Aultbea Formation, main outcrop. LAURENTIAN SHIELD, GREENLAND SECTOR: 1, Ilímaussaq intrusion, marginal syenites (1144 RS); 2, Ilímaussaq intrusion, interior fractionated rocks (1126 RS). FENNOSCANDIAN SHIELD: 1, Bamble sector, B1 and B2 remanences, two poles (1090, 1010 KA, <1139 RS); 2, Brätton-Ålgon norite-anorthosite (ca. 1030–1005 KAM); 3, post-Ålgon dolerite dyke; 4, Bamble sector A remanence, three poles (1090–1010 KA, <1139 RS); 5 Bamble sector basement rocks (1120–975 Ma); 6, Bamble-Köngsberg amphibolites; 7, Rogaland anorthosite complex, group mean and anomalous sites, two poles; 8, Egersund-Ogna and Haaland-Helleren anorthosite; 9, Haaland-Helleren massif and anomalous sites, two poles; 10, Egersund-Ogna massif; 11, Hidra body; 12 Aana-Sira massif anorthosite (940–900 KA, RS); 13, Farsundite intrusion, northern part and southern part, two poles; 14, Garsaknatt body and bordering migmatites, two poles; 15, Bjerkrem-Sogndal lopolith, phases 1, 2 + 3, and 4, three poles. AFRICAN SHIELD: 1, National Kimberlite pipe (1170 RS); 2, Montrose kimberlite pipe; 3, Waterberg dolerites (1250–600 KA); 4, Umkondo dolerites (1040 RS, 1150–650 KA); 5, Premier Mine kimberlite (1220–1170 RS); 6, Group de Char primary component (1019 RS), 7, Umkondo lavas (1115 RS); 8, Guperas lavas (<1250 RS); 9, Auborus Formation (ca. 1020 Ma); 10, Koras Group, Kalkpunt Formation (1200–1000 Ma); 11,

Kanangono kimberlite, Ivory Coast; 12, *Chela Group, Angola (1100–1050);* 13, *Nosib Group NQ1 component (<1010 UP, >844 RS);* 14, *O'okiep intrusions (1020 RS, 1090–1050 UP).* 15, *Bukoban Sandstone (>1000–990 Ma);* 16, *Abercorn Sandstone;* 17, *Klein Karas dykes (890 KA);* 18, *Kigonero Flags (890 Ma);* 19, *Mbala dolerites (940–678 KA);* 20, *Bukoban dolerites (860,810 KA);* 21, *Gagwe amygdaloidal lavas (960, 820 KA);* 22, *Manyovu red beds (<21);* 23, *Malagarasi Sandstone;* 24, *Florida Formation, secondary component;* 25, *Marico River intrusions. AUSTRALIAN SHIELD: 1, IB intrusions (1149 RS);* 2, *Giles Complex (1060 RS, 1250–1140 RS, on adjacent volcanics). INDIAN SHIELD: 1, Newer dolerites Singbhum, minor group (1600–900 Ma);* 2, *Wajrakarus Pipe 2 (1140, 840 KA);* 3, *Mundwara complex basalt (>950 Ma);* 4, *Tirupati dyke;* 5, *Veldurthi hematites (1200–900);* 6, *Chitloor dyke (1200–1100 Ma);* 7, *Cuddapah ferruginous sandstone (1400–1100 Ma);* 8, *Group 2 dykes, Dharwar craton;* 9, *Muligiripalle kimberlite pipe 5 (1140 KA);* 10, *Lattavara kimberlite pipe (1290–950 RS, 1140, 840 KA). SIBERIAN PLATE: 1, North Anabar dolerites (1135, 912, 820 KA);* 2, *Uluntui group, Northern Baikal;* 3, *Yenisei group, kuznets alatau; Yenisei Supergroup (1320–744 Ma):* 4, *Gorbilok Group, Sukhopit Series;* 5, *Uderei group, Sukhopit Series;* 6, *Pogoryui Group, Sukhopit Series;* 7, *Potoskui Group, Tingusik Series;* 8, *Shuntar Group, Tingusik Series;* 9, *Kirgitei Group, Tingusik Series;* 10, *Nizheangara Group;* 11, *Kokin Creek suite (1140 Ma);* 12, *Burovaya suite, lower Tunguska (925);* 13, *Hematite ores, lower Angara suite (925–745 Ma). EAST ANTARCTICA SHIELD: 1, Vestfold Hills, dolerite dykes (1030 RS). Lambert equal area projection centred at 200°E, 0°N, the format, symbols and abbreviations are the same as for Figure 7.1.*

---

deposition. Rb–Sr ages from sedimentary rocks may be too young if diagenetic reactions have taken place between the mineral phases, and Perry & Turekian (1974) have investigated Cenozoic sediments to show that isotopic homogenisation is not complete even at depths of 5500 m in Miocene shales. If this interpretation of Torridonian data is correct, the Lower Torridonian poles (**87–91**) correlate with the initial part of the normally-magnetised segment defined by the Keweenawan data, while the Upper Torridonian poles (**92–99**) overlap with the segment defined by the Jacobsville sediments. In common with the latter succession, it is predominantly normally-magnetised, although reversals are again encountered in the higher members of both successions; the highest members of the Torridonian also define the migration towards the succeeding Loop 8. A considerable number of poles from the African Shield fall along the return limb of Loop 7. They include results from kimberlite bodies (**1,2,5**), from basaltic lavas (**7,8**) and intrusions (**3,4**) in Southern Africa, and from sedimentary and igneous rocks of the Namaqualand belt in Namibia and Angola (**9,10,12,13**).

The data from the Laurentian Shield, covering the succeeding interval 1000–800 Ma come almost exclusively from the Grenville Province. This is a terrane bordering the SE margin of the shield which was subjected to medium- to high-grade metamorphism at 1100–1000 Ma; it is separated from the adjoining Superior Province by a narrow tectonic boundary, referred to as the Grenville Front (Fig. 9.1). The palaeomagnetic results comprise a coherent data set, separate from other Precambrian data on the shield. There appear to be two possible explanations for this observation: either

1. The Grenville Province is an integral part of the shield, where the magnetisa-

tions record a time period unrepresented elsewhere (model B in third part of Section 6.6), or

2. The province is an allochthonous terrane sutured to the shield after the magnetisations were acquired (Model A). It is recognised that the Grenville Front cannot itself define this suture, because the rocks of the Superior Province continue directly across this boundary into their metamorphosed equivalents in the Grenville Province. However, a suture might lie within the interior of the province, where the affinities of the rocks are less clear.

The remanence record in the Grenville terrane is largely a post-metamorphic phenomenon acquired during uplift and cooling; it is frequently multicomponent, and an example of one such study has been described in the concluding part of Section 6.6. The age of these magnetisations is critical to a resolution between the tectonic models. Some analyses have argued that the Grenville magnetisations are as old as the Keweenawan record, implying that the two terranes were separated by *ca.* 5000 km at 1150 Ma, and were sutured to one another at *ca.* 1100 Ma, during the peak of the Grenville episode (Palmer & Carmichael 1973; Irving, Emslie & Ueno, 1974). More recently, ^{39}Ar:^{40}Ar studies have linked the Grenville magnetisations more precisely to the cooling history defined by isotopic closure, and have suggested remanence ages in the range 980–820 Ma, and no older than 1000 Ma (Berger, York & Dunlop, 1979; Dallmeyer & Sutter, 1980). McWilliams & Dunlop (1978) have assumed that — with a few exceptions — the magnetisations in the province follow the K–Ar mica ages (although they are unlikely to be precisely correlated) and have calculated regional poles from results falling within K–Ar thermochron zones (Baer, 1976). The time sequence of poles then defines a flattened APW loop (Loop 8) with similar outward and return paths, which is later than, but follows on from, the Keweenawan record describing the latter part of Loop 7.

The fact that the remanence record is now generally recognised to be younger than the Grenville mobile episode and the development of the Front Zone does not, of course, preclude the possibility that a suturing of the province to the remainder of the shield could have taken place prior to 1100 Ma. However, the absence of a visible suture and the recently-identified presence of a relict Hudsonian-like remanence record within the Grenville terrane (Stupavsky & Symons, 1982) strongly suggest that this terrane has always formed an integral part of the shield. Nevertheless, it remains true that a palaeomagnetic record contemporaneous with the Grenville record is largely unidentified elsewhere in the shield; the possible exceptions are certain secondary magnetisations in late Keweenawan sediments (Watts, 1981) and igneous units (pole **75** in Fig. 7.8), the extension of the APWP from the Grand Canyon succession (Elston & Grommé, 1974), and possibly a meteoritic shock-induced remanence in the Lake Superior area (Halls, 1975). Attention has already been drawn to the sequences of multicomponent magnetisations in the Grenville record and, in some instances, the latter part of these sequences record the return limb of Loop 8; examples include the Haliburton A → C components (Buchan & Dunlop, 1976) and the Morin components (Irving, Park & Emslie, 1974). Since the outward and return limbs of this loop are so close (and whether it is a clockwise or anticlockwise loop has been disputed), an attempt is made in Fig. 7.8 to define the outward path by assigning only the higher coercivity/blocking-temperature components to this limb. The lower coercivity/

blocking-temperature components are assigned to the return limb, and discussed in the context of the late Proterozoic record in Section 9.2.

In the Fennoscandian Shield, a medium- to high-grade metamorphic terrane: the Sveconorwegian Belt, is situated along the margin of the shield and lies in an analogous peripheral situation to the Grenville Province in the Laurentian Shield (see Fig. 9.1). It is also separated by a narrow tectonic boundary from the adjoining shield, and it was similarly subjected to tectonothermal reworking at 1100–1000 Ma. However, a more ancient history is more widely recognised here, with the associated metamorphic events overprinting an older Svecofennian terrane (e.g. Field & Raheim, 1981). The palaeomagnetic results from this region are linked to the post-metamorphic uplift history (Stearn & Piper, 1984) and dated at ca. 1050–840 Ma. However, this body of data is no longer compatible with the primitive reconstruction of the Laurentian and Fennoscandian Shields (Patchett, Bylund & Upton, 1978) but fits a contiguous reconstruction, with the Fennoscandian Shield rotated clockwise by 60° about a local pole of rotation (Patchett & Bylund, 1977). This is the first specific indication in the Proterozoic palaeomagnetic record of relative moments between the shields, and is described in more detail in Chapter 9. With the exception of a remanence from the high-grade Bamble Sector of the belt in SE Norway (**1**) which is dated at 1090–1010 Ma, and may pre-date this movement, the Fennoscandian data are plotted in Fig. 7.8, with this shield rotated according to the modified reconstruction (Section 9.1). Insufficient K–Ar mineral ages are available to define a post-tectonic cooling history, and to analyse the data on an areal basis in the manner of the Grenville Province analysis, but a distribution of pole positions has emerged which is compatible with the flattened APW loop defined from Grenville data. Possibly the oldest remanences have been recovered from the Bamble Sector (**4 → 6**), and they are linked to ages ranging 1090–1000 Ma, while other data from near the Front Zone correlate with the earliest part of Loop 8 (**2**). The bulk of the data come from the Rogaland anorthosite and related intrusions, from the interior of the belt in SW Norway (**7–15**). These poles collectively plot along a 30° arc, and both geological relationships and blocking-temperature data define this as an arcuate anticlockwise swathe in Fig. 7.8, dated at 960–900 Ma (Stearn & Piper, 1984).

The outward limb of Loop 8 is also defined in the African Shield from the stratigraphic sequence of poles from sediments and intrusions within the Bukoban system of Tanzania (poles **15 → 18 → 20 + 21 → 22 + 23**; Piper, 1972) and Zambia (**16,19**); miscellaneous data from southern Africa correlate only with this segment of the APWP, although age data are inadequate to confirm these assignments.

A major group of poles from the Yenisei Ridge supergroup in the Siberian Shield (**3–10**; Vlasov & Popova, 1968) plot on the outward or return limbs of Loop 8. Although these poles apparently sample a succession some 5000 m thick, and have maximum K–Ar age limits ranging 1320–750 Ma, the close agreement of these poles with each other and with only a limited (< 100 Ma) segment of the APWP, implies that either these sediments were deposited within a relatively short interval of time, or that they were entirely overprinted at some later time. The former possibility is most likely, because at least four successive reversals are present, and an overprinting event might be expected to yield a narrow spectrum of age dates. It is noteworthy that palaeomagnetic data from the Proterozoic sedimentary successions (Beltian, Grand Canyon, Keweenawan, Torridonian and

Yenisei Ridge) all imply rapid deposition of several thousands of metres of sediments, within periods of no more than a few tens of millions of years (see also Sections 7.6 & 7.8).

Although a considerable number of Upper Proterozoic poles from India plot along the later part of Loop 7 and the early part of Loop 8, the age control is generally so poor that few conclusions can be drawn. However, the reconstruction appears to be valid to *ca.* 1000 Ma. Most noteworthy is the conformity of late hematite mineralisation in the Lower Vindhyan, with the path at *ca.* 1000 Ma and the agreement of *ca.* 1050 Ma kimberlite poles (**2,9,10**) with data from broadly-contemporaneous kimberlite pipes in the Laurentian and African Shields. This agreement may have a wider genetic significance noted in Section 9.1. The oldest palaeomagnetic pole from Antarctica falls on the downward limb of Loop 8, in accord with the assigned age of 1020 Ma, although the error limits on this age are large (± 200 Ma).

It remains to note the possible significance of certain data points linking Loops 7 and 8 and representing the connection between the Keweenawan and Grenville data sets in Laurentia. It has already been observed that the Jacobsville and Torridonian data imply a short southerly extension to Loop 7, while the highest Torridonian data and oldest overprints in the higher Keweenawan sediments form part of an E → W latitudinal movement. The extent of overprinting north of the Grenville Front is remarkably limited (Halls & Pesonen, 1983) but is recognised as a later overprint in the Sudbury complex (**61**) and by the remagnetisation of the *ca.* 1200 Ma Sudbury dykes as they cross the Front Zone (**62**); these poles, in common with results from some other intrusions near the Front (**63,64,72**), lie close to the beginning of Loop 8, and probably record events near the peak of Grenville tectonism. In addition, a result from an intrusion (**65**) close to the Grenville front in Labrador, and results from sediments and lavas of the Naskaupi Fold Belt (**67–69**) adjacent to the Front in the same area, are plotted here following Irving & McGlynn (1976b), because these poles are totally discordant with the APWP at their estimated ages of formation (1550–1350 Ma). Rotation about a local vertical axis during the development of this belt is a possible explanation (Roy & Fahrig, 1973), but it is most probable that these rocks were overprinted during the Grenville episode, in common with many other rock units within 30 km of the Front Zone (Irving & McGlynn, 1976b).

7.9 Reversal asymmetry and APW

Detailed investigations of the Keweenawan have highlighted a phenomenon which may both influence the definition of APW and have significant implications for the nature of the Earth's core in Proterozoic times. The inversion of field polarity identified in the lava successions here (and also in the dyke swarms) are not precisely antiparallel. There is an average inclination difference (ΔI) of $25 \pm 10°$ between the N and R directions, with the effect that one group of poles (derived from the steeper inclinations) plots closer to the sample area than the group of opposing polarity (Fig. 7.8). There are four possible causes for this anomaly (Nevanlinna & Pesonen, 1983):
1. The presence of an unremoved secondary component, superimposed on direc-

tions which were initially antiparallel: whilst studies of the Copper Harbour conglomerate have shown that a secondary component is indeed present, it has been found to have the same direction as overlying normal polarity volcanics (Palmer, Halls & Pesonen, 1981) and can therefore, not explain the observed discrepancy. Arguments based on fold tests and palaeo-intensity data can also be used to refute a secondary component as the cause in this case (Halls & Pesonen, 1983).

2. APW movement during field inversion: initially this would seem to be supported, because the direction of polar movement across the inversion is the same as that in the preceding and succeeding zones of constant polarity (Fig. 7.8). However, this case is refuted if the anomaly is maintained through more than one field inversion. Palmer (1970) originally reported two R → N polarity changes in the Mamainse lavas, but since both inversions occur in the first lavas above a major conglomerate, he proposed that the succession was duplicated here by faulting. Subsequent studies have suggested that these inversions are present in geochemically-distinct lava successions (Massey, 1979), and are separated by contrasting conglomerates (Halls & Pesonen, 1983). Hence the weight of the geological evidence favours a more complex polarity history (at least R → N → R → N) than is implied by the pattern of polarities in Fig. 7.8. Furthermore, the *ca.* 1250 Ma Gardar lava succession (**31** and **32**; Fig. 7.7) yields a reversal asymmetry of $20 \pm 5°$, which is maintained through five successive inversions of the magnetic field (Piper, 1977); the geological relationships are unequivocal here. Continental movements are therefore excluded as the principal explanation for this phenomenon.

3. A single offset dipole model: asymmetrical inversions have been observed globally in rocks of Cenozoic age (Merrill & McElhinny, 1983). Wilson (1970, 1971) has shown that this asymmetry could be produced by different offsets of the axial dipole from the centre of the Earth during the N and R epochs. To account for the Keweenawan asymmetry in terms of this model, the R dipole would need to have been located *ca.* 2000 km south of the geocentre if the N dipole was itself geocentric. On the assumption that the R and N dipoles have the same strength, this model predicts a stronger intensity for the N field than the R field, which is contrary to palaeo-intensity results (Halls & Pesonen, 1983). The single offset dipole model does not therefore appear to be applicable in this instance.

4. A coaxial two-dipole source model: according to Nevanalinna & Pesonen (1983) this is the most probable explanation of the Keweenawan and Gardar results. In one form it comprises a geocentric dipole and an offset dipole located at the core–mantle boundary. An asymmetrical inversion occurs when the geocentric dipole inverts, whilst the offset dipole retains a constant polarity. A reversal of this system causes a significant change in field magnitude: Halls & Pesonen (1983) have examined palaeo-intensity data from the Keweenawan to show that the steeply-magnetised reversed rocks appear to have been magnetised in a field with a magnitude 40% higher than the more shallowly-magnetised normal rocks. A similar difference is also present in the Gardar lavas, where the more shallow N components yield palaeo-intensities some 37% lower than the steeper R components (G. K. Taylor and author unpublished data).

One consequence of this phenomenon is that the standard dipole assumption is inapplicable in this instance and the dispersions of the N and R poles are not a true representation of APW. Nevanlinna & Pesonen (1983) suggest that the Keweenawan data have enhanced the depth of Loop 7 by some 50%, although the inclusion of data from elsewhere, and the definition of the outward limb of Loop 6 by the N polarity Greenlandic data in Fig. 7.7 imply either that this is a considerable overestimate, or that the effect is more local than predicted by the two-dipole source model. The contemporaneous Grand Canyon succession defines a smaller loop than the total data set (Elston & Grommé, 1974). This could reflect the absence of sediments spanning the interval represented by the apex of the loop, although further studies of the Unkar group identify an asymmetry in the same sense as that observed in the Keweenawan (data of Elston and Grommé reported by Halls & Pesonen, 1983). It is also possible that some of the normally magnetised studies include incompletely-subtracted Recent components which have enhanced the depth of Loop 7. The contribution of more complex dipole sources to the definition of the overall Precambrian APWP is unclear, partly because of the preponderance of a single (probably N) polarity over much of this time. In addition to the Gardar (*ca.* 1250 Ma) and Keweenawan (*ca.* 1140 Ma) examples, smaller asymmetries between N and R axes are recognised in some *ca.* 1500 Ma data. The poles derived from the separate N and R data sets are plotted in Fig. 7.6 to give a measure of this effect. They are also present in the Yenisei Ridge supergroup of Siberia (Fig. 7.8). Attention has not been drawn to this effect in Middle and Lower Proterozoic times, and some nearly-exact inversions have been described (e.g. Bingham & Evans, 1975). Hence, there is some suggestion here for buildup of a more complex source, or sources, through the later Proterozoic, to reach a maximum effect at about Keweenawan times. However, to properly evaluate this model and its effects will require the complete definition of contemporaneous segments of APW from several different shields, and much more work will be necessary to achieve this.

7.10 Tectonic implications

The most impressive feature of Figs. 7.1 & 7.3 to 7.8 is the tight conformity of the Precambrian data set to a unique APWP when only a single reconstruction, which is completely rigid in palaeomagnetic terms, is employed. Unless the palaeomagnetic investigations have completely evaluated the nature of the remanence, it has been assumed in this analysis that the remanence age is the age of the rock unit. This would be a quite unacceptable assumption in, for example, Palaeozoic sediments (Section 10.4), but it works well in the Precambrian, partly because so many of the data are derived from more suitable igneous rocks. A preponderance of data from stable cratonic interiors, and a simple diagenetic history in the sediments have contributed to the value of this data set for tectonic analyses. The analysis of Figs. 7.1 to 7.8 represents a gross average solution to the problem by demonstrating that the weight of the evidence conforms to the model. It must be stressed however, that interpretations of some individual results are unlikely to be correct, and both the positions and ages of individual poles used in this analysis will need to be revised or rejected in the light of future work. An examination of the development of Proterozoic APWPs over the past decade or so (cf. Robertson & Fahrig, 1971; Irving & McGlynn, 1976b; Piper 1976a, 1982a) demonstrates how the interpretations placed on individual poles have changed (especially in the light of new geochronological data) while the APW paths themselves have survived. In part, this is because the APW loops have degrees of repeated symmetry and radiate outwards from a common 'stable' position. This position is approximately in the central Pacific with respect to Laurentia and was repeatedly occupied by the pole (Irving & McGlynn 1976b; Morris, Schmidt & Roy 1979): the assignment of a pole lying in this general area to a particular point on the APWP is ambiguous if the age limits are broad (Fig. 7.9).

The general pattern of APW in Figs. 7.1 to 7.8 is efficiently described by a swathe *ca.* 20° in width. The actual dispersion of poles about this swathe is often defined by the data from one shield only, and may result from one or more effects including:
1. Incomplete subtraction of secondary or multiple components.
2. Non-dipole components in the geomagnetic field source (Section 7.9).
3. Broad warping of the basement terranes not evident from the geological evidence (e.g. Morgan, 1976; Piper, 1981a);
4. Complex short-term APW movements, and
5. Regional and local rotations about fault planes such as the Nipissing D example noted in Section 7.3.

For these reasons it is not possible to put constraints on the width that the swathe should have. An example of the possible influence of these factors on the definition of APW is illustrated by Halls & Pesonen (1983): by filtering pole populations according to certain reliability criteria they reduce the width of the path defined by the Keweenawan data in Fig. 7.8 from *ca.* 20° to *ca.* 10° of arc, while maintaining the length of the APWP. In view of the multiplicity of causes which can contribute to 'noise' in the data analysis, it is perhaps remarkable that the Precambrian data set shows such a close conformity to the single long APWP. It is the weight of this large and heterogeneous data set which *requires* that the continental crust behaved as a single, coherent unit throughout Proterozoic times; the details have more varied and complex interpretations which do not form part of this global synthesis, but which may nevertheless, have important regional and local geological and geophysical implications.

The major tectonic implication of this analysis is that the crust behaved as a rigid unit on the gross scale 'seen' by the palaeomagnetic data. This conclusion is reinforced by the movement of the APWP into contrasting areas with a general trend from east to west through Proterozoic times (Irving & McGlynn, 1976b and Fig. 7.9). This not only emphasises the limited importance of later thermal and tectonic events as remagnetisation events, but overcomes the limitation of short APW segments, namely that such movements can escape detection if the palaeomagnetic pole and the Eulerian pole of rotation are close together (third part of Section 6.6). The conformity of the post-2600 Ma data from Africa, Australia and Laurentia must necessarily summarise the situation for the shields in between. Model A for crustal tectonics (third part of Section 6.6) is therefore positively excluded in Proterozoic times. Model C is probably also excluded; it is not possible to be sure of this because the record is incomplete for some shields, but it is practically inconceivable that segments of crust could break apart and then come together again in the same unique reconstruction to satisfy the single APWP.

Although the palaeomagnetic data are now able to demonstrate that the con-

Fig. 7.9 Summary APWP of the Proterozoic Supercontinent between 2200 and 800 Ma. The dated points are the approximate positions of the Eulerian poles of rotation between 270 and 1000 Ma; see text for explanation.

tinental crust was constrained into a single body by large scale mantle driving forces during most of Proterozoic times, it should be stressed that there is a smaller second order of relative motions to which they appear to be insensitive. These smaller motions are of two types. Firstly, the data do not reveal the undoubted deformation of the crust along mobile belts during these times: while large scale strike slip motions according to model D are excluded, it has yet to be determined whether some smaller differences are palaeomagnetically-detectable and interpretable in the context of this model (Onstott and Hargraves 1981). Secondly, the palaeomagnetic data have not yet identified the opening and closing of small intershield ocean basins although some examples are described from late Precambrian times and there appear to be one or two older examples (section 8.5). Conventionally, the agreement of contemporaneous poles from two regions is taken to restrict the later relative motions between them to less than *ca.* 1000 km and rotational movements to less than 10° although the collective precision of the larger data sets used in the preceding analysis must render these estimates maxima.

7.11 Precambrian APW and rates of continental movement

A cursory examination of the APWP in Figs. 7.1 and 7.3 to 7.8 reveals a high degree of symmetry in the motion of the primary crust with respect to the poles. The characteristic signature is that of a closed loop, with a 'hairpin' bend at the apex, and radiating outwards from a general position which is repeatedly crossed by the path. This pattern is clear from the summary path in Fig. 7.9; it has been evident since the early analysis of Spall (1971), and the general concept has been reinforced and extended since then. The tracks and the hairpins have frequently been assigned names (mainly from Laurentian locations) in previous analyses. This no longer seems to be appropriate, both because results from the geological provinces used to name a track are seldom restricted to one track, and because the tracks and hairpins are now seen to apply to much larger areas. Furthermore no direct relationship between 'hairpins' and Precambrian orogeny can now be recognised (cf. Irving & Park, 1972b, and Section 12.8). Accordingly, it seems more appropriate to number these loops in accordance with their sequence through Proterozoic times and refer to the hairpins in terms of their ages; in common with the nomenclature of magnetostratigraphic divisions (Section 12.3), this is a simpler and more instructive procedure.

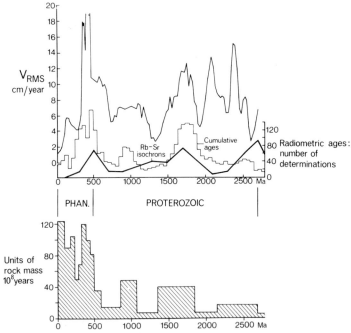

Fig. 7.10 *The root mean square velocity of the continental crust, derived from the APWPs described in this chapter and Chapter 9; the Phanerozoic section of this graph is representative of Gondwanaland. The exceptionally high velocity in Upper Ordovician to Lower Devonian times is based on the conservative interpretation of a sparse palaeomagnetic record discussed in Sections 10.2 and 10.3, which (in common with the large APW movements during Lower Cambrian times (Section 9.4)) represents the biggest uncertainty in the APW path. The V_{RMS} curve is compared with a cumulative histogram of radiometric age determinations after Dearnley (1966) and a histogram of Rb/Sr age determinations after Glikson (1983). The volumetric changes in the sedimentary rock mass are derived from Garrels & Mackenzie, (1971).*

This analysis has identified eight loops prior to 800 Ma: Loop 1 (2550–2350 Ma), Loop 2 (2350–2200 Ma), Loop 3 (2100–1950 Ma), Loop 4 (1850–1700 Ma), Loop 5 (1700–1600 Ma), Loop 6 (1600–1400 Ma), Loop 7 (1200–1000 Ma) and Loop 8 (1000–800 Ma). Although further work will no doubt necessitate a revision of the shape and time duration of these loops, it is already apparent that they all have a period of *ca.* 200 Ma. There is also a degree of symmetry, with each loop radiating from a point which is located near to the geometrical centre of the continental crust and which changes little through Proterozoic times (Fig. 7.9). Furthermore, there is a gradual temporal change in the path taken by the loops, so that they change from having a strong component of longitudinal motion in Loops 3 and 4 to a mainly latitudinal motion in Loops 5, 6, 7 and 8, accompanying a movement into a progressively more westerly sector. Finally, in late Precambrian times the loops radiate along still more westerly azimuths, contrasting with all previous paths (Fig. 7.8 and Sections 9.2–9.4).

The movements implied by the APW loops can be modelled in terms of rotation about a Eulerian pole. Assuming that the APW signature entirely reflects a movement of the continental crust, these movements must comprise a sequence of clockwise and anticlockwise rotations about the Eulerian poles, with the approximate ages and positions shown in Fig. 7.9. The hairpins at the extremity of each loop mark reversals in the sense of rotation about the Eulerian pole, or a change in the position of the pole, or both. As already noted, most of these Proterozoic loops have similar outward and return tracks, implying a reversal in the sense of rotation, accompanied by only a small change in the position of the pole of rotation. There is a major change in the style of APW motion after 2150 Ma (cf. Fig. 7.1 & 7.3–7.8) reflected in a change in the typical position of the Eulerian poles, but after 2150 Ma, until late Precambrian times, almost all of the Eulerian poles lay within the area of the continental crust, either near to its geometrical centre or towards the extremity. This observation requires that the continental crust was pivoted about a local pole of rotation, and was not translated by large distances about a remote pole of rotation. This implies, firstly that the continental velocities calculated by the method discussion in the second part of Section 6.6 will approach the true velocities, and secondly that the position of the continental crust on the globe was not a random phenomenon imparted by developments in the much more mobile oceanic lithosphere (which have gone practically unrecorded in the geological record), but was constrained by more fundamental causes (Section 12.11). This observation has important implications for the nature of Proterozoic continental motion.

The possible implications of these observations will be examined in more detail in Chapter 12. Here, the importance is noted of one other facet of Precambrian APW, namely the significance of the changing velocity of continental movement. Geochronological studies since the 1950s have suggested that geological events did not occur uniformly throughout geological time. Instead the frequency distribution of age dates identifies several well-defined peaks at 2750 ± 50 Ma, 1950 ± 50 Ma and 1075 ± 50 Ma (Gastil, 1960; Dearnley, 1966, & Fig. 7.10). This implies that heat release from the Earth's interior into the continental crust has been pulsatory rather than uniform. Although the plotting of age data as frequency histograms has been criticised because many of the ages refer to cooling and overprinting events, and because dating studies have tended to concentrate on the rocks with more suitable compositions, other assessments (e.g. York & Farquhar, 1972) have stressed that these peaks in the histogram are real effects. Their significance is now better understood: a cumulative plot of Rb–Sr isochrons identifies peaks at 2800–2600 Ma and 2000–1600 Ma (Glikson, 1983) as periods of new crustal formation (Section 8.2, and Moorbath, 1977), while the *ca.* 1075 Ma peak in K–Ar ages does not appear, and is identified as predominantly a thermal event overprinting pre-existing crust. The thermal-tectonic episodes recorded by these ages will probably promote crustal thickening, uplift and erosion. It is also observed that the volumes of sedimentary rocks in the Precambrian record correlate with the cumulative age histogram (Fig. 7.10), although it should be stressed that the changes in the sedimentary rock mass during Precambrian times are very imperfectly understood at the present time.

It is further to be expected that the broad episodes of heat release defined by the geochronological data would be reflected in a more mobile asthenosphere and more rapid movements of the continental crust. In Fig. 7.10 the minimum V_{RMS} values computed from the Proterozoic APWP (Section 6.6) show that this prediction is borne out: APW rates were generally higher during the periods of thermotectonic activity and lower during quiescent periods in between. A more specific feature of the APW rate curve is the low V_{RMS} rates at 1850 Ma; this probably represents a rather short minimum of 50–100 Ma which would not be identified in the histogram analysis of the age data. The periodicity of the age frequency/sediment volume/V_{RMS} cycle is some 800 Ma, and is four times as long as the period of the APW loops; it suggests that the mechanisms controlling the two phenomena are not directly linked. Although APWPs have been widely justified in the literature, on the grounds that they yield rates of APW comparable to the present, such arguments are clearly meaningless, because not only have rates fluctuated considerably, but the Proterozoic rates have sometimes been exceeded by Phanerozoic rates, and notably by a very rapid movement between late Ordovician and Devonian times (Sections 10.2 & 10.3). The gross features of the V_{RMS} curve in Fig. 7.10 probably reflect the changing styles of plate tectonics through geological time and are examined further in Chapters 8 & 12.

8
Growth and consolidation of the continental crust in Archaean and Proterozoic times

8.1 Introduction

Chapter 7 has shown that the Proterozoic palaeomagnetic record conforms to a single palaeomagnetically-rigid reconstruction. Although it has been argued that alternative tectonic models are now excluded by the data, they are well-known to be applicable to Phanerozoic times and are evaluated in some detail in Chapters 9, 10 and 11; furthermore, the sparse data from pre-2800 Ma suggests that an alternative tectonic régime was also applicable to these times. This chapter is devoted to testing the predictions of the palaeomagnetic data. The analysis can be approached in two ways: firstly the geochemical and isotopic signature of Proterozoic times should contrast with the preceding Archaean and the succeeding Phanerozoic eons in a way which is consistent with a dominance of Proterozoic régimes by a single coherent continental crust. Secondly, the palaeomagnetic data predict the shield configurations and the form of the primary crust, so that distributions of tectonic and magmatic features can be tested to see whether they conflict or conform with this reconstruction.

8.2 Geochemical and isotopic signatures

Recent geochemical and isotopic evidence has refuted the view that the continental crust has grown linearly or exponentially with time (e.g. Hurley & Rand, 1969) and favours the view that the bulk of the present continental crust formed in late Archaean and early Proterozoic times. Trace-element distributions in sedimentary rocks have been used to model changes in the area of continental crust, and indicate that 65–75% of the continental area formed between 3200 and 2500 Ma (Viezer & Jansen, 1979; McLennan & Taylor, 1983). Collectively, the isotopic and geochemical evidence imply that some 70–85% of the present crustal volume had formed by 2500 Ma. It is probably still growing at the present time, and a small, but disputed, fraction is no doubt returned to the mantle in subduction zones; however, the present rate of crustal growth (estimated to be in the range $0 \cdot 5 - 1 \cdot 0$ km^3/year (Brown 1977) is less than 10% of the rates of crustal growth achieved in late Archaean times.

The low-density rocks forming the continental crust are the result of partial melting and fractionation of silicic melts from basaltic and hydrous komatiitic parent materials, and their buoyancy generally precludes subduction back into the mantle. Two distinct silicic end-members of melting are observed in Archaean terranes. Na-rich tonalite-trondjhemite rocks occur with the earliest rocks in greenstone successions and are dominant in the early Archaean, while K-rich granites are most prominent in the youngest rocks of each terrane. The rare earth element (*REE*) patterns for the two suites are distinct: the *REE* evidence from sedimentary rocks produced by erosion of the Archaean terranes indicates that the K-rich granites occupied less than 10% of the area exposed to erosion in early Archaean times, but dominated the upper crust after 2500 Ma (Taylor & McLennan, 1981). Thus, the K-rich granites are interpreted as the products of intracrustal melting, and are intimately related to the growth of the primary crust. Both the Na and K-rich phases ultimately represent new additions of material from the mantle, and cannot be interpreted as the products of long-term recyling of crust of greater antiquity. The most important evidence for this contention is that gneissose terranes yield ^{87}Sr:^{86}Sr ratios coincident with, or slightly above, plausible upper mantle values, and imply that they are direct additions to the crust. Results from diverse areas of the Precambrian crust record primary additions concentrated in the intervals 3800–3500 Ma, 2900–2600 Ma and 1600–1600 Ma (Moorbath, 1977, and see also Fig. 7.10). Growth seems to have followed a two-stage evolution with addition of granodiorite and tonalite melts directly from the mantle, followed by remelting within a few hundreds of millions of years to introduce K-rich granites to the highest levels of the crust (Brown & Hennessey, 1978). The mechanism is not critical to this discussion, but it was accompanied by a spectacular degassing of the mantle (McLennan & Taylor, 1982) which was itself probably necessary for the crust to grow over such a short period of time (Campbell & Taylor, 1983). This degassing was to have an important influence on the subsequent signature of magmatism, tectonics and mineralisation in the crust (Sections 8.4 and 8.7). Essentially the same view of rapid crustal growth by 2500 Ma has emerged from Pb isotope data (Armstrong, 1981), from models for the evolution of the continental freeboard (Wise, 1974), and from seismically-determined thicknesses of the early-formed crust (Condie, 1976).

The general conclusions of these geochemical findings are in direct accord with the palaeomagnetic analysis of Section 7.2, which requires that at least 60% of the present continental *area* had stabilised as a coherent unit by 2700 Ma, and some 80–90% had stabilised by 2500 Ma (Fig. 7.1). In addition, unless the sparse pre-2800 Ma palaeomagnetic record is entirely a later overprint, it implies that relative movements took place between the early-formed Pilbara and Kaapvaal cratons (where K-rich granites appear at 3200 Ma) during the more juvenile phases of crustal development.

The gross geochemical changes through geological times are summarised in Fig. 8.1. The Sr isotope record of marine carbonate rocks has been investigated by Viezer & Compston (1976): the $^{87}Sr:^{86}Sr$ of early Archaean carbonates is similar to, or slightly lower than, upper mantle values, and implies that extensive continental areas were not then exposed to erosion. The dramatic increase at 2600–2500 Ma is best explained in terms of a large increase in the area of Rb-rich continental crust; this was accompanied by a change in the balance of sea-water Sr from predominantly mantle-derived — and related to oceanic basaltic volcanism — during Archaean times, to a largely continental source in Proterozoic times. The ratio grows consistently faster than the mantle growth line throughout Proterozoic times, and then shows a marked downturn at the beginning of Phanerozoic times, with the renewed input of mantle-derived strontium defining new ocean-floor growth between separating crustal plates (see also fifth part of Section 9.6).

The basic chemical contrast between Archaean and Proterozoic silicic igneous rocks is evident from a simple petrogenetic classification on the quartz–albite–orthoclase ternary diagram (Glikson, 1983); this shows that Archaean magmas crystallised largely in the field of plagioclase fractionation, while Proterozoic magmas had chemistries in the K feldspar–plagioclase–quartz low P_{H_2O} eutectic area. These differences reflect either advanced magmatic fractionation of the Proterozoic magmas, or derivation of the latter by partial melting from an already-differentiated source; they can be conveniently illustrated in terms of temporal variations of the K_2O/Na_2O ratios of igneous and sedimentary rocks (and their metamorphic derivatives). This is illustrated in Fig. 8.1, after Engel, Itson, Engel, Stickney & Cray (1974). This distribution again stresses the unique signature of Proterozoic times. It shows a marked downturn in Phanerozoic times, which accompanied the increasing emplacement of rocks of oceanic origin into the continental environment, and the appearance of high-P, low-T blueschists in orogenic belts; the latter are petrochemically analogous to Archaean greenstones, but they have been subducted into deeper and cooler environments beneath peripheral orogenic margins. A further expression of these chemical and petrological changes is the the ratio of quartz monzonite to quartz diorite and tonalite (Fig. 8.1).

Igneous rocks show a wide diversity of *REE* patterns, ranging from light *REE* depletion in mid-ocean ridge basalts to heavy *REE* depletion in late-stage granites. Sediments however, display much more uniform patterns, because the *REE* are completely transferred from source rocks into clastic sediments, and their solubilities and residence times in sea-water are very low. Furthermore, they do not appear to be appreciably disturbed by later post-depositional processes, and they are therefore very important indicators of crustal evolution (Taylor, 1979). The dramatic changes in the *REE* signature in early Proterozoic times are illustrated in Fig. 8.1 in terms of the ratios La/Yb, light *REE*/heavy *REE* and the total *REE* con-

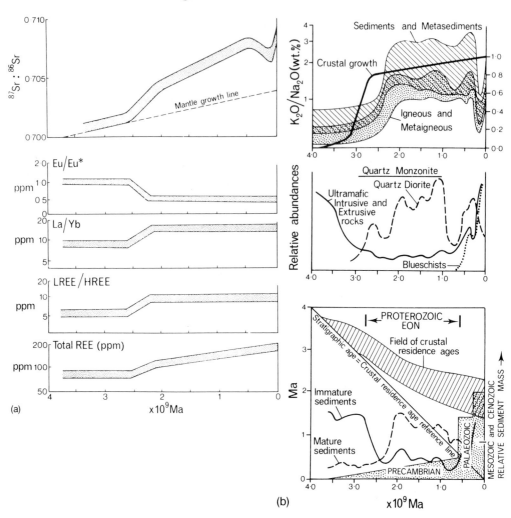

Fig. 8.1 The geological and geochemical signature of the Archaean, Proterozoic and Phanerozoic eons. (a) The evidence from the strontium isotope ratio as derived from the content of marine carbonates and compared with the mantle growth line over the same period (after Viezer & Compston, 1976), and variations in the REE patterns, after Taylor (1979). (b) Variations in the major-element chemistry as typified by $K_2O:Na_2O$ and the relative proportions of some sedimentary and igneous rock types over geological time, after Engel, Itson, Engel, Stickney & Cray (1974). The field of crustal residence ages derived from Sm:Nd studies of sediments is taken from Frost & O'Nions (1984), and is compared with the reference line to be expected for material derived directly from the mantle. This can be compared with the most probable model of crustal growth illustrated in the top diagram after McLennan & Taylor (1983) and represented as an episodic model, with most of the crust forming between 3200 and 2500 Ma. The relative mass of the sedimentary rocks over geological time is derived from Garrels & Mackenzie (1971).

tent in ppm. Europium has a signature distinct from the other *REE* because of the difference in the ionic radii of Eu^{3+} and Eu^{2+}. Archaean sediments show no appreciable Eu anomaly (expressed as Eu/Eu*, where Eu* is the theoretical value for no depletion) but Proterozoic sediments exhibit a marked Eu depletion. This cannot be explained in terms of the buildup of an oxidising atmosphere at *ca.* 2300 Ma, because that effect would favour trivalent Eu^{3+}, which is the reverse of what is observed (Taylor & McLennan, 1981). Hence the Eu anomaly is not due to superficial processes, but is an inherent property of the Proterozoic crust although its origin is not fully understood.

Although the details of these trends are obscured by the coarse stratigraphic groupings, the internal consistency of the curves, and the fact that they were obtained in different shields and sampled by different research groups, argue strongly for their general validity (Viezer & Jansen, 1979). Either the unique nature of the Archaean–Proterozoic transition, or the distinctive signature of the Proterozoic eon, is evident in each case. Without exception, these changes indicate an abrupt increase in continentality near to, or somewhat later than, the Archaean–Proterozoic transition, and a dominance of ensialic processes throughout Proterozoic times; in contrast, the Phanerozoic eon has more analogies with the Archaean.

8.3 The sedimentary record

The distribution of the sedimentary rock mass as a function of time (Fig. 7.10) exhibits a marked change at the beginning of Cambrian times, so that the mass of sediments deposited during Phanerozoic times is roughly equal to the mass of sediments deposited over the whole of the remainder of geological time: the sedimentary rocks preserved at the present time are crowded towards the last part of the geological record (Garrels & Mackenzie, 1971). In part, of course, this is because the chance of a sedimentary rock being destroyed by later erosion increases with the age of the rock, so that this distribution reflects the mass of sediments preserved. However, it is immediately apparent that no steady-state model can explain the abrupt increase in the sedimentary volume in Phanerozoic times. Garrels & Mackenzie (1971) have also shown that no models based on a progressive linear accumulation through time can explain this distribution. It must accordingly reflect the change in the style and tectonics of sedimentation near the Proterozoic/Phanerozoic transition (Section 9.6).

The *content* of this sedimentary mass has also changed (Ronov, 1968; Engel, Itson, Engel, Stickney & Cray, 1974) although these changes can currently only be described in a qualitative way (Fig. 8.1). The development of the Superior type BIF (fifth part of Section 4.2) at *ca.* 2500–2200 Ma is an early indication of broad and tectonically-quiescent continental margins by these times, but is otherwise dependent on environmental changes not directly related to the change in tectonic style. Ronov's (1968) work indicates a replacement of immature sedimentary sequences dominated by greywackes in the Archaean, by more mature sediments (initially more arkosic, later more quartz-rich) in the Proterozoic (see also Fig. 8.1). The deposition of dolomites, commencing in late Archaean times and largely displaced by limestones in Phanerozoic times, is another indicator of the establishment of widespread, stable, marine shelves (see also Section 12.8).

The character and crustal residence age of sediments can also be deduced from Sm–Nd isotopic patterns. These studies show that all sediments deposited since 2000 Ma have crustal residence ages well in excess of their stratigraphic ages; (hence they plot above the 45° reference line on the graph of stratigraphic age against residence age, shown in Fig. 8.1b). This implies that they are dominated by recycled older crustal material (Frost & O'Nions, 1984). In contrast, the limited data from Archaean rocks indicate a very short residence time. The crustal residence ages of modern, Palaeozoic, and 1300 Ma sediments all turn out to be about 2000 Ma, and although this might be an artefact of the limited data base, it is more likely to show that the input of mantle-derived material into the global sedimentary mass has been very limited since Lower Proterozoic times (Frost & O'Nions, 1984). The deposition of the banded iron formations occurred close to this transition, and Nd-isotope studies of the latter (Miller & O'Nions 1985) show that the Proterozoic (Superior-type) examples were supplied by continental weathering, and that a significant mantle contribution is excluded; in contrast the Archaean examples show ^{143}Nd:^{144}Nd ratios indicative of more or less direct derivation from the mantle.

8.4 Greenstone belts and straight belts

The earliest stages in the evolution of the continental crust are preserved in the nuclei stabilised in Archaean times as *greenstone belts* and *high-grade gneiss terranes*. They include successions, often thousands of metres thick, of mantle-derived mafic and ultramafic volcanics with immature sediments, and a component of intermediate and acid volcanics which becomes increasingly important towards the top. The contacts between the greenstone belts and the high-grade gneisses are extensively obscured by later granite intrusion, but it is clear that the high-grade gneisses in most, if not all, cases are older than the greenstone belts. The origins of the gneisses and greenstone belts are extensively discussed elsewhere (see for example, Windley, 1984) and are not directly relevant to this book. However, the palaeomagnetic reconstruction demonstrates that the greenstone belts have a very prominent linearity, both between the shields and parallel to the long axis of the primary crust (Fig. 8.2). This is an important observation because many of these belts post-date the conformity of the African and Laurentian palaeomagnetic data at 2800 Ma, and the Australian and Laurentian data after 2600 Ma (Fig. 7.1) and they therefore developed, at least in part, after the continental crust had attained an integral form. The Yilgarn examples of Western Australia formed within some 80 Ma of the initiation of magmatism at 2720 Ma (Cooper, Nesbitt, Platt & Mortemer, 1978) and the Zimbabwean belts mostly appear to be *ca.* 2670–2480 Ma in age (see summary of evidence in Goodwin, 1981). In the Laurentian Shield the Superior Province examples are mostly dated in the range 2750–2700 Ma (Krough & Davies, 1972, and Section 7.2) and the Slave craton examples are dated at *ca.* 2750 Ma. The Dharwar-type belts in India were formed at 2600–2300 Ma (Viswanatha, 1972) and the Siberian examples are <2600 Ma (Moralev, 1981). All of these younger greenstone belts have a high proportion of sediments, and are associated (although a contact cannot always be demonstrated) with high-grade gneissose crust some 600–300 Ma older (Goodwin, 1981). They have affinities with certain younger, predominantly meta-sedimentary, belts such as the Birrimian (*ca.* 2200 Ma) of West Africa, and contrast with the Kaapvaal examples of southern Africa, dated at 3426–3289 Ma (Anhaeusser, 1978) and some pre-Dharwar belts in India which are dominated by mafic and ultramafic volcanics. These old belts are essentially isotropic and do not (with the exception of the Barberton lineament in southern Africa) show the prominent linearity (Fig. 8.2). The younger belts have length to width ratios often well in excess of 10:1, and some features of the alignment of their trends were first noted on the Pangaean reconstruction by Engel & Kelm (1972); this alignment is enhanced on the Proterozoic reconstruction, which also links the belts to the form of the primary crust. The main exception to the axial trend seems to be the deflection of the Kola Peninsula and NW Scotland examples around the margin of the ancient stable nucleus in south Greenland, and two cuspate examples in central Africa.

The wide range of genetic models for these belts (Windley, 1984) includes formation as synclinal piles between large-scale fracture in older continental crust (e.g. Rutland 1973), to development within closed marginal basins behind Cordilleran-type arcs (Tarney, 1976), to development as ocean crust between uncoupled microcontinents (Talbot, 1973). The palaeomagnetic analysis now restricts the possible modes of origin: only a process approximating to the rift and sag concept of Goodwin (1977), which postulates attenuation and fissuring above hot lines in the mantle, leading to elongated basins collecting shallow-water sediments and basic volcanics, can be applicable here.

A feature of comparable, or even greater, antiquity is the emplacement of conformable and elongate layers of calcic anorthosites into the high-grade

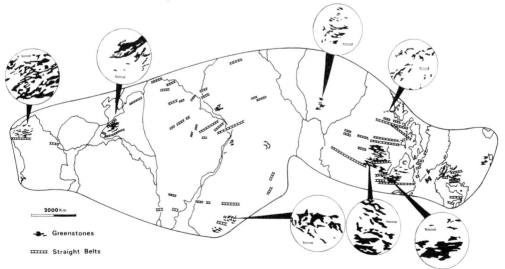

Fig. 8.2 The distribution of Archaean and Lower Proterozoic greenstone belts and straight belts within the continental crust, modified after Piper (1983b) and based on references given in the text. The regional maps show the outcrop patterns of greenstone belts (black) between gneiss terranes (white) in seven regions where this is relatively well known; the bars are equivalent to 100 km. Mollweide equal-area projection.

terranes (Windley & Bridgwater, 1971). Three possible early lineaments are: (i) Madagascar, the Queen Maud Block and India, (ii) South Africa and Zimbabwe, and (iii) east and west Greenland and NW Scotland (Fig. 8.2, and Sutton, 1971) The record is too fragmentary to determine whether the alignment of these locations has any reality (the position of Madagascar is in any case not defined, although a location near to India and Sri Lanka would seem to be appropriate: Fig. 7.1 and Kratz, 1974), but the development of thermal provinces and melting of the crust was to be repeated in a modified form by the massive anorthosite province at 1800–900 Ma (Section 8.6).

In general, the relict Archaean crust exhibits only weak anisotropy, and reconstitution by Proterozoic tectonism has only occurred freely where transfer of volatiles from the mantle and lower crust has taken place (Watson, 1973). In early Proterozoic times this effect was concentrated along narrow *straight belts*. These are the oldest pure lineaments in the continental crust, where Archaean structures are deflected into, and aligned along, well-developed planar zones. They incorporate relative movements with a strong component of strike-slip motion between adjacent, already-stabilised blocks. In character, they range from ductile shear zones several tens of km in width, of which examples are well documented in west Greenland (Bak, Sørensen, Grocott, Kortsgaard, Nash & Watterson, 1975), to long mylonite zones in West Africa, South America and the NW part of the Laurentian Shield. Examples in the Laurentian Shield can be followed discontinuously for distances > 1100 km (Watson, 1973) and lineaments at deeper levels are recognised from aeromagnetic studies. They represent a fundamental subdivision of the primary crust, because they have controlled many later tectonic and magmatic events. In the Laurentian Shield, for example, the straight belts have been reactivated later in Proterozoic times to form the boundaries of tectonic provinces (Watson, 1973, and Section 8.5).

Although their distribution is still very imperfectly understood, in common with the greenstone belts, they exhibit a prominent alignment between the shields and parallel to the long axis of the primary crust (Fig. 8.2, and Piper, 1983a). This distribution emphasises the very long-term stability of stress fields over areas of continental dimensions in Proterozoic times (Section 12.11). The straight belts appear to represent a discrete stage of crustal development (Sutton, 1978). Most were initiated between formation of the greenstone and mobile belts (Section 8.5) although this transition was diachronous on a global scale, and straight belts in some areas pre-date greenstone belts in others. Ductile shear belt development in Greenland is dated at 2900–2500 Ma (Hickman, 1979) while in NW Scotland it pre-dates intrusion of the Scourie dykes at *ca.* 2400 Ma. In the Superior Province the straight belts developed shortly after stabilisation of the greenstone belts and intrusion of the youngest granite plutons, and they are associated with the last regressive stages of deformation at deep crustal levels (Watson 1980; Park 1981). Here their formation pre-dated intrusion of the Matachewan dykes at 2690 Ma, although sinistral movements continued during intrusion of the dykes. The straight belts were also the sites of metasomatic activity, with addition of H_2O, K and Na, and loss of Fe, Ca and Mg (Beach, 1976). In every described example they appear to have been initiated prior to the widespread intrusion of dyke swarms at 2690–2400 Ma; these swarms fan out for distances of up to several hundreds of km over the shields; the latter represent a stage in the consolidation and brittle fracture of the continental crust which had not been achieved when the straight belts were initiated (Sutton, 1978; Windley, 1984; and Fig. 8.5).

Models for the development of both the greenstone and straight belts are therefore now constrained by their demonstrable alignment and their relationship to the form of the primary crust. Their areal distributions provide some clues to the scale of sub-crustal convection systems, which appear to have caused these belts: greenstone belts are now separated by distances of 50–100 km. This is somewhat less than the separations of the straight belts; in west Greenland the latter are spaced about 100 km apart and the smaller number of more widely spaced megashears in Canada (Watson, 1973) may reflect inadequate knowledge of this area. In NW Scotland two closely-spaced shear zones occur little more than 10 km apart (Beach, 1976). The mechanics of small-scale sub-crustal roll convection in the upper mantle has been widely studied in recent years (e.g. McKenzie and Richter, 1976.) There are objections to its widespread operation at the present time (e.g. Loper, 1985), but the observational evidence provided by the greenstone and straight belts shows that it must have operated in early Precambrian times, when the lithosphere was thinner and radiogenic heat production was greater. It is now possible to put some crude constraints on this convection: the episodic subsidence of the BIF basins in early Proterozoic times has been used to calculate that the lithosphere had an elastic thickness of $ca.$ 20 km at 2200 Ma (Cisne, 1984). This is less than half the contemporary thickness, and it would have been somewhat lower than this estimate during the permobile phase of crustal development represented by the greenstone and straight belts. Sub-crustal convection could have taken place at these depths (or somewhat greater depths beneath the early-stabilised granulite terranes, where mineral assemblages imply Archaean crustal thicknesses of $\leq$ 40 km (e.g. O'Hara, 1977)) as illustrated in Fig. 8.3. The general consequence of the increase in viscosity and concomitant reduction in Rayleigh number which accompanied the decline in radiogenic heat production, would have been to change the pattern of this small-scale convection from hexagonal to bimodal to longitudinal rolls for any given convecting layer (Richter & Parsons 1976, McKenzie & Richter, 1976). Longitudinal rolls are preferred at moderate Rayleigh numbers and convection will continue in this form until the viscosity becomes too high for it to be sustained. The transition from hexagonal to bimodal small-scale flow probably correlates with the change from the earliest greenstone belts (3500–3200 Ma) which exhibit little or no linearity, to the strongly anisotropic younger (< 2800 Ma) belts (Fig. 8.2). Although details are disputed, there is general agreement that these belts are the product of small-scale convection operating beneath a thin lithosphere (see Windley, 1984, and Kröner, 1981 for discussions). The stresses produced by this style of convection would have been up to an order of magnitude larger than the stresses produced by present-day convection (McKenzie & Weiss, 1975). They would appear to have been capable of disrupting the crust during formation of the greenstone belts, but not at the somewhat more advanced stage of crustal development represented by the straight belts.

The relationship of these features to each other and to the form of the primary crust implies that these small-scale rolls operating beneath the continental lithosphere were separated by a boundary layer from a large-scale convection system. The latter may have been responsible for releasing the bulk of the Earth's internal

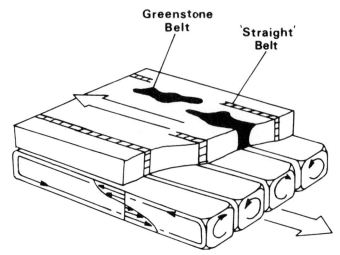

Fig. 8.3 Schematic diagram illustrating the possible relationships between straight belts and small-scale sub-continental roll convection. This small-scale motion would have been strongly affected by surface plate motions, and consisted of rolls constrained parallel to the direction of shear between lithosphere and asthenosphere. Adapted from Richter & Parsons (1976) and Davies and Runcorn (1981). The greenstone belts, which were consolidated between a few tens to a few hundreds of Ma earler, are the relics of a bimodal convection system associated with some extension and voluminous magmatism followed by shortening: motions were then tangential rather than transcurrent as in the case of the straight belts. After Piper (1983b).

heat budget within the oceanic lithosphere, which is almost entirely unpreserved in the Proterozoic rock record. The increase in the spacing of the greenstone and straight belts with time is concise evidence that the scale of sub-crustal roll convection increased with time as the continental lithosphere thickened and stabilised. After 2200 Ma, continental tectonics were dominated by large-scale thermal-tectonic mobile episodes (Section 8.5) and there is evidence only for the intermittent operation of sub-crustal roll convection on a scale up to an order larger than that implied by the greenstone and straight belts.

The transition from greenstone to straight belt tectonics is recognised as early as 2800 Ma in some areas, and as late as 2200 Ma in others, and it seems unlikely to be linked only to an increase in lithospheric thickness. The critical Rayleigh number for the onset of roll convection is increased by shear flow, so that the orientation of the small-scale cylindrical rolls is constrained by motion of the lithosphere over the asthenosphere (Fig. 8.3): rolls will only develop if the shear is sufficiently strong (Davies & Runcorn, 1981). Thus the straight belts were zones of prominent strike-slip motion, while the earlier greenstones were dominated by tensional stresses. The relationship between the straight belts and large-scale continental movements can be investigated for the belts in the Superior Province and west Greenland which are relatively well dated and mapped. They were initiated during the interval between greenstone belt formation at 2750–2700 Ma and intrusion of the Matachewan dykes at 2690–2650 Ma (Park, 1981), and relate to the initial part of the APWP defined in Fig. 7.1. The APWP defines a small circle about a Eulerian pole at *ca.* 330°E, 45°N, and implies that the Superior Province and Greenland moved southwards across the equator and rotated clockwise during this time interval. This clockwise rotation can explain the consistently sinistral sense of movement on this straight belt system (Watson, 1973; Beach, 1976), if straight belts represent zones of ductile drag between parts of the continent which were successively closer to, and more distant from, the Eulerian pole of rotation. The dry and more rigid crust in between presumably behaved as passive blocks partially uncoupled from one another by the intervening ductile belts (Sutton & Watson, 1974).

8.5 Mobile belts and accretionary orogenic belts

While the palaeomagnetic analysis has shown that the continental crust behaved as a coherent body in Proterozoic times, there is abundant evidence in the regional geology of Precambrian terranes to demonstrate that it did not behave like the rigid lithospheric plates of the present day. Instead the crust comprised a mosaic of stable Archaean massifs subdivided by the linear straight belts. After *ca.* 2200 Ma, this style of tectonics was replaced by large-scale *mobile belt tectonics*, in which much larger stable areas, essentially comprising the Archaean cratons preserved at the present day, formed rigid nuclei between broad zones of mobility. In these mobile belts, older crust was subjected to magmatism and metamorphism, and frequently to profound deformation. These belts include a wide zone of tectonic reworking, mostly at 1900–1700 Ma, which covers much of the Fennoscandian, Laurentian and Siberian Shields (e.g. Kratz, Gerling & Lobach-Zhuchenko, 1968; Stockwell, 1973; Sears & Price, 1978), and incorporates belts such as the Svecokarelian and Svecofennian in Scandinavia, the Laxfordian in NW Scotland, the Nagssugtoqidian in Greenland, and the Churchill in Canada. It can now be identified as a continuous feature in the primary crust (Fig. 8.4) showing a general alignment and continuity of structural trends. It may also have continued into wide mobile zones in west Africa and South America, which appear to be somewhat older (Hurley & Rand, 1969; Hurley, 1973) and roughly contemporaneous belts are present in India, Australia and central Africa. Appreciable, but as yet undefined areas of the Chinese and Antarctican Shields were also reworked at this time (Dearnley, 1966; Hurley & Rand, 1969; Sutton, 1978). These mobile episodes were accompanied by rapid post-tectonic uplift, linked to a large, and possibly the last major, addition of mantle-derived material to the crust (Dickinson & Watson, 1976; Condie, 1976; De Paolo, 1981).

There seems to have been a general hiatus in the development of mobile belts after 1600 Ma, which is also reflected in the minimum on the age histogram of Fig. 7.10. A younger group of narrower belts developed at *ca.* 1350–950 Ma and forms a discontinuous set of linear features running through the primary crust (Fig. 8.4). It includes the Namaqualand, Irumide and Kibaran belts in Africa (and a contemporaneous, but as yet poorly-defined, lineament in Brazil and Nigeria), the Fraser-Musgrave belt in Australia, and Aravalli belt in India and its possible continuation in the Garzon terrane (Nickerie episode) along the NE margin of the Guyana Shield, South America (Kroonenberg, 1982), and the Grenville and Sveconorwegian belts in the Laurentian-Fennoscandian sector (Fig. 8.4).

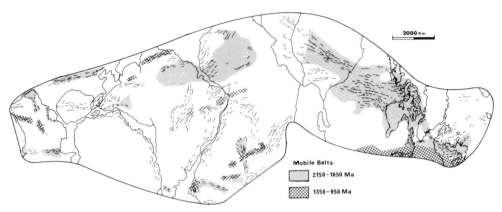

Fig. 8.4 The distribution of Proterozoic mobile belts within the primary continental crust, together with structural trends of known or probable Precambrian age. Modified after Piper (1983a). These ensialic belts are distinguished from orogenic belts which developed in continental margin settings and cordilleran-type environment; these latter belts are illustrated in Fig. 8.5.

The most significant feature of mobile belts which formed prior to 900 Ma, is the consistent way in which the structural and magmatic evidence, and the rock associations, all indicate that they originated in an ensialic environment by *in-situ* reworking or overprinting of much older crust. The evidence includes the continuity of the sialic basement, the correlation of older stuctures into and across younger belts, and the low values of integrated strain across the belts (e.g. Shackleton, 1973, and Kröner, 1977). In addition, blueschists, obducted ophiolites and indeed, any rock associations which could be linked with ancient ocean basins are absent, while the magmatic products are predominantly acid and intermediate in composition and have resulted from the melting of older crust (Glikson, 1983). The Kibaran and Irumide belts of central Africa (Figure 8.4) are typical examples, and in structural terms they have been shown to comprise a foreland fold and thrust belt dominated by thin-skinned tectonics, and an internal zone of synorogenic deep crustal shear belts (Daly, 1986). Pre-, syn- and post-tectonic magmatism is of crustal derivation, and is geochemically uncharacteristic of magmatic arc environments (Glikson, 1983). The sediments characterising these belts are of marginal basin rather than island-arc types. Nevertheless, the marginal thrust belts may exhibit several tens of km of crustal shortening, and isotopic studies of the internal zones indicate uplift rates comparable with modern Alpine belts. Hence, the foreland thrust belts may have been driven by gravity collapse following crustal thickening (England & McKenzie, 1982). However, some mobile belts are now defined by 'front' zones where the foreland thrust belts have been eroded away and mid- to deep-crustal ductile shear belts define sharp boundaries with the adjoining cratons; examples are the Grenville and Sveconorwegian Fronts (Section 7.8 and Fig. 9.1) and comparable zones are visible along the northern and eastern margins of the Zambesi belt in Zimbabwe, and the western front of the Mozambique belt in Tanzania, where they have a 'ramp and flat' geometry (Coward & Daly, 1984) and pass rapidly from flat-lying into steep structures. At greater depths they appear to change their geometry from steep structures in the amphibolite facies to shallower structures in the granulite facies, and show a combination of vertical and horizontal displacements commensurate with shuffling motions between more rigid cratons.

In addition to mobile belts with ensialic characteristics, some Proterozoic terranes contain belts with contrasting orogenic affinities, which include miogeosynclinal and eugeosynclinal assemblages, and petrochemical signatures analogous to contemporary subduction-related rock associations. Some of the late Proterozoic (notably Pan-African) belts also show evidence for consumed ocean basins in the form of ophiolite wedges. The presence of belts with both mobile and orogenic characteristics in Proterozoic times, together with other belts with unclear affinities, has led to much dispute about the true nature of tectonics during this eon; the most recent trends (see papers in Kröner, 1981) favour some kind of evolution in the style of tectonics through Proterozoic times.

The palaeomagnetic reconstruction resolves this dichotomy of opinion about Proterozoic tectonics, because the two kinds of belt are seen to have contrasting tectonic settings within the continental crust (Fig. 8.4): whilst the ensialic mobile belts largely have interior settings, a short summary of the belts attributed to subduction-related processes demonstrates that they all have a peripheral location linked to the ancient continent–ocean interface. Possibly the earliest and best-documented example is the 2000–1800 Ma Wopmay orogeny at the NW margin of the Laurentian Shield. A subsiding continental shelf along this edge of the Slave craton was the site of flysch-facies sedimentation and eruption of spilitic pillow lavas, succeeded by later molasse facies. No suture is visible, but an orogenic phase is represented by thrusting over the craton (Hoffman, 1980). Along the opposite margin of the shield, a contemporaneous succession is linked to subduction beneath the southern margin of the Superior Province; the 1900–1700 Ma Penokean orogeny was the site of deformation and eruption of calc-alkaline volcanics, with a petrology and chemistry linked to NW-dipping subduction (Van Schmus, 1979) extending from Wisconsin to possibly as far SW as Wyoming (Condie 1982). Along this ancient orogenic margin of the shield, terranes were subsequently accreted during episodes (largely magmatic in origin) at *ca.* 1800–1700, 1720–1650 and 1200–1100 Ma (Condie, 1982). In Greenland, *ca.* 1760 Ma orthogonal dyke swarms in west Greenland, the contemporaneous Ketilidian mobile belt, and the Julianahaab granodiorite batholith in south Greenland are a collective consequence of northward subduction (Watterson, 1978, and Fig. 8.5). Rocks with ophiolitic affinities occur within the Ketilidian belt, and the contact between the Ketilidian succession and the continental basement becomes progressively deformed and obscured when followed southwards into the orogenic belt (Sutton, 1978). In the Fennoscandian Shield, parts of the Svecokarelian belt contain serpentinites, apparently emplaced by NE-directed obduction at 1900–1850 Ma (Park, 1983) and the bordering Svecofennian terrain includes late- to post-tectonic plutons which can be modelled by NE-dipping subduction beneath the shield (Hietanen, 1975). The adjacent *ca.* 1100 Ma Sveconorwegian terrane in southern Sweden incorporates SW-dipping low-angle thrusts of crustal thickness piled onto one another (Berthelsen, 1976). The tectonomagmatic relationships along the strike continuation of this belt in Norway have been related to a continental margin environment, with analogies with the present-day cordilleran situation (Torske, 1977). This discussion should not imply that subduction was a continuous effect along these margins; it seems instead to have been a rather incidental phenomenon in Proterozoic times. Thus the bulk of the Penokean magmatism was confined to the interval 1860–1820 Ma, while the Wopmay orogeny seems to have occupied little more than 20–40 Ma.

Examples from the former Gondwanaland continents are less well documented. In NW India the basement gneiss complex passes into marbles and schists representing possible continental shelf and rise deposits (the Aravalli schists) and separated by mélange zones from a zone to the west which includes serpentinites, meta-gabbros and other basic rocks, comprising a possible ophiolite fragment; the collective assemblage is modelled by eastward-directed subduction

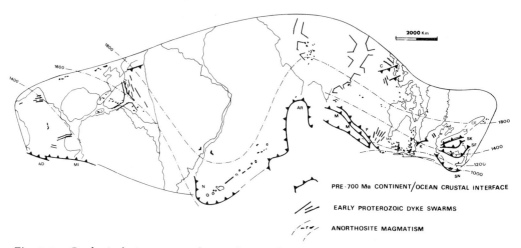

Fig. 8.5 Geological signatures of crustal growth and consolidation during Proterozoic times. The orogenic zones specifically linked to the action of intermittent subduction-related processes by geological and/or geochemical evidence are lettered: **AD**, *Adelaide;* **MI**, *Mount Isa (ca. 1800–1500 Ma);* **N**, *Namaqualand-Natal (1350–1000 Ma);* **P**, *Penokean (1900–1820 Ma);* **M**, *Matatzal (1680–1600 Ma);* **E**; *Elsonian (mainly 1480–1380 Ma granite-rhyolite terrane);* **C**; *Coronation geosyncline and Wopmay orogen (2000–1750 Ma);* **K**, *Ketilidian (1800–1650 Ma);* **SK**, *Svecokarelian (1950–1850 Ma);* **SF**, *Svecofennian (1750–1520 Ma) and* **SN**, *Sveconorwegian (1090–950 Ma).* **CE** *is the Central European miogeosyncline of Lower Proterozoic age and* **AR** *is the Afro-Arabian arc which appears to largely, or entirely, post-date this reconstruction (see Chapter 9). The 'chrontours' embracing the massive anorthosite and related magmatism are highly generalised, both because this magmatism was markedly episodic on a regional scale, and because little is known about the ages of the scattered Gondwanaland examples. Note that the early Proterozoic dyke swarms include ca. 2600 Ma examples in the Indian and Laurentian Shields which formed in a stress field perpendicular to that of the preceding greenstone belts (Section 12.11), while somewhat younger (ca. 2400 Ma) examples in W. Australia and Africa are broadly parallel to the greenstone belts. The Great Dyke and related swarms in the latter shield are part of a major magmatic lineament of Precambrian age, including dykes and central complexes shown in outline form in this figure.* **MA** *is the Matachewan swarm. The parallel bars roughly define the aulacogens, which began to develop in mid-Proterozoic times and are taken in part from Burke & Dewey (1973). Modified after Piper (1983a) mainly after sources listed in the text.*

beneath the crust, dated at least 1800 Ma and roughly contemporaneous with the Wopmay orogeny along the same margin (Fig. 8.4). In common with the Laurentian and Fennoscandian Shields, all subsequent examples are confined to the opposite margin of the supercontinent. In the Mt. Isa geosyncline (*ca.* 1800–1500 Ma), platform sediments merge continuously eastwards into a peripheral tectonic belt (Dunnett, 1976) and the Adelaide and Trans-Antarctic zones are younger belts developed along this margin. Along the margin of southern Africa, in Natal, nappes incorporating greenstone belts were emplaced northwards over the Kaapvaal foreland by marginal orogeny (Tankard, Jackson, Eriksson, Hobday, Hunter & Minter, 1982). The Afro-Arabian margin is recognised as the site of accretion of island-arc suites between *ca.* 1150 and 600 Ma. This margin includes all the signatures of subduction-related margmatism and tectonics (Greenwood, Hadley, Anderson, Fleck & Schmidt, 1976; Shimron, 1980). Recent work has extended this interpretation southwards into Kenya (Vearncombe, 1983) and has confirmed the presence here of an interface between an island-arc zone and the ancient shield incorporating the bulk of central Africa. The Bou Azzer ophiolites were emplaced in a comparable setting along the northern margin of the West African craton at 700 Ma (Leblanc, 1981).

The palaeomagnetic analysis shows that all of these occurrences are margin-related and the directions of inferred subduction are beneath the continental crust. The only major qualification necessary to many models developed in the literature is that the tectonic setting of these belts was of Cordilleran type and not a continent-continent collision as has often been inferred. There appear to be a few exceptions to this general conclusion. The best described is the Trans-Hudsonian Belt, a zone which runs SW from the margin of Hudsons Bay between the Superior and Slave cratons in the Laurentian Shield. Volcanic and intrusive igneous suites here seem to have orginated in an island arc setting at *ca.* 1800 Ma and show little sign of contamination by older crust. The accordant palaeomagnetic data from either side of this zone from 2700 Ma preclude major separation of the two margins and the absence of an obvious continuation of this belt to the NE suggests that a short term pivotal separation took place here in mid-Proterozoic times. More belts of this kind are reported from late Precambrian times and ophiolite fragments are incorporated within some of the mobile belts formed after 1000 Ma. They show that small ocean basins opened between rifted continental lithosphere and were consumed by subsequent closure. The best examples of these aborted oceans are seen in the late Proterozoic Pan African belt in the western Hoggar (*ca.* 725–670 Ma) which include a complete Wilson Cycle development (third part of Section 6.6) culminating in the emplacement of basic and ultrabasic bodies and the translation of nappes onto the West African cratonic foreland (Caby, Bertrand & Black, 1981). This belt was close to, but not parallel with, the continental margin at the time (Section 9.6) and does not appear to have involved a motion which is palaeomagnetically-detectable because the older data from the West African and Kaapvaal-Zimbabwe cratons are in general agreement (Section 7.3). The Damaran belt in Namibia is another example described by Kröner (1977): this is a NE–SW zone of intense ductile shear which probably continues into the Mwembeshi zone in Zambia, and incorporates slivers of basic and ultrabasic rocks with ophiolitic affinities at its SW extremity. The *ca.* 550 Ma tectonic movements here also seem to have incorporated a small aborted ocean basin (Downing and Coward, 1981, and first part of Section 9.4). Thus while the Pan-African belts are ensialic in palaeomagnetic terms (McWilliams and Kröner, 1981) and were characterised by an evolution spanning periods of up to several hundreds of Ma (atypical of Phanerozoic orogenic belts), they exhibit transitional characteristics. McWilliams & Kröner (1981) have developed a model for these transitional belts, which postulates that fracturing and crustal thinning resulted in ensialic geosynclines, while further stretching led to lithospheric splitting and mantle-derived igneous activity; this was followed by compressive stages represented by the interstacking of crustal segments, nappe tectonics, and partial melting to produce syn- to post-tectonic granites, so that lateral movements of $\leqslant$ a few hundred of km are involved.

The tectonic movements leading to development of mobile belts were very much smaller than contemporaneous translations of the continental crust, although continental velocities were generally higher when mobile belts were developing (Fig. 7.10). This tectonic style also correlates with two other features of the palaeomagnetic record. Firstly, loop-style APW motions were first evident when mobile belts began to replace greenstone and straight belts as the prevalent form of continental tectonics. Secondly, the V_{RMS} velocity of the continental crust increased when this transition took place (Fig. 7.10) and subsequently declined during Proterozoic times, in a manner broadly consistent with declining radiogenic heat production within the Earth. The explanation of the first correlation is not clear, and it may not be significant, because there is some evidence for loop movements in Palaeozoic and Mesozoic times (Sections 9.4 & 11.3), after these belts had ceased to form. The second correlation implies that the tectonic transition at $ca.$ 2200 Ma defines a change in the heat release within the Earth, from a system which included small-scale crustal rolls, to one which involved some form of large-scale cells only. The latter are first witnessed directly at this time, as the oldest marginal accretionary belts: e.g. the Wopmay, with their associated signatures of subduction. The transition from small-scale to large scale tectonics post-dates by $ca.$ 500 Ma the time at which crustal coherence can first be demonstrated (Fig. 7.1) and was not therefore an immediate consequence of formation of the supercontinent. A possible cause for this delay in development of large-scale crustal stability can be sought in the relationship between the declining geothermal gradient and the phase boundary of granulite stability (Tarling, 1980). The geothermal gradient in early Proterozoic times closely followed the boundary conditions for granulite stability (Tarling, 1980; Ringwood, 1975): with continuing decline in the radiogenic heat production this gradient would have begun to intersect the granulite stability field simultaneously through the entire thickness of the continental lithosphere. The formation of anhydrous granulites from the hydrous amphibolites of pre-existing crust (which would then become metastable) may have produced a vast volatile release to the atmosphere and hydrosphere, and transferred large-ion lithophile elements to the upper crust as part of the granite-forming events (Section 8.2). This dehydration would, in addition, have drastically increased crustal rigidity and have concentrated internal heat release within the oceanic lithosphere, thus decreasing the rigidity of the latter. This imbalance in the strengths of the continental and oceanic lithospheres may be one reason why the continental crust did not break up until late Precambrian times: the strength of continental lithosphere would probably change little after this event, whereas the strength of oceanic lithosphere would progressively increase as a consequence of the continuing decline in radiogenic heat.

8.6 Proterozoic anorogenic magmatism and crustal consolidation

The Proterozoic eon was characterised by the intrusion of a distinctive magmatic suite, incorporating massive anorthosites and rapakivi granites (Condie, 1976). The distribution and age relationships of the suite are discussed by several authors (Isachsen, 1969; Herz, 1969; Bridgwater & Windley, 1973; Fudao & Guanghong, 1978). This style of magmatism became progressively more restricted with time, in a way which is concisely related to the shape of the primary crust (Fig. 8.5). In the Laurentian and Fennoscandian sector, it appears to have terminated prior to 1800 Ma along one margin, but continued along the other margin at *ca*. 1650 Ma, at 1500–1300 Ma (notably in the Elsonian terrane and Labrador), and until *ca*. 950 Ma close to, and within, the Grenville-Sveconorwegian lineament. The temporal contraction is summarised by 'chrontours' in Fig 8.5 which are extended to incorporate the limited knowledge of the age relationships for anorthosite magmatism elsewhere. It is not recorded in the bulk of the African, South American and North China Shields after 1600 Ma, but appears to have continued in India and Madagascar to 1500 Ma, and in central Australia to *ca*. 1350 Ma (Hurley & Rand, 1969; Bridgwater & Windley, 1973; Sutton, 1978). The high temperature gradients necessary for the intrusion of these mushroom-shaped complexes into high levels of the already-stabilised crust (Bridgwater, Sutton & Watterson, 1974) thus contracted progressively towards one margin of the supercontinent.

Although these chrontours no doubt yield a highly simplistic view of continental crustal cooling, the impression they give appears to be fundamentally correct, for three important reasons. Firstly, the decline of temperature gradients in the crust is towards the hot margin, which incorporates all of the post-1700 Ma subduction-related orogenesis and accretion. Secondly, it was the opposite cold margin of the crust which first began to rift, with the development of deep aulacogens, notably along a sequence of triple junctions between the Laurentian and Siberian Shields (Burke & Dewey, 1973). Thirdly, the chrontours contract towards the Grenville-Sveconorwegian margin, where some 75% of exposed anorthosites are concentrated, and where a major mantle reservoir was established by 1650 Ma (Ashwal & Wooden, 1983) and was responsible for emplacing large anorthosite bodies into the crust over a period of some 700 Ma after this time.

8.7 Mineralisation in the continental crust

The fractionation processes responsible for mineralisation in the crust had not progressed far by the time that the early Archaean gneiss terranes were formed, and these zones are essentially barren. Subsequently, widespread (and mainly subaqueous) volcanism, probably of global extent, added concentrations of a number of metallic elements to the crust. (The Ni-Fe-Mn nodules forming on the present-day ocean floor are an analogue of the process). Iron, which was initially extracted by sea-water from the volcanic successions, was subsequently precipitated in supersaturated fumarolic sea-floor environments to form the Algoman-type iron formations (concluding part of Section 4.2). The komatiitic facies of the greenstone successions are the source of pods of massive Ni-Fe sulphides with low Cu and Pt contents; they are linked to magmatic partial melting in a mantle containing abundant reduced sulphide species (Hutchinson, 1981). Gold has a wider distribution in Archaean rocks, and is associated not only with the mafic-ultramafic volcanic successions, but also with acid suites in the highest parts of the successions; it appears to derive initially from the volcanic host rock by convective hydrothermal leaching, which accompanied the volcanism. With the formation of the Proterozoic Supercontinent at the end of Archaean times, there was a distinct change in the style of mineralisation. Throughout the succeeding Proterozoic eon this style reflected the predominantly ensialic processes taking place within the crust up to the Proterozoic–Phanerozoic transition (Fig. 8.6). It is, however, important to distinguish those changes which were directly linked to the style of crustal tectonics, from those caused by secular changes, unrelated (or only very indirectly so) to tectonics. The changing character of iron precipitation in the Algoman and Superior-type deposits can be linked to the oxygenation of the atmosphere at *ca.* 2200 Ma: the vast reservoir of ferrous iron in the hydrosphere could then be rapidly oxidised to the ferric state and precipitated over a short period of time in shallow, intercratonic basins. Proterozoic nickel sulphide deposits, however, became restricted to major layered igneous intrusive complexes,

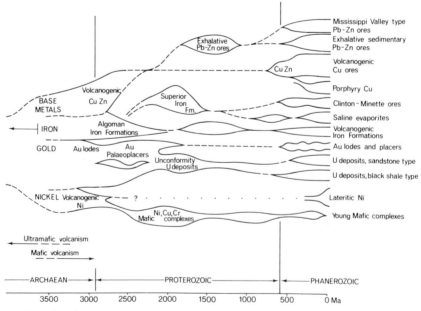

Fig. 8.6 The development and possible metallogenic relationships of some of the major types of mineral deposits over geological time. After Hutchinson (1981).

probably because by then the mantle had become largely depleted in reduced sulphur phases. Proterozoic gold lodes are discordant features redistributed by later igneous or metamorphic events, while Proterozoic placer deposits comprise the world's most important gold deposits. They are a component of thick fluviatile–deltaic sedimentary successions deposited in inter-cratonic environments at subsiding rift margins. The earliest Proterozoic sediments are mineralised with gold and uranium virtually everywhere: the gold originally derived from concordant lodes in Archaean volcanic successions, and often later mobilised and redeposited as discordant lodes, was therefore exposed to weathering and erosion in an atmosphere which was probably mildly reducing. Subsequent changes in sedimentary ores during Proterozoic times seem to be a response to changes in the chemistry of the environment but are not indicative of any fundamental changes in tectonic style (Badham, 1981). A repetition of the concordant–discordant–placer sequence is also observed in Phanerozoic orogenic environments although in this setting it took place over much more condensed time intervals (Hutchinson, 1981).

The massive base metal deposits however, change from predominantly copper- and gold-rich stratabound deposits in volcanic host rocks in Archaean times, to increasingly stratiform and lead-zinc-silver rich deposits in sedimentary host rocks in Proterozoic times. Volcanic-hosted deposits are practically absent after 1800 Ma and by *ca.* 1000 Ma, sedimentary-hosted types become very important. Ore deposits which have originated from a variety of primary sedimentary or secondary diagenetic processes are a characteristic feature of Proterozoic times; type examples of mid-Proterozoic age occur in the Mt. Isa region of Australia and late Proterozoic examples are well developed in Pan African belts of Africa; the importance of these ores reflects the widespread continentality of the environment. The other aspect of the distinctive signature of the Proterozoic times in Fig. 8.6 is the relative importance of silicic volcanism in ensialic settings. In Middle to Upper Proterozoic times the development of deep aulacogens and marginal basins where thick sequences of sediments could accumulate (Section 8.6) has produced the general association of these ores with rift environments and sometimes with basic volcanics. In Phanerozoic times there has frequently been a short term repetition of this long term Proterozoic cycle in orogenic belts, with a primitive copper-gold-rich association (which is often volcanically associated), giving way in time to lead–silver associations. The other aspect of base metal deposition correlating with the Proterozoic–Phanerozoic transition is a diversification of types of deposit in Phanerozoic times (Fig. 8.6) with the appearance of limestone-hosted Mississippi Valley type lead-zinc ores in rifted cratonic interiors and porphyry copper deposits at accreting orogenic margins. This reflects a diversification of the lateral and vertical tectonic processes plus the addition of a new tectonic element, namely the steep subduction and greater degree of partial melting of ocean lithosphere, together with the obduction of this oceanic lithosphere onto the continental crust at the sites of continent–continent collisions.

Uranium is a significant indicator of crustal evolution, since it is practically absent in Archaean rocks; it first becomes important as a placer deposit in early Proterozoic clastic deposits, and is relatively more important in rocks formed after the atmosphere ceased to be reducing. Acid igneous rocks are much richer in uranium than mafic ones, and the granites and pegmatites produced by the worldwide crust-forming episodes at the end of Archaean times, and by subsequent ensialic events (especially during the *ca.* 1800 Ma mobile episodes) provided a source of uranium which could be concentrated by subsequent weathering and erosion. Oxidation of the weathering environment was critical to this concentration, because it permitted mobilisation and transport of hexavalent uranium, which could be subsequently deposited as insoluble tetravalent uranium when it encountered more reduced environments: for example, below the ancient water table or in different rock suites. This produced the important unconformity-related uranium deposits of mid-Proterozoic age. Phanerozoic uranium is linked to the development of new kinds of reducing environments, such as the anaerobic marine black shale environments of Lower Palaeozoic times.

Possibly the most significant feature of mineralisation during the Proterozoic eon is the paucity of exhalative volcanogenic massive sulphide deposits. They are practically confined to the peripheral zones of subduction and accretion, and their more general absence may be linked to the paucity of calc-alkaline magmatism. Phanerozoic times are characterised by a return to many features of the Archaean environments, and a reappearance of all the earlier types of mineral deposits, plus a diversification of types (Fig. 8.6). Thus the evolutionary style of mineralisation in the crust is a close parallel of the tripartite geochemical and isotopic signatures discussed in Section 8.2 (cf. Figs. 8.1 & 8.6) and another indirect confirmation of the coherent crustal model derived from the palaeomagnetic evidence.

This contention can be further tested by the mineral distributions. Fig. 8.7

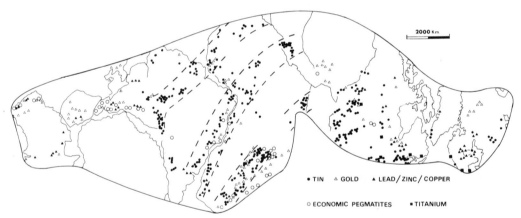

Fig. 8.7 The distribution of some metallogenic deposits within the primary continental crust, based mainly on Schuilling (1967), Petrascheck (1973) and Watson (1978). The dashed lines delimit the tin belts recognised by Schuilling (1967) and Hunter (1973). Distribution maps of this kind are controlled in part by available exposure, and by the concentration of exploration activity near to existing deposits. Several correlations do however, emerge, and include: (i) the alignment of gold deposits, (ii) the concentration of tin within several broad zones paralleling the Proterozoic tectonic trends but not coincident with mobile belts of any one age, (iii) the association of major base-metal sulphides with the Proterozoic orogenic margin and mid-Proterozoic aulacogens, and (iv) the occurrence of ilmenite and some other magmatic concentrations in major basic intrusions along the active continental margin, and especially concentrated near the mantle anorthosite source.

illustrates the distributions of several important metals and pegmatites, after the compilations of Schuilling (1967), Petrascheck (1973), Watson (1978) and Berry (1980). The available information is relatively complete for gold and tin, but much less good for other metals. Although some of the ores are of Phanerozoic age, this may be a less important reservation because they derive from the primary crustal accretions of late Archaean and early Proterozoic age, and many of the Phanerozoic examples are likely to be remobilised disseminations of Precambrian age. This is especially the case for tin, which is distributed along several major lineaments parallel to Proterozoic mobile belts; it mostly reaches economic concentrations where younger orogenic belts cross the older belts (Schuilling 1967) especially in regions such as Bolivia where the melting of older crust has taken place within a zone of black-arc thrusting.

Metallogenesis is apparently restricted to a few linear belts related to the form of the primary crust (Fig. 8.7). Although these belts are parallel to Proterozoic tectonic-magmatic lineaments, they are not constrained to lineaments of any particular age. Tin occurrences are ensialic and linked largely to the *ca*. 1100 Ma mobile belts: the linear zone between southern Africa and Egypt continues directly into a belt of uneconomic deposits in the Laurentian Shield and delineates one of the tin belts defined by Hunter (1973). Other important axial lineaments are located by gold occurrences in Africa, and between Australia and India; they typically occur towards the marginal part of greenstone belts, with basic volcanics as the host rocks. A comparable set of sub-parallel and axial belts is defined by economic pegmatite localities. As already noted, many of the Proterozoic base-metal deposits are peripheral to the supercontinent, and deposits in the Australian and Fennoscandian Shields can specifically be linked to marginal subduction-related processes in Middle and Upper Proterozoic times (Section 8.4, and Watson, 1978).

It is known that some regional distributions, such as that of nickel beneath the Laurentian and Fennoscandian Shields, and chromium beneath southern Africa were already established by mid-Archaean times (Clifford, 1966; Watson 1978) and it would appear from Fig. 8.7 that other metals with continental affinities had also been emplaced in the continental lithosphere by late Archaean crust-forming events. Presumably, the metallic elements have been subsequently concentrated and emplaced at their present sites by the succession of Proterozoic mobile episodes and Phanerozoic orogenic episodes.

9
Late Proterozoic-Cambrian palaeomagnetism: breakup and dispersal of the continental crust

9.1 Introduction

By 1300 Ma the crust had cooled and stabilised to a degree which permitted widespread rifting of continental dimensions. In the developing depressions sedimentation kept pace with rift subsidence, and led to the accumulation of great thicknesses of predominantly clastic deposits. Examples are particularly well developed along the western margin of the Laurentian Shield (Fig. 8.5) and follow patterns analogous to the triple junctions of contemporary plate tectonics (Burke & Dewey, 1973). Very large-scale brittle fracture after this time is also evident in the pattern of radiating dyke swarms which traverse the entire width of the supercontinent. In the Fennoscandian-Laurentian sector this basaltic igneous activity is represented by the Mackenzie, Sudbury and Jotnian episodes at 1250–1180 Ma (Patchett, Bylund & Upton, 1978, section 7.7 and Fig. 9.1) and, in common with alkaline magmatism which commenced at *ca*. 1300 Ma in the Gardar and Keweenawan rifts, it was the precursor of the breakup of the two shields. It cannot be demonstrated that the primitive reconstruction is applicable to the two shields after the Mackenzie and Jotnian episodes (which yield the accordant data sets plotted in Fig. 7.7) although it seems probable that it persisted until after the peak of the Grenville and Sveconorwegian mobile episodes at 1100 Ma (Baer, 1976) because the front zones and tectonic trends would then have formed a single lineament (Figs. 8.4 & 9.1). After *ca.* 1020 Ma this reconstruction is no longer applicable (Fig. 7.8, and Stearn & Piper, 1984). The Fennoscandian Shield had by then broken off from the remainder of the continental crust and rotated clockwise into a secondary configuration (Fig. 9.1) which appears to have been retained with no obvious modifications until the end of Proterozoic times (Patchett & Bylund 1977; Piper, 1982a; Stearn & Piper, 1984). This configuration is utilised here for the analaysis of post-1020 Ma data from the Fennoscandian and Ukrainian Shields (Table 9.1).

Widespread rifting was also an important feature of the Upper Proterozoic development of shields which subsequently formed the Gondwanaland continents. The Indian Shield, for example, was subdivided into the three present cratons at *ca.* 1400 Ma by developing rifts which were to be the sites of episodic sedimentation through to early Palaeozoic times (Naqvi, Divakara Rao & Narain, 1974). Within these rifts, supergroups such as the Cuddapah and Vindhyan accumulated (see also Figs. 7.7, 7.8 & 9.3). In central Africa, rifting, with concomitant development of Upper Proterozoic sedimentary sucessions, was controlled by the pattern of mobile belts and Archaean nuclei, with the former defining lines of rifting and deposition, and the latter continuing to be episodically uplifted (Piper,

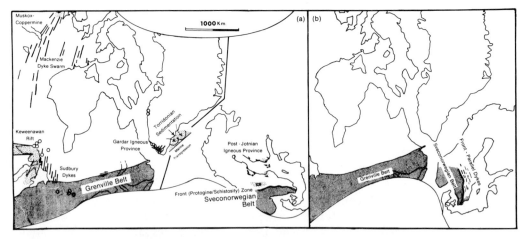

Fig. 9.1 Geological events linked to the first breakup of the continental crust in the Laurentian-Fennoscandian sector at ca. 1000 Ma (a) Prior to ca. 1050 Ma, with major magmatic and tectonic events which occurred between 1250 and 1050 Ma. (b) The situation after ca. 1000 Ma; post-tectonic uplift of the Grenville and Sveconorwegian terranes was responsible for fracturing the adjoining shield margin and led in turn to the intrusion of front-parallel dykes. Dolerite intrusions are indicated in black; circles are alkaline igneous centres. Note that the link between Torridonian rifting and sedimentation and these events is supported by the palaeomagnetic, but not the radiometric evidence. The palaeomagnetic results from the rock suites shown here are discussed in Figs. 7.7, 7.8 & 9.2. Compiled from Patchett & Bylund (1977), Piper (1983a) and Stearn & Piper (1984).

1975c). In Section 7.8 it was shown that palaeomagnetic data from Gondwanaland continents conform to the primitive reconstruction until *ca.* 1000 Ma, although this conclusion can only be regarded as a tentative one for Australia after 1150 Ma, and is dependent on many poorly-dated poles from India to *ca.* 1000 Ma. There is then a gap in the data from these continents (excepting Africa) until 750 Ma. The late-Proterozoic poles no longer conform to the primitive reconstruction (Piper, 1982a) and several analyses have suggested that they conform approximately to the Gondwanaland reconstruction of Smith & Hallam (1970) from 750 Ma onwards (McElhinny & Embleton, 1976; Klootwijk, 1979; Piper, 1982a). However, there is an important geological reason why this reconstruction cannot have been precisely achieved until Palaeozoic times: the Smith-Hallam reconstruction places India adjacent to the Horn of Africa, although it is now known that the NE margin of Gondwanaland, from the Arabian Shield as far south as the Kenya-Tanzanian region, was a continental margin zone of subduction-related magmatism and accretion from *ca.* 1000 Ma until the end of Proterozoic times (Shimron 1980; Vearncombe, 1983 and Section 8.5). This constraint will not permit the placement of the Indian-Madagascar sector north of the Mozambique-Tanzanian margin of Africa prior to Cambrian times. A reconstruction which employs this constraint, while not violating the general accordance of the palaeomagnetic data with the Gondwanaland reconstruction noted by McElhinny & Embleton (1976) and others, is utilised in Fig. 9.2. This reconstruction is referred to as 'Gondwanaland A' to distinguish it from the conventional reconstruction of Smith & Hallam (1970), which is a robust and well-tested version of the assembly of these continents in Palaeozoic and early Mesozoic times (Section 10.6). As will be shown in Section 9.4, the Gondwanaland A reconstruction actually shows better agreement with the palaeomagnetic data from 700 Ma until late Cambrian times.

Thus the peripheral parts of the primitive continental crust comprising India-Antarctica, Australia and the China Shields (collectively referred to as 'east Gondwanaland') on the one side and Fennoscandia on the other, had broken apart and moved into a new configuration by late Proterozoic times. The central portion, comprising South America, Africa (these two continents constituting 'west Gondwanaland'), Khazakhstania (?), Siberia and North America, was rifted (see Burke & Dewey, 1973; Sears & Price, 1978) but appears to have retained the coherent primitive configuration until the end of Proterozoic times (see Section 9.3 and Figs. 9.2 & 9.3). The rotational operations required to reconstruct this Late Proterozoic Supercontinent are listed in Table 9.1. It has a general shape very similar to the later supercontinent of Pangaea; the wider implications of this observation are examined in Section 12.10.

Unfortunately, the late Proterozoic palaeomagnetic record is dependent largely upon results from sedimentary formations. Most of these are weakly deformed, and the possibilities for definitive field tests are correspondingly few; it is probable too, that many of these studies have been inadequately cleaned, so that the nature and age of the remanence are often unclear. Nevertheless, with a few exceptions which are duly examined, it is shown in Sections 9.2 & 9.3 that this record shows a tight conformity with a single APWP, suggesting that the reconstruction of Table 9.1 and Fig. 9.2 was a rigid one in palaeomagnetic terms until Lower Cambrian times. After this time, the APWPs of several major blocks diverge from each other. However, since relative movements between continental plates have by now been recognised in the palaeomagnetic record (Fig. 9.1), all subsequent reconstructions

Table 9.1 The Euler rotational operations applicable to the Precambrian Shields to reconstruct the Late Proterozoic Supercontinent with the Laurentian Shield retained in present-day coordinates. Anticlockwise rotations are positive and clockwise rotations are negative.

	Euler pole °N	°E	Rotation
Australia	73·1	69·0	−170·1
East Antarctica	−63·0	281·0	164·5
India	72·1	291·0	−168·1
Africa	73·0	138·0	−146·0
Arabia	75·5	142·1	−148·8
Madagascar	76·1	155·2	−154·3
South America	−51·0	274·5	122·5
Siberia	66·0	309·0	86·0
Khazakhstania	77·0	208·0	141·5
South China	70·0	323·0	−150·0
North China	55·0	92·0	165·0
Fennoscandia–Ukraine	80·5	274·0	−66·5
NW Scotland	27·7	88·5	−38·0
Greenland	70·5	265·6	−18·0

must be regarded as transient phenomena with temporal continuities which require rigorous testing. Hence, it is likely that the gross analysis given here will need to be modified as more and better palaeomagnetic data become available.

9.2 The 900–700 Ma APW path (Fig. 9.2)

Discussion of the Proterozoic APWP in Section 7.8 finished by noting the southerly extension of the first limb of Loop 8. The loop-like form of the APWP is demonstrated in the Laurentian Shield by the distribution of the youngest magnetisations in the Grenville Province (McWilliams & Dunlop, 1978). It is implied by the sequence of blocking-temperature/coercivity components in the Morin (1), Mealy Mountain (5) and Haliburton (2 and see Section 6.6) intrusions. Few Laurentian data from outside the Grenville Province can with certainty be assigned to this segment of the APWP: certain components from sedimentary formations high in the Keweenawan succession (poles **19,20,21**) may belong here, but they are plotted as small symbols because they could equally well belong to the downward limb of the loop shown in Fig. 7.8. Similar considerations apply to results from the Windermere supergroup in the western part of the shield (**13,15,22**) and indeed, they show a more precise correlation with that path, although the later age has generally been favoured; results from sills into these sediments (**16**) are however, relevant to this path and a single pole from a low blocking-temperature component in the Grenville terrane (**23**) extends the path into a region defined by data from elsewhere.

The return limb of Loop 8 is defined in the Fennoscandian Shield by a wide range of results from dolerite and syenite intrusions into, or bordering, the Sveconorwegian Front Zone. These intrusions are linked in part to post-tectonic uplift of the bordering mobile belt which generated front-parallel fractures in the cold margin of the adjacent terrane (Patchett & Bylund, 1977). The results from these intrusions ('hyperites') correlate closely with contemporaneous cooling magnetisations in the interior of the mobile belt (Stearn & Piper, 1984). Some data (1,2,4,5) correspond directly with the remanence record from the Rogaland (3 and Fig. 7.8) complexes in the western sector while other poles are distributed along the entire return limb of Loop 8 and relate to Rb–Sr ages in the range 960–850 Ma (Patchett, 1978); other results from dolerites in this general area are plotted in Fig. 9.2 as small symbols, either because they are derived from small numbers of samples, or because they cannot with certainty by assigned to this suite. Dolerites in a comparable setting with respect to the Grenville Belt have not been specifically described from the Laurentian Shield; it is possible that the Frontenac and Grenville dykes (8 and 14) have a similar origin, although their age control is too poor to be sure of this. Similar uncertainties relate to the assignment of poles from the Egersund and Hunnedalen dykes from the interior of the Sveconorwegian belt (7,8); they may relate to this path or to its extension in Fig. 9.3.

The poles defining the youngest part of loop 8 are the Haliburton 'C' magnetisation (*ca.* 820 Ma), the Rogaland 'Y' component (*ca.* 890 Ma) and the extension of the distribution of front-parallel dyke poles. They indicate a rapid northward motion of the path at *ca.* 850 Ma, but the continuation of the path after this time can only be inferred indirectly from secondary magnetisations, and the results from sedimentary sequences in which the primary and/or univectorial nature of the remanence is in doubt. Included in this group are results from higher members of the Zil Merdak and Katav supergroups from the eastern Uralian margin of the Ukrainian Shield. These are Upper Riphean rocks linked to K–Ar ages in the range 1000–910 Ma, but they do not conform with other results of this age, and may represent a post-800 Ma extension of the path. The stratigraphic sequence of poles (10 → 11 → 12) in the Katav group is commensurate with this interpretation; the same group of sediments yields a divergent group of poles (6 → 8) which may also be interpreted in a late Precambrian context, because they are coincident with a well-defined group of Lower Sinean (800–690 Ma) poles from the Siberian Shield. The assignment of the poles from the slightly-younger Basinsk group to the continuation of this APWP is more secure, because these Vendian data are coincident with other contemporaneous data and notably with results from the *ca.* 650 Ma Franklin igneous province (next section). The small loop in the APWP at *ca.* 750 Ma (Loop 9) is also suggested by African data (2 and the sequence of remanence directions 1 → 5 in the Otavi group of Namibia) with the return limb constrained by a well-established South American data point. (2).

Two data points (18,23) from the Grenville Province do not plot on Loop 8, but seem unlikely to be relict pre-Grenville components (Section 7.8) because they correspond with no part of the pre-800 Ma APWP. They do however, plot on Loop 9, and a correlation with the curve here is supported by ^{39}Ar:^{40}Ar studies in the case of pole 18 (Baksi, 1982); they would therefore appear to record a late stage in the uplift and cooling of this terrane. The late Proterozoic APWP is defined in

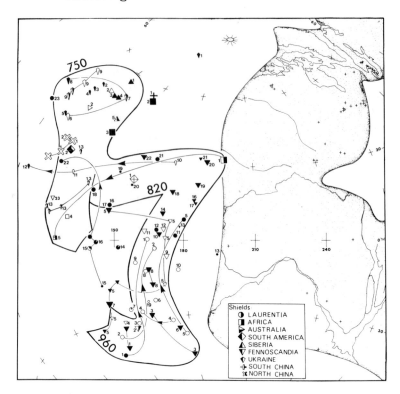

Figure 9.2 Palaeomagnetic Poles *from the* Precambrian Shields *assigned to the interval* 950–700 Ma. *The poles are numbered:* LAURENTIAN SHIELD: *1, Morin anorthosite high coercive force (M1) and medium coercive force (M2) components; 2, Haliburton intrusions HA (960 AA) and HB (820 AA) components; 3, Larrimac and Bryson diorites; 4, Thanet gabbro A2 and A1 (<A2) components; 5, Mealy Mountain E and NW components; 6, Keweenawan shock induced remanence, Slate island; 7, Borden dykes, Baffin Island (819 KA); 8, Frontenac NW dykes (817,751 KA); 9, Michigan basin borehole, igneous unit (post-Keweenawan secondary remanence); 10, Chequamegon Sandstone, primary component; 11, Rama diabase (< Grand Canyon succession, 935, 850 KA); 12, Freda and Nonesuch secondary (late Proterozoic copper mineralisation?); 13, Rapitan Formation X component (900–825 KA); 14, Grenville dykes (974–450 KA); 15, Little Dal Formation A and B components (1100–800, >769 RS); 16, Tsezoteme Formation Sills (769, 766 RS); 17, Beartooth dykes (730 KA); 18, Tudor gabbro (710 AA, 670 KAI); 19, Middle River Section, Amnicon and Orienta Formations, K1 and K2 components; 20, Eileen section (<19); 21, Saulte Ste. Marie section; 22, Little Dal lavas; 23, Cordova C remanence.* FENNOSCANDIAN SHIELD *1, Hyperite intrusions and syenite in southern part of Front Zone (1595–760 KA); 2, Hyperites, central part of Front Zone; 3, Migmatites and anorthosite complex, Rogaland; 4, Karlshamn dyke (916,853 RS); 5, Results from miscellaneous undated hyperite dykes close to Font Zone; 6, Amphibolites, Swedish sector of Svecornorwegian belt; 7, Hunnedalen dykes (<950–850 RS,KA); 8, Egersund dolerites (<950–850 RS, 850–663 KA); 9, Tuve (WNW) dolerites; 10, Rogaland Y remanence (903–850 RS, 890 KAH); 11, Årby dyke (974 RS); 12, Falun dyke (946,895 RS); 13, Bö dyke; 14, Tärnö dyke (861); 15, Masattar dolerite; 16, Ejen dolerite; 17, Väby dyke; 18, Nilstorp dyke (963 RS); 19, Listed dyke; 20, Bräkne-hoby dyke (955,862 RS); 21, Fäjö*

Australia by the stratigraphic sequence of poles from the Adelaide 'geosyncline' in the Flinders Range (McWilliams & McElhinny, 1980) where the basal Wooltana volcanics, which immediately underlie the Marinoan glacial deposits, yield a pole (2) coincident with the Lower Sinean data. Poles from higher levels in the succession are distributed along the 650–600 Ma path plotted in Fig. 9.3. The oldest palaeomagnetic results currently available from the South China Shield are poles from the Lower Sinean sediments of the Xuining District (Liu & Feng Hao, 1965). When these poles are superimposed on the Lower Sinean data from elsewhere they imply a position of this shield close to the Australia-Antarctica sector of Gondwanaland; this inference must be regarded as highly tentative because it is based on few data subjected to a.f. cleaning only. A single late Precambrian pole from the North China Shield (Lin, Fuller & Zhang, 1985) can provide no unique solution to the whereabouts of this block, but it falls on the APWP at 800–700 Ma when a position adjacent to Australia in a configuration similar to the primitive reconstruction is used (Figs. 7.1 & 9.2). There is therefore no evidence that this shield had yet broken apart from Australia.

9.3 The 700–570 Ma APW path (Fig. 9.3)

The best-defined part of this last segment of the Proterozoic APWP is represented by data from the Franklin igneous province — which includes widespread rifting and basaltic magmatism extending across northern Canada into Baffin Island and NE Greenland (Fig. 9.10); it may also extend sporadically south into the Superior and Grenville Provinces (Park, 1974). The data include results from sediments (4), lavas (8) and a wide range of dykes and sills (e.g. Fahrig, Irving & Jackson (1971) and see Palmer, Baragar, Fortier & Foster (1983) for a recent study). Unfortu-

---

dyke; 22, Bolshavn dyke A component (1000–800); 23, Kjeldsea dyke. UKRAINIAN SECTOR: Zil-Merdak Supergroup and Katav Group sedimentary formations: 1, Nugush and Biryan sequences; 2, Biryan Sequence; 3, Nugush Sequence; 4, Lemezinsk sequence sandstones; 5, Bedershinsk sequence; 3, Lower Katav sequence; 7, Middle and Upper Katav sequence limestones; 8, Podinzer sequence limestones (1000–930 KA); 9, Katavsk Group limestones, four poles (960 KA); 10, Lower Katav Group limestones and sandstones; 11, Middle and Upper Katav Group limestones; 12, Upper Katav, Podinzer Sequence limestones; 13, Basinsk Group limestones, five poles (Vendian). NORTH CHINA SHIELD: 1, Xuining Sandstones, three poles (Lower Sinean); 2, Lianta Series (Lower Sinean). SIBERIAN SHIELD: Lower Sinean sediments, eastern Siberia (800–670 Ma): 1, Omnirisk Group (maimakan suite); 2, Lakhanda Group; 3, Kakutsk area sediments; 4, Tsipanda group; 5, Malginsk Group (Malgin Creek suite). AFRICAN SHIELD: 1, Otavi Group, DCI component (<840 RS, >651 RS, 770–740 UP); 2, Pre-Nama dykes (640 RS); 3, Group de Char secondary component (Pan African?); 4, Ikorongo Group; 5, Otavi Group, DC2 + 3 secondary components. SOUTH AMERICAN SHIELD: 1, Bambui Group (886 RS); 2, La Tinta Formation (769, 723, 695 RS). AUSTRALIAN SHIELD: 2, Wooltana Volcanics (<900–850 Ma, Late Precambrian). Lambert equal area projection centred at 200°E, 0°N. Format symbols and abbreviations are the same as for Figure 7.1.

nately, the age constraints on this episode are still rather broad: K–Ar ages are mostly in the range 965–655 Ma, with a mean of 674 Ma, while an individual K–Ar isochron age is somewhat younger (Palmer & Hayatsu, 1975). With the possible exception of lavas from Quebec (13) linked by K–Ar dates to the Proterozoic–Cambrian boundary (Dankers & Lapointe, 1981), these may be the only data from the Laurentian Shield applicable to late Precambrian times. Certain other poles from sedimentary formations which were originally used by Morris & Roy (1977) to define a long APWP of this general age running from the equatorial Pacific to the central Atlantic via the present North Pole (the 'Hadrynian Track'), are probably all secondary magnetisations of uncertain age. Furthermore, the key anchor points at either end of the path were the Torridonian data (see NW Scotland sector in Fig. 7.8) which have since been shown to yield a time sequence in the opposite direction to that required by the interpretation of Morris & Roy, and to be relevant to an older part of the APWP (Fig. 7.8; Smith, Stearn & Piper, 1983). Some possible further implications of the analysis of Morris & Roy (1977) and Lapointe, Roy & Morris (1979) are noted in the second part of Section 9.4.

The immediate extension of this path to the south east is well defined by poles from African igneous rocks (6,10,11), assigned to the interval 620–600 Ma, while the remaining poles from late Precambrian sediments in this shield yield accordant data. Results from the Adelaide geosyncline in Australia continue with eight results from sediments (2 → 10) which extend upwards into rocks spanning the Proterozoic–Cambrian boundary (13,14), but with an erosional disconformity of uncertain duration between poles 10 and 14. Although connected here in stratigraphic sequence, these poles are unlikely to form an exact time sequence because the time of imprint of the characteristic magnetisations is unclear, and there are some results (2,7) which are removed from the general progression of pole positions. However, a clear history of partial overprinting by later orogenic events has been recognised (McWilliams & McElhinny 1980; Klootwijk, 1980) and the assignment of the poles to the latest Precambrian seems to be sound. In addition, the younger poles spanning the Vendian–Cambrian boundary form a sequence (6 → 7 → 9 → 10 → 13 → 14) which defines a hairpin in the APWP, marking the SE extremity of the distribution of poles along this swathe. The lowest formations in the Amadeus basin in central Australia, spanning the boundary, yield poles in general agreement with the data from the Adelaide geosyncline (Kirschvink, 1978; Klootwijk, 1980), although the hairpin extension to the path is not recognised here. The important body of recent Siberian data comes from sediments assigned to the Vendian (Upper Sinean) on palaeontological grounds; their agreement with the Franklin data and the latest Proterozoic results from Australia and Africa is particularly impressive and gives added confidence to the definition of this part of the path. The Indian results are derived from the Upper Vindhyan sediments of late Riphean or Vendian age; the Rewa and overlying Bhander sandstones yield a cluster of poles which only correspond with the APWP at 650–600 Ma, and this is therefore likely to be the remanence age.

The palaeomagnetic record of the Proterozoic–Phanerozoic transition is at present best understood in Siberia and Australia, where sedimentary successions spanning the boundary have been studied by both palaeontologists and palaeomagnetists. In the Siberian Shield, only two early Cambrian poles (6,7) plot on the APWP illustrated in Fig. 9.3. The remainder are entirely removed from this path, implying that they are either later overprints or that the Siberian Shield had broken away from the other shields by Lower Cambrian times; a case for the latter possibility is argued below. Results from Australia, however, describe a direct continuation from the latest Proterozoic results: Lower and early Middle Cambrian sediments from the Amadeus basin yield a tight cluster of dual-polarity poles plotting close to the termination of the latest Proterozoic group (20). These data are not significantly different from late Lower Cambrian (15) and Middle Cambrian (16,17,18) results from the formations in the Adelaide geosyncline, and this agreement precludes major relative movements between the peripheral and central parts of Australia in later times (Klootwijk, 1980). However, by computing separate mean poles from Lower Cambrian, Middle Cambrian and Cambro-Ordovician data from the Amadeus basin and Adelaide geosyncline, McWilliams (1981) demonstrated a small systematic difference prior to late Cambrian times. This might be explained by rifting movements between the two regions during the Delamerian orogeny, and could be accommodated by rotation about a local Eulerian pole, but a reconstruction with the Adelaide geosyncline adjacent to the Amadeus basin, as shown by McWilliams, would appear to be ruled out by the nature of the intervening crust in the Eromanga basin between them (Finlayson & Mathur, 1984).

The Lake Frome group (18,19) defines the end of sedimentation in the Flinders Range, prior to the onset of orogeny. Results from near the top of the succession in both this area (19) and in the Amadeus basin (21,23) suggest APW movement away from the Lower–Middle Cambrian group of poles sometime in Middle Cambrian times. The APWP then extends to embrace the poles from secondary overprints (25,26,27) attributed to the Delamerian orogeny (McWilliams & McElhinny, 1980; Klootwijk, 1980). The primary remanences from Australian Cambrian formations thus fall into two groups, with the suggestion of a rapid movement between the two; the stratigraphic assignments are not entirely in accord with this simple picture (cf. poles 17 and 23) although these discrepancies are more likely to be due to uncertainties of remanence acquisition during protracted diagenesis than to more complex APW movements. This latter point is emphasised by comparison with results from elsewhere in eastern Gondwanaland: poles from the Lower Cambrian Khewa (purple) sandstones and the Middle Cambrian Baghanwala (salt pseudomorph) beds of NW India (4,5) are both coincident with the second (Middle to early Upper) Cambrian group from Australia, after correction for regional rotation of the Salt Range (McElhinny & Embleton, 1976). African poles (16–18) of late Lower to Upper Cambrian age show the same accord with this position, while results from two further secondary remanences in the Pan-African belt of Namibia (15,19,23) are in general agreement with these results. An additional group of African poles from dyke swarms of NE Africa (20–22) plot 30° to the NW. They are linked to ages in the range 530–464 Ma and may be slightly younger than the Middle Cambrian group noted above; they would then imply a small loop in the APWP (of Gondwanaland) which is also suggested by second-rank African (7,8,15) and South American (3,4) data. The collective South American results applicable to this time interval comprise nine poles from sedimentary formations of Western Argentina, broadly attributed to Cambrian times because they underlie Lower Ordovician (Tremadocian) sediments. Although the South American results are only third-rank data

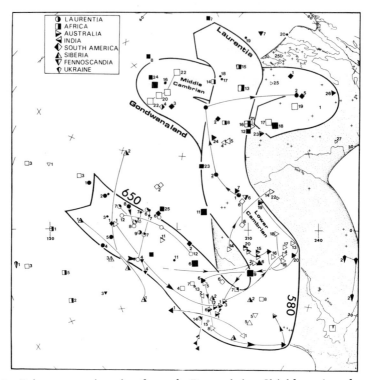

Figure 9.3 Palaeomagnetic poles *from the* Precambrian Shields *assigned to the interval* 700 Ma to Lower Cambrian times *(to Upper Cambrian-Lower ordovician times for Gondwanaland). The poles are numbered:* LAURENTIAN SHIELD: 1, *Tudor gabbro (710–640 AA, KAI);* 2, *Beartooth dykes (730 KA);* 3, *Rapitan Z component;* 4, *Reynolds Point Shales, Shaler Group (intruded by units incorporated in poles 7);* 5, *Grenville dykes (970–450 KA);* 6, *Mt. Nelson Formation, secondary component, (later Precambrian Goat River Orogeny);* 7, *Franklin Igneous Province, dykes and sills (675–625 KA, 625 KAI);* 8, *Natkusiak plateau basalts, middle reversed and upper normal lavas;* 9, *Coronation sills (800–500, 647 KA);* 10, *Cloud Mountain basalt (615 KA, 605 AA by reference to Long Range dykes);* 11, *Chequamegon Sandstone, primary component;* 12, *Johnnie Formation, corrected for local rotation (latest Precambrian);* 13, *Quebec (Buckingham area) lavas, three poles (573 KA);* 14, *Lodore Formation NRM result (late Precambrian-Cambrian);* 15, *Johnnie Formation, corrected for local rotation (latest Precambrian);* 16, *Lamotte Formation, component 2a,* 17, *Hazel Formation, New Mexico, NRM results from folded and unfolded unit (Late Precambrian-Cambrian);* 18, *Rapitan 'Y' magnetisation;* 19, *Orienta sandstone secondary components;* 20, *Chequamegon Sandstone, secondary component;* 21, *Jacobsville Sandstone, J2 component.* GREENLAND SECTOR *(underlined):* 1, *Ikertôq shear belt pseudotachylites (late Precambrian-early Palaeozoic);* 2, *West Greenland Kimberlite-lamprophyre suite, three poles (587 RS, 580–570 KA).* SIBERIAN SHIELD: 1, *Nyarovei and Kokpel Groups, Polar Urals, two poles (Vendian);* 2, *Maninsk, Nyarovei and Kopel Groups, Polar Urals, four poles (Vendian);* 3, *Sinian sediments, Patom River;* 4, *Karagasski suite, Sayan Region (Upper Sinean, 670–590 Ma).* 5, *Amphibole plagioclase gneisses (end-Baikalian metamorphism, 621–610 KA);* 6, *Podrasnotsvetnaya Group (Lower Cambrian);* 7, *Charsk Group, Olekma River (Lower Cambrian).* FENNOSCANDIAN SHIELD: 1, *Norwegian Sparagmites (Late Precambrian, Caledonian overprint?);* 2,

Vigehavn dyke; 3, *Hunnedalen dolerites (<850 KA,RS);* 4, *Egersund dolerites (850–663, for results 3 and 4 see also Figure 9.3);* 5, *Båtsfjord dolerites, Varanger Peninsula (640 KA);* 6, *Alnö Complex, shallow component (553 RS, 584 KA). UKRAINIAN SECTOR:* 1, *Lower Inzer Group;* 2, *Kuk-Karauk Group, three poles (Vendian). AFRICAN SHIELD:* 1, *Otavi Group DC2+3 component;* 2, *Upper Nama Group, N2 component (probably prefolding, 686–553 RS);* 3, *Moroccan red sandstones, three poles (Late Precambrian);* 4, *Nosib Group, NQ3 component;* 5, *Mulden Group sediments (560–550 RS on shales);* 6, *Ntonya Ring Structure (617 RS);* 7, *Nama Group, N3 component (post-folding) and two poles from secondary magnetisations in the Nama Group;* 8, *Plateau Series (Late Precambrian-Cambrian);* 9, *Sijarira Group (Late Precambrian-Cambrian);* 10, *Dokhan Volcanics (665–605 KA);* 11, *Adma diorite (616 UP, 616–590 KA);* 12, *Nosib Group, NQ2 component;* 13, *Lower Nama Group, N1 component (686–553 RS);* 14, *Mbozi Complex (750 KA);* 15, *Blaubecker Formation, NA component (<670 RS, Pan African);* 16, *Moroccan lavas (Lower Middle Cambrian);* 17, *Ben Azzer Volcanic sediments (<532 UP, Middle Cambrian);* 18, *Adras de Mauretainie CO 10 (Cambrian-Ordovician boundary);* 19, *Doornpoort Formation, Pan African overprint (500–500 Ma);* 20, *Qena-Safanga (QS) dykes (530–480 Ma);* 21, *Um-Rus (RS) dykes (497–464 KA);* 22, *Esh-el-Mellaha dykes, three poles from acid, intermediate and basic dykes (580–480, magnetisation linked to 530 Ma Pan African metamorphism);* 23, *Blaubecker Formation, NBX overprint (Pan African)* 24, *Buanji Series overprint (Pan African?). ARABIAN SHIELD SECTOR:* 25, *Jordanian Red beds (Cambrian-Ordovician). SOUTH AMERICA:* 1, *Purmamarca sediments (Cambrian);* 2, *Salta sediments, four poles (Cambrian or Ordovician);* 3, *Campanario Formation, North Tilcara (Cambrian);* 4, *Campanario Formation, South Tilcara (Cambrian);* 5, *Abra de Cajas sediments (Cambrian);* 6, *Salta and Jujuy sediments, (Late Precambrian-Ordovician). INDIAN SHIELD:* 1, *Rewa Sandstones, two poles from east and west outcrops (Late Precambrian);* 2, *Bhander Sandstone, SE outcrop (U. Vindhyan, 1);* 3, *Bhander Sandstones, NE outcrop (U. Vindhyan, late precambrian, <1);* 4, *Khewra (Purple) Sandstone (Lower Cambrian);* 5, *Baghanwala (Salt Pseudomorph) Formation (Middle Cambrian). AUSTRALIAN SHIELD:* 1, *Arumbera Sandstone (Late Precambrian-Cambrian). Stratigraphic sequence of poles from Adelaide Geosyncline (Late Precambrian <2 in Figure 9.3):* 2, *Copley Quartzite;* 3, *Merinjina Tillite;* 4, *Tapley Hill Formation;* 5, *Angepena Formation;* 6, *Brachina Formation, two poles;* 7, *Bunyeroo Formation;* 8, *Lower Marinoan Sandstone;* 9, *Upper Marinoan Sandstone;* 10, *Pound Quartzite (possibly secondary remanence);* 11, *Antrim Plateau Volcanics (Late Precambrian, <653):* 12, *Todd River Dolomite (Lower Cambrian);* 13, *Aroona Dam sediments (Lower Cambrian);* 14, *Hawker Group, basal levels and higher levels, two poles (Lower Cambrian);* 15, *Kangaroo Island red beds (late Lower Cambrian);* 16, *Billy Creek Formation, Wirrealpa Limestone- Aroona Creek Limestone (late Lower Cambrian-early Middle Cambrian);* 17, *Pantapinna Formation (late Middle Cambrian);* 18, *Lake Frome Group, basal and lower formations (Middle Cambrian);* 19, *Lake Frome Group, higher levels (Middle-Upper Cambrian);* 20, *Amadeus Basin, primary directions from Deception, Ilara, Tempe and Giles Creek (two poles) Formations, (Lower-Middle Cambrian);* 21, *Amadeus Basin, Shannon Formation;* 22, *Brachina Formation, secondary component (Cambrian-Ordovician?);* 23, *Hugh River Shale (Lower-Middle Cambrian);* 24, *Hudson Formation (Middle Cambrian);* 25, *Amadeus Basin Dalamerian orogeny overprint in Ross River section and Deception Formation (Upper Cambrian-Lower Ordovician);* 26, *Kangaroo Island, Delamerian overprint;* 27, *Merinjina tillite, Delamerian overprint (?). Lambert Equal area projection centred at 200°E, 0°N. Format, symbols and abbreviations are the same as Figure 7.1.*

they exhibit a general agreement with results from elsewhere in Gondwanaland (R. Thompson, 1972). The only Fennoscandian pole relevant to this path accords with results from similar alkaline rocks in Greenland. In common with Siberia, other Cambrian results from this plate show no agreement with the results from elsewhere on the Proterozoic reconstruction, and are discussed in the concluding part of Section 9.4.

It remains to note the late Proterozoic–early Cambrian data which do not conform directly with the APWP as illustrated in Figs. 9.2 & 9.3. Poles from the Malani rhyolites (745 Ma) of India, the YB dykes (750 Ma) of the Yilgarn craton in Australia, and Mbozi complex (710 Ma) of central Africa agree with one another on the Smith-Hallam configuration of the two shields, and provide the essential evidence that the two continents had moved together into this reconstruction by 750 Ma. They do not conform to the APWP shown in Fig. 9.3 on either the primitive or the Gondwanaland configurations, and suggest that either (i) APW movements were more complicated between loop 8 and the Lower Sinean (*ca.* 700 Ma) position, or (ii) these data record transitional movements between the two reconstructions. Utilising the data from the Damaran mobile belt and bordering zones in Namaqualand, McWilliams & Kröner (1981) develop an APWP which incorporates the antipoles of data points **2**, **4** and **5** in Fig. 9.3 and is hence very much longer than the path developed here from the wider data base. These authors admit the ambiguity in polarity of their path, and all of their data points are accommodated in this version at positions appropriate to their ages, except the Nama group N1 component. In this group only the N2 component is demonstrably pre-folding (**2**) and falls near the path at *ca.* 650 Ma commensurate with the 686–553 Ma estimate of the age of the Nama group. Some other postfolding components (**7**) fall along the path at slightly younger positions. The N1 component however, only matches the path at the Cambro-Ordovician position (**13**) in Fig. 9.3. The fold test on this component is not diagnostic, and the remanence is only present in the lower formations, in contrast to the N3 overprint which is ubiquitous (Kröner, McWilliams, Germs, Schalk & Reid, 1980). An overprinting of deeper levels by a thermal event during the Damaran mobile episode (see poles **15,19,23** in Fig. 9.3, and see also tectonic discussion relating to Fig. 9.5) is therefore a not improbable origin for this remanence.

9.4 The Cambrian–Ordovician APW paths (Fig. 9.4)

The most important feature of palaeomagnetic results from this time interval is the manner in which the data from separate continental plates become explicitly divergent and cease to conform to a rigid reconstruction. They illustrate the classic palaeomagnetic signature of continental breakup (third part of Section 6.6 and Piper, 1982a, 1985a). From Cambrian times onwards it is necessary to discuss the APWPs of each separate continental plate individually. Initially however, to demonstrate that this divergence commenced at the Proterozoic–Cambrian boundary, and is a quite unambiguous feature of the palaeomagnetic record, the Cambrian–Middle Ordovician APW records of each shield are examined with the data rotated to the late Proterozoic reconstruction of Figs. 9.2 & 9.3. It is immediately apparent from Fig. 9.4, that the data in no way conform to this reconstruction, and that it must therefore, have begun to break apart during Lower Cambrian times.

Gondwanaland

Although different assessments of the Gondwanaland* palaeomagnetic data are in agreement that the poles from the constituent continents conform to a single APWP after Middle Cambrian times (McElhinny & Embleton, 1976; Klootwijk 1979; McWilliams, 1981; Piper, 1982a), this observation is less securely based for Upper Cambrian and Ordovician times than it is for Lower and Middle Cambrian times (see discussion included in Section 9.3). First-rank data come from the Tasman geosyncline of SE Australia (Goleby, 1980) where the Molong tectonic province includes Lower Ordovician andesites (7,8) overlain by Middle to Upper Ordovician sediments (9,10). Collectively, the data imply a hairpin near the Cambrian–Ordovician boundary (defined by the secondary remanences linked to Pan-African events and orogeny in SE Australia, Section 9.3), with a southerly extension to embrace the Lower Ordovician poles, and a loop here to continue the path to the Middle Ordovician poles from Australia. As observed by McElhinny, Giddings & Embleton (1974), the summary APWP for Gondwanaland in late Proterozoic to Ordovician times accommodates the incidence of late Proterozoic to early Cambrian glaciation in West Africa; younger rocks here include Lower and later Cambrian warm-water deposits, succeeded in turn by drab-coloured Ordovician deposits, and culminating in deposits formed during the great Saharan glaciation of Middle–Upper Ordovician times (Beuf, Biju-Duval, de Charpal, Rognon, Gariel & Bennacef (1971) and first part of Section 10.3).

The timing of the transition from the Gondwanaland A to the Gondwanaland B configuration can be determined by calculating mean poles along segments of the APWP illustrated in Figs. 9.2–9.4 separately for the eastern segment of Gondwanaland (the data here come solely from Australia and India), and comparing this path with the mean APWP for the remainder of the Late Proterozoic supercontinent. This test is illustrated in Fig. 9.5c. It is not a very definitive one, but the Gondwanaland A configuration gives a consistently better match to the APWP until late Cambrian or early Ordovician times, when the Gondwanaland B (Smith-Hallam) reconstruction yields a better fit. The latter reconstruction is well established for subsequent times (e.g. Klootwijk, 1979, and Section 10.6). This analysis implies that 3000 km of sinistral strike-slip motion took place between eastern and western Gondwanaland in Cambro-Ordovician times. The motion was synchronous with a hairpin in the APWP (defining a change in the direction of motion of Gondwanaland and rotation about a new Eulerian pole, third part of Section 6.6) and it correlates with major tectonism in the Pan-African belts and the Tasman orogenic belt in Australia. This model is comparable with that of McWilliams (1981), in that it implies relative movements between East and West Gondwanaland at this time, but in contrast to the latter

---

* The breakup of the late Proterozoic supercontinent created a number of continental plates which comprise not only the major shields of late Archaean origin, but also peripheral accretions of Middle and Upper Proterozoic age. The shield names are therefore no longer appropriate, and following Ziegler, Scotese, McKerrow, Johnson & Bambach (1979) the Lower Palaeozoic plates are referred to here by their area names, such as Gondwanaland, Baltica, Laurentia and Siberia.

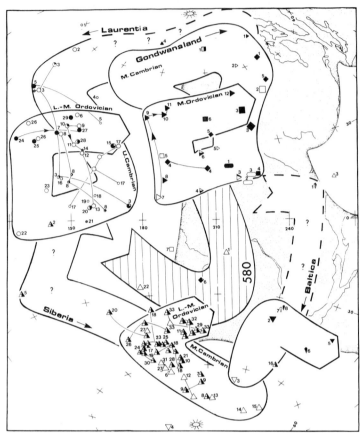

Figure 9.4 A summary of palaeomagnetic data for Late Cambrian-Mid Ordovician times from Gondwanaland (B reconstruction) and late Lower-Upper Cambrian data from the Laurentia, Baltica and Siberia. Although these data are plotted on the Late Proterozoic Supercontinent reconstruction, this is merely intended to demonstrate the explicit divergence of the APWPs for these shields in Cambrian times and is not otherwise relevant to Lower Palaeozoic times. AUSTRALIAN SHIELD (right directed triangles): 1, Ross River Section Delamerian overprint (C-O); 2, Deception Formation Delamerian overprint (C-O); 3, Tumblagooda Sandstone (Ol); 4, Dundas Group (Ol); 5, Brachina Formation, secondary component; 6, Jinduckin Formation (O); 7, Mt. Pleasant andesite (O); 8, Walli andesite (Ol); 9, Cliefden Limestone (Om-u); 10, Malongulli Formation (Om-u); 11, Angullong Tuff (Om-u); 12, Stairway Sandstone (Om, Upper Llanvirn-Llandeilo). AFRICAN SHIELD (squares): 1, Blaubecker Formation overprint (Pan Africa); 2, Doornpoort Formation (Pan African folding remanence); 3, Adras de Maurentainie CO 10 (C-O boundary); 4, Tassili sediments (C-O); 5, Table Mountain Series (O-S); 6, Hasi-messaud sediments (C-O); 7, Hook intrusives (500 RS). SOUTH AMERICA (diamonds): 1, Bolivian sediments (O); 2, Abra de Cajas sediments (C-O); 3, Suri Formation (Om, Llanvirn); 4, Urucum Formation (O-S); 5, Jacaido Series sediments (O-S). ANTARCTICA SHIELD: 1, Charnockites, Mirny'y station (Cu-Ol, 502 RS); 2, Lamprophyre dykes, Taylor Valley (Ol, 470 KA); 3, Ongul Gneiss (470, Pan African overprint). LAURENTIAN SHIELD (circles): 1, Abrigo Formation (Cm-u); 2, Muav Limestone (Cm); 3, Colorado intrusive suite (530–511 RS); 4, Covey Hill Formation, Potsdam Group (Cl-m); 5, Chateauguay Forma-

model it requires shear motion in place of continental collision. In both models the implied suture lies within the Pan-African Mozambique Belt, and is now concealed beneath post-mid Triassic rocks which accumulated here when this zone was rifted apart during the breakup of Gondwanaland B. The transition from Gondwanaland A to Gondwanaland B however, is part of a wider set of movements which link the collective tectonics of the Pan-African belts in central and eastern Africa with the single sinistral motion of eastern Gondwanaland around the SE margin of Africa (Fig. 9.5d). In Namibia the Damaran belt is dominated by an

---

tion, Potsdam Group (Cu); 6, Rome Formation (late Cl); 7, Lamotte component 1; 8, Wichita Granites and gabbros, four poles (525 RS). Llano Uplift, Texas, five poles in stratigraphic sequence: 9, Riley Formation, Hickory Sandstone member (Cm-u); 10, Riley Formation, Cap Mountain Limestone member (Cu); 11, Riley Formation, Lion Mountain limestone (Cu); 12, Wilberns Formation, Welge Sandstone-Morgan Creek limestone (Cu); 13, Wilberns Formation, Point Peak Member (Cu); 14, Caldwell Group and Thetford Mines (Cu-Ol); 15, Waynesboro Formation (Cl-m); 16, Unicoi Formation, in situ intermediate blocking temperature component; 17, Bradore Formation, W. Newfoundland (Cl); 18, Carrara-Bonanza King Formation, corrected for tectonic rotation (Cm); 19, Carrara Formation (Cm); 20, Tapeats Sandstone (Cl); 21, Wood Canyon Formation, corrected for tectonic rotation (Cl); 22, Quebec dykes (497 KA); 23, Grenville front dykes (492–401 KA); 24, Chapman Ridge Formation (Ol-m); 25, St. Georges dolostone, Newfoundland (Ol); 26, St. Georges limestone (early O); 27, Mocassin Formation (Om, Llanvirn-Llandeilo 24); 28, Bays Formation (Om, Llanvirn-Llandeilo, <24); 29, Martinsburg Formation prefolding remanence (Om). SIBERIAN SHIELD (upright triangles): 1, Podrasnotsvetnaya Group, Lena River (Cl); 2, Sediments, Lena River Section (early Lower Cambrian); 3, East Aldan sediments (Cl); 4, Charsk Group, Olekma River (Cl); 5, Ust'kunda Group, Kuznets Alatau (Cl); 6, Ust'Agul'sk Group (Cl, 609 KA); 7, Emyaksa Group, River Olenek (Cl); 8, Chernoles group, Middle and Upper Aldan River (Cl); 9, Ust'Maisk Group (Anga Group, Cm); 10, Olenek River Group (Cm); 11, Silisir, Dzhakhtar and Olenek Groups (Cm); 12, Ust'Maya group (Cm); 13, Ust'Maya and Chaya groups River Maya (Cm); 14, Amga River sediments (Cm); 15, Ust'botoma Group, River Lena (Cm); 16, Izluck'ye Suite (Late Precambrian-Cambrian); 17, Verkholensk Group, River Lena (Cu); 18, Verkholensk Group, Irkutsk area (Cu); 19, Verkholensk Group, Angara and Oka (C); 20, Verkholensk Group (Cu); 21, Evenkiisk Group, River Angara (Cu); 22, Gornaya Altai Group (Cu-O); 23, Verkholensk Group, River Nepa. 24, Ilginsk Group, River Lena (Cu). 25, Chukuk and Markha Groups, River Olenek (Cu); 26, Upper Lena Group sediments, three poles (Cl); 27, Lena River Sills (Cu-Ol); 28, Ust'Kutsk Group, Lena River (Ol); 29, Chrertovsk group (Ol); 30, Ust'Kutsk Group and Kazimirovsk Group, three poles (Ol); 31, Limestones, River Alakit (Ol); 32, Lena River sediments (Om); 33, Makarovskoe Group (Om-u). FENNOSCANDIAN-UKRAINIAN SHIELD (inverted triangles): 1, Volyn, Sokolets and Yaryshev sediments (C); 2, Nexö Sandstone (Cl); 3, Fen complex rødberg (600–530 KA, RS); 4, Alnö Complex, steep component (553 RS, 554 KA); 5, Vanern Limestones (O); 6, Obolus Limestones (Ol, Tremadoc); 7, Leningrad district limestones (O); 8, Obolous Sandstones (Ol, Tremadoc). In the age assignments C – Cambrian, O – Ordovician and l, m, u – Lower, Middle and Upper divisions; otherwise format, symbols and abbreviations are the same as Figure 7.1.

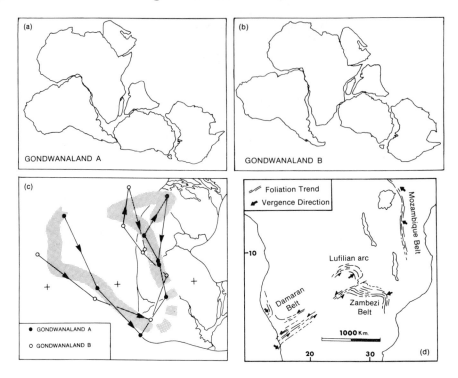

Fig. 9.5 (a) and (b) show the Gondwanaland A and B reconstructions. (c) shows a comparison of mean poles calculated from the data of Figs. 9.2 & 9.3 from Eastern Gondwanaland, with the mean path calculated from data from the remainder of the Late Proterozoic Supercontinent. The dashed segment of the path is defined by data from eastern Gondwanaland only. (d) illustrates tectonic elements in the Pan-African domains of southern Africa, linked to the transition from the Gondwanaland A to the Gondwanaland B configurations, (after Coward & Daly, 1984).

intense NE–SW sinistral ductile shear zone (Downing & Coward, 1981) which continues directly into the Mwembeshi Zone of Zambia. Associated movements along mobile zones between the Kasai, Zimbabwe and Bangweulu-Tanganyika blocks produced the Zambezi and Lufilian mobile belts which include overthrusting onto the adjacent cratons, crustal thickening, and high-pressure metamorphism (Coward & Daly, 1984). In the Mozambique Belt, a contrasting NW to WNW overthrusting direction is deduced from shear zones and mineral elongations (Coward & Daly, 1984) but is consistent with the continuing northwards convergence of east Gondwanaland against this margin of Africa (Fig. 9.5d).

Laurentia

The Cambrian palaeomagnetic data from North America have been summarised and critically assessed by Watts, Van der Voo & Reeve (1980). More than half of these poles are third-rank data, with problematic ages of remanence or uncertain tectonic control. They include results from the perimeter of cratonic North America, including thrust slices within the Desert Ranges of Nevada (Gillett & Van Alstine, 1979) and the Appalachian Belt (Watts, Van der Voo & French 1980); although well-tested, the possibility of local rotations cannot be excluded here. The minority of igneous data assigned to this period from the Colorado intrusives (3) and Wichita granites (8) help to constrain the overall APWP; other first-rank data include results from the Llano uplift of Texas which yields the stratigraphic sequence 9 → 13. In spite of the reservations noted, all of these poles fall on a streaked distribution running from *ca.* 155°E, 0°N to 90°E, 60°N (Fig. 9.4). Several explanations for this distribution are possible (Watts, Van der Voo & Reeve, 1980) but the most probable one is that they represent a single APW swathe. A tight loop has been proposed to accommodate these data in their entirety, but it seems unlikely that an outward path is recorded (a possible exception being pole 6 from the late Lower Cambrian Rome formation). The extremity is defined by two poles from sediments of Middle or early Upper Cambrian age (1,2). Although these poles are coincident with younger North American poles of Permo-Triassic age, this seems to be a real Cambrian pole position because it is defined by a range of igneous data from the Colorado alkaline intrusive suite (3) with an assigned age of 530–511 Ma. The stratigraphic sequence of Upper Cambrian poles from the Llano uplift defines the latter part of this path.

The first-rank poles, and possibly all of the data, incorporated in this swathe are late Lower Cambrian or younger in age, and therefore refer to the interval *ca.* 540–500 Ma (or possibly less, see Patchett, Gale, Goodwin & Humm, 1980). There is thus a gap of 50–100 Ma between the Franklin igneous episode and late Lower Cambrian times during an interval of rapid APW which is practically unrepresented by data from Laurentia. The possible outward limb of a Cambrian APW loop applicable to this plate is recorded in the *ca.* 590–570 Ma West Greenland alkaline province (Greenland poles 1–2 in Fig. 9.3) where the characteristic remanences in pseudotachylytes and igneous intrusions have inclinations ranging from shallow to steep (Piper, 1981a). A continuation of this early Cambrian path into the central Atlantic, in order to reconcile the probable polarities of the Precambrian and Phanerozoic APWPs (McElhinny, Giddings & Embleton 1974; Piper, 1981a; see first part of Section 10.3) is not a valid interpretation, because the shortest APWP connects the late Proterozoic–early Cambrian data from Greenland in Fig. 9.3, with the late Lower to Upper Cambrian data in Fig. 9.4 via the NW Pacific (Loop 11). It may however, be correct to include the secondary magnetisations considered by Morris & Roy (1977) in the early Cambrian APWP, for three reasons: (i) The steep components of magnetisation in North American sediments of late Precambrian age are more likely to be linked to thermal-tectonic events such as rifting (Section 9.6) of similar age to the diagenesis, than to the Recent field direction of no particular significance to the geological history of these rocks. To this extent these steep components would be analogous to the widespread overprinting of Ordovician to Lower Carboniferous sediments by a Permo-Carboniferous remanence, during protracted diagenesis or later thermal-tectonic events (Section 10.4). (ii) An early Cambrian age for these components would explain why they are frequently present in rocks of this age, but are not observed in younger Palaeozoic sediments of comparable lithologies; where a shallow Permo-Carboniferous overprint is ubiquitous (see also Lapointe,

Roy and Morris, 1978). (iii) The APWP carries North America into polar latitudes consistent with evidence for glaciation in early Cambrian times (first part of Section 9.6).

Poles from Lower Cambrian rocks at the periphery of the plate (Fig. 9.4) from the Bradore (17), Tapeats (20) and Wood Canyon (21) formations yield poles coincident with the Upper Cambrian extremity of Loop 11, and distinct from the Lower Cambrian position predicted from the Gondwanaland data in Fig. 9.3. This discrepancy is not alleviated by reviewing the tectonic corrections applied to these poles, and it seems likely that either these rocks were magnetised after their stratigraphic ages, or that the Laurentian APWP had become distinct from the Gondwanaland APWP by Lower Cambrian times, with the implication that these plates had begun to separate by then.

The Laurentian APWP is continued in Fig. 9.4 into the fields of Lower–Middle Ordovician poles, via two igneous units (22,23) of probable early Ordovician age. It incorporates Lower Ordovician poles (24,25) and then moves to Middle Ordovician poles (27 → 29) with the general motion defined by one stratigraphic sequence (24 → 27 + 28). These are possibly the only data which can be used to define the Lower Ordovician path, because the characteristic magnetisations of many North American Lower Palaeozoic sediments are known, or suspected, overprints. This problem is discussed in Section 10.4. It will be observed that Middle–Upper Cambrian poles coincide with Middle Ordovician poles, although the primary origin of the remanence in some of these Cambrian rocks has been cogently argued. Hence it would appear that the Ordovician APWP crossed over the Cambrian path to define a small loop in the same way as the contemporaneous path from Gondwanaland; indeed the similarity of these Cambro-Ordovician loops suggests that they may record a common movement of true polar wander (Section 12.7).

Siberia

The scatter of earliest Cambrian data from this shield is so large that the nature of the path connecting the Vendian and Lower Cambrian poles remains speculative. Probably the best-defined pole comes from the earliest Cambrian sediments in the Lena River section, studied by Kirshvink & Rozanov (1984; 2 in Fig. 9.4). By connecting this pole with the Vendian poles of Fig. 9.3 (see also Lower Cambrian poles 1 in Fig. 9.4) and then continuing the path to the compact distribution of late Lower to Upper Cambrian poles via the shortest arc length, the track illustrated in Fig. 9.4 is derived. However, since this route requires a path of 75° which is totally unrepresented by data points, it must be regarded as no more than one possibility; in addition it fails to incorporate two aberrant poles from Lower Cambrian rocks (3 and 4). The succeeding poles form a tight cluster, which includes results from the Middle Cambrian Ust 'Maya group (12,13) and the Upper Cambrian Verkholensk group (17 → 20,23). There is a tentative suggestion of an outward and return motion of the APWP here. The Upper Cambrian distribution is essentially identical to the Lower Ordovician distribution defined by the Ust 'Kutsk group (28,30), but there is a small shift between these poles and the poles from Middle Ordovician rocks from higher stratigraphic levels in the same area (32,33).

This path is based almost entirely on sedimentary studies (excepting poles numbered 27) from the Lower Palaeozoic successions exposed in the Lena River. The possibility of later overprinting is suggested by experience elsewhere, and by the observation that these poles plot in a similar field to Upper Palaeozoic poles from the same area. Arguing against this possibility is the finding of dual polarities in many of these studies (Lower Palaeozoic rocks from Laurentia, which have experienced later overprinting, possess a predominant reversed polarity acquired during the Permo-Carboniferous Reversed superchron: Sections 10.4 & 12.2). In addition, the Lower–Middle Ordovician passage of the APWP across the Cambrian path is also observed in the Gondwanaland and Laurentian records.

Baltica

The paucity of data from this plate precludes the proper definition of a Cambro-Ordovician APWP. From early Cambrian results 1 and 2, this path should probably be continued to include divergent poles from Lower Cambrian units 3 and 4 and then be retraced to incorporate poles 5–7 from Ordovician sediments.

9.5 Palaeomagnetism of late Precambrian–early Palaeozoic microplates and accretionary terranes

Orogenic processes around the periphery of the Precambrian shields were continuing to add new crust to the continents in Upper Proterozoic and Cambrian times. They have produced terranes which now possess isotopic signatures with little or no vestiges of an Archaean origin, and appear to be underlain by basement no older than late Precambrian in age. These terranes originated in the vicinity of rifted margins, and have been dismembered into small structural elements which have subsequently become incorporated into the orogenic belts formeed by continental collisions in Phanerozoic times. In their present situations they are suspect terranes, and need to be considered as microplates until their final emplacement during the consolidation of the younger orogenic belts.

The incidence of later thermal and tectonic events means that these areas represent particularly difficult subjects for palaeomagnetic study: a history of partial or complete magnetic overprinting is always present, and the tectonic complexities are usually too inadequately understood to be corrected for completely. These difficulties are partially offset by the opportunities for fold tests and the abundance of igneous material. Currently, the palaeomagnetic records of only three such terranes (the Armorican, Bohemian and Anglo-Welsh microplates) have been studied in any detail. Each one subsequently became incorporated into the Caledonian and Hercynian orogenic belts in Europe, and they are now exposed as isolated massifs separated by sediments deposited in Upper Palaeozoic and younger times in rifted zones between (Ziegler, 1984).

The Armorican microplate

The late Precambrian–early Cambrian Cadomian orogeny (which was predominantly magmatic in nature) led to the stabilisation of a low-grade terrane

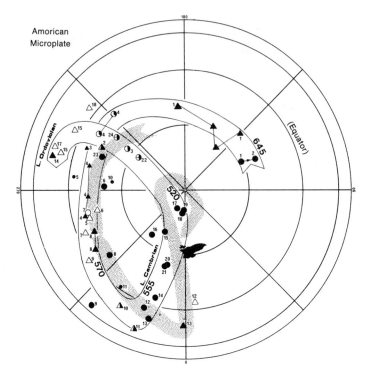

Figure 9.6 Late Precambrian palaeomagnetic poles from the Armorican and Bohemian micro-plates. The poles are numbered: ARMORICAN SUB-PLATE (circles): 1, Treguier, Brioverian keratophyre tuffs, pre-folding magnetite component (640 RS); 2, Spilites, Paimpol (640 RS, <1); 3, Treguier, Brioverian keratophyre tuffs, post-folding hematite component (<640); 4, St. Peter Port gabbro, in situ and dip corrected poles; 5, Bordeaux diorite contact, Guernsey (660–570); 6, Gabbro, Kerlain, primary component late Precambrian-Cambrian); 7, Mainland rhyolites, high blocking temperature component (546, 531 RS); 8, Microgranite dykes and contacts (573–570, 557 RS, 515–510 KA); 9, Diorite, St. Quay, primary component; 10, Diorite St. Quay, secondary component; 11, Gabbro Kerlain, secondary (Later Precambrian-Cambrian?); 12, Longuivy microgranite dykes, primary component (548 RS); 13, Mainland rhyolites (546 531 RS); 14, Porzscarff Granite, altered facies (557 RS); 15, Carteret Red beds fa, primary component, revised pole (Lower Cambrian); 16, Jersey Volcanics (552 RS); 17, SE Granite, Jersey (520 RS); 18, NW Granite, Jersey (490 RS); 19, SW Granite, Jersey (565 RS); 20, Erquy Spilite, in situ post folding component (<482 Ma); 21, Moulin de Chateaupanne Formation (Ol, Arenig, Ou diagenetic remanence?); 22, Paimpol-Bréhac, high blocking temperature component (472 Ma); 23, Jersey dolerites, group C NE dykes (<520 Ma); 24, Site k Erquy-Cap Fréhal, high blocking temperature component. BOHEMIAN SUB-PLATE (triangles): 1, Barrandian Porphyries, C, D, E and F groups (late Precambrian-Cambrian?); 2, Late Precambrian sediments; 3, Porphyrites (Lower Cambrian); 4, Barrandian porphyries, A1 to A4 groups (Late Precambrian to Cambrian, 562 Ma); 5, Glubshsky conglomerates (Lower-Middle Cambrian); 6, Earliest Barrandian; 7, Barrandian sediments (Cl); 8, Hlubos sediments, two poles (Cl); 9, Sadecky beds (Cl); 10, Jince beds (Cm, Upper Palaeozoic remagnetisation?); 11, Barrandian porphyry (Ol, 475 Ma); 12, Barrandian porphyries (ca. 11 or upper Palaeozoic remagnetisation); 13, Oolitic ores 1 (Om); 14, Oolitic ores 2 (Om); 15, Krusne

comprising the northern part of the Brittany Peninsula and adjacent areas in the English Channel. In detail this region comprises a larger number of structural units (Cogné, 1974) although motions between them do not appear to be palaeomagnetically-detectable, and for these purposes the terrane can be treated as a single entity (Hagstrum, Van der Voo, Auvray & Bonhommet, 1980). Cessation of magmatic activity in Lower Cambrian times was followed by deposition of red beds in Cambrian and early Ordovician times (Duff, 1979b). The ensuing Hercynian orogeny included several phases, beginning in Upper Devonian to Lower Carboniferous (Tournasian) times, and continuing into late Carboniferous times (Perroud, Bonhommet & Robardet, 1982; and concluding part of Section 10.3).

The results of a variety of recent palaeomagnetic studies are summarised by Duff (1979b, 1980), Hagstrum, Van der Voo, Auvray & Bonhommet (1980) and Perigo, Van der Voo, Auvray & Bonhommet (1983). The collective data are plotted in Fig. 9.6. There appears to be a large, and as yet undefined, APW shift between the oldest poles from the Armorican microplate (1,2) assigned to ca. 640 Ma, and the larger body of late Precambrian–Cambrian data; this shift may be defined by a secondary magnetisation (3) and data from early gabbros (4–6). The APWP continues via a range of poles from acidic igneous rocks, with assigned ages close to the Vendian–Cambrian boundary (8,12–14). The presence of a remanence (8) contrasting with the magnetisation in host spilites (2) implies that a later remagnetisation is unlikely here, and there appears to be a hairpin in the APWP at this point, because the subsequent sector of the path shows a movement from low to high latitudes, via a remanence pre-dating the Cadomian orogeny (16) to poles from post-tectonic granites (17–19 and Duff, 1980). A probable primary component in the Cartaret red beds (15) could post-date these poles, but the continuation of the APWP beyond this point is unclear. The path could continue to poles 20 and 21 of post-Cambrian and probable Upper Ordovician age (third part of Section 10.3), or it may continue to poles 23–24 and possibly the secondary pole 10. A range of contrasting poles from post-tectonic dyke swarms in the Channel Islands, and secondary components in the Cambro-Ordovician red beds correlates with remanences acquired in the Hercynian orogeny; these results are discussed in the third part of Section 10.3 (Duff, 1980; Perroud, Bonhommet & Robardet, 1982).

Hagstrum Van der Voo, Auvray & Bonhommet (1980) note the close similarity between this APWP and the contemporaneous path defined by the Gondwanaland data. There is a good correspondence between the post-640 Ma segments of the two paths, when the Armorican microplate is rotated by 140·5° about a Eulerian pole at 173°E, 76°N into the reconstruction of Fig. 9.3. The rotated position of this segment of the Gondwanaland path is shown as the stippled swathe in Fig. 9.6. The Armorican data are observed to conform closely to the

---

Hore beds (Ol, Tremadoc); 16, Zahorany beds (Om-u, Upper Caradoc); 17, Sediments (Ou, Caradoc-Ashgill); 18, Diabase and red beds (Ol). Pole centred Lambert Equal Area projections of the whole globe sectored at 45° intervals. The stippled swathe is the contemporaneous Gondawanaland APWP for these times rotated according to the inset reconstruction of Figure 9.7b.

Gondwanaland APWP until late Lower–Middle Cambrian times. The rotational operation places the Armorican fragment between the Khazakhstanian, Siberian and African Shields, in a position (Fig. 9.7) which conforms with the regional correlations of Upper Proterozoic microfossils (Chauvel & Schopf, 1978), and links the Cadomian orogeny to the events which immediately preceded continental breakup and dispersal (second part of Section 9.6). The Armorican microplate subsequently moved by a series of unspecified movements to become incorporated in the Hercynian Belts, following the collision of Gondwanaland and Laurasia (Section 10.3).

The Bohemian microplate

The Bohemian (Barrandian) massif of Czechoslovakia experienced a geological development comparable to the Armorican massif, with acid volcanic activity in late Precambrian and Cambrian times, succeeded by terrestrial deposition. Renewed volcanism in Upper Cambrian times was followed by deposition of sediments spanning much of Ordovician times. Early palaeomagnetic studies, incorporating some partial a.f. demagnetisation and field tests, are summarised by Bucha (1965); more modern and better-constrained studies are unfortunately sparse (Krs & Vlasimsky, 1976).

The late Precambrian and Cambrian poles from this area appear to plot on the same APWP as the Armorican data (Hagstrum, Van der Voo, Auvray & Bonhommet, 1980, and Fig. 9.6). Poles from porphyries of probable late Precambrian or early Cambrian age (**1,3,4**) and late Precambrian sediments (**2**), define a path which continues to poles from Lower–Middle Cambrian sediments (**5–7**). There is no evidence here for the hairpin at the Vendian–Cambrian boundary, or the subsequent movement of the pole from low to high latitudes. Furthermore, the supposed poles of Lower Cambrian age plot further back along the APWP than would be predicted from Armorican and other data, and the movement to the present pole from the Middle Cambrian Jince beds (**10**) should be no younger than earliest Cambrian if the correlation with the Armorican APWP is correct. Pole **10** and nearby poles **11** and **12** from Lower Ordovician rocks plot in the region of Hercynian remagnetisations, and it is likely that they are not relevant to definition of the early Palaeozoic path (Burrett, 1983). Accordingly, a correlation between the Armorican and Bohemian massifs, in approximately their present relative positions, is the simplest interpretation of the palaeomagnetic data, but it should not be regarded as established until more is known about the remanences and ages of the Bohemian rocks. The correlation between the two regions is reinforced by the continuation of the APWP to the Bohemian Ordovician rocks (**13–17**) via the Armorican poles **22–24**.

The main geological implication of this correlation is the recognition of an elongate microplate, extending from Brittany in the west, to the suture with the Fennoscandian-Ukrainian shield in the east. This suture is defined by the Holy Cross Mountains in Poland, where an orogenic belt of Caledonian age with a WNW–ESE trend is part of a lineament which extends into the northern North Sea, and has controlled much of the subsequent tectonic development of this region. The approximate location of this microplate prior to Middle Cambrian times, as predicted by the palaeomagnetic correlations, is illustrated in Fig. 9.7b.

The Anglo-Welsh microplate

The Precambrian basement of England and Wales, between the Iapetus suture in the north (Section 10.3) and the Hercynian belt in the south (Fig. 10.2b) comprises igneous rocks and meta-sediments metamorphosed to low grades in late Precambrian times (*ca.* 700—650 Ma), with isotopic characteristics indicating derivation from a mantle source no more than 300 Ma before this (Thorpe, Beckinsale, Patchett, Piper, Davies & Evans, 1984). This microplate may extend westwards to incorporate the Rosslare complex of SE Ireland (where the basement appears to have a more complex origin, although the late Precambrian–Cambrian history is similar), and the Avalonian terrane in the NE sector of the Appalachian orogenic belt; there are few palaeomagnetic data bearing on this possibility. It is separated by a suture of Devonian age from the Armorican microplate to the south, with a regional significance discussed in the concluding part of Section 10.3.

Palaeomagnetic data from this microplate are summarised by Piper (1979a, 1982b) and Thorpe, Beckinsale, Patchett, Piper, Davies & Evans (1984). The sequence of higher to lower blocking-temperature/coercivity components in the low-grade metamorphic basement yield a short APW segment dated 700–650 Ma (poles **1–5** in Fig. 9.7); the link between these results and early Lower Cambrian data is not yet defined. The younger path commences with poles assigned to earliest Cambrian times, and runs via pre- and post-folding remanences, from widespread calc-alkaline volcanics (**9,11–13,15**) which were strongly folded in Lower Cambrian times. The continuation of the path is poorly defined by secondary magnetisations linked to this folding (**16,18–21,23**) and post-tectonic dolerites (**22**). All of these events took place prior to the onset of quiet shelf sedimentation in a contrasting environment, during late Lower Cambrian times. The collective data define a long APWP which is dated by the stratigraphic evidence, and the sequence of primary to secondary magnetisations as entirely applicable to Lower Cambrian times (Patchett, Gale, Goodwin & Humm, 1980; Piper, 1982b).

The regional distribution of Anglo-Welsh data encompasses a triangular area, ranging from the border of a major NE–SW early Palaeozoic dislocation in Anglesey (to the NW), to the Pembrokeshire inlier bordering the later Hercynian belt in the SW, and the English Midlands in the east (Fig. 10.2c). Although Caledonian structural dislocations within this area have been considerable, they have evidently not been on a scale sufficient to destroy the gross features of this APWP, because the low to high latitudinal movement is recognised separately in three areas, namely in North Wales and Anglesey (**8 → 14 → 20**), in SW Wales (**6,7** and **19**) and in the English Midlands (**9 → 11 → 15 → 17 → 18**).

The APWP from this microplate is distinct from the APWP of the Armorican and Bohemian microplate, on present-day coordinates (Piper, 1982b; Burrett, 1983). However, the feature common to the Anglo-Welsh, Armorican and Gondwanaland APWPs is the long, low- to high-latitude movements in Lower Cambrian times, following the hairpin near the Vendian–Cambrian boundary. When these features are matched they imply a rotation of the Anglo-Welsh

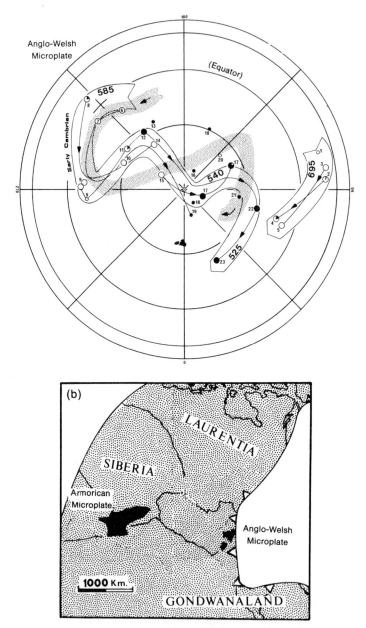

Figure 9.7 (a) *Late Cambrian to Cambrian palaeomagnetic poles from the Anglo-Welsh micro-plate. The poles are numbered: 1, Rushton Schist (>536 KA); 2, Malvernian diorites, M2 component (681 RS, 670 UP); 3, Stanner-Hanter Complex (702 RS); 4, Malvernian M3 component (<2); 5, Stanner Hanter B1 component (<3); 6, St. Davids Granophyre (587 UP); 7, Johnston Complex, secondary remanence (<643 UP); 8, Gwna Group pillow lavas (earliest Cambrian); 9, Eastern Uriconian Volcanics, three poles from two lavas successions (558 RS); 10, Old Radnor sediments, uplift and folding remanence; 11, Western Uriconian Volcanics; 12, Hope Bowlder (Uriconian) Volcanics, pre-folding remanence; 13, Batch Volcanics (<11, >23); 14, Arvonian ignimbrites and tuffs (Lower*

microplate by 81·5° clockwise about a Eulerian pole at 156°E, 43°N. This operation places the microplate in the vicinity of the Afro-Arabian arc until Lower Cambrian times (Fig. 9.7b); there are close similarities between the later Precambrian and early Cambrian calc-alkaline and transitional alkaline volcanic histories of the two regions (Greenwood, Hadley, Anderson, Fleck & Schmidt, 1976; Thorpe, Beckinsale, Patchett, Piper, Davies & Evans, 1984). The 750–650 Ma data are not accommodated by this reconstruction and imply either (i) that relative movements took place between formation of the metamorphic basement and the later calc-alkaline magmatism (and were possibly linked to the emplacement of obducted ocean floor (unit 8 in Fig. 9.9) and contemporaneous metamorphism at the boundary of the microplate in Anglesey); or (ii) that APW relative to the continental crust was more complex than that defined by Figs. 9.2 & 9.3 (see also Section 9.3). The contrasting locations of the Armorican and Anglo-Welsh microplates in the late Precambrian crust are also consistent with the presence of *ca.* 2300 Ma (Pentavrian) basement in the former area, and the apparent absence of crust older than *ca.* 1000 Ma in the latter area.

9.6 Geological implications

The late Precambrian glaciations

The evidence for very widespread glaciation in late Precambrian times was assembled by Harland (1964). In this review, he noted the palaeomagnetic evidence (obtained before proper cleaning and analysis of magnetic components were routine) which suggested an origin within *ca.* 10° of the ancient equator. Subsequently, there has been much debate on this issue, with one school arguing that the late Precambrian glaciations were distinguished from Phanerozoic glaciations by their exceptional extent, so that the tillites (diamictites) define globally-synchronous events (e.g. Dunn, Thompson & Rankama, 1971). The opposing view is that these glaciations were essentially circum-polar effects comparable to the Phanerozoic glaciations, and that the continents were glaciated and deglaciated as they drifted into and out of polar latitudes (Crawford & Daily 1971). What is certain is that much of the late Precambrian to early Cambrian interval (560 Ma extending back to perhaps as much as 950 Ma) was a period of exceptionally low temperatures. Although the continental crust frequently moved across the pole in

---

Cambrian, <8); 15, Nuneaton Volcanics (Lower Cambrian); 16, Malvernian M1 component (early Cambrian overprint?); 17, Charnwood diorites, two poles from intermediate (dual polarity) and steep inclination components (540, 535 RS); 18, Uriconian overprint (pre-late Ordivician); 19, Pebidian Volcanics, post-folding remanence; 20, Bangor red beds (Lower Cambrian); 21, Caer Caradoc, post-folding remanence; 22, Post-Uriconian dolerites, main group (<9 + 12, 553 Ma); 23, Longmyndian sediments, folding and uplift remanence (529 RS, 530–520 FT). Pole centred Lambert Equal Area projection of the whole globe sectored at 45° intervals. The stippled swathe is the contemporaneous Gondawanaland APWP for these times rotated according to the inset reconstruction (b) discussed in the text.

earlier times (Chapter 7), there are no occurrences of glaciogenic rocks which can be positively attributed to the interval between 2200 and 1000 Ma. In contrast, the late Precambrian record is replete with rocks of certain or suspected glacial origin. The climatic cooling appears also to be recorded in the biological record, because the diversity of the planktonic fauna experienced a marked drop in early Vendian times (*ca.* 700 Ma) and only began to recover again in the *Holmia* stage of Lower Cambrian times at *ca.* 550 Ma, (Vidal & Knoll, 1983). It has also been observed that continental volcanism was low during this interval (as shown indirectly by the content of the palaeomagnetic record) while dolomite deposition was high; collectively these effects would have produced a CO_2 depletion, leading to a thinner atmosphere and a steeper vertical temperature gradient (Schemerhorn, 1983). A possible additional explanation for the extraordinary characteristics of these glaciations is provided by G. E. Williams (1975); he proposed that the obliquity of the Earth's ecliptic was much higher ($54°-126°$) during late Precambrian times. This would reverse the climatic zonation, and could produce glaciation only in low to intermediate latitudes.

Both geochronological and palaeomagnetic studies have contributed to this discussion in recent years. Many more age estimates fom sedimentary sequences incorporating glacial deposits have suggested that these glaciations occurred over a period of *ca.* 400 Ma; the ages concentrate very roughly into groups, and suggest four main periods, distinguished as the Lower Congo (950–850 Ma), the Sturtian (820–770 Ma), the Varangian (700–630 Ma) and the Lower Sinean (600–520 Ma) by Harland (1983). Unfortunately, radiometric, and specifically Rb–Sr, studies of glacial horizons are unlikely to produce a definitive answer to this problem, both because they may date de-watering events (first part of Section 6.4) and because they have associated errors which incorporate appreciable APW movement (Figs. 9.2–9.4). A number of palaeomagnetic studies have attempted to infer the palaeolatitude in which the glacial rocks were deposited. However, direct studies of ancient glacial rocks are fraught with difficulties, because the nature of the remanence in these drab-coloured rocks is difficult to decipher and is unlikely to be primary. Tarling (1974) reported low ($<10°$) palaeolatitudes from early Cambrian tillites of Scotland, but the remanence here has since been shown to be secondary, and of Caledonian age (Stupavsky, Symons & Gravenor, 1982). Morris (1977b) subsequently identified a characteristic remanence acquired at low ($14°$) palaeolatitudes in the Rapitan group (Fig. 9.2), but the primary nature of this remanence could not be demonstrated; the Nama formation (Fig. 9.3) is another example of a rock group containing glaciogenic rocks, which yields a remanence record with low inclinations. McElhinny, Giddings & Embleton (1974) compiled a composite APWP for Gondwanaland (comparable with the path illustrated in Fig. 9.3) and observed that the sequence of late Vendian–early Cambrian and Ordovician glaciations within this landmass, could be accommodated by movement across a polar ice cap (see first part of Section 9.4). Subsequently, however, McWilliams & McElhinny (1980) noted that deposition in the Adelaide geosyncline, which includes a number of well-developed glacial horizons, appeared to have taken place at low latitudes. Recognising that the few palaeomagnetic studies of late Precambrian glacial deposits have all identified a low-inclination remanence record, they suggested that low-latitude deposition was a feature of these glaciations. In fact, the Wooltana volcanics yield a steep remanence direction, while the overlying Merinjina tillite yields a range of directions, of which the majority have low inclinations; a subordinate group of steep inclinations appears to be secondary.

In view of these difficulties, it is more profitable to examine the distributions and estimated ages of the glacial deposits in the context of the APWP; they are compiled together in Fig. 9.8. The record of ancient glacial deposits has been compiled by Harland over many years (1964, 1983; see also Harland & Herod, 1974); the ensuing discussion is based largely on these references, and the reader is referred to them for more discussion of individual occurrences. The absolute age estimates of the late Precambrian glacial rocks show a frequency distribution which peaks at *ca.* 950, 800, 710 and 610 Ma (see Williams, 1975 and Piper, 1983a) and at each of these times a large area of continental crust lay within polar latitudes; furthermore, there is a paucity of estimates in the time interval 680–640 Ma, when the crust lay mostly in middle to low latitudes and the APWP explains in part,

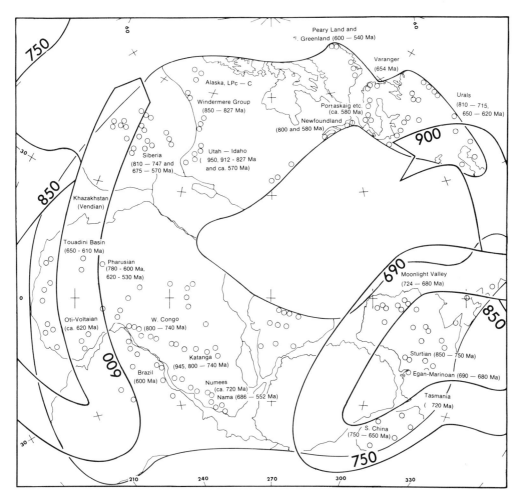

Fig. 9.8 The distribution of the late-Precambrian–early Cambrian glacial deposits, with the estimated ages (mainly after Harland, 1964, 1983; Harland & Herod, 1974; Williams, 1975; and Chumakov, 1981). The APWP is derived from Figs. 9.2 & 9.3.

why a number of areas such as Australia, Siberia and western North America experienced glaciation at two stratigraphic levels between late Riphean and early Cambrian times.

The evidence for the earliest (Lower Congo) glacial period is slender, and relies mainly on an estimate of 950 ± 50 Ma for the Lower tillite in Lower Zaïre; the Upper tillite here is estimated to be 820 ± 50 Ma in age. The region moved no higher than 40° latitude throughout this time period (Fig. 9.8). Rocks of glacial origin in Utah, Idaho and Washington appear to have an age of *ca.* 900–800 Ma, while the Toby formation in British Columbia is reckoned to be *ca.* 850–820 Ma in age, and the earlier Siberian glaciation at 810–715 Ma appears to belong to this episode. These occurrences correlate with the extension of the APWP at *ca.* 850–800 Ma in Fig. 9.3, but the North American locations would appear to have reached latitudes no higher than 50°–60° at this time. It is the glacial rocks, assigned to the interval 800–670 Ma, that show a concise relationship to the APWP. The loop (9) defined by Sinean-age poles passes close to Australia twice during this period (Fig. 9.8) and accommodates both the Sturtian glacial deposits (800–750 Ma) and the later Marinoan glacial deposits (*ca.* 690–680 Ma) in the Adelaide geosyncline. The Moonlight Valley tillites (740 ± 30 Ma) and the Egan tillites (655 ± 45 Ma) in the Kimberley district and NW Territories are also accommodated by the two limbs of this APW loop. The China Shields were in proximity to Australia at this time, and experienced widespread glaciation during up to four separate episodes assigned to the interval 720–660 Ma (except possibly the Huishan episode, which may be as old as 950 Ma). The same polar ice cap could have accommodated the glacial horizons in the Vindhyan supergroup of India. All of these localities moved within 30° of the pole during this time interval, as did Namibia, where the Nosib and Numees (*ca.* 720 Ma) tillites are of about this age (and presumably older than the Nama N2 pre-folding remanence in Fig. 9.3).

After 700 Ma there is clear evidence that the ice sheets attained a much wider distribution. The APWP moved no closer than 50° from the type area of the Varangian glaciation in northern Norway, where the Nyborg formation, which occurs between two tillites of Vendian age, has been dated at 654 ± 23 Ma. Extensive correlative tillites in the western USSR are of Vendian age, with radiometric ages in the range 650–630 Ma, and two tillites overlie the Eleonora Bay group in Greenland, which has yielded Vendian acritarchs;* the Visingö beds in Sweden, with a similar fauna, may be as old as the Riphean–Vendian boundary (710 Ma) and contain possible glacial deposits. None of these latter localities would appear to have moved to latitudes higher than 50° at this time. However, the passage of the APWP across Africa and South America at *ca.* 620–590 Ma, and then across the Siberian and Laurentian margin of the continent, can be correlated with the very wide range of late Vendian to early Cambrian glacial deposits in these areas. They include the glaciations in Brazil (*ca.* 600 Ma), in Namibia (the Nama formation, immediately underlying the Lower Cambrian Fish River formation), in Ghana (the Oti tillite dated 620 Ma), in Mauretainia (650–610 Ma), in Siberia (675–570 Ma) and NW Canada (immediately below fossiliferous Lower Cambrian rocks). However, this period also includes deposits in Newfoundland, Peary Land,

---

* *Acritarchs* are (single celled) marine microplankton of uncertain affinity. They are widely preserved as fossils typically 10–20 μm in size.

Svalbard, Scotland and Norway with a late Vendian to earliest Cambrian age. None of these localities appear to have moved into latitudes higher then 40° at this time, although it is presently difficult to define the Lower Cambrian APWP for these areas (second part of Section 9.4) and this conclusion may need to be modified in the light of further work.

Thus the APWP suggests that the sequence of late Precambrian glaciations can be largely explained in terms of the movement of continental crust across, or close to, one of the poles. Although formation of glacial rocks in equatorial latitudes cannot be ruled out, it is not well supported by the palaeomagnetic evidence, and is not required by the APWP. High-latitude deposits of appropriate age can be found near to the APWP between 750 and 650 Ma, and between 620 and 570 Ma, and imply that the ice caps at this time were pole centred. This would rule out an increase in the obliquity of the ecliptic as an explanation of these events. At the same time it is also possible to identify glacial deposits in areas which do not appear to have moved closer than 40°–50° from the ancient pole during the interval 700–600 Ma when the biological record identifies a decline in faunal diversity (Vidal & Knoll, 1983). Hence these glaciations would appear to have been more extensive than their Permo-Carboniferous and Pleistocene counterparts, and some kind of non-uniformitarian model may be applicable to them (Harland & Herod, 1974, Schemerhorn 1983). A unique and unexplained feature of the 700–570 Ma tillites is their tendency to be interbedded with sediments formed in warm environments, including dolomites, red sediments of possible lateritic origin, iron formations, and sediments containing glauconite.

Passive margins and the timing of breakup

The most important conclusion to emerge from the analysis of Sections 9.2–9.4, is that a large supercontinent broke up at the beginning of Cambrian times to form a number of separate plates and microplates. This is explicitly demonstrated by the way in which the single APWP splits up into several divergent paths (c.f. Figs. 9.3 & 9.4), and is direct palaeomagnetic confirmation of an episode of rifting and breakup which has long been suspected from the geological record. In general terms, it has been noted in the transition from restricted deposition in developing aulacogens, to deposition of mature sediments in circum-shield seas (Stewart, 1976). In regional studies, for example, Rankin (1976) describes the evidence for opening of failed arm salients along the Appalachian margin of Laurentia, where the aulacogens were subsequently filled and replaced by passive marine margins (miogeoclines) of Cambrian age. Anderton (1980), noted the southerly derivation of parts of the latest Vendian–Dalradian succession in Scotland (presumably from the bordering Fennoscandian region, Fig. 9.1), and inferred that the Iapetus suture did not open here until the beginning of Cambrian times; widespread basaltic magmatism accompanied rifting in both areas (Mt. Rogers, Blue Ridge and Tayvallich volcanics). Another example is provided by the Cordilleran margin of western North America, where a protracted (perhaps >200 Ma) rifting history is defined by rift-controlled fluviatile sedimentation with great lateral and vertical variability, and a provenance in the craton to the east. A transition to post-rift subsidence and marine sedimentation took place here at the beginning of Cambrian times (Bond, Christie-Blick, Kominz & Devlin, 1985).

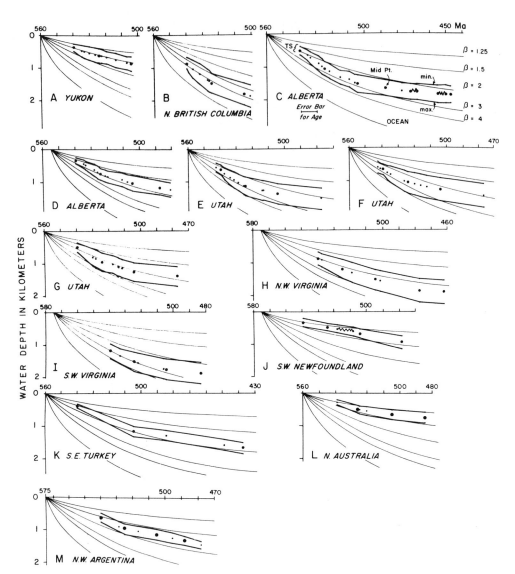

Fig. 9.9 Tectonic subsidence inferred from Cambrian–Ordovician successions which developed along the passive continental margins formed after continental breakup in Lower Cambrian times. After Bond, Nickeson & Kominz (1984). The light solid curves are the post-rift thermal subsidence curves of McKenzie (1978) for different amounts of stretching (σ). The thick curves are tectonic subsidence curves for maximum (max.) and minimum (min.) delithification factors and the dots are the mid points between the two curves. The large dots are stratigraphic boundaries with ages, and the small dots are boundaries postulated without ages, and located by assuming constant sedimentation between dated points; the zig-zag lines are unconformities. See Bond, Nickeson & Kominz (1984) and text for further explanation.

Recently, this event has been documented in detail by reconstructing the temporal evolution of the early Palaeozoic marine basins (Bond, Nickeson & Kominz, 1984). This approach first corrects measured stratigraphic sections for the effects of compaction and cementation; it then compares the implied subsidence histories with the tectonic subsidence curves derived from a model for the subsidence of modern passive margins (McKenzie, 1978). Tectonic subsidence of this kind has a form independent of the rifting mechanism and model parameters after *ca.* 15 Ma, and it reaches equilibrium after *ca.* 200 Ma. Since most of the sedimentary sequences are best understood from Middle Cambrian times onwards (some palaeontological complexities of Lower Cambrian times are noted below), it is this part of the subsidence history which is best matched with the theoretical curves. The results from widely separated localities of North America, South America, Australia and the Middle East show that *in every case* the initiation of passive margins and marine sedimentation occurred between 625 and 575 Ma (Fig. 9.9).

The Vendian–Lower Cambrian alkaline province

Alkaline igneous activity is almost exclusively found in an intraplate setting, in high heat flow environments. Many alkaline rocks are also demonstrably rift-controlled, and their relationship to a thick lithosphere is suggested by their temporal evolution. Highly evolved types such as carbonatites are absent in Archaean greenstone sequences, apart from the isolated Phalabowra carbonatite within the early-stabilised Kaapvaal craton (Fig. 7.3), they are absent prior to *ca.* 1850 Ma. They first become important in association with the Upper Proterozoic aulacogens (Section 7.7 and Fig. 9.1). Alkaline rocks are widely associated with prolonged updoming and rifting of continental crust since Triassic times. They reach their most evolved development (e.g. East Africa and the Rhine Rift) where the rifting has not proceeded to continental separation, apparently because the rift geometry could not be accommodated into the evolving global plate geometry.

A discrete episode of alkaline magmatism is now also recognised as a distinctive Vendian–early Cambrian event linked to the very widespread rifting of the continental crust at this time (Doig, 1970); it appears to have preceded breakup and separation of the continental plates by between 0 and 70 Ma (Bond, Nickeson & Kominz, 1984; Piper, 1985a). The distribution of the rock suites included in this province is shown in Fig. 9.10. It includes 570–560 Ma carbonatites and syenites distributed along the St. Lawrence graben, and along incipient rifts running from Lake Superior to Hudson's Bay, and along the margins of the Labrador Sea (Doig, 1970). In Greenland this activity is represented by intrusion of kimberlite dyke swarms at 585–565 Ma along reactivated older Proterozoic lineaments. Contemporaneous rifts in Scandinavia were associated with intrusion of central complexes (Alnö, Fen, Fig. 9.4), and alkaline intrusion at this time is also recognised in northern Norway and the Kola Peninsula. Carbonatites, alkaline granites and syenites of this age also occur along the southern rifted margins of the Siberian Shield. Representatives of the province occur in NE Africa, Iran and Turkey, although they appear to be rare beyond this northern periphery of Gondwanaland, where the influence of this breakup event was not widely felt.

An areal distinction, but a causal relationship, are identified between this alkaline province and widespread basaltic magmatism during these times. The

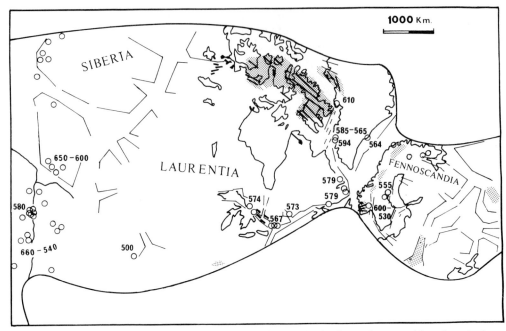

Fig. 9.10 The distribution of the igneous provinces linked to rifting and breakup in late Vendian and Lower Cambrian times, within the N African–Fennoscandian sector of the Late Proterozoic Supercontinent. The open circles refer to alkaline rocks of this age, and include alkali granites and syenites, carbonatites, kimberlites, potassic lamprophyres and alnöites, and the sources are based mainly on Doig (1970) and Harris & Gass (1981). The thin lines are aulacogens and other rift zones of known or probable, Upper Proterozoic age. The stippled areas are Vendian basaltic provinces with the thick lines indicating dyke trends, and the irregular black outcrops indicating lavas or sills. The Franklin Province extends from the Coronation Gulf to NW Greenland. Modified after Piper (1985a).

latter includes the Franklin episode at *ca.* 675–625 Ma, but other basaltic magmatism occurs in Australia (Antrim plateau basalts), southern Siberia, the lesser Himalayas of Pakistan, Morocco, NW Argentina and in British Columbia (the early Cambrian Hamill group). Extensive late Vendian dolerites occur along most of the length of the Scandinavian Caledonides, and also post-date the late Precambrian tillites in Newfoundland (Roberts & Gale, 1978). The distribution of magma types within individual regions is a sensitive indicator of the evolving mantle conditions beneath the continental crust, and shows that a fundamental change in the upper mantle conditions beneath these areas accompanied continental breakup at the beginning of Cambrian times. This is currently most clearly described in the Afro-Arabian Shield (Harris & Gass, 1981) where an accreting continental margin was the site of calc-alkaline magmatism between 1100 and 650 Ma linked to one or more eastward-dipping subduction zones (Section 8.5). After about 600 Ma subduction ceased here and the high P_{H_2O} calc-alkaline magmatism was replaced by low P_{H_2O} conditions responsible for alkaline magmatism throughout Phanerozoic times.

Environmental changes and marine transgression

Marine transgression is the most typical consequence of continental breakup, because the growth of new ocean ridges, accompanied by the development of hot elevated ocean crust, reduces the oceanic volume. An ancillary factor contributing to transgression is the flexure-controlled subsidence which results from sediment loading; this latter effect can extend some 200–300 km from the hinge zone onto the craton. As predicted by the palaeomagnetic evidence, Cambrian times were characterised by one of the most distinctive episodes of marine transgression in the geological record: the Vendian rifts contain volcanics and sediments characterised by large lateral and vertical facies changes, and were displaced in Cambrian times by marine sediments with great lateral continuity, deposited on passive marine shelves. They may include shallow-marine limestones, shales and sandstones, but they typically commence with a pure quartz sandstone of distinctive appearance. Although the transition from syn-rift to marine shelf environment cannot be dated accurately by the palaeontological record, it is apparent that it generally conforms with ages estimated more precisely from the marine subsidence data (Fig. 9.9).

There appears to have been a large measure of continuity between the limited seas around the periphery of the supercontinent in late Proterozoic times: the Upper Riphean and Vendian acritarch assemblages can be widely correlated between the Ukraine, Finnemark, East Greenland and North America for example, (Vidal 1977, 1979). A large measure of correlation is also possible with the Gondwanaland successions. It is in Lower Cambrian times that widespread diversity first becomes evident in the palaeontological record as the continents separated and deep sea-ways developed between them. This is especially apparent in *Holmia* and *Protolenus* times (Cowie 1971; Holland 1974); discrepancies in both lithofacies and faunas developed in the late Lower Cambrian between the continents bordering the Atlantic and suggest that both regions were well apart by these times (Fig. 9.12). The disconformity caused by this marine transgression is well developed in Baltica and Laurentia. Other areas flooded earlier, so that there was a broad continuity of sedimentation across the Vendian–Cambrian boundary. This applies to parts of Australia, Africa and China, which appear to have been remote from the zones of breakup, but it also applies to Siberia; evidently the sea-boards of these continents were lower at the time of rifting and breakup (Bond, Nickeson & Kominz, 1984). The characteristics of the Lower Cambrian unconformity also vary considerably from area to area (Holland, 1974) and this may be because the continental breakup was a sequential event; this is suggested by the divergence of the APWPs in Fig. 9.4, but cannot be confirmed by the present Lower Cambrian palaeomagnetic data. The link between continental breakup and the appearance of hard-bodied organisms is no doubt linked to the convergence of several factors, including the great extension of the marine shelf environment accompanying transgression, the increase in the length of the marine coastline by nearly 50% and a more favourable balance in the sea-water chemistry (see next subsection).

The early Palaeozoic transgression reached its maximum in late Cambrian or early Ordovician times (510–500 Ma) or some 70–90 Ma after the initial breakup. (Fig. 9.9). This is to be anticipated from the time lag between the time of initiation of the new ocean basins and the time of maximum reduction in the ocean basin volume. The older ocean floor on the ridge flanks is preferentially subducted as new spreading develops, and this effect causes the maximum reduction in ocean volume to be achieved on a time scale comparable to the mean age of the ocean floor (Parsons, 1982). In practice this is *ca.* 70 Ma after spreading begins, and a delay of this order is also observed in Mesozoic times between the Jurassic rifting and the peak of transgression in late Cretaceous times (Section 12.8)

The isotopic and chemical signature

Continental breakup in Lower Cambrian times coincided with the largest single increase in ^{34}S in the marine environment in the geological record (Holser, 1977). The concentration of this isotope is linked to sulphate removal from the oceanic system, by three possible mechanisms, namely (i) evaporite formation, (ii) bacterial reduction and (iii) cycling through the mid-ocean ridge system. However, the catastrophic increase at the beginning of Cambrian times can probably be attributed only to two sequential events. The first was the accumulation of brines in enclosed basins in rifted continental environments, and the second was continental fragmentation. The accumulation of evaporites may have proceeded over the entire interval of aulacogen development in appropriate climatic conditions, but particularly thick evaporites of late Vendian and early Cambrian age are found along the margin of Gondwanaland incorporating Arabia, Iran and Pakistan (Bond, Nickeson & Kominz, 1984). A comparable but smaller amplitude response followed the Mesozoic fragmentation of Pangaea (Fig. 9.11 and Chapter 11).

The progressive increase of the ^{87}Sr:^{86}Sr isotopic ratio of sea-water is also interrupted in early Cambrian times, reflecting a marked increase in the upper mantle contribution to the strontium balance (Fig. 9.11 and Section 8.2). It is also the response anticipated from the formation of new ocean basins in Cambrian times, with the concomitant development of new ocean ridge systems: for the first time in 2000 Ma, this contribution offset the ^{87}Sr input from erosion of older con-

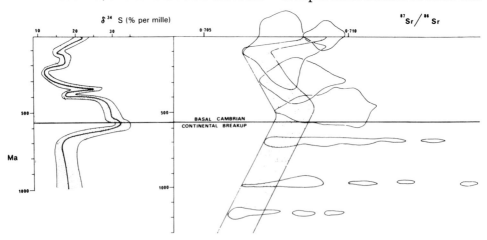

Fig. 9.11 The isotopic response of the oceans to the continental breakup in Lower Cambrian times, indicated by (i) a sharp rise in ^{34}S caused by the catastrophic mixing of brines and the surface ocean, initiated by the destruction of enclosed basins, and (ii) a reduction in the ^{87}Sr:^{86}Sr ratio, reflecting a relative increase in input of mantle-derived strontium. After Piper (1985a) from data in Holser (1977) and Viezer & Compston (1976).

tinental crust and the subsidiary effect of limestone recycling. Additional chemical and isotopic changes linked to the change in tectonic style at the Proterozoic–Phanerozoic boundary are noted in Sections 8.2 and 8.7.

Some of the chemical and isotopic changes may have been critical to the development of hard-bodied organisms at the beginning of Cambrian times. Holser (1977) speculates that the later stages of the mixing cycle, associated with the sulphide–sulphate transfer, restored nutrients in solution to surface waters and enhanced the deposition of organic carbon. Probably, nutrient productivity also increased as a result of upwelling along the new continental margins. Another unique signature of Lower Cambrian times, favourable to the apatite-secreting metazoa, was the widespread deposition of phosphorites at this time. The term 'phosphorite' is generally applied to rocks containing $>10\%$ P_2O_5, which is present predominantly in the form of carbonate fluorapatite (francolite). These rocks developed at intervals through geological times, in response to a unique set of oceanic/climatic conditions on stable marine shelves <500 m in depth (e.g. Northolt, 1980). They are characteristic of sediment-starved environments, where prolonged diagenesis during 'stillstand' conditions and sediment reworking are able to form and concentrate the phosphate minerals. The developing shelf environments in front of peneplained hinterlands no doubt explain why the Lower Cambrian phosphate deposits are the most important in the geological record (Cook & McElhinny, 1979); the next most significant deposits are Upper Cretaceous to in age and are also partly linked to marine transgression. Present-day phosphate development in the shelf seas off Peru, Namibia and elsewhere is related to the upwelling of nutrient-rich ocean currents at low latitudes. The proximity of the Lower Cambrian shelves to developing deep oceans also links these phosphorites to dynamic upwelling, (although the link is an indirect one, because phosphorite formation results from diagenetic and authigenic processes acting on accumulated organic material in a nutrient-rich environment (Marshall & Cook, 1980). Three major environmental belts are recognised in the Lower Cambrian sedimentary successions bordering Laurentia and Baltica, namely (i) an inner detrital belt at the ancient coastline, (ii) a belt of oolite shoals and archaeocyathid* bioherms, and (iii) an outer coastal environment (Robinson & Rowell, 1976). Apatite and glauconite developed partly behind the carbonate barriers, but mostly on the ocean-facing shelves (Brazier, 1979). At least three transgressive–regressive cycles are recognised in the Lower Cambrian (and are related in part to repetitive phosphorite deposition) prior to more general submergence in late Lower and Middle Cambrian times (Brazier, 1979 and Figure 9.9). Since these cycles are both contemporaneous and equal in number to the stages of breakup required to form the four major early Palaeozoic continents (Gondwanaland, Laurentia, Baltica and Siberia), they are a further suggestion that the dismemberment of the continental crust took place sequentially, as suggested by the bifurcation of the APWPs in Figs. 9.3 & 9.4. The development of multicellular animals had evidently reached a stage at which it was able to capitalise on the range of chemical changes associated with these events.

---

* The Archaeocyathidae were a group of organisms which secreted calcareous skeletons, and resembled the sponges. They evolved rapidly over the expanding marine shelves of Lower Cambrian times, and are an important component of many limestones of this age. They became extinct in Middle or Upper Cambrian times, and their ecological niche was not occupied again until the corals appeared in late Ordovician times.

The palaeogeography of Vendian–Cambrian times

Some of the palaeogeographical implications of Vendian and Cambrian APW are summarised in Figs. 9.12(a) to (c) as palaeogeographic maps relevant to *ca.* 625 Ma (a), *ca.* 575 Ma (b) and Middle-Upper Cambrian times (c); the orientation of the supercontinent in (b) must approximate to the last position of the continental crust before it fragmented later in Lower Cambrian times. The Middle-Upper Cambrian configuration (*ca.* 530–520 Ma) should be regarded as highly tentative, both because of the uncertainties in palaeolongitude, and because the polarities of the Baltican and Siberian APWPs are uncertain (Sections 9.4 & 9.5 and Fig. 9.4).

Evaporites are virtually absent in Vendian to earliest Cambrian successions of Laurentia and Baltica (where tillites of this age are common) with the exception of limited deposits in north and east Greenland, but they are common along the eastern margin of Gondwanaland (Bond, Nickeson & Kominz, 1984). This region straddled the equator during these times (Fig. 9.12a). The earliest multicellular animals (metazoans) are soft-bodied, and are collectively referred to as the 'Ediacaran' fauna from the classic locality in South Australia. The base of the Cambrian has traditionally been located at the point where diverse hard-bodied forms, including trilobites, brachiopods and archaeocyathids first appear. However, this does not appear to represent an instantaneous radiation, because assemblages of shelly fossils have since been discovered below the first trilobites (e.g. Brazier, 1979) and used to define a basal Cambrian stage called the Tommotian.* It now appears that an Ediacaran fauna *ca.* 650–600 Ma in age was displaced by a pre-trilobite hard-bodied assemblage at *ca.* 600–570 Ma, and then by a trilobite fauna after *ca.* 550 Ma (McMenamin, 1982). The Ediacaran fauna appears to be confined to localities in Gondwanaland and Baltica, which all lay within about 30° of the equator in Upper Vendian times (Fig. 9.12a). In common with the tillite and evaporite evidence noted above, it implies that temperature gradients between poles and equator were steep at this time, and that the climatic belts were correspondingly compressed towards the equator. The partially-mineralised fossil *Cloudina* may, in part, be as old as the Ediacaran fauna (McMenamin, 1982) in which case its distribution shows that it flourished at somewhat higher latitudes (30°–60°). The earliest hard-bodied faunas of Tommotian times are defined by the fossils *Protohertzina* and *Anabarites*. They are found in localities (McMenamin, 1982; Jiang Zhiwen, 1984) which are seen to comprise well-defined provinces in each hemisphere (Fig. 9.12c). The first incorporates occurrences ranging through South China, Antarctica, Australia and North China, and the second incorporates Siberia, Khazakhstania, NW Africa, Britain, Armorica, Newfoundland, and the western margin of Laurentia (Jiang Zhiwen, 1984). Both were located at intermediate to high latitudes, although the near-polar location of the second province may be caused by uncertainties in locating the appearance of this fauna within a period of rapid APW. It is apparent however, that the earliest hard-bodied metazoa did not have an equatorial distribution like the Ediacaran fauna; their development appears to have been stimulated by the marginal environments near contracting ice sheets.

---

* Following the geological time scale of Harland and co-workers (1982), the base of the Tommotian is regarded here as 590 Ma, although some radiometric studies of this stage in China suggest that it may be as old as 610 Ma (Zichao, Guogan & Huaquin, 1984). It is possible of course, that this boundary is diachronous.

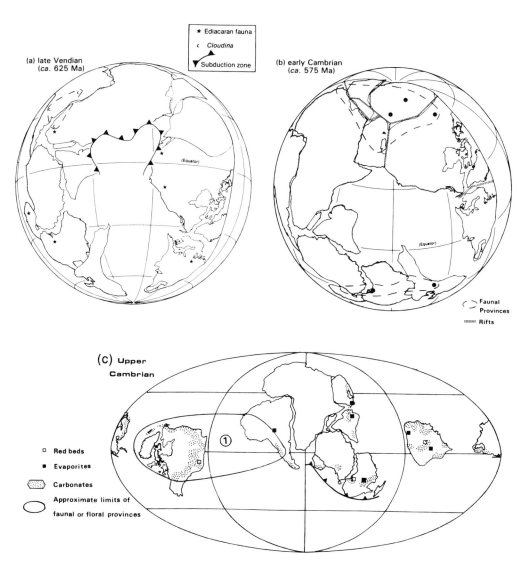

Fig. 9.12 Palaeogeographical maps for the interval between late Vendian and Middle Cambrian times derived from the palaeomagnetic data. (a) Late Vendian times, showing the distribution of Ediacaran fauna. (After McMenamin, 1982, plus occurrences in the Carolina Slate Belt and Iran (with the approximate location of the latter area estimated)). (b) Early Cambrian reconstruction showing the locations of the earliest shelly metazoan localities (circles), after McMenamin, (1982) and the boundaries of the provinces indicated by the dashed lines, after Jiang Zhiwen (1984). (c) A possible reconstruction for middle Cambrian times, with some palaeoclimatic indicators (mainly after Ziegler, Scotese, McKerrow, Johnson and Bambach, 1979. (a) and (b) are orthographic projections and (c) is a Mollweide projection.

The transition to the Middle-Upper Cambrian distribution of Fig. 9.12c is entirely speculative, and it can only be noted that in the interval between early and middle Cambrian times (*ca.* 570–520 Ma) the dispersing continents moved rapidly away from the North Pole into low latitudes leaving wide polar oceans. Extensive carbonates then developed in Middle and Upper Cambrian times on the new marine shelves, and appear to have formed within 40° of the equator. The widely distributed archaeocyathidae are generally believed to have occupied tropical latitudes, and a broad belt of evaporites of this age extended from the Arabian peninsula to India and possibly to China.

The provinciality of the faunas at this time has already been noted. The ordering of the continents around the equator in this figure can be varied to preserve the orientation defined by the palaeomagnetic data. Thus, Bond, Nickeson & Kominz (1984) favour a position of Laurentia closest to the western sea-board of Gondwanaland, because of similarities between the trilobite Ollenelid fauna of the Appalachians and NW Argentina. Single Lower Cambrian poles from North and South China place these shields in positions little different from their Proterozoic locations (Section 9.2) and it cannot be demonstrated unambiguously that they had moved relative to the remainder of Gondwanaland by this time. Lin, Fuller & Zhang (1985) place the South China plate against Northern Australia in Cambrian times, noting that this accords with the trilobite affinities of the two areas, and yields a continuity between the miogeosyncline along eastern Australia and the South China Fold Belt (Fig. 9.12c). The North China plate may have been located either *ca.* 30° north or south of the equator (depending on the interpretation of the polarity of the pole). The former possibility places it close to its Proterozoic position, but either possibility is consistent with the trilobite provinces; the Cambrian evaporites of the North and South China plates are then linked to the evaporite belt extending latitudinally eastwards from Arabia.

10
Palaeozoic palaeomagnetism and the formation of Pangaea

10.1 Introduction

The Palaeozoic palaeomagnetic data have presented some of the most difficult and intriguing problems of the whole geological record. In part this is because the continental crust had by now broken up into a mosaic of plates and microplates moving relative to one another, and because there is no remaining ocean floor of this age to provide a second constraint on the motions. But, in addition, the rocks, and in particular the sediments, of this age developed in a higher Eh condition than during any previous times within environments which supported an abundant terrestrial/marine flora and fauna. This factor seems to have contributed to a complex diagenesis, and led to the widespread alteration of primary magnetic phases to secondary phases (second part of Section 4.2). As a result, the primary palaeomagnetic record has proved particularly difficult to unravel. Almost all Laurentian pole positions from rocks of mid-Ordovician to Lower Carboniferous age, for example, lie within 20° of the Permo-Carboniferous poles (Section 10.4). These positions imply either that the plate moved much less than other crustal plates (such as Gondwanaland) during this interval, or that the remanences in the Lower Palaeozoic rocks are dominated by Permo-Carboniferous overprints. In addition to the palaeomagnetic arguments outlined in Section 10.4, the latter view is supported by the observations that (i) the palaeomagnetic record is dominated by sedimentary studies (ii) the region formed part of a vast, emergent supercontinent which moved into low latitudes characterised by tropical weathering in

Permo-Carboniferous times, and (iii) these times were also characterised by tectonic activity, marking the culmination of suturing into this supercontinent.

The Laurentian plate is not therefore, a good starting point with which to begin discussion of Palaeozoic APW. This subject is addressed instead by considering two regions where much of the primary remanence record appears to have survived these events. The first region is Gondwanaland, which lay mainly in middle to high latitudes throughout this era, and remote from the belts of profound chemical weathering. Some systematic investigations here, notably in SE Australia, have elaborated a long APWP which is divergent from the younger path and cannot therefore be readily confused with later overprinting. The discussion is then extended to the British microplate where the record is dominated by igneous lithologies and a long Lower Palaeozoic APWP is now documented in some detail. A brief comparative discussion is then extended to the Laurentian, European and Siberian plates, where there is considerable evidence for variable amounts of late Palaeozoic overprinting and the value of the palaeomagnetic evidence for tectonic and palaeogeographical analyses is correspondingly suspect.

10.2 Gondwanaland

Assessments of the palaeomagnetic data from Gondwanaland have all suggested that the constituent continents record a single APWP when assembled into the Smith-Hallam (1970) configuration (McElhinny 1973, McElhinny and Embleton 1974, Hailwood 1974, Schmidt and Morris 1977, and Goleby 1980). The complexity of the path has however, increased progressively: the main features of the earlier analyses were a loose grouping of Ordovician poles situated near northern Africa, followed by a path moving to successively younger poles across Africa, then moving across Antarctica, and reaching the eastern extremity of the supercontinent by Permo-Carboniferous times. This path correlated in a general way with the incidence of glaciation (McElhinny 1973). More recently, this picture has been complicated by a range of results from Silurian and Devonian rocks in the Tasman geosyncline. McElhinny & Embleton (1974) and Embleton, McElhinny, Crawford and Luck (1974) proposed a two-plate model to accommodate these result which required an anticlockwise rotation through 90° and a welding onto the craton of a SE Australian subplate in Upper Devonian times. The accretionary nature of the Tasman Zone is suggested by a number of serpentinite belts between the Adelaide geosyncline and the east coast, which become progressively younger towards the east. The large rotation required is more problematic and other workers have preferred to regard this terrane as essentially integral with the remainder of Australia when the studied rock formations were formed. Schmidt and Morris (1977) integrated these results into a single APWP with the remainder of the Gondwanaland data. Their model postulated that the Cambrian and Ordovician poles which plotted near northern Africa were reversed with respect to the post-Ordovician data, so that all of the Silurian and Devonian poles could be incorporated into a single track. Although this interpretation required a very rapid APW movement in Upper Ordovician-Lower Silurian times, it had the advantages of incorporating all of the reliable Australian poles in their correct stratigraphic order, and all movements between successive poles in the data set at that time were <90°.

Subsequently Goleby (1980) studied the Tasman Geosyncline in more detail. By extending the stratigraphic record back to Lower Ordovician times he confirmed that this area was essentially integral with the remainder of Gondwanaland in Ordovician times (poles 4–9 in Fig. 10.1), but a better stratigraphic coverage of Silurian and Devonian times has shown that the APWP now appears to incorporate a loop, and the direction of APW is, in part, in the opposite sense to that inferred by Schmidt & Morris (1977). The data are summarised on the Smith-Hallam reconstruction of Gondwanaland in Fig. 10.1. Much of the record from elsewhere relies on data from unfossiliferous sediments with only loosely-defined ages; Examples from South America include poles 2–7, which accord with the general distribution of poles from the other continents but cannot be used to substantiate any part of the path. The separation of a discrete sequence of primary and secondary components of magnetisation in the Tasman orogen and the re-valuation of some older data in the context of new structural information (Goleby 1980) gives added credence to the path defined from this area. Discussion of the pre-Carboniferous record from Gondwanaland must accordingly be focused on these results, because it is only here that a set of high–quality stratigraphically-controlled data is available.

The late Cambrian to Middle Ordovician poles comprise a monotonic age sequence accommodated by a clockwise loop which has already been discussed in Section 9.4. The continuation of the path to the NW in Fig. 10.1 extends to three Middle-Upper Ordovician poles from the Tasman geosyncline (7–9) and probably includes two poles from Antarctica (1,3) and one from Africa (8) the latter of which, although constrained by a fold test, could be younger than Ordovician in age. These data points lie a little over 90° from the Upper Ordovician (Caradocian) pole from the Alcaparrosa lavas of Argentina (6, Vilas & Valencio, 1978); this latter pole appears to pre-date Lower Palaeozoic folding, and is sufficiently removed from Carboniferous data to preclude an obvious explanation in terms of later remagnetisation. In the context of the implied NW motion of the APWP in Middle Ordovician times, it is appropriate to regard the Cambro-Ordovician poles, plotted in Fig. 10.1, as of normal polarity and to continue the APWP as a reversed path beginning with South American pole 6. This interpretation would mean that a rapid APW movement is present in the path, as proposed by Schmidt & Morris (1977), although it took place slightly earlier than they suggested. There are clearly too few data points to define the connection between the Middle Ordovician and Silurian poles, and thus fix the polarity of the older APWP with certainty but, as is discussed in Section 10.3, the contemporaneous record from the British microplate lends support to the interpretation of Fig. 10.1.

The swathe continues via the stratigraphic sequence of Lower–Middle Silurian (10–14) to Middle–Upper Silurian (15–19) and to Lower–Middle Devonian (27,30,31) Australian poles. Only single South American (7) and Antarctican (3) poles can suggest that this path is applicable to the remainder of Gondwanaland, but since the pre-Upper Ordovician data from the Tasman geosyncline accord with the other Gondwanaland data (Fig. 10.1 and Section 9.4), they are relevant to the whole supercontinent. In addition, new Devonian data from the Iranian plate favour this interpretation of the APWP (Wensink, 1983). The data are still sparsely

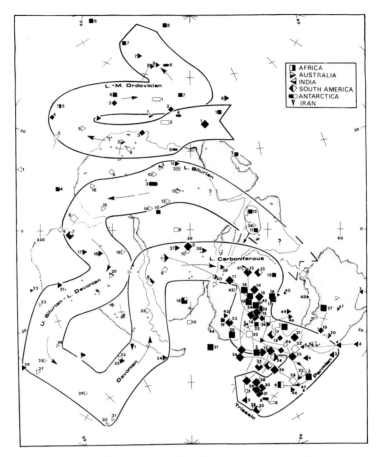

Figure 10.1 Palaeomagnetic data from Gondwanaland *assigned to the interval between* Lower Ordovician *and* Upper Permian *times, ca. 500–250 Ma. The poles are numbered:* AUSTRALIA: 1, *Jinduckin Formation (Ol);* 2, *Dundas Group (Ol);* 3, *Tumblagooda Sandstone (Cu-Ou);* 4, *Walli Sandstone (Ol);* 5, *Mt. Pleasant Andesite, primary component (Ol);* 6, *Stairway Sandstone (Om, Llanvirn-Llandeilo);* 7, *Cliefden Limestone (Om-u);* 8, *Malongulli Formation (Om-u);* 9, *Angulong Tuff, primary component (Om-u);* 10, *Rockdale Formation (Sl-m);* 11, *Narragual Limestone (Sl-m);* 12, *Mereenie Sandstone (S-D);* 13, *Millambri Formation (Sl-m);* 14, *Bebubula Shale (Sl-m);* 15, *Canowindra porphyry (Sm-u);* 16, *Upper Avoca Shale, primary component (Sm-u);* 17, *Ghost Hill Formation, primary (Sm-u);* 18, *Mumbil Formation (Sm-u);* 19, *Mugga Mugga Porphyry, tilt corrected pole (Su, early Ludlow);* 20, *Canberra igneous rocks (Su-Dl);* 21, *Catombal Group, (Du);* 22, *Millambri Formation overprint;* 23, *Belubula Formation overprint;* 24, *Angullong Tuff overprint;* 25, *Ghost Hill Formation overprint;* 26, *Locheil Formation (Du);* 27, *Cunningham Formation primary (Dl-m);* 28, *Laidlaw and Douro Volcanics (S, Wenlock-early Ludlow, 412 RS);* 29, *Upper Avoca Shale primary (Sm-u);* 30, *Cowra granodiorite (Dl-m);* 31, *Dolerite intrusion (Dl-m);* 32, *Tenandra Formation (Dl-m);* 33, *Ainslie Volcanics (Sm, Devonian overprint);* 34, *Isismirra and Gilmore Volcanics (Carbl);* 35, *Yetholme adamellite (Carbu, 318 RS);* 36, *Yalwal Stage basalts (Du);* 37, *Browning Group Volcanics (Dl; Gedinnian);* 38, *Ross River Secondary directions, two poles (D-Carb);* 39, *Gilmore (Lower Kuttung) andesites (Carb, 300 KA);* 40, *Housetop granite and aureole (Dm, 375);* 41, *Mulga Downs Group (Du);* 42, *Cleifden Limestone secondary;* 43,

Dotswood Red beds (Du, Carbu remagnetisation); 44, Paterson Toscanite (Carbu 298 KA); 45, Amadeus Basin sediments, Alice Springs orogeny overprint, two poles; 46, Upper Kuttung sediments and lavas (Carbu); 47, Percy Creek Volcanics (Carbu); 48, Moonbi lamprophyre (Pm); 49, Upper Marine Latites (Pm-u, 248 KA); 50, Milton monzonite (Pu, 204 KA); 51, Gerringong Volcanics (Pm). SOUTH AMERICA: 1, Suri Formation (Om, Llanvirn); 2, Jacadigo Series sediments (O-S); 3, Urucum Formation two poles (O-S); 4, Salta and Jujuy sediments, four poles (C-O); 5, Nunorco igneous rocks (416–310 KA); 6, Alcaparrosa Formation (O, Caradocian); 7, Bolivian sediments, two poles (O-D, Siluro-Devonian overprint?); 8, Picos and Passagem Series (Dl-m); 9, Salta and Jujuy sediments (D); 10, Taiguati Formation, N component (Carbl); 11, Piaua Formation, two poles (Carbu); 12, La Colina Formation red beds, five poles (Carbu-Pl); 13, Pipiral Formation (Carb-P); 14, Tubarao Group, Itarara Subgroup, (Carbu); 15, Middle Paganzo Formation, three poles (Carbu-Pl); 16, La Colina basalt (295–266 KA, in the middle of section yielding poles numbered 12); 17, La Mendieta sandstones (Dl, Carb-P remagnetisation?); 18, Tupambi Formation (Carbl); 19, Grey Tuffs, Peru (Carbl); 20, La Rioja sediments (Carbl); 21, Los Tunas and Bonete Formation (Pu); 22, Irati Formation dolomites (Pu); 23, Amana Formation (P-T, <18); 24, Lagares Formation (Carbu); 25, Violaceo Formation (Carbl); 26, Middle Paganzo Formation (Pu, 290–266 KA); 27, Uruguay sediments (D, Carb-P remagnetisation?); 28, Corumbatai Formation, Passa Dou Group (Pu); 29, Cerro Bola-Cerro Colorado Formation lavas (Pl); 30, Cerro Carrizalito Group lavas (Pm, 263 KA); 31, Sierra de Laventana red beds (Pu); 32, Mitu Formation (Pm). INDIA: 1, Rudraprayag Formation (S-D); 2, Talchin Beds (Carb-P); 3, Durgapipal Formation (Pm); 4, Wardha (Speckled) Sandstone (Pl); 5, Kamthi Beds; 6, Wardha Valley red beds (Pu-Tl). MADAGASCAR: 1, Sakoa Group glacial deposits (Carbu); 2, Sakoa Group red beds (Pl); 3, Series Rouge Inférieure (P); 4, Lower Sakamena Group (Pu). ANTARCTICA: 1, Mirnyy Station Charnockites (Cu-O, 502 RS); 2, Lamprophyre dykes, Taylor Valley (Ol, 470 KA); 3, Sør Rondane intrusions (Ol, 400 RS) 4,5, Ongul Gneiss and Lutzow-Hohn Bay rocks, Pan African magnetisation; 6, Hawkes Rhyodacite, Gambacorta Formation (500 RS); 7, Wright Valley dykes (>D); 8, Beacon Group sediments, two poles (D-T). AFRICA: 1,2, Adras de Mauretainie CO10 and CO8 divisions (C-O boundary); 3, Tassili sediments (C-O); 4, Hook intrusives (C-O, 500 RS); 5, Jordanian red sandstones (C-O); 6, Table Mountain Series, two poles (O-S, possible recent component incorporated in these data); 7, Hasi-Messaud sediments (C-O); 8, Sabaloka Ring Structure (500–330 KA, 383 RS); 9, Msissi norite (Du); 10, Dwyka Varves, Zimbabwe, two poles (Carb); 11, Djebel Hadid red beds (Carbl, Visean); 12, Qued Draa Aftez limestone (Carbl, Visean); 13, Basalts, diorites and Contact (Carb, 297–294 KA); 14, Ain-ech-Chebbi Formation (Carbl, Muscovian); 15, Hassi Bachi Formation (Carbl); 16 and 17, Djebel Tarhat red beds (Pl); 18, Glacial Varves, Dwyka (Carb); 19, Gneiguira Supergroup (Dl-m); 20, Mejeria Sandstone (C, Carb remagnetisation); 21, K3 beds Galula coalfield; 22, Taztot trachyandesites (Pl); 23, Ain-ech-Chebbi/Hassi Bachir Formations, secondary component (Pl); 25, Serie d'Abadla (Carbu-P); 26, Chougrane red beds (Pl); 27, K3 Songwe-Kiwira and Ketewaka-Mchuchuma coalfields (Carb-P); 28, Maji-ya-Chumvi beds, Kenya (P-T). IRAN: 1, Central Iranian Red beds. The bracketed ages employ the abbreviations: C – Cambrian, O – Ordovician, S – Silurian, D – Devonian, Carb – Carboniferous, P – Permian, T – Triassic, l, m, u – Lower, middle and upper divisions. Lambert Equal Area projection centred at 30°E, 0°N with Africa retained in present day coordinates and the reconstruction employing the operations of Smith and Hallam (1970). Note that results from Madagascar use the African symbol with the numbers underlined. The crosses refer to study areas.

distributed along this path, and although the simplest integral path is an open loop, as shown by Goleby (1980) and Figure 10.1, there are certain data points (21,26) with stratigraphic ages which are too young for their positions on this path and suggest that it may be more complicated in detail. The latter point has some support from data elsewhere (Section 10.3 and Fig. 10.2b) and is suggested by the Australian palaeoclimatic data, because Lower Carboniferous coralline limestones were rapidly succeeded here by glacial deposits near to the Permian–Carboniferous boundary.

The path can be continued to a group of poles of Devonian or Carboniferous age, from South America (8–10), Africa (9,11,12) and Australia (34–39), although the position of the Devonian-Carboniferous boundary within these data points is unclear, and the path is not tightly constrained until the group of Carboniferous poles plotting across Antarctica is reached. The N–S swathe here incorporates the bulk of the Permo-Carboniferous data from Viséan times onwards. There is little obvious systematic trend in the individual age assignments along this path and their distribution hints at a complex history of diagenetic magnetisation or later overprinting. The overall trend in late Carboniferous–early Permian APW appears however, to be recorded in South America by the stratigraphic sequence in the La Colina red beds (12) and basalt lava (16) with a possible return loop of the path in Upper Permian times defined by the overlying Amana formation (23; and Valencio, Vilas & Mendia 1977). The earlier part of this trend is also identified by the sequence of Lower (39) and Upper (46) Kuttung poles from Australia (Irving 1966). Some well-defined Lower–Middle Permian poles from India are slightly removed from this main distribution and suggest either that further complexities are present in the APWP or that relative movements between the constituent blocks of Gondwanaland had begun in Permian times (Klootwijk 1979). The more complex APWP of Fig. 10.1 agrees better with the incidence of glaciation in Gondwanaland than the earlier assessments because the Table Mountain group in the Cape system of South Africa includes a tillite of late Ordovician or earliest Silurian age, whilst the glaciogenic deposits of South America include both the Siluro-Devonian deposits of Brazil and early Carboniferous glacial deposits of the Parana basin and the Andean Belt (Crowell & Frakes 1970). Central and southern Africa were glaciated *again* during the Dwyka episode, which is variously estimated to have taken place between Upper Devonian and Upper Carboniferous times (see McElhinny 1973), although varves belonging to this episode in South Africa are of early Lower Permian age (Anderson & Schwyzer, 1977).

African pole 9 from the Mssissi norite of Morocco has often been used to derive a Devonian palaeolatitude for Gondwanaland (eg Van der Voo & Scotese, 1981), owing to the paucity of other Devonian data from the supercontinent. The implied reconstructions have suggested a wide ocean between Gondwanaland and Laurentia-Baltica at this time, in conflict with the faunal evidence which suggests a close proximity of these continents by mid-Devonian times (e.g Boucot, Johnson & Talent, 1969; Burrett, 1983). It is doubtful however, whether pole 9 and the aberrant Carboniferous poles 11 and 12 from Morocco should be incorpoated into the integral APWP. Martin, Nairn, Noltimer, Petty & Schmidt (1978) detected no significant difference between the Upper Palaeozoic palaeomagnetic results from the High Atlas, Anti Atlas and Meseta structural divisions of Morocco, although these data collectively differ from Carboniferous data from the Saharan region. The systematic discrepancy of Upper Devonian and Carboniferous poles from Morocco, evident in Fig. 10.1 could be caused by later dextral movements along the South Atlas Fault Zone. The Middle Devonian Gueiguira supergroup of the western Sahara yields a pole (**19**) contrasting with the Mssissi result. Since pole **19** plots in the vicinity of Carboniferous poles it could be a later overprint. Alternatively, it may be a representative Devonian pole for cratonic Gondwanaland (Kent, Dia & Sougy, 1984); the latter interpretation would imply that the NW margin of Gondwanaland had moved back to within 20° of the equator at this time consistent with the development of a carbonate fringe here on the first occasion since early Cambrian times (Section 10.7, and Ziegler, Scotese, McKerrow, Johnson & Bambach, 1979).

10.3 Palaeozoic palaeomagnetism of Western Europe

The British region

The British microplate, recognised as a geological and palaeomagnetic entity in Lower Palaeozoic times, incorporates the late Precambrian basement discussed in the fourth part of Section 9.5, together with WNW-to-NE trending volcanic arcs and miogeosynclines which formed along the southern margin of the Iapetus suture in Britain and Ireland from Ordovician–Devonian times (the paratectonic Caledonides of Fig. 10.2c). The suture defining this ancient ocean is now buried and extends through the Solway line of southern Scotland and central Ireland to separate the 'Pacific' graptolite province found in the Girvan and County Mayo districts of Scotland and Ireland respectively, from the 'Atlantic' province found in the Lake District of northern England, and County Meath in Ireland; a contrast is also observed in the shelly faunas from either side of this suture in Cambrian (Cowie, 1971) and Lower Ordovician (Williams, 1973) times. Critical problems to which palaeomagnetic data are relevant include the timing of the final closure across the suture and the welding of the Paratectonic Caledonides to the Southern Uplands accretionary wedge and metamorphic orthotectonic Caledonides of the Grampian and Northern Highlands of Scotland (Dewey, 1969 and Fig. 10.2c). In addition, they constrain the scale of post-closure strike-slip motions which were an important feature of the belt in Devonian and early Carboniferous times (Pitcher, 1969; Watson, 1984).

This small region is covered by the longest and most intensive study of Lower Palaeozoic palaeomagnetism. The APW record was first described in detail by Briden, Morris & Piper (1973), and has been refined by a number of later studies (see Piper, 1979b, and Briden, Turnell & Watts, 1984 for reviews). The collective data are plotted in Figs. 10.2 (a) and (b) and 10.3.

Stratigraphically-constrained studies from igneous rocks of Arenig to late Llandeilo age (*ca.* 485–460 Ma) define a W → E APW movement through about 40° (poles **1–11**). Although some of these results are defined by fold tests, they are not immune from the effects of overprinting by later thermal-tectonic events;

by careful thermal demagnetisation studies, Briden & Mullan (1984) have separated an Hercynian (late Carboniferous) component from unit 2. Their results suggest that overprints may be incompletely removed from a number of older studies plotted in Fig. 10.2(a). This would have the resultant effect of shifting the APWP here away from the sampling area and towards the Carboniferous poles. Briden, Morris & Piper (1973) and Briden, Turnell & Watts (1984) considered that some palaeomagnetic results from the orthotectonic Caledonides are of this age and can be used to deduce the separation across the Iapetus Ocean at this time. However, there are good reasons discussed below for now believing that the latter record is entirely younger than late Ordovician, and therefore is not relevant to this problem. Two poles come from *ca.* 475 Ma units post-dating the Arenig emplacement of the Ballantrae ophiolite complex at the nothern margin of the Southern Uplands block (**2,3**); they plot 30° to the east of most Llanvirnian-age poles from south of the suture and show a large shift from a single result of probable Arenig age (**1**) from the NW margin of the orthotectonic Caledonides. A similar APW shift is suggested by data from north of the suture in western Ireland between results from Arenig (**1,2**) and Llanvirn (**3**) units. Deutsch (1980) has compared pole **3** with results from a single contemporaneous unit from south of the suture in Ireland (**1**, Tramore Volcanics) and integrated the data with a geological model for closure and stabilisation of the suture (Phillips, Stillman and Murphy (1976); he infers a separation across the suture of 3300 ± 2200 km in Llanvirnian times. Direct comparison between the British and Irish data is complicated by the possibility of regional rotations (Morris 1976); the trend of the late Proterozoic–early Palaeozoic Dalradian rocks in the orthotectonic Caledonides of Scotland is NE–SW, while in western Ireland (Connemara), the same stratigraphic divisions strike WNW–ESE. Relative rotation between these areas cannot be tested by the existing Ordovician data, but, as discussed below, is a possible explanation of Silurian data from Connemara (Fig. 10.2b)..

The Upper Ordovician (Caradocian and Ashgillian) palaeomagnetic results from the British sub-plate comprise an extraordinary data set, dominated by dual polarity and steep inclination results, which contrast with both the older (Fig. 10.2a) and younger (Fig. 10.2b) data. Recently, Perroud, Van der Voo & Bonhommet (1984, 1985) have also recognised a steep dual-polarity axis of magnetisation in late Ordovician rocks from the Armorican massif (Fig. 10.3). The best-constrained result here comes from the Thouars massif of southern Brittany, dated 444 Ma (Ashgill) in age; further steep magnetisations of this probable age have also been identified in the Iberian Peninsula (pole **1** in Fig. 10.3, corrected for opening of the Bay of Biscay) and they are present in some early results from the Barrandian massif (Bucha, 1965). Steep inclination data have been recognised in Britain for a number of years. Thomas & Briden (1976) found steep negative inclinations to be exclusively present in intrusions of the Welsh Igneous Province (poles **8,9,15–21** in Fig. 10.3); although the age of these intrusions is not known with certainty, a similar remanence is present in the Cader Idris basalts (**10**) of Caradoc age. Piper, McCook, Watkins, Brown and Morris (1978) and Piper (1979b) subsequently reported steep directions, mostly of opposite polarity, from mid-Ashgill (**13**) and otherwise late Ordovician (**6,11,12,14,22**) intrusions from the Lake District. Re-evaluation of some other British data has also helped to clarify the significance of these data: remapping of the Shelve Inlier in the Anglo-

Welsh Caledonides has shown that igneous intrusive suites here are of late Ordovician age and predate Llandovery age folding (Lynas, Rundle & Sanderson 1985). Correction of existing palaeomagnetic data from this area (Piper 1978) for the effects of regional folding, marginally improves the overall precision and radically alters the pole positions (2,4). The time sequence of intrusion here, and in a Caradoc age lava (3), and a pre-mid Ashgill age intrusion (5) from the nearby Breidden Hill Inlier provide additional data.

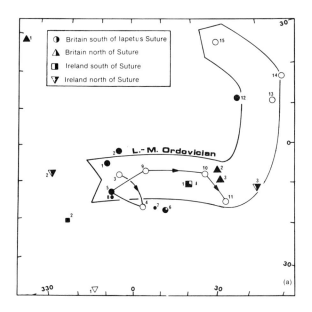

Figure 10.2 (a) *Palaeomagnetic results from* Britain *and* Ireland *assigned to* Lower and Middle Ordovician *times. The data are separated according to the following tectonic subdivisions:* BRITAIN SOUTH OF THE IAPETUS SUTURE (CIRCLES): 1, *Trefgarn Andesitic Series (Ol, Arenig);* 2, *Builth Volcanic Series (Ol, Llanvirn);* 3, *Eycott Group (Ol, early Llanvirn);* 4, *Carrock Fell Complex (Ol);* 5, *Round Knott dolerite (Ol-m, <4);* 6, *Stapely Hill Volcanic Series (Ol, Llanvirn);* 7, *Caerfai Series (Cl, O diagenetic remanence);* 8, *Caerbwdy Sandstone (Cl, O magnetisation as for 7?);* 9, *Lowerr Borrowdale Volcanic Series (O, Llandeilo-Lower Caradoc);* 10, *Upper Borrowdale Volcanic Series (O, Llandeilo-Lower Caradoc);* 11, *Upper Borrowdale Volcanic Series, highest levels at Kentmere;* 12, *Haweswater dolerite (O, Upper Llanvirn-early Caradoc);* 13, *Dolerite intrusions, Shelve Inlier, tilt corrected (O, Caradocian);* 14, *E-W dolerite dyke swarm, Welsh Borderlands (O, as for 13);* 15, *Moel-y-Golfa andesite (O, Caradoc);* 16, *E-W Kersantite dyke swarm, Northern England (Ou?);* 17, *Breidden Hill dolerite (Ou, Caradoc-mid Ashgill).* BRITAIN NORTH OF THE IAPETUS SUTURE (*upright triangles*): 1, *Canisp Porphyry (Ol, probably Arenig);* 2, *Ballantrae serpentinites;* 3, *Byne Hill gabbro (Ol-m, 475 KA).* IRELAND SOUTH OF THE IAPETUS SUTURE (*squares*): 1, *Tramore Volcanics (O, Llanvirn-early Caradoc);* 2, *Grangegeeth Volcanic Series (O, Llanvirn).* IRELAND NORTH OF IAPETUS SUTURE (*inverted triangles*): 1, *Loch Nafooey Spilites (O, Arenig);* 2, *Glensaul Felsite (O, Arenig);* 3, *Mweelrae ignimbrites (O, mid-Llanvirn).*

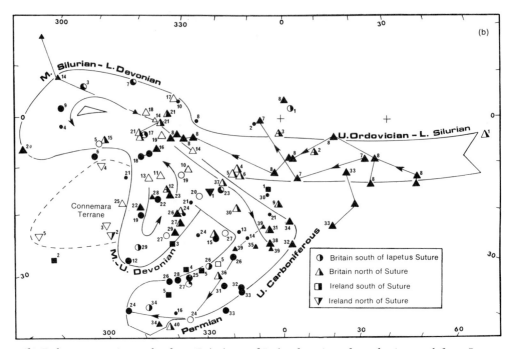

(b) Palaeomagnetic results from Britain and Ireland assigned to the interval from Lower Silurian to Upper Permian times. The data are separated according to the following tectonic sub-divisions: BRITAIN NORTH OF IAPETUS SUTURE (upright triangles): 1, WNW-ESE components, micro-diorite – dolerite suite (450–420 KA, RS); 2, Borrolan Ledimorite; 3, Ben Loyal Complex; 4, Dolerite suite, main group (450–420 KA,RS); 5, Microdiorite suite (450–420 KA, RS); 6, Ratagan complex (415 RS); 7, Aberdeenshire gabbros, post-folding uplift remanence, chrontour zone means based mainly on a.f. demagnetised results for >468, 468–443, <443 K-Ar mica ages; 8, Aberdeenshire gabbros, uplift remanence, site means based on thermal demagnetisation; 9, Borrolan pseudoleucite (430 UP); 10, Achmelvich dyke; 11, Alkaline dykes, NW Scotland; 12, Borrolan syenite; 13, Loch Ailsh complex; 14, Ratagan and Dalradian metasediments, five poles; 15, Morar Division metasediments (480–410 KA); 16, Arrochar complex (Su, 418 KA); 17, Glencoe complex lavas (Lower ORS, Dl?); 18, Lorne Plateau lavas (Su-Dl, 409 KA, 401 RS); 19, Garabal Hill-Glen Fyne complex (412 KA, Su-Dl); 20, Comrie diorite (408 RS); 21, Lower Old Red Sandstone lavas and sediments, Strathmore region (Su-Dl); 22, John O'Groats Sandstone (Dm-u); 23, Orcadian Basin, Lower to Middle Old Red Sandstone (Dl-Dm); 24, Orkney lavas (Dm, 370 AA); 25, Esha Ness ignimbrite (Dm); 26, Duncansby volcanic neck, B component (258–239 KA, Du remanence?); 27, Caithness Sandstone (Dm); 28, Unst ophiolite, B component; 29, Foyers Old Red Sandstone sediments (Dm); 30, Foyers complex (400 KA); 31, Helmsdale granite, A group (420 UP, 400 KAB); 32, Mainland dykes, Shetland (370 AA); 33, Port Askaig tillite, Caledonian overprint, three poles; 34, Gamrie outlier red sediments, two poles (Dl-m?); 35, Unst ophiolite C component (Dm-u uplift remanence?); 36, Northern Highlands lamprophyre dykes (Dm-u); 37, Kinghorn lavas (Carbl); 38, E-W dolerite dyke swarm, NW Highlands (320 KA); 39, Mauchline lavas, sediments and intrusions (P); 40, Stornoway Formation, Isle of Lewis (P-T age inferred from remanence direction). BRITAIN SOUTH OF THE IAPETUS SUTURE (circles): 1, Dolerite intrusions, Builth Inlier (latest O-early S); 2, Diorite dyke, Stile End (Ou-Sl); 3, Somerset and Gloucestershire lavas (Sl-m, late

Llandovery-early Wenlock); 4, Fishguard Volcanic Group (S-D remagnetisation); 5, NW-SE dykes, Anglesey (Ou-Du); 6, Cautley Fell dolerite (>Carbl); 7, Cheviot Volcanic Series (394–374 KA); 8, Cheviot granite (383–372 KA); 9, Anglo-Welsh Cuvette, Old Red Sandstones (Su-Dl); 10, Lizard Complex, tilt corrected (Dl?) remanence; 11, NE-SW Minnette dyke swarm, Northern England (Dm-u); 12, Shap granite (Su-Dl, 394 RS, 399–394 KA); 13, Hendre dolerite (D); 14, Blodwell keratophyre (D); 15, Upper Old Red Sandstone, Bristol District (Du-Cl); 16, Ashprington Volcanic Series (Dm-u, P-Carb remagnetisation?); 17, Derbyshire (Miller's Dale) lavas (Carbl); 18, Derbyshire lavas (Carbl); 19, Cumbrian hematite ore (Carbl?); 20, Cockermouth lavas (Carbl); 21, Clee Hill sills and contacts, three poles (Carb); 22, Draughton Limestone (Carbl, Visean); 23, Coal Measures sills and lavas; 24, Skomer-Volcanic Group (Sl, Carbu remagnetisation?); 25, Carboniferous limestones, Craven Basin (Carbl); 26, Carboniferous Limestones, Craven Basin, six poles in stratigraphic order (Carbl, Tournasian-Visean); 27, Carboniferous Limestones, Askrigg Block (Carbl, Upper Tournasian-Visean); 28, Pendleside Limestone (Carbl, Visean); 29, Shatterford intrusion (Carbu); 30, Wakerfield dyke (303 KA); 31, Whin Sill (295, 281 KA); 32, Devonshire sediments (Pl); 33, Exeter lavas (Pl, 280 KA); 34, St. Bees Sandstone (Pu-T). IRELAND NORTH OF THE IAPETUS SUTURE (inverted triangles): 1, Derry Bay Felsite (<S, U. Llandovery); 2, Basal Keratophyre, Lough Nafooey (S, U. Llandovery); 3, Salrock Group (Sm, Wenlock); 4, Knocknaveen Group (S, Wenlock); 5, Microgranodiorite intrusion (<Sm, post-Wenlock). IRELAND SOUTH OF IAPETUS SUTURE (squares): 1, Chair of Kildare Volcanics (O-S, late Ashgill-early Llandovery) 2, Balbriggan-Sherrick's Island volcanics (O, Hercynian remagnetisation?); 3, Portrane andesite (O, Hercynian remagnetisation?); 4, Limestones, Limerick-Tipperary (Carbl); 5, Carboniferous Limestones, five poles (Carbl).

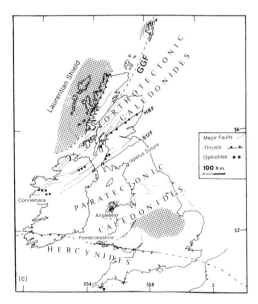

(c) Outline tectonic map of the British region with localities and geological divisions referred to in the text. The shaded areas are regions of known basement stabilised in earlier times. North of the Iapetus Suture this basement is of early Precambrian age, and once formed part of the Laurentian Shield (Sections 7.5 and 7.8), while the basement south of this suture is of late Precambrian-Cambrian age and has an origin discussed in the second part of Section 9.5.)

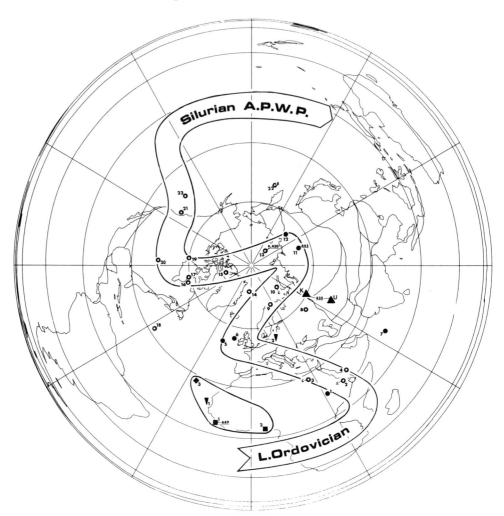

Figure 10.3 The "anomalous" Upper Ordovician palaeomagnetic poles from Britain, Ireland and Armorica which define a rapid APW movement during Caradocian and Ashgill times. The preceding and succeeding APWPs are illustrated in Figures 10.2(a) and 10.2(b) respectively. The poles are numbered: BRITAN SOUTH OF THE IAPETUS SUTURE (circles): 1, Haweswater dolerite (Om-u); 2, Dolerite intrusions, Shelve Inlier (O, Caradocian); 3, Moel-y-Golfa andesite (O, Caradoc); 4, E-W dolerite dyke swarm, Welsh Borderlands; 5, Breidden Hill dolerite (Ou, Caradoc-mid Ashgill); 6, E-W kersantite dyke swarm, N. England (Ou); 7, Corndon Hill phacolith (Ou, Caradoc); 8, Gurndu granodiorite; 9, Foel Fras granodiorite; 10, Cader Idris basalts (Ou, Caradoc); 11, Threlkeld-St. John's microgranodiorite (445 RS); 12, WNW dyke swarm, N. England; 13, Stockdale rhyolite (Ou, mid-Ashgill, 430 RS); 14, Lower Borrowdale volcanics, overprint (post-Llandeilo); 15, Penmaenmawr granodiorite; 16, Carreg-y-Llan granodiorite; 17, Penmaenbach rhyolite; 18, Yr Eifl microgranite; 19, Garnfor granodiorite; 20, Bodeilias granodiorite; 21, Myndd Nefn granodiorite; 22, Cautley Fell rhyolite (Ou, Ashgill). IRELAND SOUTH OF IAPETUS SUTURE (diamond): 1, Lambey Island andesite (O). ARMORICAN MASSIF (squares): 1, Thouars massif rocks (Ou, 444 RS); 2, Moulin de

The overall data set is plotted in Fig. 10.3. The continuity with the Lower–Middle Ordivician APWP is illustrated by the rapid shift from the late Llanvirn-Llandeilo age poles (9 → 10 → 11) to a NW motion via Caradocian poles 13,14,15 and 17 in Fig. 10.2a. The remaining poles comprise a scattered distribution in high latitudes, and no clear movement can be discerned between them, except that there is a clear time shift to poles 13 and 22 from mid-Ashgillian rocks and pole 11 which is dated 445 Ma (early Ashgill) in age. Thus, there is positive evidence for continuing the NW APW motion towards the present north pole in Fig 10.3, but, pending radiometric investigations of poles 15–21 from the Welsh Province, the continuity of the APWP can only be speculated upon. The shorter path (<90°) to the Silurian data continues into the central Pacific as a *north* APWP. This interpretation would imply that the Lower–Middle Ordovician APWP plotted in Fig. 10.2a is of normal polarity, while the post-Ordovician APWP plotted in Fig 10.2b is of reversed polarity. The generalised path drawn through the British data in Fig 10.3 should only be regarded as tentative, both because the time interval involved is no more than 20 Ma, and because some of the apparently aberrant points might reflect a multipole field source and not simply record the movement of the landmass with respect to a geocentric dipole source. An example of a complex field source in Lower Devonian times is noted elsewhere (Section 12.6).

The aberrant poles from the Armorican massif and Iberian peninsula comprise a discrete group, marginally removed from the British data, and imply that the Armorican-Bohemian sub–plate was still separated from the British sub–plate at these times. A magnetisation from the Shetland ophiolite complex, linked to emplacement at *ca*. 435 Ma corresponds closely with the results from south of the Iapetus suture in Britain, as does a result from the Karmøy ophiolite along the strike continuation (but opposing foreland) of this belt in Norway. There is also evidence that these 'anomalous' Ordovician field directions are recorded within Baltica: a large body of results from Lower–Middle Ordovician limestones of the Baltic platform in Sweden yield steep directions remote from Ordovician field directions for Europe (Claesson, 1978). The mean remanence direction is removed from the present field direction, and there is no plausible reason why the Recent field should be so widely recorded as a single component; a diagenetic origin in Upper Ordovician times is more likely.

There are several lines of geological evidence which point to the wider significance of the rapid APW movement recorded by the data of known, or probable, Upper Ordovician age in NW Europe. Firstly, the APWP of Fig. 10.3 implies that the British and Armorican sub-plates moved close to the North pole in

---

Chateaupanne Formation (Ol, Ou diagenetic remanence?). IBERIAN PENINSULA (inverted triangle): limestone (Ol, Arenig-Llanvirn). Pole marked U is the Shetland (Unst) ophiolite A component (435–425 AA). This unit is emplaced onto Dalradian rocks at the northern margin of the Iapetus suture and may define the first closure of the ocean here; its tectonic relationship to the tectonic block south of the suture at this time is, however, unclear. Pole K is a pole from the Karmøy ophiolite of SW Norway with the magnetisation dated ca. 445 Ma. This unit is of similar age and tectonic setting to the Shetland ophiolite but is emplaced onto the opposite foreland of the Caledonian orogen.

Caradocian to early Ashgillian times (Perroud, Van der Voo and Bonhommet, 1984) and were then translated rapidly into middle latitudes of the southern hemisphere, while the APWP of Figure 10.1 suggests that the northern sector of Gondwanaland also moved across the same pole at about this time. These movements explain the distribution of widespread late Ordovician glacial deposits. Glacial dropstones occur within the Upper Ordovician deposits of the accretionary wedge in southern Scotland (Harland, 1972) and in probable late Ordovician rocks of W Ireland (Williams, 1980). In eastern North America, conglomerates and varves of probable glacial origin extend from Quebec to the Gaspé peninsula and include the Caradocian White Rock formation of Nova Scotia. In Normandy, late Ordovician glacio-marine deposits occur in three localities and are correlated with similar deposits in Celt-Iberia in Spain and Thuringia in Germany; the latter are of established Ashgill age. Within Gondwanaland, Ashgillian-age tillites occur in the Anti-Atlas of Morocco (Harland, 1972) and are an extension of the more widespread Upper Ordovician glaciation of the central Sahara.

Secondly, late Ordovician times were characterised by a major episode of faunal extinction. This was one of the four major episodes of faunal extinction in Phanerozoic times (the others occurred in Permian, Triassic and Cretaceous times, see also section 12.8). It was probably a composite event made up of waves of extinctions in late Caradoc/early Ashgill and in late Ashgill times. The latter event is correlated with rapid lowering of sea level during the last (Hirnantian) stage of Ashgill times which displaced first the deep benthic faunas, and then the more shallow benthic faunas, towards the margins of the continental shelves (Brenchley, 1985). The entire episode probably took place within 2 Ma and while several factors may have contributed to these eustatic changes, only the rapid growth of ice sheets at this time can apparently account for the rapidity of the sea level fall (Brenchley op cit.). The growth of the polar ice at this time is further suggested by the evidence for enhanced oceanic circulation promoted by higher temperature gradients (Force, 1984 and Section 12.8) because the black shale facies of Ashgill times is appreciably less reduced than the older Ordovician and younger Silurian examples.

The rapid rise in sea level and flooding of the continental shelves in earliest Silurian times is another remarkable feature of this episode. It seems that both the near-elimination of shallow marine environments, and the rapid movement of a large part of the continental crust into polar latitudes could be invoked to explain the Hirnantian extinction, but the rapid translation of the continental areas into high latitudes, could itself have precipitated the climatic deterioration responsible for the short-lived late Ordovician (mid-Caradoc(?)–late Ashgill) glaciation by drastically changing oceanic and atmospheric circulation patterns (see for example, Robinson, 1973). This inference is further suggested by the lack of evidence for repeated climatic cycles in the Hirnantian (Brenchley 1985), and that the Pleistocene glaciations (which were not associated with rapid continental movements) correlate with no such comprehensive faunal extinctions.

The recognition of rapid APW movement in late Ordovician times may also reconcile one of the outstanding problems of the APWPs prior to Upper Palaeozoic times. The most conservative assessments of the Laurentian Shield APWP described in Chapters 7 and 9 continue this path via the Phanerozoic poles plotting in the Pacific area, and then towards the present north pole, and define this APWP as one of N polarity (eg. Irving & McGlynn 1976b). The paths for the constituent shields of Gondwanaland continue via the Phanerozoic poles towards the present south pole and apparently define this APWP as of R polarity (eg. McElhinny 1973). This contradiction can only be reconciled if there is a rapid (and as yet unrecognised) polar shift applicable to either one or the other of these shields. McElhinny, Giddings and Embleton (1974) recognised this problem and proposed that the polarities could be reconciled by a rapid shift in the early Cambrian Laurentian path. This argument was developed by Piper (1981b) but, as discussed in the second part of Section 9.4, it now appears to be precluded by the Cambrian palaeomagnetic record from Laurentia. An alternative solution is presented if the large shift identified in the British sub-plate is applicable to either the Gondwanaland (see discussion in Section 10.2) or the Laurentian plates. (The Siberian and Baltica APWPs in Lower Cambrian times (Section 9.4) are too poorly defined to suggest the continuities between Proterozoic and Phanerozoic paths). Three points suggest that the APWP defined from Britain (and probably closely applicable to the Armorican and Baltican plates) is to be linked to Gondwanaland rather than Laurentia: (1) their faunal provinces are more closely linked, (ii) a large and similar movement is recognised between the Middle Ordovician and Lower Silurian poles in each plate, and (iii) the two paths can be closely matched on the conventional reconstruction from mid-Silurian times onwards (Fig. 10.5); by this time welding between them is also apparent from geological evidence.

The post-Ordovician Palaeozoic results from Britain are summarised in Fig. 10.2b. The results from within the metamorphic (orthotectonic) Caledonian belt present a similar problem to the data from many Precambrian terranes because the source rocks were uplifted and cooled over an interval of 50–100 Ma following the climax of tectonic activity (Dewey and Pankhurst 1970, Watson 1976) the Aberdeenshire younger gabbros (7) in NE Scotland yield a distributed post-folding remanence which has been interpreted in the context of the K–Ar mineral age chrontours (Dewey and Pankhurst op cit.) to represent an E → W APW swathe recorded during regional uplift and cooling (Watts & Briden, 1984). Although these authors have also interpreted this path as recording an appreciable interval of Ordovician APW, this remanence is regarded here as a TCRM of Silurian age, for three reasons: (i) only one (R) polarity is recognised throughout the swathe, although Ordovician times include a number of reversals and are represented by a considerable number of results of opposite polarity (Section 12.2 and Figure 10.2b); (ii) the direction of APW is opposite to the stratigraphically-constrained Ordovician data of Fig. 10.2(a), and (iii) the Aberdeenshire swathe incorporates results from the Northern Highlands of Scotland (the sector of the orthotectonic Caledonides north of the Great Glen Fault, Figure 10.2c) assigned to the interval 450–420 Ma (1,4,5, Fig. 10.2b) as well as results from the same structural divisions from post-Ordovician, and probable Silurian, age alkaline intrusions (2,3,9–13, Turnell & Briden, 1983) and from the ca 430–400 post-tectonic granites (see also Torsvik 1985a). Some other results from Siluro-Devonian rocks of these regions have also yielded distributed directions with palaeomagnetic poles plotting along this swathe (Kneen 1973, Esang & Piper 1984). Most of this APW movement may have taken place between latest Ordovician and Middle Silurian (Wenlock) times because pole 3 of this latter age from south of the suture plots towards the younger end of the swathe and is constrained by a fold test to the Llandovery–Wenlock boundary (Piper 1975d); furthermore, result **15** from north of the suture is derived from Moinian psammites (Watts 1982) with uplift K–Ar mineral ages ranging 500–430 Ma (although, as in the case of the Aberdeenshire gabbros, the relationships between the remanence and isotopic closure is uncertain here). This position also includes poles from a wide range of Upper Silurian–Lower Devonian rocks, which are post-orogenic with respect to the Caledonian belt and have a distribution ranging from the orthotectonic Caledonides in the north to the Hercynian Front in the south (McClelland-Brown 1983). They have long been central to the arguments for and against widespread remagnetisation of European Lower Palaeozoic rocks (Section 10.4). Their primary nature is now generally accepted (Torsvik 1985b) while the preceeding E → W swathe is a feature of most recent assessments (Briden, Turnell & Watts 1984); some versions have a southerly extension to incorporate results from some Caledonian granites (**30,31**, Torsvik op cit.) but to accommodate the Silurian data discussed above would introduce another loop into the path, and it seems more likely that these intrusions record a post-Lower Devonian remanence. A westerly extension of the path shown in Fig. 10.2b to 290°E, is suggested by a few data points notably pole **10** (408 Ma) from north of the suture.

A shift from these poles to the Middle–Upper Devonian group is identified by data points north of the suture (**22–25,27,29** and **32**) and poles **11–15** from south of the suture. The evidence from Lower Carboniferous data is conflicting: poles from Lower Carboniferous volcanics south of the suture (**17,18**) plot close to the Upper Silurian–Lower Devonian group, while poles from contemporaneous limestones plot 30°–50° further to the south (**25–28**). Although igneous-derived poles numbered **18** are untreated, they are supported by a variety of field tests and a later demagnetised result (**17**, see also Rolphe & Shaw, 1985 for confirmatory data). A SE shift from these poles is suggested by igneous results **37 → 38** from north of the suture, with the latter result assigned to the Lower-Upper Carboniferous boundary. The collective data would imply that the APWP was a small double loop in Upper Devonian–Lower Carboniferous times. Although the Lower Carboniferous limestones have been widely studied in Britain and Ireland (see Turner & Tarling, 1975), the significance of their remanence is far from clear; the poles from this facies are smeared towards, or are coincident with, the Upper Carboniferous–Permian path, and there is a strong suspicion that they are dominated by later diagenetic remanences. This is also a likely possibility in the context of recent North American experience (McCabe, Van der Voo & Ballard, 1984 and Section 4.2), although the remanence in British limestones cannot have been acquired entirely in the Permo-Carboniferous superchron. (Section 12.2.) because up to 25% of these studies have N polarity. The remaining Upper Carboniferous–Permian data from Britain are consistent with a NE → SW APW motion recording this superchron, although this period has not been as well studied in Britain as in continental Europe (Fig. 10.4); poles **34** and **40** include data with N polarity and therefore were probably magnetised in Triassic times (Section 12.2).

Palaeomagnetism and the Caledonian orogeny

Although the results have been separated here on the basis of their origins either north or south of the Iapetus suture, it will be evident from Fig. 10.2b that there is no distinction between these two data sets after Ordovician times. The E → W swathe is recorded in the orthotectonic Caledonides both north (Esang & Piper, 1984) and south (Watts & Briden, 1984) of the Great Glen Fault Zone (Fig. 10.2b), and from late Ordovician–early Silurian (1) and mid-Silurian results from south of the suture. Palaeomagnetic arguments for large scale Carboniferous movements along the Great Glen Fault (Storetvedt, 1974, Van der Voo & Scotese, 1981) have been assessed and refuted by Briden, Turnell & Watts (1984). Thus it is implicit in Figs. 10.2b & 10.3 and this discussion that the final closure of the Iapetus Ocean took place during late Ordovician times and is recorded by ophiolite emplacement of this age in the Shetland Islands and the Norwegian Caledonides. This palaeomagnetic conclusion generally accords with palaeontological evidence, (see also Section 10.7) which recognises a reduction in faunal provincialism in Upper Ordovician times, reaching a minimum in Ashgillian times (Burrett, 1973) and its disappearance within the Caledonian orogen in Silurian times. The sequence of geological events in the Caledonides during Ordovician times can be modelled in terms of oblique collision of the two continental margins accompanying SW migration of an unstable triple junction (Phillips, Stillman & Murphy, 1976); major SE-directed and subduction-related volcanism on the south side of suture terminated progressively between late Llandeilian and Ashgillian times.

Silurian results fom the Connemara terrane of western Ireland are significantly removed from the remaining data in Figure 10.2b; and more recent studies have shown that these data points are influenced by Devonian overprinting; a dual polarity axis of magnetisation of mid-Silurian age, defined from lavas and red sediments plots 30–50° to the west of this group (author, in preparation). Hence the Connemara region is quite clearly a suspect terrane. An anticlockwise rotation of 45° about a local pole of rotation brings the poles from this terrane into general agreement with contemporaneous poles from Britain, and closely aligns the strikes of Caledonian trends in each area. This motion could have been caused by rotation during dextral movements along the orogenic belt during Devonian times because probable Hercynian-age results from this region correspond to the data from elsewhere. However, the motions of this terrane were probably more complex in detail because a small latitudinal discrepancy remains; this may be related to the observations that Connemara lies to the south of the Ordovician volcanic arc, along the northern margin of the Iapetus suture in Britain, and rocks correlative with the Dalradian of Connemara are found only on the north side of the suture in Scotland (fig. 10.2c). The E–W trending Ireland-Newfoundland sector of the Caledonian Belt is anomalous with respect to the NE-SW trending remainder of the orogen, in terms of the absence of both large-scale recumbent folding and thrusting of Silurian age, and of major strike slip faults. Phillips, Stillman & Murphy (1976) interpret these contrasts in terms of a collision misfit, or gap, along this sector. Although closure of the Iapetus Ocean was completed here by Silurian times, dextral strike-slip motions within the orogen in Devonian times (Phillips, Stillman and Murphy op cit., Watson, 1984) could have been responsible for rotating (and possibly translating) the Connemara terrane because it was appropriately sited to accommodate block rotations (Morris, 1976). It is possible that the effects of larger amounts of smaller block rotations within the orogenic belt are represented by the distribution of palaeomagnetic poles in Fig. 10.2b; before such movements can be identified, more refined palaeomagnetic studies will be required to separate tectonic effects from the APW collectively applicable to the orogen and its stable forelands.

Palaeozoic palaeomagnetism of the remainder of Europe

The bulk of post-Ordovician palaeomagnetic record from the Armorican massif is represented by known, or suspected, overprints acquired during the Hercynian orogeny. There is a suggestion that the first five poles in Fig. 10.4 may record the Siluro-Devonian swathe described by British data, but only one pole (4) has an age positively confirming this. The remaining data comprise a distribution plotting between British Middle Devonian to Lower Carboniferous poles and commensurate with a primary or secondary remanence of this age (6–17). Results from elsewhere in France, notably from the Vosges, define a concordant group of Lower Permian poles. Permian data from the French sector of the Alpine belt are aberrant; they imply a $30°-45°$ anticlockwise rotation of this arc during the Alpine orogeny (Westphal, 1973) and are excluded from Fig. 10.4. The results fom the Iberian peninsula are plotted in this figure, after correction for the opening of the Bay of Biscay (first part of Section 11.4) and come from rocks with stratigraphic ages as old as Silurian. They all plot along the same NW–SE swathe incorporating the Middle Devonian to Lower Permian results from Britain and the Armorican massif. They would therefore appear to be dominated by overprints acquired during the Hercynian orogeny; this interpretation is further suggested by the predominant R polarity of the data, linking them to the Permo-Carboniferous superchron (Section 12.2). In the Iberian arc of NE Spain the remanence in pre-Upper Carboniferous rocks has been shown to pre-date the F1 folds which form the arc structure and are contiguous with folds in the southern part of the Armorican massif (Reis, Richardson & Shackleton, 1980 and Fig. 10.4). The remanence here is largely, but not entirely, corrected for by unfolding the Iberian orocline, implying that it was imparted (in this instance by the redistribution of hematite) during the Upper Carboniferous folding of the arc.

The Silurian swathe is recorded by two further results (1,2) from central Europe, but as in the Armorican area, the bulk of the remaining data are either suspected overprints plotting on a 'Hercynian remagnetisation' swathe (3–5) or are Upper Carboniferous sediments (7,9 → 15); the latter are exclusively reversely magnetised and comprise a tight group correlating with the Upper Carboniferous–Lower Permian results from elsewhere. Systematically aberrant data comprise the pre-Lower Carboniferous results from the Vosges (16,17) the Harz Mountains (18–20) and the Franconian Forest (21–23); these discordant results derive from rocks of Lower to Upper Devonian age, and can be matched with results of this age from further west in Europe by a $30°$ clockwise rotation. Bachtadse, Heller & Kröner (1983) interpret this rotation in terms of a sinistral shear in late Devonian–early Carboniferous times; this movement could also be relevant to the Cambrian correlations between the Armorican and Bohemian massifs (Section 9.5), although data from the latter area are presently insufficient to test this proposal. The Permian results include a volcanic component from central Europe which conforms closely with the sedimentary results in this reversely-magnetised distribution and with the Upper Carboniferous data. It should be noted that there are uncertainties in the age assignments of many European poles in this general group: many rocks traditionally considered to be Lower Permian are probably latest Carboniferous in age (see Van der Voo, Peinado & Scotese, 1984). The European plate defined by these data extends to the SE as far as the Carpathian Front. Permian results from the Carpathians plot to the east of the data illustrated in Fig. 10.4, and are compatible with anticlockwise rotations during the Alpine translations (Krs, Muska & Pagac 1982).

Some data from the Norwegian Caledonides may also record the Siluro-Devonian swathe (Fig. 10.4), but the diagenetic or overprint origin of the remanence in Baltic Platform sediments is illustrated by the tight group of poles from Lower–Upper Silurian limestones (4), and the characteristic remanences in Siluro-Devonian red beds (5–7). Both sets accord most closely with Upper Devonian–Lower Carboniferous data from elsewhere. Thus, while there is some evidence for the E $\rightarrow$ W Silurian APW swathe in the palaeomagnetic data from both the Baltic platform (Fig. 10.4), and the European platform of the USSR (second part of Section 10.5), it is not very compelling and gives only qualified support to geological evidence which implies that this plate was also welded into the Pangaean configuration by Silurian times. Continent–continent collision between the Laurentian and European plates is related to intense deformation in the Scandinavian Caledonides leading to the recumbent folding, thrusting and nappe emplacement already noted. Less is known about the continuation of the collisional belt in E Greenland, although the main thermal and tectonic events are linked to the westward translation of nappes over the Laurentian foreland and they are also dated as late Silurian (Haller, 1970). Harland (eg 1978) has addressed the problem of the position of Spitsbergen in this orogenic espisode. Although, following the original work of Du Toit (1937), this region has been conventionally placed off the north coast of Greenland, the geological similarities between the two regions are not close. Spitsbergen incorporates a continuation of the Caledonian belt, which was the site of substantial strike-slip movement in late Devonian times. Harland (1978) recognises at least three structural provinces separated by two major fault zones with estimated movements of 200–1000 km. Hence, in mid-Devonian times, the eastern part of Spitsbergen may have been as far south as central east Greenland, with the western segment distributed further along the NE margin of the Greenland Caledonides. Post-tectonic granites here pre-date Lower Devonian molasse sediments and identify an orogenic climax compatible with the timing of events in Scandinavia and Britain. Upper Devonian and Lower Carboniferous sediments of central Spitsbergen yield an APW segment which coincides with the British APWP of this age (Torsvik, Lovlie and Sturt, 1985) and precludes major strike slip displacements between the two regions since these times.

The collision between the Anglo-Welsh sub-plate defined in the concluding part of Section 9.5. (which may have extended westwards to include the Avalonian terrane in the Appalachian sector) with the Iberian-Armorican-Bohemian sub-plate (second and third part of section 9.5) in the south, and the Baltica plate in the east, is defined by an intersection of orogenic belts in the vicinity of the present North Sea. The NE–SW Caledonian trend in western Britain swings around to merge into a poorly-exposed NNW–SSE (Charnoid) trend in eastern England which includes calc-alkaline magmatism (e.g. Turner, 1949; Wills, 1978). Further to the east, the site of collision may include the Tornquist lineament in the North

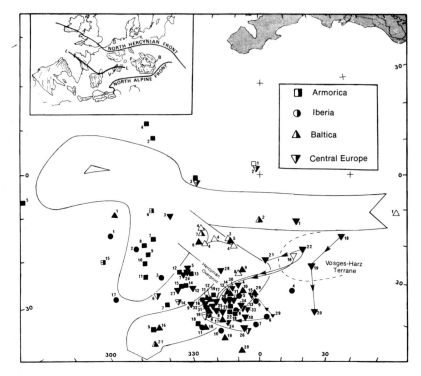

Figure 10.4 Palaeomagnetic south poles from Continental Europe assigned to the interval between Lower Silurian and Upper Permian times (438–250 Ma). ARMORICAN MASSIF AND REMAINDER OF FRANCE (squares): 1, Crezon Peninsula dolerites (Ou); 2, Amorican red beds C component (O, S-D remagnetisation); 3, Jersey dolerites group B (<565 RS, S-D remagnetisation?); 4, Lamprophyre dykes, Jersey (427 KAH); 5, Roanne and Morvan region volcanics (Du-Carbl); 6, Armorican red beds B3 component (O, Du-Carbl remagnetisation?); 7, Red beds, Montmartin syncline, two poles, (Du, D-Carb remagnetisation?); 8, Moulin de Chateaupanne Formation, Hercynian overprint; 9, Crezon peninsula dolerites, two poles, Hercynian overprints; 10, Paimpol-Bréhac (Plouviro) red beds (472, D-Carb remagnetisation); 11, Brittany dykes (Carb); 12, Flamanville granite (340–300 Ma); 13, Jersey dolerites, group A ENE dykes (C, Carbm-u remagnetisation?); 14, Trégastel-Ploumanac'h granite (300 RS,KA); 15, Zone Bocaine red beds, two poles (D-Carb remagnetisation?); 16, Erquy-Cap Fréhal red beds (Ol, D-Carb remagnetiation?); 17, Laval syncline volcanics (Carbl, Hercynian overprint); 18, Loedeve Sandstone, two Poles (Pl); 19, Montcenis Sandstone, Morvan (P); 20, Nidek-Donon volcanics (Carbu-P); 21, Lower Nidek volcanics (Carbu-P); 22, Nidek Porphyry. IBERIAN PENINSULA (circles): 1, Iberian arc of northern Spain, Palaeozoic rocks, Hercynian remagnetisation; 2, San Pedro Formation (Su-Dl); 3, Cabo de Pinas Sandstones (Du-Carbl); 4, Almaden igneous rocks (Su-Dl, Hercynian remagnetisation?); 5, Atienza igneous rocks (Carbl-m); 6, Atienza sediments, two poles from lower and upper formations Carbu, 287 KA); 7, Viar red beds (Carbu-Pl); 8, Viar dykes and sills (Carbu-Pl); 9, Bucaco red beds (Carbu-Pl); 10, Red beds and volcanics (Pl); 11, Seo de Vogel andesites (Pl). CENTRAL EUROPE (inverted triangles): 1, Diabases, Czechoslovakia, two poles including one Hercynian remagnetisation (Sm-u); 2, Basalts, Czechoslovakia (Sm-u); 3, Claystones, East Germany (O, Hercynian remagnetisation?); 4, Barrandian red limestones (Dl-m, Hercy-

Sea and continues into the Caledonian-age Holy Cross Mountains in Poland. The suture is also largely hidden within the Appalachian sector: mafic bodies of Ordovician–Upper Devonian age occur in five localities between Armorica and New England; together with some more circumstantial evidence, they suggest the presence of a Taconic to Acandian suture (*ca.* 440–400 Ma). The ages of these sutures are all older than the mid Carboniferous age required by interpretations of the palaeomagnetic evidence before the significance of widespread magnetic overprinting in the Palaeozoic palaeomagnetic record was understood (Section

---

nian remagnetisation?); 5, Granites and sediments (O, Hercynian remagnetisation); 6, Diabase, East Germany (D); 7, Plenzen Basin sediments, Czechoslovakia (Carbu); 8, Eifel Sandstones (Dl-m, Hercynian remagnetisation); 9, Blanice graben red beds (Carbu) 10, Krakonese basin red beds (Carbu); 11, Boskovice graben red beds (Carbu); 12, Kladno-Radovin basin red beds; 13, Lower Silesia volcanics (Carbu); 14, Mineralised hematite veins, Czechoslovakia (Carb-P); 15, Sediments and Igneous rocks, Inner Sudetic Basin (Carbu); 16,17, Vosges greywackes, Upper Devonian and Lower Carboniferous units; 18,19,20, Harz sediments, Lower Devonian, Upper Devonian and Lower Carboniferous units; 21,22,23, Franconian Forest sediments, Middle Devonian, Middle-Upper Devonian and Lower Carboniferous units; 24, Schopfheim Basin sediments, W. Germany (P); 25, Krakow District volcanics (P); 26, Volcanics and sediments, W. Germany; 27, Sandstones, Czechoslovakia (P); 28, Porphyry and sediments, E. Germany (Pl); 29, Rotliegende sediments and lavas, W. Germany; 30, Rotliegende red beds, Czechoslovakia, two poles for lower and middle-upper Rotliegende formations; 31, Boskovice graben red beds, two poles, Czechoslovakia (Pl); 33, Lower Silesia volcanics, E. Germany (Pl); 34, Nahe region igneous rocks, W. Germany (Pl, 278–253 KA); 35, Kosice area red beds (Pu). BALTIC PLATFORM: 1, Fongen-Hylingen gabbro, NW and SW components (Su-Dl, 405 RS); 2, Sulitjelma gabbro (Su-Dl); 3, Sarv Nappe dolerites (750–700 KA, Caledonian remagnetisation (Su-D); 4, Gotland Limestones (Sl-u); 5, Kvamshesten sediments (Dm-u); 6, Ringerike Sandstone (Su); 7, Roragen Sandstone (D); 8, Vanern limestones (O, Carb remagnetisation?); 9, Mt. Billingen sill (Carb-Pl, 287 KA); 10, Skåne dolerite dykes, group A (Carb-Pl, 294 KA); 11, Oslo igneous rocks (Carbu; 290 KA); 12, Ny-Hellusund dolerite (Carb-Pl, 270 KA); 13, Skåne dolerite dykes, group E (carbu-Pl, 294 KA); 14, Arendal dykes, A and C systems (Carbu-Pl); 15, Mt. Henneberg Sill (Carbu-Pl, 279 KA); 16, Arendal dykes, B and D systems (Carbu-Pl); 17, Oslo graben lavas (Pl); 18, Oslo region igneous complex (Pl, 270 KA)); 19, Bohuslan multiple dykes (Pl); 20, Mellaphyre dykes, Skåne, (Carbu, 300 KA); 21, Brummudal lavas (Pl, later overprint?). Stereographic projection centred at 0°E, 0°N; the outline of the British APWP covering the same time interval (Figure 10.2(b)) is shown by the swathes. The location of the central European block from which these data are derived is shown stippled in the top left hand part of the figure.

The inset diagram shows the distribution of the pre-Permian Hercynian inliers of central Europe with generalised structural trends; Corsica and Sardinia are shown in their pre-drift configurations (first part of section 11.4). Localities referred to in the text are lettered: B, Bohemian massif, F, Franconian Forest and V, Vosges. L-L is the Ligerian Front of mid-Devonian age.

10.4). However, they are compatible with the collisional history described above and implied by Fig. 10.5.

Palaeomagnetism and the Hercynian orogeny

With the exception of some data which exhibit an anomaly in declination, and suggest regional rotational movements (second part of Section 10.3), the palaeomagnetic data from the orthotectonic Caledonides in the north, to central Europe in the south conform to a single APWP from the beginning of Silurian times onwards. Furthermore, when the Ordovician to Permian APWP defined from Britain, in Fig. 10.2b is rotated with Britain into the Pangaean configuration (Bullard, Everitt & Smith, 1965) it is similar to the Gondwanaland APWP described in Section 10.2 and Fig. 10.1 after Lower Silurian times (Fig. 10.5). It is implicit in this observation that the crust between Britain and Gondwanaland comprising the Iberian-Armorican-Bohemian microplate, was practically continuous with, if not actually welded into, the configuration by these times. Hence the timing of continental collision must now be assigned to a Silurian rather than a Carboniferous age (*cf.* LeFort & Van der Voo, 1981). This agrees better with geological evidence, because it links the climactic Caledonian deformation and metamorphism to this event. In the British, Irish and Scandinavian Caledonides, polyphase deformation linked to this collision took place in Middle–Upper Silurian times, while sequential accretion of the Armorican and Gondwanaland plates into the mosaic is suggested by the manner in which deformation occurred progressively later towards the south and west (Roberts & Gale, 1978). In southern Baltica for example, the peak of activity occurred in late Silurian (Downtonian) times, and in southern Ireland it was reached in Lower Devonian times; in the Ligerian zone of high pressure metamorphism and nappe development, which extends across central Europe from Brittany in the west (Figure 10.4), deformation occurred in Middle Devonian times (Zeigler 1984). The former events were contemporaneous with the Acadian orogeny in the Appalachian sector which shortened and thickened the crust here and developed tight vertical folds followed by intrusion of granites in Upper Silurian and Lower Devonian times (Roberts and Gale 1978).

A varied range of tectonic models has been proposed to accommodate the ensuing Hercynian orogeny of late Devonian–Upper Carboniferous times. They include closure of an ocean between Armorica and Britain culminating in continent–continent collision (e.g. Dewey & Burke, 1971, Burrett, 1972); other workers have proposed that Cordilleran environments were followed by ocean closure in southern Europe (Floyd, 1972), or that the Hercynian belts originated as a series of microplates between stable Europe and Gondwanaland (Riding 1974; Badham & Halls, 1975). In the context of the palaeomagnetic evidence, the Hercynian orogeny must now be regarded as an ensialic phenomenon. The counter-arguments which have stressed the faunal and lithological unity of the terranes spanning the orogen (Ager, 1975) and the ensialic nature of magmatic activity (Krebs & Walschendorf, 1973) are supported. This conclusion also accords with several peculiar signatures of the belt, including the absence of high-P–low T metamorphism, mélanges and nappe structures. It also helps to explain the great width of the orogen, which exceeds 800 km in Europe (Fig. 10.4) and extends southwards to include the Meseta and Anti-Atlas belts of Morocco and the Mauretainides in W and NW Africa. The Hercynides evidently represent the consequence of an extensive thermal event, responsible for intraplate melting, and generation of granites and migmatites (Read, 1960, Zwart 1967). Widespread folding in arcuate zones is linked to this episode, but the isolated ophiolite occurrences (e.g. Kirby, 1980) can represent no more than incipient developments in regional tensional régimes, perhaps emplaced during strike-slip movements (Badham, 1982, Zeigler 1984). The extent of the thermal effects associated with

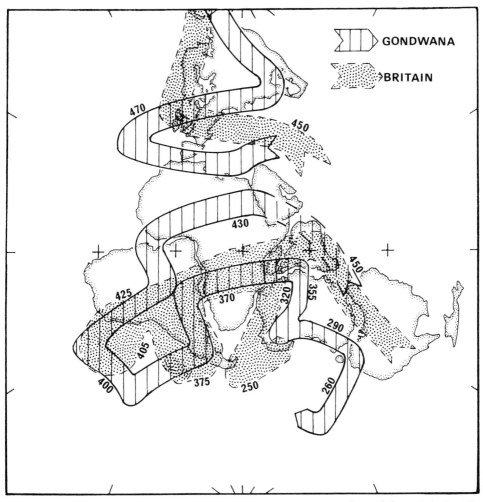

Fig. 10.5 The correlation of the Ordovician–Triassic APWP derived from Britain (Fig. 10.2), with the contemporaneous APWP from Gondwanaland (Fig. 10.1), following rotation into the Pangaean A configuration of Bullard, Everitt & Smith (1965), see also Section 10.6. The close correspondence between these paths implies that the Lower Palaeozoic oceans between the Caledonides in the north, and the Atlas region in the south, had closed (in palaeomagnetic terms) by early Silurian times. The later residual differences between these paths are likely to result from an incomplete understanding of APW rather than from tectonic causes. Stereographic projection centred at 30°E, 0°N with respect to Africa.

this orogeny is remarkable. It is identified in the palaeomagnetic record from the widespread magnetic overprinting of this age, which extends will beyond the orogenic forelands (section 10.4) although the relative contributions of tectonic, diagenetic and climatic effects to this phenomenon have not yet been clearly resolved.

Thus, while the Caledonian and Acadian orogenic belts can be explicitly related to consumption of ancient ocean basins and continent–continent collision, the tectonic significance of the Hercynian belts is less clear. Many workers have drawn attention to the importance of strike-slip faulting along, or near to, the older sutures. These faults comprise a group of NE–SW lineaments, characterised by sinistral and dextral movements which are probably mainly Devonian in age (Pitcher, 1969, Webb, 1969); prominent E–W faults are dominated by Carboniferous dextral movements, and transect the older Caledonian and Acadian trends. Following the models of Tapponier & Molnar (1976) and Molnar & Tapponier (1977), linking strike-slip tectonics to the impingement of India against Asia during the Alpine-Himalayan orogeny (see Fig. 10.9 and Section 11.4), Lefort & Van der Voo (1979) relate these fault systems to the movement of the rigid West African cratonic margin of Gondwanaland into the Appalachian-Maritime margin of Laurentia. The resultant effect was to force segments of the intervening zones sideways: Hercynian Europe moved laterally to produce the E–W dextral faults north of Africa and the SW–NE sinistral movements along the eastern margin of Laurentia. This model could accommodate such effects as the dextral rotation of central Europe noted in section 10.3 and may have contributed frictional heat to produce granite magmas by intracrustal melting.

10.4 Palaeozoic palaeomagnetism and the remagnetisation hypothesis

Before proceeding to examine the Palaeozoic palaeomagnetic record of the remaining plates, it is necessary to consider a problem which affects them all to an appreciable extent. The Gondwanaland, and British (and in part the Armorican) palaeomagnetic data define long APWPs throughout the Ordovician–Carboniferous interval, which are largely divergent from post-Carboniferous paths. The data from Laurentia, Siberia and the eastern sector of the European plates appear, by contrast, to define much shorter paths, with many of the Ordovician–Lower Carboniferous poles plotting in similar positions to the Permo-Carboniferous poles. This could indicate either that these continents experienced much smaller motions than Gondwanaland, and Britain–Armorica (an unlikely scenario over such a long period of time), or that the magnetisations recorded by the Palaeozoic rocks of these plates are largely Permo-Carboniferous overprints. This discussion has a long pedigree: Creer (1968a,) first recognised that the Lower Devonian field axis derived from the Strathmore (Midland Valley) lavas of Scotland (Fig. 10.2b) and Lower Old Red Sandstone sediments of the Anglo-Welsh Cuvette (Chamalaun & Creer, 1964) differed from all other European pole positions obtained from Devonian rocks. Creer argued that *only* the British results were representative of the Lower Devonian field and that all other results were remagnetisations. There were, of course, counter-arguments which considered the British results to be anomalous and the European results to be true records of the Devonian field; these were prompted by studies of Norwegian Siluro-Devonian red sediments which failed to detect the field direction recognised in the British rocks (Storetvedt, Halvorsen & Gjellestad, 1968). However, since the British results are removed from all younger field directions (with the possible exception of Lower Carboniferous ones, see Fig. 10.2b and the first part of Section 10.3), these results should represent some complex mixture of components (Storetvedt & Halvorsen, 1968). This has been effectively refuted by a range of fold, contact and conglomerate tests and the observation of multiple reversals in the Midland Valley lavas and red beds (Sallomy & Piper 1973; Torsvik, 1985b).

The proposal that widespread remagnetisation took place in Palaeozoic times is referred to as the *remagnetisation hypothesis*. Creer (1968b) suggested that the dehydration and oxidation of iron hydroxides and oxyhydroxides at near surface levels in tropical latitudes produced the secondary magnetic phases (second part of Section 4.2). These processes are associated at the present times with laterisation, a process which operates between about latitudes $30°N$ and S of the equator; they could formerly have been important within regions such as Laurentia and Europe, which formed vast emergent areas straddling these latitudes during Permo-Carboniferous times, while they are unlikely to have been important in Gondwanaland, which lay in intermediate to high latitudes throughout this period of continentality. Parallel arguments for and against widespread remagnetisation have also considered the Laurentian palaeomagnetic record. McElhinny (1973) argued that Creer was interpreting the data within a preconceived notion of the Pangaean reconstruction and should instead be interpreting differences between poles in terms of relative plate movements. McElhinny & Opdyke (1973) utilised the (Ordovician) Trenton limestone to demonstrate that what they considered to be a primary Ordovician field axis was also similar to the Upper Carboniferous one. Although this study did not permit a fold test, the recognition of a principal susceptibility axis close to the cleaned remanence direction suggested that the magnetic grains were aligned in the ancient magnetic field and that the remanence was therefore of primary depositional origin. Subsequent results tended to endorse this view (e.g. Van der Voo & French 1977). Studies in the Appalachians, in particular, were able to distinguish a widespread post-folding overprint from a pre-folding remanence. The post-folding remanence here correlates with the Permo-Carboniferous palaeofield direction, and is linked to Hercynian folding (Van der Voo 1979).

It now appears however, that refutations of the remagnetisation hypothesis were premature. In its broadest sense, remagnetisation can include thermal and chemical effects occurring during protracted lithification and diagenesis, as well as superficial weathering. In this latter context, the hypothesis has enjoyed a new lease of life in the past few years. By careful work in the Maritime Provinces Roy (1969) has recognised a sequence of magnetisations in Carboniferous sediments which both predate and post-date late Carboniferous (Hercynian) folding. Subsequently, Roy & Morris (1983) have re-examined the remagnetisation hypothesis in the context of these and other data from cratonic Laurentia. They note that even a remanence which is demonstrably pre-folding can be secondary; they identify a further criterion which can be applied here: the magnetic field reversed its polarity frequently in Lower Carboniferous times but beginning in early Upper Carboniferous times, it retained R polarity almost exclusively, for 70 Ma (Section 12.2). Thus, in addition to the results of the standard field tests, the occurrence of both polarities is a strong indication that the remanence predates Upper Carboniferous times, while the presence of uniform R polarity suggests that the remanence is of Permo-Carboniferous age. Roy & Morris (1983), show that the actual APW record for North America in mid-Carboniferous times comprises a 25° swathe (Fig. 10.6); the older segment of this swathe is defined by pre-folding and/or dual polarity remanences while the younger R segment includes Permian poles, post-folding remanences in Carboniferous rocks, and some remanences in Devonian and even older rocks (see also Fig. 10.7).

This problem has specifically afflicted Laurentian palaeomagnetism, because the record here is derived predominantly from sedimentary material; it became evident in Europe rather sooner (Briden, Morris & Piper 1973), because the largely igneous-based record from Britain contrasted with the sedimentary-based record from continental Europe. Paradoxically, the Laurentian rocks which possess a

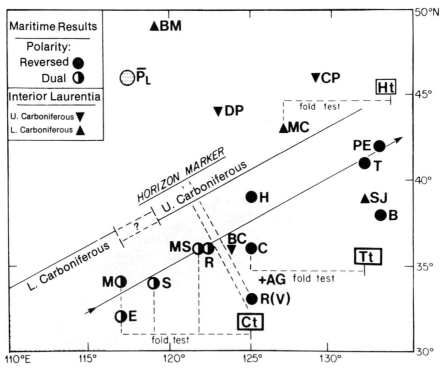

Fig. 10.6 Palaeomagnetic poles from Carboniferous rocks of North America. The horizon marker is the location of the transition from the Lower Carboniferous interval of mixed polarity into the CN superchron (see Section 12.2). Poles derived from remanences acquired prior to folding are shown, with the upper time limit of the folding event as Ct (Cumberland time), Tt (Tormentine time) and Ht (Hercynian time). The square is the mean Lower Permian pole from Irving & Irving (1982). The age limits of the Austell gneiss (AG) are 365–325 Ma. Note that the triangles in this figure refer to the age of the rock unit, which is not necessarily the age of the remanence. After Roy, Tanczyk & Lapointe (1983).

demonstrably syn- or early post-depositional remanence come from the terranes in the Maritime Provinces affected by Acadian or Hercynian tectonism, whilst sediments from stable cratonic environments, sometimes as old as Cambrian in age, have been widely overprinted in Permo-Carboniferous times. It appears that the sediments in the orogenic zone have been lithified and indurated by tectonism within a few millions of years of deposition, whilst diagenetic processes continued within the sedimentary rocks in more stable environments for at last 50–100 Ma, after deposition. As a typical example, a further study of the Trenton limestone has found that the magnetisation recognised by McElhinny & Opdyke (1973) is a mixture of two components, with the characteristic remanence probably of Permo-Carboniferous age, and carried by diagenetic magnetite (McCabe, Van der Voo & Ballard, 1984).

Subsequently, other workers of the Canadian palaeomagnetic school have endorsed the interpretation of Roy & Morris (1983). In a study of the Burin Peninsula in the Avalonian terrane of SE Newfoundland, Irving & Strong (1985) separate a Permo-Carboniferous component from a probable primary Devonian component with an inclination some 30° steeper. They show that there is little or no overprinting in the late Precambrian basement, here and attribute the overprinting in the Palaeozoic rocks to groundwater circulation. These authors also assess other Devonian data from Laurentia, and demonstrate that these are all statistically identical to the Permo-Carboniferous field direction; small divergences in declination, from the Colorado Plateau, may be the consequence of the Mesozoic–Cenozoic dextral shear along the western Cordillera (concluding part of Section 11.4). It is strange that Devonian rocks from cratonic Laurentia have been so thoroughly overprinted 50–100 Ma after deposition, and particular interest centres on the Middle–Upper Devonian Catskill formation, a well studied cratonic unit (Kent & Opdyke, 1978, Van der Voo, French & French, 1979). The bulk of the remanence here pre-dates the Hercynian folding, but the conformity with the Permo-Carboniferous field direction implies that this remanence could not have been acquired long before the folding took place (Irving & Strong, 1985); thermal demagnetisation of these rocks close to the Curie point of the hematite remanence carrier indicates steeper directions although their significance has not yet been evaluated (Van der Voo, French and French, 1979.).

The recognition of widespread overprinting in the Palaeozoic palaeomagnetic record has necessitated the re-evaluation of certain tectonic hypotheses. Noting the 10°–15° difference in the palaeolatitudes indicated by data sets from the Devonian and Lower Carboniferous rocks of the Maritime Provinces and cratonic Laurentia, and the disappearance of this discrepancy in Permo-Carboniferous times, Kent & Opdyke (1978) proposed that an allochthonous terrane (the 'Avalonian microplate') was emplaced against the remainder of Laurentia by *ca*. 1000 km of sinistral strike-slip movement in mid-Carboniferous times. This concept was subsequently extended to incorporate the whole of the European plate as far east as the Uralian suture, which was inferred to be placed some 2000 km south of its position in the Pangaean reconstruction before the mid-Carboniferous (Van der Voo, French & French, 1979). Van der Voo & Scotese (1981) were specific about the NE continuation of this suture; they noted differences in the magnetic inclinations on either side of the Great Glen Fault in Scotland and developed a palaeomagnetic case for sinistral-strike slip motion in mid-Carboniferous time.

Unfortunately, this case ignored the extensive geological evidence for only a few tens of km of movement along this fault in post-Lower Devonian times (Donovan, Archer, Turner & Tarling 1976); it also appears that much of the remanence in the Devonian sediments of this region has a complex diagenetic origin (see, Cisowski, 1984). In the context of more recent studies, the APWP of Fig. 10.6 must be interpreted in terms of a real northward movement of the whole of Pangaea during Carboniferous times, rather than in terms of relative motions between constituent parts of the supercontinent.

10.5 Palaeozoic palaeomagnetism of the remaining continental plates

Laurentia

In addition to the Precambrian basement discussed in previous sections, the Laurentian plate in Palaeozoic times comprised a sedimentary-volcanic wedge of cordilleran affinities (the 'Andean margin') along the SE perimeter, succeeded eastwards by the suture zone noted above (p. 273) and the Avalonian microcontinent (which was sutured to the plate in the Acadian orogeny, following consumption of the Iapetus Ocean, Williams, 1978). This Avalonian microplate comprises a basement of late Precambrian volcanic and sedimentary rocks metamorphosed to low grade. Still further to the east, segments of the Appalachian Piedmont and Nova Scotia have geological relationships not readily linked to the Avalonian microplate, and it has been speculated that they represent fragments from the NW African margin of Gondwanaland, transferred to the Laurentian plate following the Mesozoic rifting (e.g. Hatcher, 1978).

The Palaeozoic palaeomagnetic poles from Laurentia are plotted in Fig. 10.7. The symbols here refer to the age of the rock unit inferred from radiometric or stratigraphic studies. In both cases there is sometimes an uncertainty in this allocation and it is appropriate to consider the Silurian and Devonian rock units together because a distinction between Upper Silurian and Lower Devonian ages cannot often be made. The Permian poles come only from cratonic Laurentia, and plot along a NE → SW distribution that includes poles with a 'Permo-Carboniferous' assignment, as well as poles from older rocks — now considered as Hercynian remagnetisations. The general SW migration of the APWP is however, identified in some individual studies, as for example in the Deer Lake group of Newfoundland (Irving & Strong, 1984) and the Casper formation (Diehl & Shive, 1981). The 'elbow' in the Carboniferous path extends the APWP back along the track identified by Roy & Morris (1983) in the Maritime Provinces. This track coincides with the 'Hercynian remagnetisation' swathe identified in European data on the Pangaean reconstruction (Figs. 10.4 & 10.5). A large number of results plot near this change in APWP direction, and although many of them are overprints, they include results of defined late Lower–early Upper Carboniferous age, and date this bend near the transition between the two divisions of Carboniferous times.

The results from pre-Carboniferous rocks which do not coincide with the younger path, are more dispersed and are considered by Roy, Tanczyk & Lapointe

(1983) to fall into two groups. The first is strung along latitude 30°S and has a latitudinal distribution which could logically be interpreted in terms of block rotations during the Acadian orogeny. The second group includes poles plotting in higher latitudes (20°S–60°N) and is not explicable in terms of later remagnetisation or (at least in a major way) in terms of regional rotations. There is some support for this two-fold separation of the data in a comparison with the British APWP (Fig. 10.2c). The second group conforms in a general way with the E–W Siluro-Devonian swathe. It is mainly recorded by a range of secondary and possible primary components in Ordovician–Silurian igneous and metamorphic rocks from the Andean margin of the plate in Quebec (e.g. Seguin 1981) and in the Avalonian terrane of Nova Scotia (Rao, Seguin & Deutsch, 1981). It is also identified in Devonian sediments from the opposite passive margin of the plate (Dankers, 1982) and agreement with the British APWP is defined by results from the Lower Silurian Botwood group rhyolites (423 Ma), the Mt Peyton diorite and the high-stability component in the Upper Silurian Mascarene group intrusions and sediments (Roy & Anderson, 1981). This agreement with the British path is to be anticipated from the intergrity of the British Silurian data from the orthotectonic Caledonides bordering the Laurentian foreland in the north, to the Hercynian Front in the south (Section 10.3 and McClelland-Brown, 1983).

The correlation with the British APWP also identifies the possible significance of the first group of poles recognised by Roy, Tanczyk & Lapointe (1983). Certain pre-mid Devonian poles plot to the west of the APWP in the same area as poles from contemporaneous rocks from the Connemara terrane (second part of Section 10.3). This westerly group of poles includes results from the Andean margin and Avalonian divisions of Newfoundland and eastern Nova Scotia; the sample locations yielding this group do not appear to extend west of latitude 293°E in SW New Brunswick (Roy & Anderson, 1982) although there is some evidence for block rotations in the extreme NE of New England (Traveller felsite in Figure 10.7 and Sparisou & Kent, 1983) and the inclusion of data from the Andean margin in Quebec (Compton Metasediments, Seguin, Rao & Pineault, 1982) in this group shows that this anomaly must post-date full closure of the Iapetus Ocean. The possible geological implications discussed in the context of Connemara in the second part of Section 10.3 may also be applicable here. As in the case of the British data, it would be unwise to exclude tectonic causes for the dispersions of some other data in Fig. 10.7. Irving and Strong (1985) for example, note a declination difference of 30° between late Devonian units on opposite sides of the Avalonian terrane in Newfoundland which might be a consequence of regional rotations during sinistral shear movements.

Two general observations emerge. Fistly, very widespread overprinting has taken place during the Acadian and Hercynian episodes, in both the orogenic belt and in the superficial cover rocks of the stable craton, although the mechanisms of this overprinting probably vary from area to area (Section 10.4). Secondly, the extent of Hercynian overprinting becomes predominant towards the south and west of the Appalachian sector. The igneous and metamorphic rocks of the SE Piedmont for example, possess exclusively a remanence of Upper Carboniferous or Permian age (e.g. Ellwood, 1982 and Fig. 10.7). However, at the NE extremity of the Piedmont rocks, last metamorphosed during the Taconic orogeny (*ca.* 440 Ma) have steep inclination remanences (Brown & Van der Voo, 1980; Rao & Van der

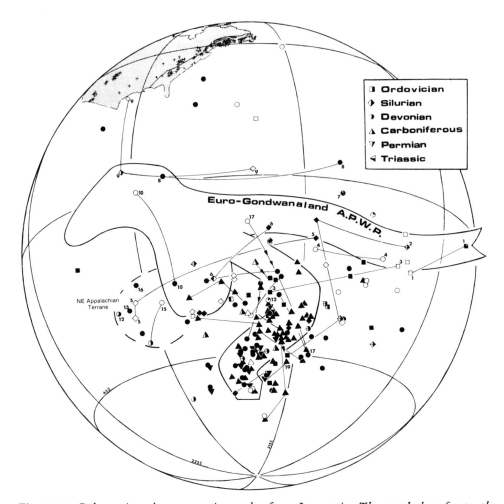

Fig. 10.7 Palaeozoic palaeomagnetic results from Laurentia. The symbols refer to the geological age of the rock unit which is not necessarily the magnetic age. The lines connect poles derived from the same rock unit. The APW swathe drawn in solid lines is the path deduced from the British data (Fig. 10.2b) rotated to the Laurentian plate according to the Pangaean reconstruction of Bullard, Everitt & Smith (1965). Although not all of the data are referenced here, the poles of relevance to points discussed in the text are numbered: **1**, Quebec volcanics, three components (Om, 450, 376 AA); **2**, Mt. Aylmer association (Cm–Dm), four components; **3**, Weedon-Gould area rocks, three components (Ol–m); **4**, St. Cécile–St. Sebastian granite, FG, MG and G components (362 KA, 365–360 on related granites); **5**, Cape Breton granites, SE, NE1 and NE2 components (587–460 RS); **6**, Botwood group rhyolites (Sl, 423, 422 KA); **7**, Bonavista Bay alkaline dykes, groups 1 and 2 (400–344KA, 375 AA); **8**, Avalon Peninsula mafic dykes, SE and SW groups; **9**, Mascarene Group intrusions, greywackes and siltstones (S, 420–400 RS); **10**, Peel Sound formation (Dl); **11**, Mt. Peyton diorite (480 RS, 410 KA); **12**, Compton meta-sediments (Dl–m); **13**, Traveller Felsite (Dl); **14**, Hartwicke diorite (S); **15**, St. Georges pluton (D, 394 RS, 404–377 KA); **16** Dockendorff Group (Dl, 406, post-Acadian remanence); **17**, St. Lawrence granite and dykes (360 RS); **18**, Deer Lake group (L. Carboniferous, Viséan–Namurian); **19**, Casper formation (Carboniferous–Lower Permian). The crosses refer to the locations of the palaeomagnetic studies.

Voo 1980). These data points should logically be compared with other late Ordovician data, and on the Pangaean reconstruction they conform closely to British and Armorican data of this age (Fig. 10.3). If the Piedmont was then an integral part of the Laurentian plate, this agreement could mean that the Iapetus Ocean was already nearly closed by Upper Ordovician times (see also Fig. 10.5 and preceding discussion). Alternatively, it may link the Piedmont to the British or Armorican microplates, and imply that this region was attached to the Laurentian Plate following the Acadian orogeny (e.g. Hatcher, 1978). In common with the Grampian orogeny, which deformed and metamorphosed the terrane within the orthotectonic Caledonides (Fig. 10.2c) at 510–480 Ma, the Taconic orogeny cannot yet be linked to a specific area of the palaeomagnetic evidence.

The European plate (USSR sector)

The results from the numerous palaeomagnetic laboratories in the USSR generally only appear in the West as tabulated group mean data, and it is difficult to assess their reliability. Although thermal and a.f. demagnetisation are now employed routinely, many published studies have evaluated magnetic stability from studies of coercivity and steady field demagnetisation, and tests have not usually been as rigorous as those employed elsewhere (McElhinny, 1973). The Palaeozoic data from the Russian sector of the European plate are plotted in Fig. 10.8. The late Lower Carboniferous–Upper Permian results are based largely on extensive studies of the Moscow and Donbass basins, and comprise a distribution coincident with other European data of this age; there is not however, a clear distinction between the data sets for these two periods, and the nature of the remanence in these rocks is unclear. The group also includes poles from some Ordovician–Devonian rocks. Other more scattered data points appear to record older magnetisations. Results from twelve igneous and sedimentary units plot along the Siluro-Devonian track of Figs. 10.2 and 10.4, in accordance with the inferred structural continuity of this region with western Europe, although seven of these results come from close to the Uralian (1,2,3,4) of Caucasus (7,11,12) margins, and may have been influenced by tectonic effects. Three units of Upper Devonian–Lower Carboniferous age (10,13,14) support an implication of the British data which suggests that the Lower Carboniferous and mid-Silurian–early Devonian poles occupied similar positions (first part of Section 10.3).

Siberia

Although Phanerozoic rocks of the Siberian plate have been quite extensively studied, the data show considerable scatter (Fig. 10.9) and do not justify the calculation or definition of a pre-Carboniferous APWP. The possible reasons for this scatter have been discussed by McElhinny (1973). They relate in part to problems of defining the area of this plate. The West Siberian lowlands bordering the Urals are largely covered by Mesozoic rocks, but Mussokovsky (1973) defines crust of oceanic affinities here, some 200 km to the east of the Uralian chain (Fig. 10.10). The southern margin of the plate is generally located at the Irtysh zone, a WNW–ESE trending lineament of Devonian age, which parallels a number of

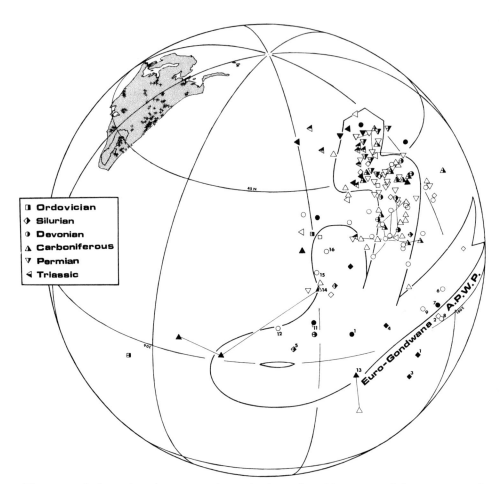

Fig. 10.8 Palaeozoic palaeomagnetic results from the USSR sector of the European plate. The outline of the APWP deduced from the British data (see also Fig. 10.5) is shown for comparison. Not all data are referenced, but the poles relevant to points discussed in this section are numbered: 1, Polyakovka group tuffs and dolerites (Sl); 2, Polyakovka group tuffs and sandstones (Sm); 3, Shemakha and Kubinsk limestones (Sl–Sm); 4, Orthophyre group igneous rocks (Sm–Dl); 5, River Dneister sediments (Sm); 6, North Timan sediments (Du); 7, Basaltic porphyrites, North Caucasus (Dm–Du); 8, Volongi River sediments (Dl–Du, Famennian–Givetian); 9, Travyansk and Nadezhda groups (Dm, Frasnian); 10, North Timan sediments (Du, Frasnian); 11, Bakhmuta group red sediments, North Caucasus (Dl–Dm); 12, Basaltic porphyrites, North Caucasus (Dl–Dm); 13, South Urals limestones (Carbl, Tournasian); 14, Zilairsk group (Du–Carbl, Tournaisian–Famenian); 15, Lower Kartdzhurt group, North Caucasus (Dm–Du); 16, Domanik beds (Du, Frasnian). The crosses refer to the locations of the palaeomagnetic studies.

ancient volcanic arcs and actually lies to the south of a zone of Hercynian deformation with probable ophiolitic ultrabasic bodies (Burrett, 1974). South of the Lake Baikal region, the SE perimeter of the plate includes several concentric zones of Phanerozoic age. These include ophiolites and become progressively younger towards the south. Collectively, they comprise a broad zone of accretion to the Siberian plate prior to fusion to the North China plate. With a width of some 800 km this Central Asian fold belt is the largest orogenic belt in the world. The margins of the Siberian plate are better defined in the NE, where the Precambrian

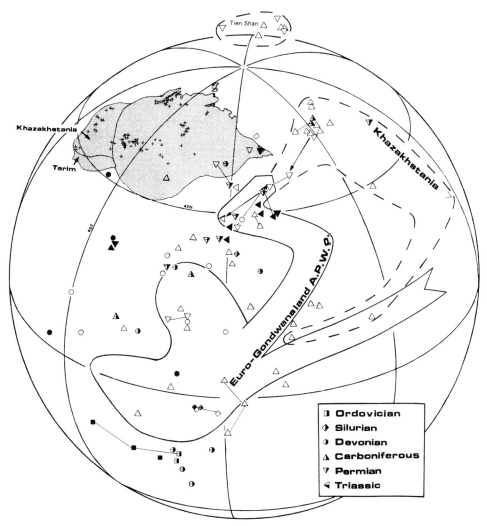

Fig. 10.9 *The palaeomagnetic poles from rock units of Palaeozoic age in the Siberian and Khazakhstanian plates. Note that the palaeomagnetic data from Khazakhstania plot in a region removed from the Siberian data until Permo-Triassic times, and are enclosed here by the dashed line. The APWP from the European plate as deduced from British data in Fig. 10.2 is shown in outline for comparison. The crosses within the plate outline refer to the locations of the palaeomagnetic studies.*

platform borders the Verkoyansk orogenic belt formed by the welding to the Kolyma plate in Mesozoic times, (third part of Section 11.4).

The palaeomagnetic results from Siberian Palaeozoic rocks plot largely to the east of the European APWP, although the two data sets are not widely removed, and suggest that the two plates were in close proximity in post-Ordovician times (see also Irving, 1977). Some results from rocks as young as Permian plot in discretely different regions and would be commensurate with final welding of the two plates along the Uralian Orogenic Belt in Permian or Triassic times (Hamilton, 1970).

Khazakhstania and the China plates

The palaeomagnetic results from the Khazakhstanian plate are no older than Carboniferous in age. Four poles from Carboniferous rocks are widely divergent from one another while the remaining poles are closely grouped near to a Permian result (Fig. 10.9). Irrespective of whether or not these latter points are Permian overprints, all of these data are removed from the Siberian results. There appears to be a systematic movement towards other Permian, and then to Permo-Triassic poles, defining a convergence with the Siberian data. This is consistent with the history of the Altai Sayan Zone separating the two plates, which was a faunal barrier until Permian times, when terrestrial conditions replaced marine environments (Burrett, 1974). The Tien Shan orogen forms the SE margin to this plate and defines the suture with the Tarim massif (Fig. 10.10); this suture includes Devonian and Carboniferous ophiolites, but orogeny appears to have terminated in Permian times, implying a collision of the Tarim and Khazakhstanian plates at this time. These plates shared a common Angaran flora with the Siberian plate in Upper Permian times (Section 10.7). Middle Carboniferous–Lower Permian poles from the Tien Shan belt are divergent (Fig. 10.9), but there are inadequate data from younger rocks to confirm this age for the suturing. However, according to Guiliang, Quigge, Yhang, Yonfan, Yuje & Dongjiang (1985) recent palaeomagnetic data imply that the Tarim block had collided with Khazakhstania by late Carboniferous or early Permian times (Fig. 10.10).

Three Upper Permian poles from the North China plate plot at points remote from the data of Fig. 10.9, and suggest that this plate had not become welded into the mosaic of Asian plates by this time. These data imply an equatorial location, possibly still in contact with the NE margin of Gondwanaland. The translation of this plate to the Eurasian agglomeration may have taken place shortly after this, because folding of the Central Asian orogenic belt took place in late Permian times to be followed by deposition of molasse facies sediments and granite intrusions during Triassic and early Jurassic times (Burrett, 1974). This orogenic belt continued to be a zone of strike slip translation during Cenozoic times with the result that APWPs of this age from the China Shields diverge from the Eurasian APWP until Cenozoic times (Lin, Fuller & Zhang 1985 and Section 11.4). The North and South China plates are welded along the Tsinling (Qinling) orogenic belt which formed a faunal barrier and zone of ophiolite emplacement throughout Palaeozoic times (Burrett, 1974). A history of contrasting orientations in Palaeozoic times, followed by motion as a unified plate in Mesozoic times is supported by the limited paleomagnetic data from these blocks (Lin, Fuller & Zhang 1985.). In late Permian

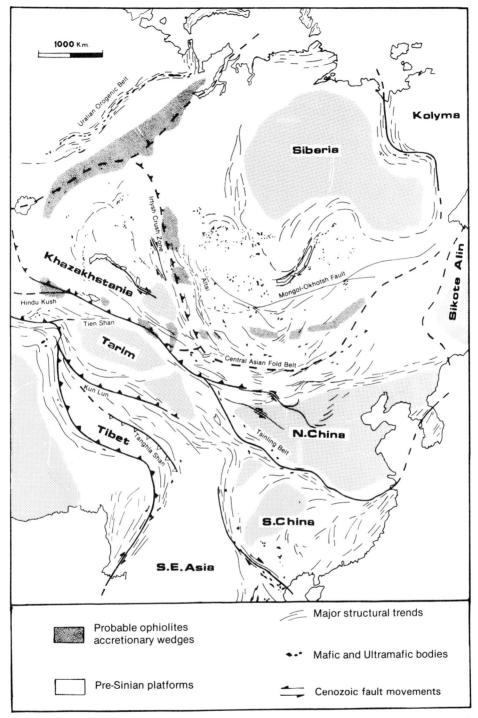

Fig. 10.10 *The ancient plates of Asia and features of the suture zones between them. The Tarim region is shown as a separate microplate, but may have constituted a single unit with Tsiadam and Tibet. Simplified after Burrett (1974) with additions from Molnar & Tapponier (1977) and Boulin (1982).*

times, they lay in equatorial latitudes still well separated from the Siberian plate (McElhinny, Embleton, Ma & Zhang, 1981); relative rotation through 120° was followed by suturing along the Tsinling belt in late Triassic or early Jurassic times.

The SE Asian plate is separated from the South China plate by a suture (the Song Ma/Song Da zone) running along the line of the Red River in Vietnam (Fig. 10.10). It appears to have its western margin at a Lower Palaeozoic geosyncline, which runs along the western side of the Malaysian peninsula and then into Thailand, Burma and the Yunnan Province of southern China. This plate now comprises at least four ancient terranes: the Sibumasi (parts of Burma, Thailand and Malaya), Indo-China (parts of E Thailand, Laos, Kampuchea and Vietnam), E Malaya-Sumatra, and the Borneo block. Lower Palaeozoic faunal and stratigraphical evidence is compatible with a position within the Gondwanaland assemblage close to NW Australia (Burrett & Stait 1985). In contrast, warm water Upper Palaeozoic facies imply a location remote from India and Australia during these times, because the latter areas were then experiencing glaciation. Rocks assigned to the interval Lower Carboniferous–Permo-Triassic from the Malayan peninsula yield similar magnetisations (McElhinny, Haile, & Crawford, 1974), and although they pre-date late Triassic folding, these magnetisations are probably younger than the rock ages (cf. Fig. 10.1). A mean palaeolatitude of 15° is compatible with a position to the east of northern Africa in Permo-Triassic times as suggested by geological evidence (Stauffer, 1974). Ordovician and Silurian limestones (Haile, 1980) yield similar magnetisations, to each other and are certainly younger than the rock ages, although an inferred palaeolatitude of 43° would be compatible with a similar position in the Gondwanaland reconstruction at some stages of the Ordovician–Carboniferous interval (Fig. 10.1). An Upper Triassic (Carnian or Norian) collision between the SE Asia and South China plates (Burrett, 1974) is demonstrated by the general conformity of Triassic–Jurassic data from Thailand with those from South China (Barr, MacDonald & Haile, 1978, Haile, 1981). A plate comprising Iran and western Afghanistan was also a segment in this assemblage. It was part of a passive margin removed from the influence of the Caledonian events. Palaeomagnetic data from Devonian–Permian rocks suggest that it was located at the edge of the supercontinent between Arabia and NE India (Wensink 1979). Tectonic events beginning in late Triassic times seem to have been responsible for moving this sub-plate into a position close to its present one. The post-Carboniferous data set from the Arabian plate, extending up to the North Anatolian Fault zone, conforms to the African APWP following closure of the Red Sea (McElhinny, 1973) and this observation defines the probable northerly extension of Gondwanaland. To the east, this margin probably extends to the Caucasus suture, and to the west it includes a segment (referred to as the Adriatic promontory) underlying part of the present Mediterranean, although the present degree of coupling between this segment and Africa is in dispute (Section 11.4).

It is now apparent that the re-entrant in the Pangaean reconstruction, generally considered to be the site of the Tethyan Ocean, was partially occupied by a number of plates during Upper Palaeozoic times, including SE Asia, North and south China and probably Kolyma and Sikote Alin (Fig. 10.10, Section 11.4 and McElhinny, Embleton, Ma & Zhang, 1981). Collectively, crustal fragments extending westwards as far as the Black Sea (and referred to as the 'Cimmerian mosaic' by Sengor, 1979) were transferred from Gondwanaland to Eurasia and were accom-

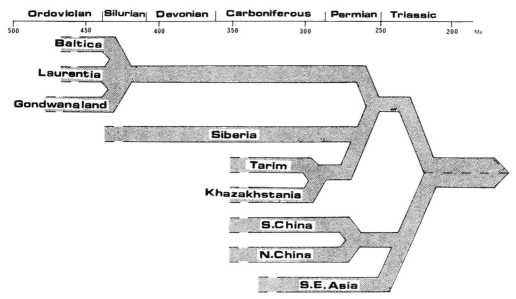

Fig. 10.11 Summary of the accretionary history of the Eurasian wing of Pangaea, from geological and palaeomagnetic evidence (see text). Note that although the China Shields and SE Asia were probably coupled to Eurasia by early Mesozoic times, the palaeomagnetic evidence requires that large-scale strike-slip motions between these plates and the remainder of Asia took place in later times (Sections 11.3, & 11.4).

panied by the closure of an ocean basin (Palaeo-Tethys) between Triassic and mid-Jurassic times. This was complemented by the opening of a new ocean basin (Neo-Tethys) which is now recorded by obducted ophiolite fragments which have been incorporated into the Alpine-Himalayan orogenic belt by the relative movements between Gondwanaland and Eurasia, following the breakup of Pangaea; these latter ophiolites are derived from ocean crust which is no older than Triassic in age (Sengor, 1979. and Section 10.7).

10.6 Palaeomagnetism and the reconstruction of Pangaea

The preceding analysis has demonstrated that most of the continental plates had clustered together into the single landmass known as *Pangaea*, by the end of Palaeozoic times, in general accord with the case first developed by Wegener in 1923, and since supported by a wide range of other geological evidence (Section 10.7). The reconstruction of Wegener as quantified by Bullard, Everitt & Smith (1965) and referred to as Pangaea A, does not bring the Upper Carboniferous–early Triassic poles from the northern continents (Laurasia) into precise conformity with data from the southern continents (comprising Gondwanaland). The Upper Triassic and Lower Jurassic poles from these two regions are, however, statistically identical on this reconstruction, and imply that Pangaea A is a faithful representation of the continental crust during these times. There appear to be three possible causes of the Upper Carboniferous–Upper Permian discrepancy: (i) the geomagnetic field may have included non-dipole components during these times thus invalidating the dipole assumption used to calculate the poles (Briden, Smith & Sallomy, 1971); (ii) the APWPs are still inadequately defined, and (iii) relative motions took place between the Laurasian and Gondwanaland sectors of Pangaea, so that a different reconstruction from Pangaea A was applicable prior to mid-Triassic times. The latter possibility has been most widely investigated. Van der Voo & French, (1979) proposed a modification to Wegener's Pangaea A referred to as Pangaea A2, in which Gondwanaland is rotated anticlockwise by 20° relative to Laurentia, so that the NW margin of South America fits into the Gulf of Mexico. Irving (1977) proposed a third reconstruction known as Pangaea B, requiring a 35° anticlockwise rotation of Gondwanaland with respect to Laurasia; in this reconstruction the NW margin of South America is placed against the east coast of North America. A still more extreme reconstruction with Gondwanaland sited well to the east of Laurasia has been proposed by Smith, Hurley and Briden (1980); this is referred to as Pangaea C. Van der Voo, Pienado & Scotese (1984) have re-evaluated the A2 and B reconstructions, and conclude that both are compatible with the present data base between late Carboniferous and Lower Permian times, although the A2 fit has higher statistical precision. The Pangaea B reconstruction, however, provides a better fit to the small Upper Permian data set.

Taken at its face-value, this conclusion would require a sinistral motion, followed by a larger dextral one to achieve the Pangaea A reconstruction. There are however, severe geological problems with both the A2 and B reconstructions (Hallam, 1983). The A2 reconstruction provides no space for a number of microcontinents which now comprise the Nicaraguan-Yucatan-Campeche group, and the continental basement between the Canary Island group and North Africa (Ager, 1975). The former group is generally considered to have lain immediately to the south of the Ouachita-Marathon belt, and within the Gulf Coast region in Permo-Triassic times (White, 1980). According to Hallam (1983), structural, stratigraphic and faunal considerations argue strongly in favour of the classical A reconstruction, and the absence of real evidence for 3500 km of dextral shear in the Tethyian region renders both the A2 and B reconstructions unlikely.

More consideration should be given to the possibilities (i) and (ii) noted above. The reality or otherwise of non-dipole components in the geomagnetic field in Permo-Triassic times cannot be rigorously tested until more well dated magnetisations of undoubted primary origin have been recovered from each of the larger plates. However, it is probable that the calculation of mean poles for divisions of geological time which has been used in the assessments of the A2 and B reconstructions is not a valid procedure. Examination of the data plotted in Figs. 10.1, 10.2 & 10.4 (see also Fig. 10.5 which uses the Pangaea A reconstruction) for example, identifies appreciable dispersion within the late Carboniferous–late Permian interval (*ca.* 295–235 Ma) which is lost in the mean pole calculations. Specific problems have been discussed by Van der Voo, Peinado & Scotese (1984) and include two possibilities. Firstly, small secondary overprints may have been incompletely removed in some studies. Secondly, differences in the interpretation of the geological time-scale, and uncertainties in the stratigraphic ages critically affect the assignment of some data points to divisions used for mean pole calculations. The comparison of mean APWPs from different continents on well established reconstructions, such as the North Atlantic fit of Bullard, Everitt & Smith (1965), shows that there are important discrepancies, even in Mesozoic times, which cannot be accommodated by uncertainties in the reconstructions (Tarling 1979 & Section 11.2). It is therefore, clearly unwise to consider the palaeomagnetic data to the exclusion of geological considerations, until the nature of APW and the geomagnetic field source are better understood.

The assembly of the southern continents into the reconstruction referred to as Gondwanaland B in section 9.1 was first proposed in full by Du Toit (1937), and quantified by Smith & Hallam (1970). The west Antarctica peninsula laps over the Falklands Plateau in this reconstruction, although both areas are continental crust. The significance of this overlap has been widely discussed. The attempts to avoid superimposing the present continental outlines by placing East Gondwanaland some 2000 km further south have not enjoyed marked success: the variants arguing for a southerly position for Madagascar adjacent to Mozambique are not supported by the palaeomagnetic data (e.g. Embleton & McElhinny, 1975) and most recent discussions have accepted a position adjacent to Kenya and Tanzania. Irving & Irving (1982) have tested the Barron, Harrison & Hay (1978), the Norton & Sclater (1979) and the Powell, Johnson & Veevers (1980) modifications of the reconstruction, which favour the more southerly position of East Gondwanaland; in every case they find that the Du Toit-Smith-Hallam reconstruction provides a better fit to the palaeomagnetic reconstruction. Klootwijk (1979) and Embleton, Veveers, Powell & Johnson, (1980) reach a similar conclusion. Furthermore, the modifications to the Du Toit reconstruction are no longer necessary. West Antarctica comprises a number of suspect terranes which have probably been attached here since the breakup of Gondwanaland (Section 11.4). In addition, recent palaeomagnetic studies have demonstrated that the Falkland Islands are part of microplate rotated through 120° from SE Africa anticlockwise into its present position relative to South America, following the dispersal of Gondwanaland (Mitchell, Taylor, Cox & Shaw, 1986). One qualification to this discussion is necessary; Klootwijk (1979) and Irving & Irving, (1982) find that reconstructions with a southerly position for Madagascar satisfy the palaeomagnetic data better over the short interval from late Permian to mid-Triassic times, better than the northerly position. Klootwijk integrated this observation into the supposed anticlockwise rotation of Gondwanaland with respect to Laurentia, and proposed that East Gondwanaland first lagged behind West Gondwanaland and then caught up again in late Triassic times. Irving & Irving suggest that this observation is an artefact of the inadequate data base, and reflects poor definition of an interval of rapid APW which is recognised in all of the major plates at this time (Chapter 11 and Fig. 12.12).

10.7 The palaeogeography of Palaeozoic times

Since Smith, Briden & Drewry, (1973) published their set of Phanerozoic world maps based on the predictions of the palaeomagnetic data, a number of workers have made more refined reconstructions of Palaeozoic palaeogeography, and have integrated the palaeomagnetic constraints with data from palaeoclimatology, biogeography and tectonics (see for example, Seyfert & Sirkin (1973), Irving 1977,

Ziegler, Hanson, Johnson & Kelly, 1977, Kanesewich, Havskov & Evans, 1978, Ziegler, Scotese, McKerrow, Johnson & Bambach 1979). In this section we examine the general implications of these studies, and note the revisions which are necessary in the light of more recent palaeomagnetic studies.

As a result of the diversification of fauna and flora in Palaeozoic times, more varied and definitive parameters are available to test the palaeomagnetic predictions during these younger times (second part of Section 6.5). In post-Cambrian times the flora and fauna, for example, constrain the east to west order of the continents along the same latitudinal belt and partially overcome the indeterminacy of palaeolongitude. Faunal and floral provinces are important indicators of major oceanic barriers, but they provide no indication of the width of such barriers. The Wallace's Line in SE Asia is a Cenozoic example of a major faunal barrier involving only a small lateral separation. Thus the faunal, floral and palaeomagnetic evidence should be assessed collectively to reconstruct the ancient geography of the continental crust in Phanerozoic times. The rock facies formed in the terrestrial and shallow-marine environments are the most sensitive to climatic changes affecting the atmosphere and hydrosphere, and are therefore most directly linked to insolation. They include:

(i) *tillites*: the unsorted products of the decay of regional ice sheets; they incorporate scratched boulders and may have been directly dumped on striated pavements, scratched and polished by the glacier movement. Regionally-associated rocks may include marine or lacustrine sediments with dropstones deposited from melting icebergs, and varves; the latter result from episodic influx of the products of spring melting into water bodies peripheral to the ice sheets. These deposits are collectively characteristic of polar latitudes, but they may have formed in relatively low latitudes during glacial epochs. Sediments with tillite-like aspect (tilloids) may result from alternative mechanisms, including mud-flowing and sliding, slumping, and turbidity flow; these alternatives need to be positively excluded before a deposit can be classified as a tillite and assigned a palaeoclimatic significance. The association of a tillite with a striated pavement is the most positive indication of a glacial origin, while interbedding with marine sediments is the best indication of glaciation reaching down to sea-level and hence recording low mean temperatures, and, in particular, low summer temperatures.

(ii) *Evaporites* require a large excess of evaporation over precipitation for their formation, where salt-charged marine waters can either be periodically evaporated in coastal lagoons or drawn inland by capillary action through porous coastal deposits of blown sand and reef debris. These conditions are typical of the arid trade wind zones between the warm temperate and sub-tropical belts. They may occur between $10°$ and $50°$ latitude and their distribution seems to be greatly influenced by the degree of continentality.

(iii) *Carbonates* are preferentially deposited by marine organisms where marine waters are nearly saturated with calcium carbonate. At the present time carbonate production is greatest within $30°$ of the equator and falls off dramatically outside this belt. The activity of carbonate-secreting organisms is confined to the euphotic zone or the top 100–120 metres of the marine environment, while direct secretion of carbonate can only occur where the ocean waters are warmed in intertidal and immediate sub-tidal environments. During periods of equitable climate when the polar ice sheets were small, or absent, and the tropical and temperate belts were broad, the limits of the carbonate belt appear to have been much wider. The Upper Cretaceous chalks for example, were deposited in latitudes as far as 60° from the equator.

(iv) *Coals* can form in a wide variety of environments, where the rate of vegetation accumulation exceeds the rate of oxidation or physical removal. The environments include tropical rain forests, where growth throughout the year causes growth rings to be small or absent, and higher latitudes where decay is inhibited by cold winters and seasonal changes are recorded by prominent growth rings. In general, the presence of trees can be taken to indicate that mean summer temperatures were above freezing. Land plants first became established in Silurian times, and the oldest fossil coals and peats are of Devonian age. By Carboniferous times forest trees and ground cover vegetation had developed; this flora comprises the oldest major coal deposits. Flowering plants had developed by Cretaceous times and the bulk of the extensive Cenozoic coals are derived from flowering plants.

(v) *Red beds*: warm climates favour the diagenetic alteration of iron minerals in sediments to hematite, and the formation of lateritic soils (second part of Section 4.2.). Their distribution is wider than that of evaporites, because they also characterise moist environments. In addition diagenetic changes producing red colouration can take place long after deposition, by which time the region may have moved with respect to the pole

(vi) *Large-scale dune-bedding* and *sun-cracked surfaces* are features associated with desert climates, and are indicative of a low seasonal rainfall in a landscape bare of vegetation for most of the year. Provided that the sedimentological characteristics define an aeolian origin, the cross bedding directions can be used to indicate the direction of transportation. At the present day these sands occur between 18 and 40° from the equator and are deposited by trade winds originating in zones of high barometric pressure at latitudes *ca.* 30° north and south of the equator; the Coriolis force imparts a westerly component to these winds. Appreciable amounts of aeolian sands were deposited in Permian and Triassic times, and the palaeowind evidence relates primarily to these times (Fig. 10.12 (e)).

(vii) *Hydrocarbons* result primarily from the bacterial degradation of planktonic algae which follows their accumulation in sediments at shallow depths of burial; this must in turn be followed by thermal maturation at somewhat greater depths of burial equivalent to temperature in the range 50–150°C (Stonely & Bailey, 1981). Algal production is favoured by warm waters and an abundant influx of nutrients, while preservation of the dead algal requires the proximity of a reducing environment. This latter condition is provided by a stratification of the water body; it is achieved in rather few environments at the present time although it seems to have operated in the open ocean basins during some intervals of geological time, notably during the Cretaceous (Section 12.8), when oceanic circulation was minimal. Hydrocarbons are not very precise indicators of the ancient environment because they have normally migrated upwards from the source rock into younger reservoir rocks. It is often assumed that they are mobilised shortly after formation prior to complete lithification of the source rocks; this would mean that the reservoir rocks are generally only marginally younger than the source rocks. There are no doubt important exceptions to this rule, but analysis of the palaeomagnetic information using this premise suggests that most hydocarbons have been formed within 40° of the ancient equator (Irving, North & Couillard 1974, Tarling 1983).

The first five of these palaeoclimatic indicators are plotted on a series of palaeogeographical maps covering Palaeozoic and early Mesozoic times, in Fig. 10.12; this evidence is based primarily on Nairn (1964), Seyfert & Sirkin, 1973 and Ziegler, Scotese, McKerrow, Johnson & Bambach, 1979. Fig. 10.12a is a map satisfying the approximate palaeomagnetic constraints for Lower to Middle Ordovician times prior to the rapid APW motions described in Section 10.3. This does not differ in a major way from other assessments, and shows that the bulk of the continental crust still lay within the low latitudes occupied during Middle and Upper Cambrian times (Fig. 9.12c). Carbonates, evaporites and red sediments all occur within the latitudinal limits of their present distributions. Subduction of ocean crust was taking place during these times along the Appalachian, Norwegian and British sectors of the Iapetus Ocean. A tectonic pulse in early Ordovician (Tremadoc and Arenig) times resulted in the Taconic and Grampian orogenies while by Llanvirnian and Llandeilian times renewed subduction was taking place along these margins; it culminated in further ophiolite emplacement and final ocean closure in Middle–Upper Ordovician times (Section 10.3). The Iapetus Ocean was a faunal barrier during Lower and Middle Ordovician times, and it is possible that the width was greater than indicated in this figure. In addition Arenig–Llanvirn trilobites from Siberia belong to a Bathyurid Province, which incorporates Laurentian faunas (Whittington & Hughes, 1974) and an alternative position for Siberia between Laurentia and Gondwanaland is also possible. In contrast the Olenellid-Ceratopygid trilobite fauna occupied a sector including Baltica, Gondwanaland and Armorica-Bohemia suggesting a proximity between these areas, although there are some faunal differences between Northern and Southern Europe and Gondwanaland (see Ager, 1975; and Burrett, 1973) which might reflect the different latitudinal belts occupied by these localities during the earlier part of Ordovician times.

The Silurian reconstruction of Fig. 10.12b is approximately applicable to the latter part of Silurian times and to Lower Devonian times. It differs from earlier assessments by illustrating that the ocean which had separated Laurentia, Baltica and Gondwanaland had closed up by these times, as shown by the most recent palaeomagnetic data (see Sections 10.2 & 10.3 and Fig. 10.5). It is also possible that the oceans separating Baltica, Siberia and Khazakhstania had also nearly closed by these times (see also Irving, 1977) and they are shown in that situation here. These proximities are implied for example, by the evidence from the Siluro-Devonian agnathid fishes (and especially by the armoured versions of these fishes, the thelodonts); although strongly constrained to specific ecological niches, these fossils have a distribution which requires either continental continuity, or the implausible postulate of a number of temporary land bridges (Turner & Tarling, 1982). Extensive shallow seas covered the continents, and the sites of carbonate deposition seem to have extended nearly 60° from the equator; in addition, evaporites in Siberia may have formed in latitudes as high as 50°N. Tillites (probable Silurian and Lower Devonian examples are plotted in this figure) were confined within 20° of the South pole, and although the glacial movements are uncertain, their distribution correlates with the passage of Gondwanaland across the South pole described in Fig. 10.1. It would be compatible with the existence of only a small polar ice cap at this time, and corresponds to a more equitable climate accompanying the sea level rise following the Hirnantian glaciation discussed in the first part of Section 10.3.

Faunal relationships are unclear in Silurian times, but barriers to migration seem to have been small by late Silurian times, and the faunal provinces show strong latitudinal control. This figure (Fig. 10.12b) assigns several blocks, including North China, SE Asia, Tibet and Iran-Afghanistan to the perimeter of Gondwanaland, in the approximate positions which they had probably occupied since Cambrian times although the evidence for this is faunal rather than palaeomagnetic (concluding part of Section 10.5). The mid-Carboniferous reconstruction of Fig. 10.12c differs from earlier assessments in illustrating a complete closure between Laurentia-Baltica and Gondwanaland (Sections 10.3 & 10.4); the Gondwanaland A reconstruction is preferred here for the reasons discussed in Section 10.6. A long latitudinal belt of coal deposits formed in tropical rain forests is concentrated within 20° of the ancient equator and is largely applicable to Lower Carboniferous times; these coals contain fossils indicating minimal seasonal effects, and now comprise the bulk of the northern hemisphere Palaeozoic deposits. Evaporites are also widely distributed within 25° of this equator and seem to have formed mostly during the period of increasing continentality during Upper Carboniferous times. Carbonate deposition may have extended as far as latitudes 60°N and S of the equator although (in common with the wide distribution of the evidence for glaciation) this conclusion is no doubt influenced by the considerable APWP during the earlier half of Carboniferous times: the APWPs of Figs. 10.2 & 10.3 are equivalent to latitudinal movements of at least 25° during Upper Carboniferous and Permian times. The South polar ice cap appears to have been more extensive than during Silurian and Devonian times, but probably did not extend further than 30° from the pole; the pattern of ice movement (Crowell & Frakes, 1970) indicates a dispersal away from this pole. Several blocks had probably broken off from the perimeter of Gondwanaland by these times, and South China and SE Asia may have moved towards equatorial positions.

Khazakhstania was fully sutured to the remainder of Pangaea by Permian times (Fig. 10.12d), while the China blocks. Tibet and SE Asia were in the process of being translated towards Eurasia by the closure of the palaeo-Tethys Ocean (section 10.5). They seem to have become essentially coupled to Eurasia during Triassic (Fig. 10.12e) or early Jurassic times. Worldwide regression accompanied an extension of continental environments to produce a broad belt of arid and semi-arid environments between 25°N and S of the equator; palaeowind directions link aeolian sands here to the ancient Trade Wind Belt. Equatorial rain forests seem to have contracted and may only have had a significant distribution in the eastern Tethyan region (Fig. 10.12d). Vegetation was more extensive in the temperate regions above latitudes of 30° and cold-resistant plants (the Glossopteridae) flourished close to the perimeter of the ice sheet. Coal beds are found in this region and extend to the margins of the South polar ice cap, where some thick coal beds are now found associated with tillites. The trees in these deposits have prominent growth rings implying slow growth in an environment comparable with the present-day muskeq swamps of Siberia (Seyfert & Sirkin 1973). These coals range

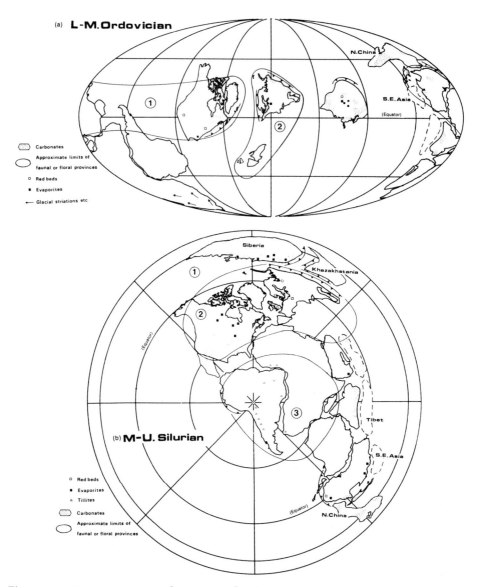

Fig. 10.12 Representative palaeogeographical maps for Palaeozoic and early Mesozoic times, with palaeoclimatic indicators taken mainly after sources in Nairn (1964) and from Ziegler, Scotese, McKerrow, Johnson & Bambach (1979). Sites of active subduction are indicated by the same symbol as Fig. 9.12. (a) Mid-Ordovician times; the faunal provinces are numbered, in circles: 1, North American realm and 2, European realm. (b) Middle–Upper Silurian times; this figure also includes glacial evidence from Lower Devonian times and faunal provinces of this age which are numbered: 1, Uralian realm, 2, American realm and 3, Malvinokaffric realm. The locations of the Iranian, Tibetan and SE Asian blocks which are dashed here are suggested by geological evidence only. (c) Mid-Carboniferous times; three faunal provinces are numbered: 1, Angaran flora, 2, Euramerican flora and 3, Glossopteris flora. (d) Permian times: this map includes glacial evidence of probable Upper Carboniferous age; the floral provinces are numbered: 1, Angaran, 2, Euramerican, 3, Cathaysian and 4, Glossopteris. (e) Triassic times: the faunal provinces are numbered as for part (d).

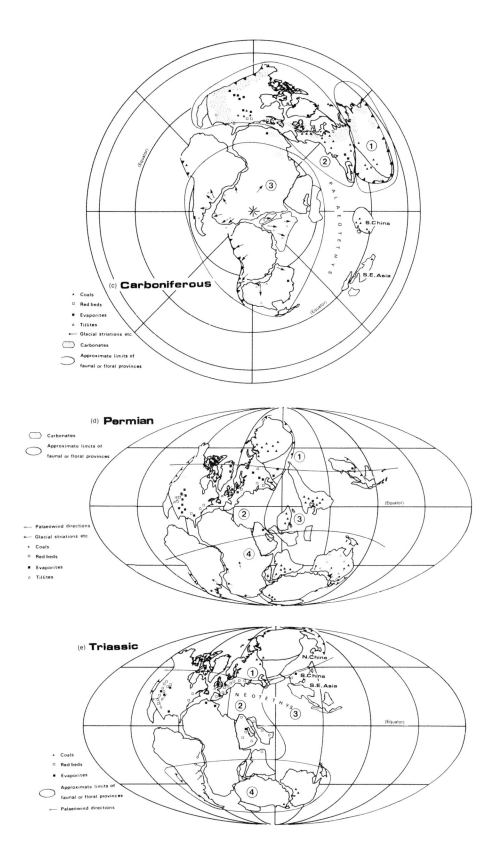

from Upper Carboniferous to Triassic in age and comprise the bulk of the southern hemisphere deposits. Note that the extension of the Permian tillites across Australia provides some support for the palaeomagnetic evidence, suggesting a Permian migration of the APWP across this continent (Section 10.2). Faunal barriers were by now largely unrelated to plate boundaries and exhibit a strong latitudinal control.

The effects of continentality remained strong during Triassic times, as the entire continental crust moved northwards and rotated clockwise through some 30°. A broad belt of red beds and evaporites extended from near the ancient equator to 40°N and S. The northward movement of Pangaea seems to have translated the crust out of the influence of the South polar ice cap during this period (Fig. 10.12e).

It is evident from Fig. 10.12 that the palaeoclimatic indicators correspond closely with the predictions of the palaeomagnetic evidence, and reinforce the gross validity of the axial dipole assumption during Phanerozoic times. The main difference between the Palaeozoic and present-day distributions applies to the continental or epicontinental climatic indicators (red beds, evaporites and desert sandstones) which were concentrated substantially nearer to the equator during the earlier part of the Palaeozoic era. Carbonates (which are the only marine facies of well-defined climatic significance considered here) do not show this effect. One possible reason is that continental conditions were enhanced by the agglomeration of the continental crust during the latter part of Palaeozoic times (Briden & Irving 1964). A second possible reason is related to the faster rotation rates of the Earth at this time (Rosenberg & Runcorn 1975): Coriolis forces acting on the atmosphere and hydrosphere would then have been stronger and would have tended to compress the dry trade wind belts towards the equator. Ziegler, Scotese, McKerrow, Johnson & Bambach (1979) have examined this effect, and conclude that the magnitude of this change would not have exceeded 20% since the beginning of Palaeozoic times.

11
Mesozoic and Cenozoic palaeomagnetism and the breakup of Pangaea

11.1 Introduction

The palaeomagnetic analysis in the preceding chapters has shown that APWPs can be regarded as continuously evolving phenomena. For this reason and because of inadequacies in the data, mean pole calculations have not been presented and are unlikely to be very meaningful. The palaeomagnetic record for the last part of geological time, which is described in this chapter, has generally been covered in considerably greater detail than earlier times, and it is not possible within the bounds of a single book to list and evaluate all of these data. Instead, this youngest part of the record is examined using a procedure similar to that adopted by Irving & Irving (1982, see also Section 6.6). The palaeomagnetic poles from each plate are first assigned an age by reference to the geological time scale and/or radiometric studies, and they are then ordered in time sequence. This is generally a straightfoward procedure over this youngest part of geological time because later remagnetisations can usually be identifed. However, there may still be uncertainties in the age of remanence acquisition (especially in the case of sediments) and the age may then not be known to better than ± 10–25 Ma. Results from the oldest interval, or *time window*, are averaged, and the window is moved forward in time by a specified increment. A new mean is calculated, and the procedure is repeated forward to the last time interval covered by the data set. The mean APWP is then defined by joining together the successive mean palaeomagnetic poles. Each pole refers to an interval of time, and has an associated circle of confidence which yields an error envelope about the mean path.

The palaeomagnetic record is obviously highly variable from continent to continent, and from time interval to time interval within the same continent, and the width of this error envelope gives some measure of this variability. It is constructed using the standard error ($P = 0.63$), rather than the 95% error, because this is more useful for evaluating differences between the mean poles. The screening procedure adopted by Irving & Irving is used here, and excludes poles with errors exceeding 25° and those based on < 10 oriented samples. Results which are clearly younger overprints, or which have age uncertainties of more than ± 25 Ma are also excluded. When more than one investigation has been made of a rock unit, the most complete study is used. Calculations of plate velocities in terms of V_{RMS} are analysed separately (concluding part of Section 12.8).

11.2 Laurentia and Eurasia

The data points from Laurentia are derived from Greenland and North America, and yield the mean APWP listed in Table 11.1. Results from suspect terranes in the Western Cordillera are excluded because these terranes have experienced tectonic displacements relative to the Laurentian block since the breakup of Pangaea (concluding part of Section 11.4). There remain some important time gaps in the data, notably in Middle Jurassic (*ca.* 170 Ma) and Upper Cretaceous (*ca.* 90 Ma) times, but the time window is not very critical to the analysis: a 20 Ma window is used in Fig. 11.1, although the general S-shaped form of the APWP is still evident with a 100 Ma window (Irving & Irving, 1982). Quasi-static intervals are defined by the grouping of mean poles during Upper Carboniferous–Lower Permian (300–260 Ma), Lower Jurassic (200–180 Ma) and mid-Cretaceous times. They are separated by rapid movements, of which the Middle Jurassic–Lower Cretaceous one is still poorly defined.

Eurasia is taken to include the component parts of the European block discussed in Sections 10.3 and the second part of 10.5 (the Siberian block, Section 10.5) and the other plates sutured to this block during the latter part of Palaeozoic times are discussed in the concluding part of Section 10.5. The Kolyma and Sikote Alin blocks appear to have a different origin, discussed in Section 10.5; together with the Caucasus area (Transcaucasus) which may have moved relative to Eurasia in Palaeogene or earlier times. The data from these latter areas are excluded from the reference path plotted in Fig. 11.2 and listed in Table 11.2. Although relatively well-defined during the Upper Carboniferous–Lower Triassic interval, this path is poorly constrained over much of the Mesozoic era, espcially during Upper Triassic and Lower Jurassic times. The APWP describes a small loop in Cretaceous times, with hairpins at *ca.* 120 and 80 Ma; this features persists when a broader window is used, and in common with a contemporaneous signature in the North American path (Fig. 11.1), it appears to be a real effect.

Table 11.1 Mean palaeomagnetic poles for the North American plate from 300 Ma to the present.

Age (Ma)	Pole Position °E	°N	N	Alpha 95	Alpha 63
300	125	39	14	4	2
290	127	41	22	3	2
280	124	44	21	3	2
270	122	45	16	3	2
260	124	47	3	16	10
250	118	52	14	7	4
240	119	54	12	8	5
230	105	62	27	8	5
220	95	66	36	7	4
210	71	70	25	6	4
200	81	71	19	6	4
190	80	74	14	8	4
180	120	72	4	13	8
170	144	78	6	14	8
160	139	72	9	11	6
150	133	67	6	10	6
140	158	66	2	—	—
130	168	73	2	—	—
120	175	72	4	11	7
110	181	70	2	14	8
100	187	65	1	—	—
90	179	66	3	14	8
80	182	70	7	6	3
70	182	70	6	7	4
60	188	81	10	5	3
50	189	82	13	4	2
40	168	84	11	4	2
30	154	86	14	3	2
20	121	88	17	3	2
10	101	88	18	3	2

These calculations utilise a 20 Ma window for the data set incremented in steps of 10 Ma. The results for 60–10 Ma are taken from Irving & Irving (1982) and utilise a window of 30 Ma. N is the number of palaeomagnetic studies.

302   *Palaeomagnetism and the Continental Crust*

The North American and Eurasian APWPs were first interpreted in the context of the opening of the North Atlantic Ocean by Runcorn (1956). He showed that the North American path plots systematically to the west of the European path. Hospers & Van Andel (1968) subsequently showed that the palaeomagnetic difference is closely compatible with the morphological fit of Bullard, Everitt & Smith (1965). The agreement of the two paths as defined by

Fig. 11.1 APW path for North America from late Palaeozoic times, expressed in terms of the mean poles averaged over time windows of 20 Ma, and plotted as closed circles connected by the thick line. The Cenozoic results in this and the next three figures are taken from Irving & Irving (1982). The ages of the mid-points of each window are given in Ma. The APWP shown by the open circles and connected by the thin line is the Eurasian APWP for the same time interval, after rotation to close the North Atlantic, according to the reconstruction of Bullard, Everitt & Smith (1965). Note that the two paths differ in a systematic way between 310 and 200 Ma, and show little correspondence ater 200 Ma, for reasons which do not appear to have a tectonic explanation.

Table 11.2 Mean palaeomagnetic poles for the Eurasian plate from 300 Ma to the present.

Age (Ma)	Pole position °E	°N	N	Alpha 95	Alpha 63
310	169	29	18	6	3
300	167	34	37	4	3
290	165	39	39	3	2
280	165	41	40	3	2
270	159	41	36	7	4
260	160	44	50	5	3
250	160	47	62	4	2
240	154	51	28	6	3
230	134	57	7	11	6
220	124	58	6	13	7
210	103	70	10	14	8
200	70	67	12	14	8
190	86	68	13	13	8
180	110	69	3	33	19
170	111	63	5	18	10
160	122	67	6	12	7
150	129	72	5	9	5
140	132	76	3	23	13
130	170	80	10	7	4
120	176	80	9	7	4
110	170	80	5	15	9
100	177	73	5	16	9
90	174	70	5	11	7
80	165	74	3	22	13
70	156	72	4	15	7
60	153	75	14	6	3
50	155	75	14	6	3
40	159	78	13	6	4
30	198	83	15	5	3
20	196	85	35	3	2
10	192	86	52	2	1

The data are processed as for Table 11.1. Results for the interval 70–10 Ma are from Irving & Irving (1982) and use a 30 Ma window.

the present data when rotated according to this fit is illustrated in Fig. 11.1; it demonstrates that the two APWPs show only a poor correspondence after 200 Ma. In addition, there is a residual shift of the Eurasian path to the east of the North American path prior to 200 Ma, which cannot be eliminated by any plausible adjustments to the morphological reconstruction (Tarling, 1979). These differences between the two APWPs cannot be interpreted in a tectonic context prior to Cretaceous times, and possible causes include poor definition of the paths in Jurassic and Lower Cretaceous times, variable overprinting of the late Palaeozoic data, and departures from the geocentric dipole assumption (Sections 10.4 & 10.6).

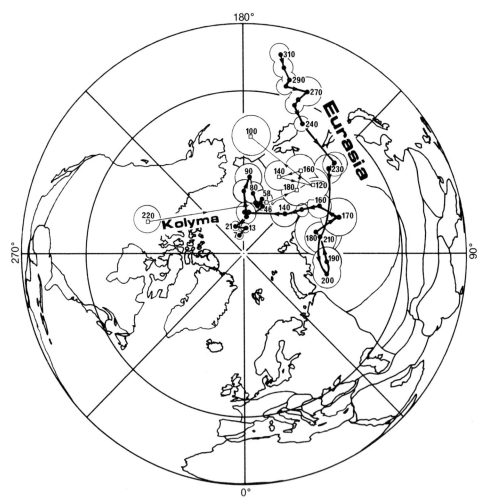

Fig. 11.2 APWP for the Eurasian plate, defined here in terms of mean poles calculated through a 20 Ma window and plotted as closed circles connected by the thick line. The mean APWP for the Kolyma block (see third part of Section 11.4) is calculated through a 40 Ma window, incremented in 20 Ma steps, and plotted as open squares connected by the thin line. The palaeomagnetic record from this plate is largely pre-Upper Cretaceous in age and does not accurately define the time of suturing with the Eurasian landmass.

11.3 Gondwanaland

Fig. 11.3 and Table 11.3 give a mean APWP for Gondwanaland, when the poles are rotated together according to the reconstruction of Smith & Hallam (1970, see

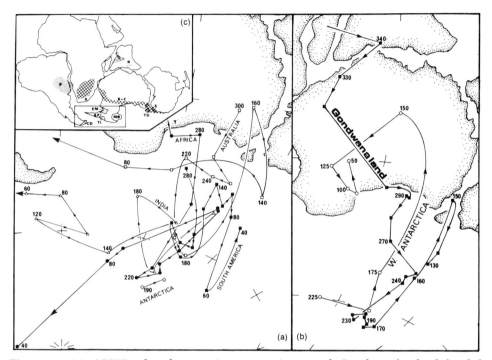

Fig. 11.3 (a) APWPs for the constituent continents of Gondwanaland, defined by calculating mean poles over time windows of 40 Ma and increments of 20 Ma. The paths are rotated towards Africa, according to the reconstruction of Smith & Hallam (1970). Note that the APWP for cratonic eastern Antarctica is only currently described by results from the Ferrar-Beacon volcanism of early Jurassic age, and the Indian APWP is sparsely represented by data prior to eruption of the Deccan Traps in latest Cretaceous times. The poles are indicated as closed circles (Africa), closed squares (South America), open squares (Australia), open triangles (India) and diamonds (Antarctica). (b) Mean APWP for Gondwanaland (closed squares), calculated by averaging the collective palaeomagnetic data from the Gondwanaland continents over 20 Ma windows incremented in intervals of 10 Ma. The APWP shown as small open circles is calculated from data for the Cordilleran margin of Western Antarctica (see also fifth part of Section 11.4). (c) The distribution of volcanic activity linked to the breakup and dispersal of Gondwanaland between late Triassic and early Cenozoic times. Locations are lettered: K = Karroo basin, B-F = Beacon-Ferrar dolerites, TD = Tasmanian dolerites, S = Sydney basin (all mostly of early Jurassic age), P = Parana Basin, R = Rajmahal Traps (both early Cretaceous) and D = Deccan (latest Cretaceous or earliest Cenozoic). The terranes comprising Western Antarctica are lettered; AP = Antarctic peninsula, TI = Eights Coast–Thurston Island, MB = Marie Byrd Land, EM = Ellsworth Mountains and CD = Cordillera Darwin. These terranes were not in their present positions relative to cratonic Eastern Antarctica during the lifetime of Gondwanaland, and are enclosed by the box.

Table 11.3 Mean palaeomagnetic poles from Gondwanaland for the interval 320–150 Ma relative to Africa. Smith-Hallam reconstruction.

Age (Ma)	Pole Position °E	°N	N	Alpha 95	Alpha 63
320	223	21	6	16	9
310	242	33	9	14	8
300	245	36	14	8	5
290	244	35	18	5	3
280	243	40	14	7	4
270	245	44	9	9	5
260	263	52	14	5	3
250	258	55	36	4	2
240	255	56	31	5	3
230	246	64	18	5	9
220	243	62	12	8	8
210	253	62	15	6	5
200	272	62	18	5	3
190	274	65	22	4	4
180	253	64	15	5	5
170	256	64	14	5	3
160	258	50	9	10	2
150	257	45	12	8	3

Data processed as for Table 11.1

also Section 10.6). Mean paths for the constituent continents in present-day coordinates are listed in Tables 11.4–11.8. Data from both Africa and South America define in general terms the southerly transition between 330 and 230 Ma, followed by a quasi-static interval between ca. 230 and 180 Ma. Following a hairpin, the path executes a loop in Jurassic and Cretaceous times (140–80 Ma) which is recognised in both continents, but is best described from South America (Vilas & Valencio, 1979). The African data are biased towards results from the Karroo volcanics, which were probably all erupted close to the Triassic–Jurassic boundary. The Australian path shows a departure from the African and South American paths in Triassic times (Irving & Irving, 1982). Mesozoic data from India are few; an extension to the APWP may be defined here by the Permian record, but the Triassic data are entirely consistent with those from elsewhere.

Some measure of the complexity of APW relative to Australia in Mesozoic times first became evident from the discrepancy between Jurassic results from the Tasmanian dolerites and results from the other continents, and notably, from the Ferrar dolerities of Antarctica (Fig. 11.3). This was first noted by Irving & Robertson (1969) and has been discussed in detail by many authors since then. More recent data from rock units in Australia (Embleton & Schmidt, 1977), and complementary DSDP results (McElhinny & Embleton, 1974) have established the reality of this difference; these data imply that it represents real APW. Subsequent studies in SE Australia (Schmidt & Embleton, 1981) have elaborated the form of an APW loop executed between *ca*. 170 and 100 Ma, which moved across eastern Australia and New Zealand. This loop is presumably also relevant to eastern Antarctica but, in part, post-dates the fragmentation of Gondwanaland.

The paths of the Gondwanaland continents diverge sharply at *ca*. 140 Ma, and define the palaeomagnetic signature of continental breakup (Fig. 11.3a). Note that the African and South American paths diverge between 120 and 100 Ma, which is *ca*. 10 Ma after the age of the oldest ocean crust between them (Maxwell, Von Herzen, Hsu, Andrews, Saito, Percival, Mitlow & Boyce, 1970) and post-dates, by some 50 Ma, the initial (late Jurassic) sedimentation in developing rifts along the margins of South Atlantic-bordering continents (Funnell & Smith, 1968). A range of geological evidence shows that a marine margin first began to develop along the edge of eastern Africa in Lower and Middle Jurassic times (McElhinny, 1973), and accompanied the widespread Karroo rifting and volcanism (Fig. 11.3c). The APWP of Australia, and the combined path of South America and Africa diverge in mid-Cretaceous times at 120 Ma, while sparse data suggest that India may already have separated from these continents and began its northward movement by these times (Figure 11.3a). The history of subsequent movements is more precisely defined by the ocean-floor record, which is not covered by

Table 11.4 Mean palaeomagnetic poles for Africa, from 320 Ma

Age (Ma)	Pole Position °E	°N	N	Alpha 95	Alpha 63
320	230	22	8	13	8
300	223	34	5	14	8
280	238	33	7	7	4
260	244	57	12	15	9
240	256	72	26	7	4
220	260	72	20	7	4
200	256	67	14	5	3
180	258	71	19	8	5
160	263	68	11	16	9
140	256	53	6	14	8
120	255	58	4	18	11
100	228	62	28	5	3
80	226	63	29	5	3
40	95	80	3	23	10
30	152	84	7	10	5
20	164	84	12	6	3
10	187	87	10	5	3

These poles are calculated through a 40 Ma window, incremented in steps of 20 Ma. The post-80 Ma results are from Irving & Irving (1982) and use a 30 Ma window.

this book. It is sufficient to note that India seems to have separated from Australia in Lower Cretaceous times (140 Ma), and Antartica separated from Australia in late Palaeocene times (53 Ma), following rifting which began in Middle Jurassic times (McKenzie & Sclater, 1971; McElhinny & Embleton, 1974). The Tasman Sea developed by separation of the New Zealand Plateau and Lord Howe Rise from Australia in late Cretaceous times (82 Ma, Veevers & McElhinny, 1974). This event is recorded in the Sydney basin as a widespread remagnetisation episode, linked to rapid uplift and erosion following burial heating along this SE margin of Australia (Schmidt & Embleton, 1981). Embleton & McElhinny (1982) have used palaeomagnetic data from lateritic soil profiles in Australia to develop a smoothed APWP for Cenozoic times. They show that the APWPs of the former Gondwanaland continents are entirely consistent with the model defined by sea-floor spreading since 150 Ma (Norton & Sclater 1979), when the ocean basins between them are progressively closed up.

The perimeter of Gondwanaland in the north is considered to have extended into central New Guinea in Palaeozoic times (e.g. Burrett, 1974). Terranes beyond this are part of the SE Asian block discussed in the last part of Section 10.5 and the fifth part of 11.4. Mesozoic studies from east Antarctica have concentrated on the Ferrar volcanics, and little information is available for other times. The Australian data, however, show that this continent moved close to the South Pole between 100 and 80 Ma, and has remained within 20° of this pole ever since (Fig. 11.3b). Results from western Antarctica come from allochthonous terranes, and are discussed in the fifth part of Section 11.4. The palaeomagnetic record from India

Table 11.5 Mean palaeomagnetic poles for South America from 280 Ma.

Age (Ma)	Pole Position E°	°N	N	Alpha 95	Alpha 63
300	170	65	6	10	6
280	156	75	11	9	5
260	55	83	27	6	4
240	48	82	26	6	4
220	61	82	11	10	6
200	106	82	9	10	6
180	63	82	6	12	7
160	134	89	6	12	7
140	178	84	7	9	5
120	333	90	12	6	3
100	227	88	12	7	4
80	310	85	8	17	10
60	333	88	3	17	7
20	358	85	3	3	2

Pole calculations use a 40 Ma window incremented in steps of 20 Ma. Post-80 Ma results are from Irving & Irving (1982).

Table 11.6 Mean palaeomagnetic poles for Australia from 320 Ma.

Age (Ma)	Pole Position °E	°N	N	Alpha 95	Alpha 63
320	312	45	6	18	10
300	312	45	6	18	10
280	323	48	4	9	5
260	335	43	3	27	15
240	328	51	5	17	10
220	322	59	3	15	9
200	354	49	6	6	3
180	357	49	8	3	2
160	312	48	4	20	12
140	338	37	8	15	9
120	339	42	11	11	7
100	330	57	7	10	6
80	334	65	2	54	31
70	311	68	2	15	4
40	309	69	3	11	5
30	294	73	4	12	6
20	293	77	5	12	6
10	267	81	3	17	7

Pole calculations use a 40 Ma window incremented in steps of 20 Ma.

Table 11.7 Palaeomagnetic poles for India from 260 Ma.

Age (Ma)	Pole Position °E	°N	N	Alpha 95	Alpha 63
260	309	−2	7	8	5
240	305	3	10	10	6
220	299	15	3	31	18
180	292	11	4	24	14
160	300	12	7	14	8
140	305	17	4	16	9
120	287	34	5	17	10
100	293	28	6	6	4
80	285	29	16	6	4
60	281	34	21	6	3

Pole calculations use a 40 Ma window incremented in steps of 20 Ma.

Table 11.8 Palaeomagnetic poles from Antarctica.

Age (Ma)	Pole Position °E	°N	N	Alpha 95	Alpha 63
(i) East Antarctica					
190	45	51	3	11	6
170	38	53	4	9	5
(ii) West Antarctica					
225	55	63	1	—	—
200	38	59	7	9	5
175	35	67	7	24	14
150	264	78	2	—	—
125	155	86	4	12	7
100	93	88	7	10	6
75	45	86	4	18	10

The calculations for East Antarctica use a 20 Ma window incremented in 20 Ma steps. The West Antarctica calculations employ a window of 50 Ma, incremented in steps of 25 Ma; they assume that the West Antarctica peninsula can be treated as a single plate in palaeomagnetic terms.

is heavily biased towards results from the Deccan Traps, erupted within a short interval of late Cretaceous or early Cenozoic times. India was still situated at a latitude of *ca.* 30°S at this time, and a subsequent movement at velocities of *ca.* 18 cm/year, between 75 and 55 Ma, carried this continent away from Antarctica into a collision with Eurasia (Norton & Sclater, 1979).

By utilising data from DSDP cores Klootwijk & Pierce (1979) have defined an APWP for the Indian plate during Cenozoic times. Comparison with the Eurasian data in general, and results from the NW Himalaya north of the Indus suture, in particular, (Klootwijk & Radhakrishnamurty, 1981; Klootwijk, Nazirulla, de Jong & Ahmed, 1981) show that collision between the Indo-Pakistan block and an island arc along the southern margin of Eurasia, commenced in late Palaeocene or early Eocene times in the west, and continued to Eocene times in the east. The Himalayan orogenic belt resulted from this collision and now comprises a thrust zone of stacked continental and oceanic crust, with a foreland of folded and thrusted molasse sequences. By comparing palaeomagnetic results from the southern margin of the Tibet Plateau above the Main Central thrust at the site of the collision, with contemporaneous data from the Indian subcontinent, Bingham & Klootwijk (1980) have recognised an excess anticlockwise rotation of India of 10°–15° with respect to its leading edge. This suggests that India has rotated anticlockwise about a Euler pole in the NW Himalaya since Oligocene times. The resultant underthrusting of the Tibetan plate is predicted to increase from zero in the west to 650 km at the longitude of western Nepal (Klootwijk, Conaghan & Powell, 1985). This model correlates well with the distribution of calc-alkaline volcanism, and explains the crustal thickening to the extent of some 70 km, which has in turn produced the uplift of the Tibetan plateau to altitudes averaging 3,500 m; this model also predicts that the Indian continental block (Greater India) extends northwards as far as the Kun Lun fracture zone (Fig. 10.10).

The palaeomagnetic declinations in the Himalayan arc and southern Tibet change systematically in an anticlockwise sense along the length of the belt, from 45° to the NE, in the NW Himalayas, to slightly west of north, in the Lhasa region. This pattern is consistent with continued post-collision convergence of Greater India into southern Asia until early Miocene times, followed by rotational underthrusting of Greater India (in an anticlockwise sense) beneath the Asian block. This latter motion has accompanied an oroclinal bending of the arc in Neogene times (Klootwijk, Conaghan & Powell, 1985). The southern margin of the Eurasian landmass has responded to the impingement of the rigid Indian block over a period of some 50 Ma, following ocean closure, by yielding along systems of strike-slip and conjugate faults; these systems are associated with regional extensions and compressions, and can be concisely linked to the geometries of the colliding blocks (Molnar & Tapponier, 1977 and Fig. 10.10).

11.4 Sub-plates, microplates and suspect terranes

The Mediterranean region

The Mediterranean region incorporates a number of microplates in a complex mosaic which has resulted from a sequence of strike-slip and closure movements between the African and Eurasian margins of the Tethyan Ocean. The African margin appears to have presented a rigid promontory during these movements, whilst the Eurasian margin behaved as a semi-rigid and semi-plastic margin which deformed into oroclinal belts (Tapponier, 1977). The present situation (Fig. 11.4) is a tight one in the east and west, where larger numbers of narrow, folded belts are separated by broader, weakly-deformed zones, and an open one in the centre, where two wider oroclinal belts are separated by oceanic and semi-oceanic crust. The palaeomagnetic evidence for the component plate movements will be discussed here from west to east.

The Bay of Biscay has resulted from the anticlockwise rotation of the Iberian peninsula about a nearby Eulerian pole. Bullard, Everitt & Smith (1965) found that a rotation of 32° was required to fit the adjacent coasts at the 500 fathom isobath; this is close to the 36° angular rotation required to match the Permo-Triassic palaeomagnetic results from the Iberian peninsula with the data from central Europe (McElhinny, 1973). From study of the (Eocene) Lisbon volcanics, Watkins & Richardson (1968) suggested that this rotation was in part post-Eocene. Van der Voo & Zijderveld (1971) recognised a pre-folding component in these rocks, and argued that the rotation occurred entirely between late Jurassic and late Cretaceous times. This interpretation is supported by the presence of Cretaceous deposits in the Bay of Biscay, and the probable identification of (Upper Senonian) anomaly 34 here (C. A. Williams, 1975). From palaeomagnetic study of Mesozoic sediments VandenBerg (1980) was able to show that opening post-dated Barremian–Aptian times. However, Storetvedt (1972) identified a more complex history of remanence in the Lisbon volcanics and disputed the interpretations of earlier studies. He regarded the rotation of the peninsula as taking place largely after the eruption of these rocks, and during a few Ma of late Eocene and early Oligocene times (*ca.* 40 Ma ago). This interpretation is supported by the northerly derivation of Eocene-age turbidites along the northern Galician coast (Crimes, 1976) which seems to preclude a major opening in the Bay of Biscay by these times, and by the correlation of this event with the main phase of the Pyrenean orogeny at *ca.* 40 Ma. Thus, the collective evidence is most compatible with a two-stage model for the rotation of the Iberian peninsula.

The pole of rotation cannot be identified from the palaeomagnetic data alone, but geological evidence suggests that it cannot have lain within the Pyrenean orogenic belt, because a pure rotation here would result in a progressive increase in crustal shortening to the east. This rotation would be incompatible with the distribution of Mesozoic and Cenozoic sediments and with the comparable degrees of crustal shortening along the length of the belt (Mattauer, 1969). It seems more likely that Iberia originated as a western extension of the Armorican-Barrandian plate (Sections 9.5 and 10.3) and that rotation was accompanied, or followed by, several hundreds of km of sinistral displacement along the North Pyrenean fault zone, about a Euler pole close to Paris (Le Pichon, Bonnin, Francheteau & Sibuet, 1971). The magnetic anomalies in the Bay of Biscay are partially disrupted in a way which would be compatible with movements of this kind (Storetvedt, 1972).

Corsica and Sardinia are surrounded by regions of Cenozoic extension, and are practically unaffected by the Alpine orogeny; there is a general continuity of Permian and younger dykes between Sardinia and Corsica, implying that the two islands have moved as a single microplate (Arthaud & Matté, 1977). Permian palaeomagnetic studies in Corsica (Nairn & Westphal, 1968) and Sardinia (Zijderveld, de Jong & Ven der Voo, 1970) suggest a 60° anticlockwise rotation from a position adjacent to southern France (Arthaud and Matté, 1977). A range of palaeomagnetic data constrain some 30° of this rotation to a 1 or 2 Ma period in the Burdigalian stage of Miocene times (Van der Voo & Channell, 1980, Montigny, Edel & Thvizat, 1981); this movement is linked to the opening of the North Balaeric Basin (Bellon, Coulon & Edel, 1977). The significance of the residual rotation is obscured, in part, by the effects of magnetic overprinting, but is presumed to have taken place in unison with the rotation of Iberia in Cretaceous and/or Eocene times.

The Peri-adriatic belt is a double arc, extending from northern Sicily through the Apennines into the Southern Alps, and thence SE via the Dinarides and Hellenides into Greece (Fig. 11.4). It is generally regarded as a deformed continental margin, which originated at the southern side of the Tethyan Ocean, and it comprises nappe piles and deformed sedimentary troughs of Mesozoic rocks, resting, in part, on Triassic evaporites. The latter have formed a mobile discontinuity with the underlying Hercynian basement (Section 10.3). Since contrasting deformations are involved in at least two crustal levels, the tectonic evolution was a complex one. It has presented a considerable challenge to palaeomagnetic investigations, and the predominance of carbonate lithologies in the pre-Cenozoic cover rocks has required the use of cryogenic magnetometers. In palaeomagnetic terms, the Adriatic region has generally been considered to be a promontory of the African continent, and therefore as a autochthonous unit which has moved with Africa throughout Mesozoic times. This view is supported by the conformity of

Cretaceous results from the Iblean (southern Sicily) and Apulian (SE Italy) autochthonous inliers, with the late Cretaceous poles from Africa (Channell, D'Argenio & Horvath, 1980), and by the correlation of results from Permo-Triassic rocks of the Southern Alps with data from African rocks of this age (Van der Voo & Channell, 1980; and Fig. 10.4). The northern margin of this crustal block is difficult to define in palaeomagnetic terms, but probably lies beneath the Northern Calcareous Alps (Tarling, 1983 and Fig. 11.4).

Results from the cover rocks are more difficult to interpret; the wider range of declinations and inclinations appear to reflect the movements (often inadequately understood from field observations) of individual allochthonous units on the scale of individual nappes (Channell & D'Argenio, 1980). The evolution of the Calabrian arc, linking Sicily and the Apennines, has been investigated in some detail by a comparison of results from the cover rocks with data from the Iblean and Apulian autochthonous regions (e.g. Channell, Catalano & D'Argenio, 1980). A tectonic picture emerges which involves an anticlockwise rotation of the southern Apennines, accompanying a clockwise rotation of the northern Alpine margin of Sicily during Cenozoic deformation. Hence, the Calabrian arc is a secondary feature, which presumably developed during the opening of the Tyrrhenian Sea (Van der Voo & Channell, 1980); the Calabrian nappe pile formed by the stacking of nappes, derived from the north during collision with the Adriatic promontory of the African plate (Van der Linden, 1985).

The Mesozoic carbonates of the Umbrian sector of the central Apennines have been investigated in some detail by palaeomagnetism. Some of these results have important implications to the geomagnetic time scale discussed in Section 12.2. By correlating the sequence of directional changes from this area with the African pattern, VandenBerg (1979) identified a 27° clockwise rotation. He interpreted this in terms of a decoupling of the Adriatic promontory from Africa in post-early Cenozoic times. Channell, Lowrie, Medizza & Alvarez (1980) disputed this analysis and preferred to interpret changes in palaeomagnetic directions in terms of regional tectonic deformations. They observed that the palaeomagnetic directions from the Umbrian region are in part dependent on geographical position around the tectonic arc and they considered that different amounts of horizontal transport (décollement) provide the best explanation for this. There remains a more westerly mean declination in this area, compared with the Apulian sector, which these authors attribute to an allochthonous origin. However, VandenBerg (1983) has since demonstrated a common 25° anticlockwise rotation at three separate localities around the Adriatic, ranging from Gargano in SE Italy to the Istrian peninsula of Yugoslavia. Support for relative movements between the Adriatic block and the African plate comes from studies of marine sediments from the Transdanubian Central Mountains of Hungary (Marton & Marton 1981). Results from mid-Triassic to mid-Cretaceous rocks define a clockwise APW loop which is analogous to the Gondwanaland path, but is removed from it by an amount commensurate with a 35° anticlockwise rotation; this movement is dated as post-Senonian. Other palaeomagnetic studies in this area define a positive link between the Pannonian sector and the Italian peninsuala (Marton & Veljovic, 1983), and faunal data from the Transdanubian area correlate this region with the southern margin of Tethys in Jurassic times (Marton & Marton, 1981.). Marton & Marton (1983) have since refined the correlation with the African APWP, and confirmed a rotation similar to the one identified by VandenBerg in the Adriatic. This anomaly does not extend eastwards of the arc formed by the Carpathian Mountains and the Transylvanian Alps, because the Mesozoic palaeomagnetic data from western Bulgaria correlate directly with the Eurasian APWP (Nozharov, Petkov, Yanev, Kropacek, Krs & Pruner, 1980).

The concept of a single Adriatic promontory, extending northwards as far as the Carpathian collisional zone has not received unequivocal support, because development of the Aegean and Ora regions, in a back-arc spreading mode behind the Hellenic-Cyprian subduction zone, requires an Adriatic suture zone. The

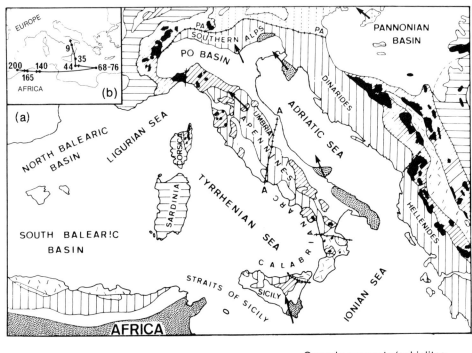

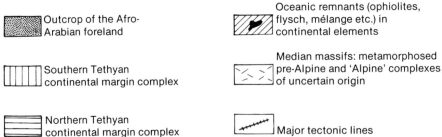

*Fig. 11.4 (a) Major structural divisions of the Peri-adriatic region, after Van der Voo & Channell (1980) with additions. The tectonic boundaries lettered **AA** and **PA** are the Anzio-Ancona and Peri-adriatic lines respectively. The arrows are mean declinations derived from Cretaceous rocks overlying the autochthonous inliers and from contemporaneous rocks from elsewhere. (b) Movement of Africa relative to Europe since early Mesozoic times, derived from magnetic anomalies in the Atlantic Ocean, after Biju-Duval, Dercourt & Le Pichon (1977) and Van der Linden (1985).*

oroclinal concept of Carey (1958) suggests that the evolution of the Adriatic arc would have required differential movement along the Adriatic, with anticlockwise movement of Italy accompanied by clockwise rotation of the Dinarides and Hellenides (Van der Linden, 1985). These latter zones comprise NNE–SSE-trending structural units separated, in part, by ophiolite wedges composed of Triassic and younger age sea-floor. Preliminary palaeomagnetic investigations suggested large rotational motions between these blocks during emplacement, but more recent results (Horner & Freeman, 1983) have defined a *ca.* 38° clockwise rotation of the nappe pile since early Cenozoic times, commensurate with the oroclinal model. These results imply that there is unlikely to be a direct continuity of ancient basement between Hungary and Yugoslavia, although differential rotational movements between the Italian peninsula and the Transdanubian region have clearly been minimal. Laj, Jamet, Sorel & Valente (1982) attribute the rotation in W. Greece to a combination of the rotation of nappe sheets during Miocene deformation and the extensional tectonics of the Aegean basin during the last 13 Ma. Their palaeomagnetic studies of Mio-Pliocene formations suggest that the Cretian arc has not been significantly deformed during the last 5 Ma, although the anticlockwise rotation of the Hellenides has occurred at a rate of *ca.* 5°/Ma to give a net rotation of 26° over this time period.

Palaeomagnetic studies in the Turkey, Syrian and Lebanon Sectors of Asia Minor, on rocks of Permian and younger age (Gregor & Zijderveld, 1964; Van Dongen, Van der Voo & Raven, 1967; Van der Voo, 1968) yield poles which largely follow the Gondwanaland APWP, after anticlockwise rotation to close the Red Sea and Gulf of Aden (McElhinny, 1973). The exceptions are certain results from the Dead Sea fault system in Israel, which appear to reflect the local rotation of individual fault blocks (Freund & Tarling, 1979). The regional data imply that the Arabian sector of the African plate probably extends as far north as the Caucasus Mountains, and identify the latter as a collisional belt, formed by northward movement of the African plate and closure of this sector of the Tethyan Ocean. The wider extent of the African plate into Iran and Afghanistan has been noted in the concluding part of Section 10.5. This region is now separated from the Arabian area by the Zagros thrust zone, where subduction and obduction of ocean crust have partly accommodated crustal separation in the Red Sea, and sinistral movements along the Dead Sea Fault since Miocene times (Quennell, 1983).

The complex tectonic development of the Mediterranean region can be concisely linked to relative movements between the African and Eurasian plates since early Jurassic times (Channell & Horvath, 1976: Biju-Duval, Dercourt & Le Pichon, 1977; Van der Linden, 1985). Between 200 and 76 Ma, movement was predominantly in a strike-slip sense (Fig. 11.4b) with a component of tensional and later compressional movement. This stage included initial emplacement of the Dinaride and Hellenide nappes (beginning in late Jurassic times), rifting the Bay of Biscay, and the earlier rotation of Corsica and Sardinia. A reverse motion between *ca.* 76 and 44 Ma relates to deformation of the Calabrian arc, and opening of the Balaeric, Tyrrhenian and Pannonian basins. Since *ca.* 44 Ma, the African plate has moved directly northwards, and the interaction of the Adriatic promontory with the collage of belts formed during earlier episodes of deformation, has produced the deformation of the Alps (especially during the Eocene– Oligocene interval) and the Apennines. Rotation of the Hellenides has led to back-arc growth in a developing extensional régime to the east of this orogenic belt, possibly linked to the rotation of the complex Adriatic promontory.

The Japanese arc

Kawai, Ito & Kume (1961, 1962) identified a striking difference of 60° between the magnetic declinations of pre-Cenozoic rocks in the northern and southern parts of mainland Japan (Honshu Island). Cenozoic magnetisations from the two areas appeared to show similar directions, in conformity with the present field. These workers interpreted their results in terms of bending of the Japanese mainland during late Mesozoic or Cenozoic times — the present angle between the axes of these two segments of the mainland is *ca.* 60°. More recent studies have clarified and extended this early work, but have failed to resolve a pre-Cretaceous evolution of the Japanese arc, because most older rocks are strongly overprinted (Tosha & Kono, 1984), probably largely by widespread Cretaceous magmatism. The island group as a whole has moved north by *ca.* 30° since Cretaceous times (Fujiwara & Morinaga, 1984), while study of the Upper Cretaceous Shimanto belt in SW Japan suggests rapid northward movement at that time, and identifies this area, at least, as an accreted terrane (Kodama, Taira, Okamura & Saito, 1983).

Studies of Cenozoic volcanics in SW Japan have identified a change in the mean declination from NE to N between 21 and 12 Ma, which corresponds with an anticlockwise rotation of SW Japan in early Miocene times (Tosha and Kono, 1984; Otofuji & Matsuda 1983). Otofuji, Matsuda & Nohda (1985) have studied contemporaneous rocks from NE Japan, and identified a change from W to N declinations between 21 and 11 Ma. The radiometric and palaeomagnetic data are best-constrained in SW Japan, and imply that these rotations were concentrated between 15 and 12 Ma. Hence, the bend in the Japanese arc does not appear to be an example of gradual oroclinal bending. Instead, it seems to have resulted from easterly-directed stresses in the zone between Korea and Sikote Alin, which split an ancient volcanic arc. The resultant rotations accompanied the opening of a region of back-arc spreading (the Japan Sea), in Lower–Middle Miocene times (Tosha & Kono, 1984). Prior to these motions the Japanese region appears to have formed a volcanic arc paralleling the Asiatic mainland. Ancient strike-slip faults along the axis of this zone (presumably linked to oblique subduction, as in the example from Sumatra noted in the fourth part of Section 11.4) were reactivated by these rotations, so that Palaeogene–Miocene motions were in a dextral sense in NE Japan, and in a sinistral sense in SE Japan, These regional movements can be likened to the slips within a pile of planks pivoted at either end and stressed at the centre (Otofuji, Matsuda & Nohda, 1985). The junction between the two rotated segments now coincides with the Fossa Magna zone, a N–S dislocation sited close to the change in direction of Honshu Island from NNE to ENE–WSW. The zonal structure of paired metamorphic belts, which is well developed in SW Japan, is interrupted at this point, although it can be recognised further to the NE (Tosha and Kono, 1984). At present there are inadequate palaeomagnetic data from the Islands of Kyushu to the SW, and Hokkaido to the NW, to extend this interpretation outside the Japanese mainland, although some early data from Okinawa (Sasajima & Shimada, 1964) suggest that the Ryukyu arc may have been partly involved in the rotation of SW Japan.

North and South China, Kolyma, Sikote Alin and Korea

Geological evidence suggests that the North and South China blocks were sutured during Triassic times, and palaeomagnetic evidence is consistent with this view from mid-Triassic times (Lin, Fuller & Zhang, 1985 and concluding part of Section 10.5). The APWP of the China blocks is displaced fom the Eurasian APWP until Cenozoic times although the shapes of the two paths are similar from 250 Ma (Fig. 11.5) and they predict that the Eurasian and China blocks were in continuity, but

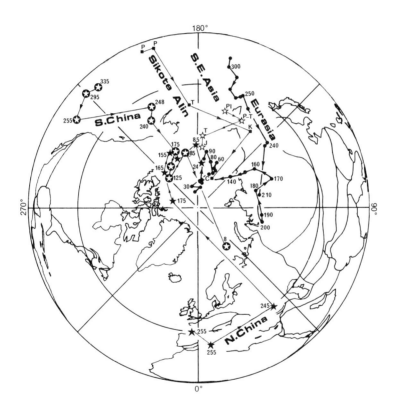

Fig. 11.5 The APWP for Eurasia since Upper Palaeozoic times (closed circles connected by the thick line) compared with APWPs from the North (closed stars) and South (open and circled stars) China plates, SE Asia (open stars) and Sikote Alin (closed squares). The APWPs from the China Shields are based mostly on single poles (Lin, Fuller & Zhang, 1985), and are assigned ages in Ma, commensurate with the source-rock stratigraphic ages. The SE Asia path is based on single poles or small numbers of poles referred to Permian (P), Triassic (T), Jurassic (J), and Cenozoic (C) ages.

in different orientations from the present. They can be matched by a 50° clockwise rotation of the China path about a Euler pole near to the present pole. This rotation brings the China blocks into the region now occupied by India and Pakistan, and suggests that a sinistral motion of 4000 km has taken place between the Chinese and SE Asian blocks on the one hand, and the Eurasian block on the other. It is tempting to correlate this movement with the Palaeocene–Eocene collision between India and Eurasia, especially since the required motion is equal to the width of northern India (3000 km) plus the width of the transverse ranges between India and the South China blocks (Lin, Fuller & Zhang, 1985), but it can only be constrained by the present data to the interval 70–10 Ma. The relative movements have presumably taken place along the Central Asian fold belt, which currently separates the seismically-active Chinese blocks from the aseismic northern Eurasian plate. Collectively, this motion is another facet of the deformation of an orogenic terrane between two colliding plates (Molnar & Tapponier, 1977 and Section 11.3).

The data from Sikote Alin (see Fig 10.10 for the location of this block), only converge with the Eurasian block in Cenozoic times (Fig. 11.5). The sparsely-defined older path is similar in form to the Eurasian path, but rotated relative to it, by an amount which is about half that of the South China APWP. This suggests that Sikote Alin shared in the sinistral displacement of the China Shields, but was translated by a smaller amount. The APWP from the Kolyma block remains discrete from the Eurasian APWP until at least 100 Ma (Fig. 11.3), and commensurate with the view that the Verkhoyansk Mountains separating the Siberian and Kolyma blocks define an early Cretaceous collision zone (Burrett, 1974). Cretaceous palaeomagnetic data from South Korea conform to the China APWP, and support a continuity of the basement between North China and Korea (Ito & Tokieda, 1980, see also Fig. 7.1 and Chapter 8).

South East Asia

The Indian plate is currently underthrusting the continental margin of the SE Asia plate from the south. To the NW of the Sunda Straits, the trench bordering Sumatra has a NW–SE trend, and is therefore highly oblique to the direction of subduction. The continental plate has responded to this stress system by developing major dextral fractures running along the magmatic arc parallel to the trench (Fitch, 1972). This is the situation which is most favourable to the lateral transport and emplacement of suspect terranes (see also concluding part of Section 11.4 and Harland, 1971). An exposed fracture (the Semangku fault) runs along the axis of Sumatra although reconnaissance palaeomagnetic studies from either side of this zone (Sasajima. Otofuji, Kirooka, Suparka & Hehuwat, 1978; Haile, 1981) have identified no relative movements on a palaeomagnetic scale here. It is possible that appreciable movements have occurred along fractures to the NE, because the mean palaeolatitude calculated from Permo-Triassic data from Sumatra is *ca*. 9° less than the corresponding palaeolatitude for Malaysia and Thailand (Haile, 1981).

A mean result calculated fom the data for Thailand, the Malayan peninsula, W Kalimantan and SW Sulawesi — which may collectively belong to a single SE Asian plate — yields a Permo-Triassic pole — based on four studies — which plots between the Eurasian and South China APWPs (Fig. 11.5). It suggests that the SE

Asian plate has experienced somewhat less than half of the sinistral displacement of the China blocks already referred to, and is compatible with a SE extrusion of this block in Eocene – early Miocene times, as a consequence of the impingement of India and Asia (Molnar & Tapponier, 1977). The most significant feature of the data from Sumatra is that they plot consistently to the east of the other SE Asian data, and imply a *ca*. 40° clockwise rotation since Triassic times (Haile, 1979). In contrast, sparse palaeomagnetic results from W. Kalimantan (Borneo) and SW Sulawesi (Celebes) suggest that these areas formed part of a plate which has rotated anticlockwise by 30°–50° between Cretaceous and Miocene times. The data hint at a general closure of accreted terranes around the eastern and western margins of the SE Asian plate since Upper Cretaceous times. Unfortunately, data are still so few that it is impossible to be clear whether individual studies record the gross movements of microplates, or rotations about individual fault planes (Sasajima, Otofuji, Kirooka, Suparka & Hehuwat, 1978.).

Along the eastward extension of the Banda arcs, contrasting results from Seram (late Triassic) and Timor (Permian) link the former to arc rotation and the latter to the Gondwanaland plate (Chamalaun, 1977, Haile, 1981). None of the terranes in this area have participated in the rapid, post-Jurassic, nothward movement of the Japanese arc. The SE Asian region appears to have remained at about the same palaeolatitude, or to have moved slightly southwards since Permo-Triassic times, although the data coverage is too poor to monitor movements over much of Mesozoic and Cenozoic times.

West Antarctica

At least four separate microcontinents are included within the West Antarctic Cordilleran margin. They include the Ellsworth Mountains, Eights Coast–Thurston Island and Marie Byrd land blocks, and the West Antarctic peninsula (Dalziel, 1982 and Fig. 11.3c). All are potentially suspect terranes, in common with the E–W-trending segment of the Andes in Tierra del Fuega (Cordillera Darwin). Palaeomagnetic data from Cambrian rocks of the Ellsworth Mountains suggest little relative translation between this block and East Antarctica (Watts & Bramall, 1981) although no precise correlation is possible (cf. Fig. 9.3); a large rotation about a local rotation axis has taken place here, and is responsible for disrupting the trend of Palaeozoic structures between the Ellsworth block and the nearby Pensacola Mountains. Studies in the West Antarctic Peninsuala are more varied, and include plutons and volcanics of Jurassic and younger age. When an integral APWP is calculated from these data (there is no certainty that this is a valid procedure), an APWP is derived which plots close to the Gondwanaland path between 225 and 150 Ma (Fig. 11.3b); since that time West Antarctica has remained close to its present location near the South Pole. Thus, the present results suggest that the West Antarctic peninsula has remained close to its present situation, with respect to East Antarctica, since Jurassic times. Longshaw & Griffiths (1983) argue for a location south of the base of the present peninsula, with a northward translation and a small clockwise rotation in post-Jurassic times.

Because all of the studied areas of this region have occupied high latitudes since Jurassic times, the palaeomagnetic inclinations are steep, and the definition of palaeodeclination is correspondingly poor. For this reason, and because the present data comprise relatively few studies, based mostly on small numbers of samples, it is not yet possible to establish that oroclinal bending has occurred around the southern Andes and the South Shetland arc (Kellog & Reynolds, 1978; Van der Voo & Channell, 1980), although the data suggest that this is the case (Burns, Rickard, Belbin & Chamalaun, 1980).

The Andean margin of South America

Subduction appears to have begun in the zone between North and South America, in late Mesozoic or early Cenozoic times, some 100 Ma after their initial separation (Malfait & Dinkelman, 1972). The Greater Antilles, from Cuba to Puerto Rico, developed subsequently in an island-arc setting, linked to SW-directed subduction. After Miocene times, the contemporary Caribbean plate formed by a decoupling from the Pacific plates at the site of the Middle American trench and island arc. The new plate changed its direction of movement to easterly, with respect to North America, to produce the Lesser Antilles island arc (Malfait & Dinkelman, 1972). It is to be anticipated that terranes now lying at the opposite margins of the Caribbean plate will have been deformed by sinistral shear motions in the north, and dextral shear motions in the south. Discordant palaeomagnetic directions have been reported from Cretaceous and Cenozoic rocks in the northernmost Andes of Colombia and Venezuela (MacDonald & Opdyke, 1972; Skerlec & Hargraves, 1980; Stearns, Mauk & Van der Voo, 1982). The declinations are roughly E–W in this area, and are consistent with a clockwise rotation of $ca.$ 90° between the east-moving Caribbean plate and the west-moving South American plate. Westerly declinations along the northern margin of the Caribbean plate in the Greater Antilles (e.g. Vincenz & Dasgupta, 1978) are compatible with anticlockwise rotation along this northern margin of the plate. It is possible that these rotated terranes in the nothernmost Andes and the Greater Antilles are the separated segments of a former single arc, which were rotated apart during formation of the Caribbean plate (Skerlec & Hargraves, 1980).

Large-scale strike-slip motions within the Andean margin are not recognised south of Ecuador, where the strike of the margin is nearly normal to the subduction direction of the Nazca plate. In Peru, the Precambrian basement extends to the coast. Geological data suggest that this region has been essentially continuous with the Brazilian Shield since early Devonian, and probably since Ordovician, times; some palaeomagnetic data from the Arequipa massif on this margin support the correlation (Shackleton, Ries, Coward & Cobbold, 1979). However, magnetic declinations from Cretaceous rocks in coastal Peru yield poles incompatible with the existing APWP for South America (Fig. 11.3a). These declinations have been interpreted in terms of anticlockwise rotation of this terrane by 20°–40° in Cenozoic times (Heki, Hamano, Kinoshita, Taira & Kono, 1984). These authors identify the anomaly as far south as the bend in the Andean orogenic trend in northernmost Chile (Palmer, Hayatsu & MacDonald, 1980) and consider it to be a consequence of the oroclinal bending of the Andes, as proposed by Carey (1958). This interpretation is supported by the identification of an apparent clockwise rotation in early Cretaceous sediments from further south along the Andean belt, in Chile (Turner, Clemmey & Flint, 1984). Alternatively, although these poles do not correspond with Cretaceous poles from cratonic South America, they plot along the *ca.* 150 Ma extension of the Gondwanaland path defined by Australian data in Fig. 11.3b, and cannot be considered as unambiguous evidence for oroclinal rotation until the Cretaceous APWP is completely defined.

Mexico

The Mexican region is separated from the North American plate by a tectonic discontinuity (the Texas lineament) which has a long history dating from mid-Proterozoic rifting and aulacogen formation. The SW margin is the destructive plate boundary of the Cocos plate and the origin of extensive Cenozoic volcanics, which now cover much of this sector. Of the several small Precambrian nucleii in the south of Mexico, the Oaxacen terrane is correlated with the Grenville province, but is overlain by Cambro-Ordovician rocks containing a fauna with Gondwanaland affinities (Whittington & Hughes 1974), and a link with the Grenville-age basement at the NW margin of South America is more likely (Fig. 8.4). Geological evidence suggests that other terranes in south Mexico are allochthonous, and welded to this area during the development of Pangaea in Devonian times (concluding part of Section 10.3).

The bulk of Mexico has been regarded as a single plate, which is usually positioned close to the cratonic margin of SW North America in conventional reconstructions (Scotese, Bambach, Barton, Van der Voo & Ziegler, 1979), but the shape of this unit has been modified by a succession of movements (mainly sinistral) on a set of major NNW–ESE trending (arcuate) faults. The region should therefore, be considered as a collage of tectonic blocks (Urrutia-Fucugauchi, 1983). Palaeomagnetic results from this region first start to diverge from the APWP for North America in early Cenozoic times, when they are systematically rotated by *ca.* 20° anticlockwise, and indicate an anticlockwise movement of northern Mexico, and/or sinistral movements on the major faults. A somewhat larger rotation is indicated by a *ca.* 50° rotation of the Trans-Mexican volcanic belt in the south (Urrutia-Fucagauchi, 1981). This early Cenozoic motion seems to have been associated with development of the Caribbean plate and the dislocation of the Mesozoic batholith belt within Mexico; it may also have resulted in the bending of the Sierra Madré Oriental orogenic belt.

Cretaceous results from the same region show no systematic deviation from the contemporaneous field directions of North America (Urrutia-Fucugauchi, 1982) and suggest that components of pre-Cenozoic clockwise rotation have been largely cancelled out by Cenozoic anticlockwise rotations. The Jurassic poles, too, appear to show no significant divergence from the North American APWP, and suggest that this block moved in unison with North America, during initial opening of the North Atlantic Ocean (Urrutia-Fucugauchi, 1984). However, this event was associated with major sinistral movements along some of the major faults in Jurassic times, these movements have not yet proved to be palaeomagnetically detectable. Lower Jurassic and older results from Mexico are not amenable to simple interpretation; some may be later remagnetisations, while others have distributions compatible with local rotations of small crustal blocks. In NE Mexico, for example, progressive anticlockwise rotations by as much as 130° seem to have taken place between Middle Permian and Middle Jurassic times, within small

The Cordilleran margin of North America

Aberrant palaeomagnetic directions were first identified in this area by the pioneer studies of Cox (1957) in Oregon; at that time they were interpreted in a geomagnetic context. Regional studies have since shown that the majority of the results from within a zone extending from Alaska to California, and as much as 600 km inland, deviate by variable amounts from the reference APWP for the North American plate. More than 50 distinct suspect terranes have since been recognised in this region, and it has become an important area for the palaeomagnetic study of such terranes. When considered collectively, the majority of Cordilleran poles plot to the right-hand side of APWP from cratonic North America, in a way which is compatible with variable degrees of clockwise rotation of the sampled areas about local vertical axes. A considerable number of the poles also plot on the far side of the APWP (Fig. 11.6), and indicate that some blocks have been translated northwards relative to the craton, by distances ranging 1000–5000 km (Irving & McGlynn, 1981; Van der Voo & Channell, 1980). Thus, in general, the palaeomagnetic directions have declinations which deviate in a clockwise sense, and inclinations which are shallower, than would be expected if the sampled area had not moved with respect to the adjacent continent. A comparison of the palaeolatitudes calculated from these aberrant results — with the palaeolatitudes of the craton since Carboniferous times — emphasises that rotations (which are not apparent in such a comparison) are the predominant component of movement, while northward translations are apparent in a smaller number of cases (Irving & McGlynn, 1981).

It would be unwise to interpret all data from this region in terms of this tectonic premise, because results from some of the large plutons are possibly due, in part, to local tilting (see for example, Symons, 1973). However, Beck (1980) has argued that local tilting is unable to explain the magnitude of the differences between the remanence directions in large plutons and the reference directions from the adjoining shield. The geotectonic solution is strongly supported by the analyses of Beck (1980) and his co-workers in Washington and Oregon, and by Yole & Irving (1980) in coastal British Columbia. Furthermore, sedimentary and lava sequences, from which the original orientation can be recovered unambiguously, yield results supporting the tectonic arguments. A considerable number of examples, constrained by fold tests and dual polarities, are now described from terranes ranging from Alaska to California (see Beck, 1980 for review). In the coastal segment of Washington and Oregon, west of the Cascade Range, rotations are consistently clockwise and range from $25°-70°$ in rocks varying from 55 to 30 Ma in age; the largest rotations are found in the oldest rocks. Unlike the situation in California, outlined later in this section, this region is not one of widespread contemporary seismicity and faulting. Simpson & Cox (1977) favour a model in which the Coastal Range (Willamette terrane) was rotated and 'docked' adjacent to the craton in Eocene times (55–40 Ma) and then rotated away from the cratonic margin — in a second phase beginning 20 Ma ago — as the crust was stretched and thinned in the Basin and Range Province to the east. The complete explana-

Mesozoic and Cenozoic palaeomagnetism and the breakup of Pangaea 323

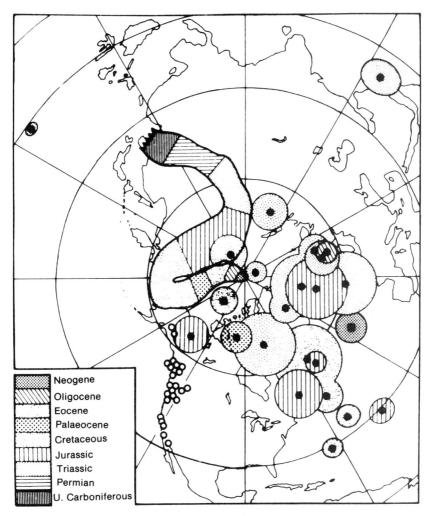

Fig. 11.6 Palaeomagnetic poles from the Cordilleran margin of North America, with their assigned stratigraphic ages, compared with the contemporaneous APWP for the North American plate. The open circles are the locations of the palaeomagnetic studies. From Van der Voo & Channell (1980) after Irving & McGlynn (1981).

tion of these data is more complex: by integrating palaeomagnetic and structural studies in western Washington, Wells & Coe (1985) have recognised two major stages in the tectonic evolution of this region. Mid-Eocene volcanics yield variable declinations, pre-dating Middle–Upper Eocene folding. By subdividing the region into small (<30 km) structural domains, they identify a correlation between the mean directions of the fold axes and the remanence directions. These distributed directions are interpreted in terms of the convergence of the structural divisions with the continental margin in mid-Eocene times. Upper Eocene volcanics have a coherent remanence, rotated clockwise by 23° with respect to the reference Eocene direction from North America. This coherent rotation involves post-Eocene movements along fault planes separating many small crustal blocks. The rotation

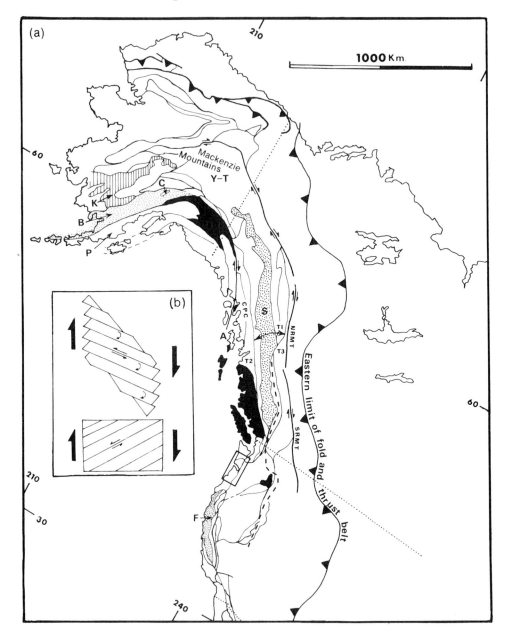

Fig. 11.7 (a) Terrane map of the Cordilleran margin of western North America. The regions in black are the fragments of Wrangellia. Other terranes are lettered: **F**, Franciscan, **S**, Stikine, **Y-T**, Yukon-Tanana, **P**, Peninsula and Chugach, **K**, Kuskokwim, and **C**, Chutlina; **B** is an Upper Mesozoic flysch belt. The zones delineated as **T1**, **T2** and **T3** are terranes 1, 2 and 3 respectively. **NRMT** and **SRMT** are the North and South Rocky Mountain troughs, and **CPC** is the Coastal plutonic complex. Mainly after Jones, Cox, Coney & Beck (1982), Chamberlain & Lambert (1985) and Stone, Panuska & Packer (1982). The western limit of the Laurentian block is most effectively delineated (dashed line) by the transition from $^{87}Sr/^{86}Sr$ values of <0.704 in the west (indicating an underlying crust of

can be integrated with the tectonics of the Pacific plate–North American plate contact, in terms of a model involving sinistral movements along WNW–ESE faults, within a broad zone separating the coastal ranges from the stable continental margin. The latter two elements are continuing to move relative to one another in a dextral sense (Fig. 11.7b).

An elonaged terrane in the Wrangell Mountains of eastern Alaska, yields palaeomagnetic data compatible with an ancient location near present-day California (Hillhouse, 1977). This region has strong geological similarities with a number of isolated terranes, ranging from Alaska in the north to Oregon in the south; Jones, Silberling & Hillhouse (1977) have proposed that they once formed a single crustal block referred to as 'Wrangellia,' which has since been dismembered by Cordilleran tectonic movements (Fig. 11.7). The south-easternmost fragment of Wrangellia, in the Seven Devils region of Oregon, yields primary and secondary remanence directions which have both been rotated with respect to the North American plate. When the primary remanence in Triassic rocks is corrected for the rotation of the (Jurassic–Cretaceous) secondary remanence, it accords with results from the fragment of Wrangellia comprising Vancouver Island, thus supporting the former correlation between the two regions (Hillhouse, Grommé & Vallier, 1982).

Unfortunately, the incomplete definition of the APWP introduces a problem here, common to many suspect terranes which have experienced large rotations, i.e. because these terranes have experienced large rotations (and therefore show large changes in declination) over short intervals of time, it is often unclear whether palaeomagnetic directions representing short periods of geological time are of N or R polarity. In the case of Wrangellia, either a post-Triassic northward movement of *ca.* 3000 km or *ca.* 6000 km would be compatible with the palaeomagnetic data, with accompanying rotations taking place in an anticlockwise or clockwise rotation sense, respectively. Palaeomagnetic investigations in the Alexander terrane of SE coastal Alaska however, have accommodated rocks ranging over a long term period — from Ordovician to Triassic times — and have permitted the definition of a long segment of APWP (Van der Voo, Jones, Grommé, Eberlein & Churkin, 1980). This path exhibits similarities to the APWP from Laurentia; a comparison between the two paths suggests a former location some 20° further south with respect to North America, and adjacent to central California prior to Triassic times.

To the east of the dismembered segments of Wrangellia lies a broad terrane, referred to as 'terrane 1' by Canadian workers (Monger, Price & Tempelman-Kluit, 1982). It was welded to a sliver of cratonic North America in mid-Jurassic times. Subsequently, movements have occurred along a major dislocation some 100–200 km to the east, referred to as the Southern Rocky Mountains trench (Fig. 11.7); Chamberlain & Lambert (1985) define this silver as terrane 2. The palaeomagnetic record of terrane 1 (see for example, Symons, 1973, 1983) is

---

oceanic or island arc origin) to $^{87}Sr/^{86}Sr$ values of >0.706 in the east (which indicate a deep crust of continental origin). (b) Tectonic model incorporating both dextral and sinistral fault movements, which may explain the coherent rotations of some Cordilleran terranes following their 'docking' against Laurentia; see text for explanation.

amenable to several interpretations, but is compatible with a northward translation of this terrane through some 14–20° of latitude and a clockwise rotation of 45°–55° since mid-Cretaceous times (Chamberlain & Lambert, 1985.). This is similar to the inferred motions of terrane 3 (which includes segments of Wrangellia) to the west, and these authors speculate that terranes 1, 2 and 3 have

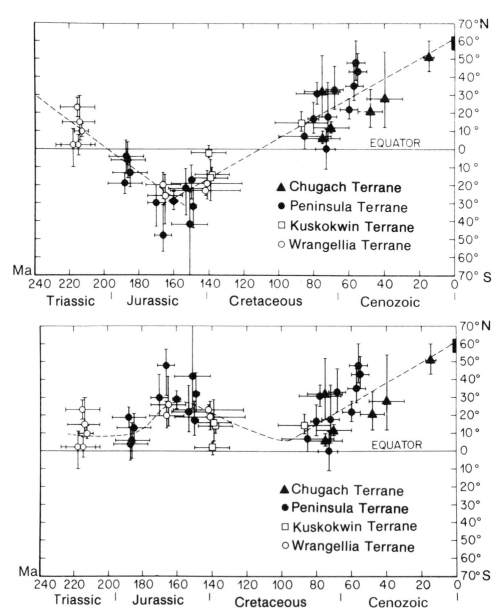

Fig. 11.8 Changes in palaeolatitude of four Alaskan terranes (see Fig. 11.7 for locations) since Triassic times according to two models, using both polarity options for the pre-Cretaceous data; several lines of evidence suggest that model (a) is the most likely one. The dashed line is equivalent to a motion of 6 cm/year. After Stone, Panuska & Packer (1982).

moved in unison, as a single microcontinent, since Cretaceous times. The palaeomagnetically-defined motion may have been accommodated by a combination of oblique subduction in a back-arc basin beneath the NE margin of this block, together with strike-slip motion; final collision with the craton appears to be defined by deformation and magmatism in the Mackenzie Mountains at *ca.* 95 Ma (Fig. 11.7). Clearly, much more study will be necessary to determine the unity, or otherwise, of groups of terranes during their northward motion. Present evidence does not seem to favour a common motion of all the displaced terranes, and it also suggests that movements have been irregular rather than continuous. Study of the Sanpoil terrane at the cratonic margin of NE Washington, for example, illustrates a 25° clockwise rotation of 55–48 Ma Eocene rocks, but no rotation in post-40 Ma rocks (Fox & Beck, 1985). South-west Alaska, at the extremity of the collage of Cordilleran terranes, predictably incorporates the furthest-travelled blocks. Palaeomagnetic data here define a northward movement, beginning south of the equator and probably continuing northwards through most of late Mesozoic and Cenozoic times. By considering the palaeomagnetic data from four terranes between western Alaska and the Wrangell Mountains, Stone, Panuska & Packer (1982) derive the change of palaeolatitude with time (Fig. 11.8). The Cretaceous and younger results almost certainly refer to the northern hemisphere, because a location south of the equator would require a very rapid motion in Cenozoic times. The incomplete time coverage of the older data set however, leaves an uncertainty in the polarity of these results, and two possible graphs are possible. Limited data from rocks formed during the PC Superchron (Section 12.2), together with the derivation of a graph illustrating a progressive northward movement with an average velocity of 6 cm/year, seem to favour the first of the options illustrated in Fig. 11.8b. Note that the movement of these terranes seems to have been southwards between Triassic and Jurassic times, prior to the breakup of Pangaea.

In California, well-constrained palaeomagnetic studies of the Franciscan formation indicate an origin some 20° south of its present location; they support the view that these sediments and volcanics were formed in an oceanic environment on the Farallon plate, and were emplaced by obduction (Alvarez, Kent, Silva, Schweichert and Larson, 1980). Study of the Salinian terrane, on the west side of the San Andreas fault, suggests that it had amalgamated with several other terranes by late Cretaceous times, and has since moved northwards by 2500 km (Champion, Howell & Grommé, 1984). In the vicinity of the San Andreas fault zone, palaeomagnetic studies have identified aberrant declinations, some in excess of 60°, in post-Miocene rocks. Luyendyk, Kamerling & Schopf (1980) interpret these results in terms of variable clockwise rotations linked to dextral movements near the fault. Considerable interest centres on the size of blocks which have rotated as individual entities; some of these are evidently no more than 10–20 km across, and it would appear that the faults bounding them must be confined to the top $\leqslant 15$ km of brittle crust.*

While the combination of faunal and palaeomagnetic evidence shows that many of the Cordilleran terranes were transported on ocean crust, they have clearly

---

* The suspect terranes along this Cordilleran margin range from <10 to >1000 km in size, and is clear that the smaller ones cannot be directly attached to lithosphere. Since, by definition, a microplate must be an integral segment of lithosphere, only some of the larger terranes can be regarded as true plates.

survived subduction, and the paucity of subduction-related effects and the small degree of internal deformation are noteworthy features of this region. The terrane margins are bounded by thrust faults or strike-slip faults. Following the docking against the continental margin, the motions along these faults have greatly telescoped and elongated the original terrane shapes. The distinctive characteristics of this margin are a consequence of the interaction between the plates of the Pacific Ocean basin and the North American continent since Jurassic times; they have probably been concentrated in those intervals when the convergence of the Pacific plates was strongly oblique to the continental margin. An oceanic plate (the Farallon) was adjacent to western North America during the northward translation of the terranes, and since late Mesozoic times this plate appears to have had a component of northward motion relative to the continental margin (Coney, 1978). A mid-ocean ridge (the Farallon–Pacific rise) approached the ancient subduction zone along the margin in late Cenozoic times, and a tranform-type margin incorporating the San Andreas fault zone subsequently developed here. The NE motion and former extent of the Farallon plate are indicated by the number and extent of accretionary terranes between Mexico and Alaska, by the oceanic origin of many of them, and by the fusilinid (foraminiferal) fauna which contrasts with cratonic North America, but corresponds closely with the eastern Tethyan domain (Jones, Cox, Coney & Beck, 1982, see also Fig. 12.10). The APWP of cratonic North America during Mesozoic and Cenozoic times comprises three major segments (Fig. 11.1): a northerly motion to *ca.* 170 Ma was interrupted in Jurassic times, and then resumed (following an anticlockwise rotation) between 150 and 120 Ma. Following a quasi-static interval between *ca.* 130 and 90 Ma, a further rapid motion takes the path in an easterly direction, and is largely accommodated by a clockwise rotation of the North American plate. This rotation has accompanied the opening of the North Atlantic, and seems to have changed the convergence between the Farallon plate and North America from E to NE, and thus facilitated lateral transport. It was during this episode, and particularly during the episode of rapid sea-floor spreading at *ca.* 80 Ma, that most of these terranes seem to have been docked and translated (Sections 12.8 & 12.10).

12
The geomagnetic field, continental movements and configurations, and mantle convection since Archaean times

12.1 Introduction

In Chapters 7, 9 10 and 11, the palaeomagnetic data are used to describe the movements of the pole with respect to the continental crust over geological time. This description provides a measure of the velocities of the continental plates, together with their directions of movement and their orientations. The latter information forms the basis for reconstructing the palaeogeography of the continental crust, and has already been discussed in some detail in the preceding chapters. The velocities and directions of continental motion defined by the APWPs are the response to a variety of tectonic processes. In addition to defining the style of these tectonics over geological time (Chapter 8 and Sections 9.6, 10.3 & 10.7), they help to constrain the possible mechanisms driving the plates. This knowledge can be related to the other facet of the palaeomagnetic data, namely the nature of the geomagnetic field through time. Although this book is concerned with the effects of the geomagnetic field rather than its origin, temporal changes in the field are important, because they are the only direct signature of processes occurring within the core of the Earth. They provide the principal evidence for linking together core, mantle and crustal phenomena.

12.2 The polarity time scale

Cenozoic and Mesozoic times

Although self-reversal is an established and reproducible phenomenon, it only seems to take place in magnetic minerals of rather unusual composition (section 2.3), and a vast number of direct and circumstantial observations shows that geomagnetic field inversions are a characteristic feature of the palaeomagnetic record since the end of Archaean times (Section 6.5 and Chapter 7). The picture to emerge from more than three decades of investigation is of periods of time, punctuated by frequent inversions at irregular intervals, separated from periods when inversions were infrequent. The reversals during the last part of geological time have followed a chronology crucial to the recognition of sea-floor spreading (Vine & Matthews, 1963; Vine, 1966) and to its subsequent integration with continental drift (inferred in part from land-based palaeomagnetic studies) to define the tenets of plate tectonics (Morgan, 1968).

Reversals are not symmetrical phenomena: the time spent in the normal (N) state is not generally equal to the time spent in the reversed (R) state. It is commoner for the field to be strongly biased towards one polarity or the other. The chronology of reversals is referred to as the *polarity time scale*. It comprises intervals referred to as *subchrons*** (events), typically of the order of 10^5 or 10^6 years duration, of N or R polarity. Viewed over longer time intervals ($10^6 - 10^7$ years), the polarity subchrons group into *chrons* (epochs) in which one polarity or the other predominates. Over long time periods ($10^7 - 10^8$ years) the polarity time scale comprises much longer intervals, biased towards one or other polarity. Periods of very strong bias, when there were few or no inversions, are described as 'quiet' intervals, while times of weak bias and frequent reversals are described as 'disturbed' intervals. Quiet intervals are equivalent to time periods known as *superchrons* (sometimes formerly referred to as hyperzones), while disturbed intervals comprise a sequence of many short N and R subchrons. Successive quiet and disturbed intervals usually have the same N or R polarity bias, and groups with the same bias are referred to as *bias intervals* (Irving & Pullaiah, 1976). The equivalent rock divisions, or segments of ocean-floor, mapped on the basis of their polarity, are defined as quiet, disturbed and biased zones.

---

* The nomenclature of the time periods identified by their magnetic polarity, has been defined by the International Subcommission on Stratigraphic Classification (ISSC). They recommend (Anon., 1979) that 'chron' should replace the word 'epoch' for a period of one predominant polarity, so that the Brunhes epoch becomes the Brunhes chron. An event such as the Jaramillo then becomes a subchron. The equivalent stratigraphic divisions are defined as a chronozone and a subchronozone respectively.

The chronology of reversals over the last 4 Ma was established during the 1960s, from joint palaeomagnetic and K-Ar studies of lava flows. These early investigations identified four intervals of polarity bias named the Brunhes, Matuyama, Gauss and Gilbert chrons. The time scale to emerge from these studies is shown in Fig. 12.1 from a recent assessment by Mankinen & Dalrymple (1979). It does not differ substantially from the classic chronology of geomagnetic reversals published by Cox in 1969, and the main uncertainties are the reality or otherwise of the short subchrons (Laschamp & Blake) within the Brunhes chron as global rather than regional events, and the existence of a short N polarity ('X') event in the Matuyama chron, which has been recognised as a signal in marine magnetic anomaly profiles (Hiertzler, Dickson, Herron, Pitman & Le Pichon, 1968), but has not so far been identified on land. For a recent discussion of this time scale, refer to Jacobs (1984).

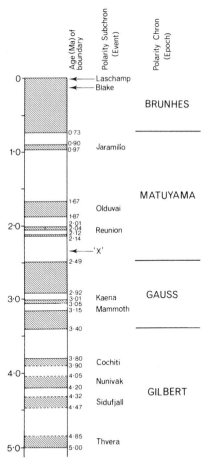

Fig. 12.1 *The polarity time scale for the last 5 Ma of geological time. This is based primarily on K-Ar and palaeomagnetic investigations of subaerial volcanics, but includes information from marine magnetic profiles and deep-sea sediment cores. Shaded intervals are of normal polarity and blank intervals are of reversed polarity. After Mankinen & Dalrymple (1979).*

The error on K–Ar age determinations older than 4–5 Ma (which is typically between 2 and 5%), becomes greater than the length of the shorter subchrons, and it is no longer possible to construct a geomagnetic time scale from the straightforward procedures employed to derive the scale of Fig. 12.1. Greater precision may be obtained from a number of K–Ar determinations on lavas near a magnetic boundary. By using thick and continuous sequences of lavas in western Iceland, McDougall, Saemundsson, Johannesson, Watkins & Kristjansson (1977) have extended the polarity time scale back to 7 Ma. Comparable studies of the lava pile in eastern Iceland by McDougall, Watkins & Kristjansson (1976) and McDougall, Watkins, Walker & Kristjansson (1976) have extended the scale back to 12 Ma. Both sequences can be correlated with the time scale inferred from marine magnetic anomalies; short events may however, have been missed because they did not coincide with the eruption of a lava. In addition, the stratigraphic thicknesses of subchrons are not reliable indicators of their duration, because the stratigraphic sections change in thickness along the strike in a way which is related to the locations of the volcanic centres (Gibson & Piper, 1972). The oldest part of the lava succession in Iceland is *ca.* 16 Ma in age and there seems to be little prospect of extending the polarity time scale beyond this by combined K–Ar and palaeomagnetic studies of lava successions on land, except for isolated intervals of time. Over approximately the same interval as that represented by the lava pile in Iceland, the polarity time scale can be reinforced by the study of deep-sea sediment piston cores. Sedimentation rates in the abyssal plains have been just sufficient for closely spaced sampling of these cores to discern most of the subchrons. Using sedimentation rates deduced from the known polarity time scale covering the last 4 Ma, Harrison, McDougall & Watkins (1976) have employed these cores to independently produce a geomagnetic time scale back to 13 Ma. This scale is in close agreement with the time scale inferred from marine magnetic anomalies. The sedimentation rates have been very uniform here (this assumption can be checked by correlating with nearby cores and with the alternative polarity time scale), this approach has the advantage that the thickness of each subchron is proportional to its duration.

From 5 Ma ago, back to the age of the oldest surviving ocean crust of Upper Jurassic (Oxfordian) age, marine magnetic anomalies have provided the most important record of the geomagnetic polarity time scale. Complexities are introduced into the direct interpretation of these anomalies by topographic irregularities, faulting, and the occasional shifts of the ridge axis to new locations. In addition, the rate of growth of the ocean crust changes from time to time, and the crust beneath the anomalies must be calibrated from the biostratigraphy of sediment cores and radiometric dating of sea-floor basalts. However, even with these complications, the nature of sea-floor spreading is such that it provides a remarkably continuous and high-fidelity record of magnetic field polarity. A single profile from the South Atlantic was used by Hiertzler, Dickson, Herron, Pitman & Le Pichon (1968) to derive the first polarity time scale from this record. They assumed a constant spreading rate over the last 80 Ma, and identified the polarity chrons from the most prominent magnetic anomalies. The chrons were numbered in sequence outwards from the mid-ocean ridge, so that the positive anomaly over the axial zone of present-day accretion is number 1. A subsequent revision of this scale by LaBreque, Kent & Cande (1977), used two calibration points applied to a composite profile, namely an age of 3·32 Ma for the Gilbert–Gauss chron boundary, and an age of 64·9 Ma for the beginning of anomaly 29, close to the Cretaceous–Palaeocene boundary. Ness, Levi & Couch (1980) revised the polarity time scale by incorporating the geological time scale with the absolute ages recalculated to the revised constants (Steiger & Jäger, 1977); they then employed four calibration points of 3·40 Ma for the Gilbert–Gauss boundary, 10·30 Ma and 54·90 Ma for the older boundaries of anomalies 5 and 24 respectively, and 66·70 Ma for the Cretaceous–Palaeocene boundary just prior to anomaly 29. Lowrie & Alvarez (1981) integrated the record from the marine magnetic anomalies with magnetostratigraphic sections through the Gubbio pelagic limestone succession in Italy, which is palaeontologically dated by an abundant microfossil fauna and crosses the Cretaceous–Cenozoic boundary without a stratigraphic break. These workers were able to link the magnetic boundaries to the appearances and extinctions of planktonic foraminifera and coccoliths. They correlated the polarity zones with the marine magnetic anomaly sequence of LaBreque, Kent & Cande (1977), and located the Eocene–Palaeocene boundary within the negative polarity zone, just younger than anomaly 25, with the Cretaceous–Cenozoic boundary between anomalies 29 and 30. The time scale of Lowrie & Alvarez has been incorporated, with slight modification, into the polarity time scale of Harland, Cox, Llewellyn, Picton, Smith & Walters (1982 and Fig. 12.2).

Anomaly 24 is the oldest magnetic anomaly identified over the oceanic crust in the North Atlantic, and is overlain by sediments with an age close to the Eocene–Palaeocene boundary (Sclater, Jarrard, McGowran & Gartner, 1974). Furthermore, sediments spanning this boundary are interbedded with the early Cenozoic plateau lava sequence in East Greenland, where the lavas are dated at 53–48 Ma by the K–Ar method (Tarling & Mitchell, 1976). These lavas are entirely of R polarity, and are therefore likely to have been erupted during a prolonged reversal, such as that preceding anomaly 24 (Faller, 1975; Tarling, 1983). On the revised time scale this anomaly is regarded as *ca.* 53·5 Ma in age, which is some 8 Ma younger than the original estimate of Heirtzler, Dickson, Herron, Pitman & Le Pichon (1968).

It is probable that all the chrons have now been recognised in the record back to Upper Jurassic times. There remains some scope for the interpretion of small-scale anomalies in the marine magnetic profiles (with wavelengths of *ca.* 10 km or less, and amplitudes of <50 nT.) in terms of more subchrons, but pending more magnetostratigraphic studies on land, most workers have erred on the side of caution, and have preferred to regard these features as the consequences of intensity fluctuations in the dipole source. The use of the chron numbers derived from marine magnetic anomalies has now achieved international acceptance, and they are more useful than chronozones identified from magnetostratigraphic sections on land, for two reasons: firstly, due to variations in the rates of deposition, the thicknesses of the chronozones vary from place to place, even between sections through potentially ideal recorders such as marine pelagic limestones (Lowrie & Alvarez, 1981). Although episodic on a scale of *ca.* 10^2–10^3 years, the addition of igneous material to the ocean crust is very regular on the 10^5–10^6 year time scale of the subchrons, and contrasts with the episodic growth of volcanic provinces in continental environments. Secondly sedimentary successions are liable to be influenced by pauses in deposition, which cause the total magnetic record to be incomplete.

Global magnetic anomalies are absent over the ocean-floor formed between Aptian and Santonian times (*ca.* 118–83 Ma), and this is interpreted as a record of the Cretaceous Normal (CN) quiet interval. It was first identified in Japan by Sasajima & Shimada (1964). Subsequently, Helsley & Steiner (1969) recognised this superchron in the North American record, and Irving & Couillard (1973) synthesised land-based observations to estimate that it spanned the interval 110–83 Ma. This was in close agreement with the marine-based estimate of Larson & Pitman (1972) and it has been little affected by more recent studies (Fig. 12.2).

The preceding interval between Upper Jurassic and Lower Cretaceous (Oxfordian–Barremian) times was characterised by frequent inversions of the geomagnetic field. The evidence from the marine magnetic record is summarised by Cande, Larson & LaBreque (1978) and Larson & Hilde (1975). The end of this disturbed interval is identified as the beginning of Aptian times, from the magnetostratigraphic studies of Lowrie, Channell & Alvarez (1980) in the Apennines, although an age a few millions of years younger cannot be excluded from the DSDP results (Larson, Golovchenko & Pitman, 1981). The chrons in this interval are numbered M0 to M29 by Harland, Cox, Llewellyn, Pickton, Smith & Walters (1982), following the marine anomalies, and they are assigned to the negative anomalies (This contrasts with the numbering of normal anomalies in the Cenozoic–Upper Cretaceous interval in the order 1 to 33, (see Fig. 12.2)). Since the ocean-floor beneath these old anomalies has not yet been drilled, their age cannot be directly determined, and the dating of this part of the scale is therefore less precise.

The notable feature of the oldest part of the marine record is the absence of magnetic anomalies in Middle Jurassic times. This has been identified as a Middle Jurassic quiet zone of N polarity (Larson & Helsley, 1975). However, paleomagnetic studies of Callovian sedimentary formations in North America suggest that this was a period of predominantly R polarity (Steiner, 1978), in accordance with results from Upper Callovian sediments of the USSR (Khramov, 1973). Since ocean-floor of this age is currently close to being subducted into contemporary subduction zones, this conflicting evidence might be explained if the ocean crust has been overprinted by a Recent field component as a result of low-grade thermal metamorphism. A magnetostratigraphic study of the Lower–Middle Jurassic limestones of the Breggia gorge in Italy (Horner & Heller, 1983) has indicated that the geomagnetic field inverted frequently (*ca.* 5 times/Ma) during parts of this interval. These data are supported by results from Hungary and the Umbrian section, and provide the basis for extending the polarity time scale in Fig. 12.2 back to the middle of Pliensbachian times at 198 Ma. Magnetostratigraphic studies of the Umbrian limestones, by Channell, Lowrie, Pialli & Venturi (1984) identify two intervals of mixed polarity in Jurassic times, separated by a quiet interval which was probably restricted to the Callovian and Oxfordian stages crossing the Middle–Upper Jurassic boundary. These authors refer the youngest disturbed interval to the oldest part of the M-sequence observed in the ocean basins, with the quiet interval equivalent to the oldest preserved segments of the ocean crust, where anomalies are either absent or poorly defined.

Not enough is known about the magnetostratigraphy of Triassic times to

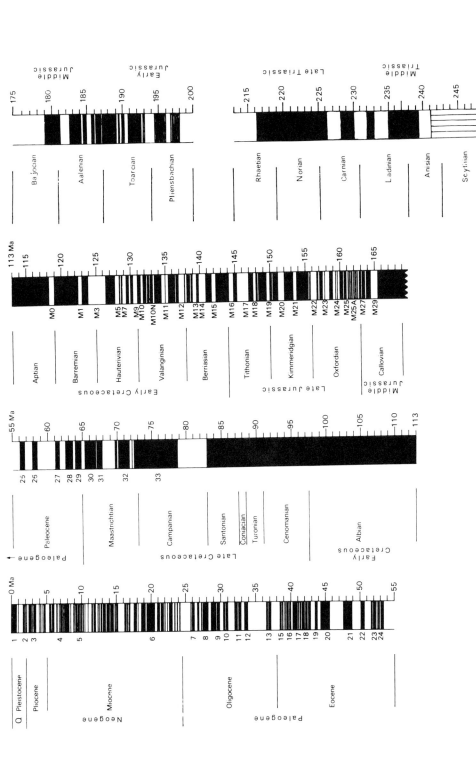

Fig. 12.2 The geomagnetic polarity time scale for Cenozoic and Mesozoic times. The record back to Upper Jurassic times is based on a combination of land and marine studies, as described in the text. The Lower–Middle Jurassic section is based on magnetostratigraphic studies in the Breggia gorge (Horner & Heller, 1983) and the Triassic section is based on the assessment of McElhinny & Burek (1971). White intervals have R polarity and black intervals have N polarity; vertical lines indicate frequent reversals of unknown number. Reproduced in part from Harland, Cox, Llewellyn, Pickton, Smith & Walters (1982).

extend the record of Fig. 12.2 back beyond 200 Ma. This is because these studies have focused on the non-marine red sediments which are characteristic of this period. The faunal content is typically inadequate for the purposes of precisely dating and correlating the successions. Furthermore, it is difficult to discriminate the PDDRM record in these rocks from the CRM one (Section 4.2). An important early study was made by Picard (1964) on the Chugwater formation of Wyoming, and it was followed by the work of Helsley (1969) on the Moenkopi formation of Colorado; the latter studies were subsequently extended by Helsley (1969), Steiner & Helsley (1974) and Elston & Purucker (1979). Frequent reversals were noted in the Lower Triassic, while the Upper Triassic was found to be predominantly of N polarity, with a single R subchron near the Triassic–Jurassic boundary. The origin of the remanence in these rocks has been questioned by Larson & Walker (e.g. 1985) and it is still uncertain to what degree they record the history of the Triassic geomagnetic field. However, the lateral persistence of the magnetostratigraphy, and the record of polarity transitions strongly suggest that the diagenetic remanence here was imposed very shortly after deposition (Herero-Bereva & Helsley, 1983, and the second part of Section 4.2). Comparable studies of European Triassic rocks were made by Burek (1964, 1970) who also identified a large number of reversals in the Lower Triassic Bunter sandstones. Unfortunately, in the absence of a fossil record, there is no way of correlating the European and North American records.

Palaeozoic magnetostratigraphy

The most comprehensive studies of Palaeozoic magnetostratigraphy have been made in the USSR. Khramov & Rodionov (1980) have synthesised many sedimentary palaeomagnetic studies, linked to type sections and palaeontologically-dated sections, to produce a composite polarity time scale for this era. This cannot, of course, be accepted uncritically, because there is a suspicion that these successions in part, may have been remagnetised in later times (Sections 10.4, 10.5). However, it forms a basis for discussing the polarity history of Palaeozoic times, and is reproduced in Fig. 12.3, with modifications which have emerged from studies elsewhere in the world.

The Cambrian–Silurian record is compiled mainly from stratigraphic sections of the Siberian plate (Khramov, Rodionov & Komissarova, 1965). Kirschvink & Rozanov (1984) have correlated the magnetostratigraphy of the late Precambrian–Lower Cambrian section in the Lena River with archaeocyathid zones (see page 249), to produce a detailed magnetostratigraphy for the Tommotian–to Lenian interval. In addition, Gillett & Van Alstine (1979) studied the Lower–Middle Cambrian succession of the Desert Range in Nevada, linked to trilobite zones, and recognised a sequence of zones of probable reversed and mixed polarity.

Lower Ordovician times (Tremadocian–Arenig) are regarded by Khramov & Rodinov as an interval of R polarity, although a higher frequency of reversals is implied by global data (Section 12.4). The Llanvirn epoch includes at least two N → R inversions (Briden & Morris, 1973; Piper, 1978; Briden & Mullan, 1984). The Borrowdale volcanic series of northern England may span much of the Lower Llandeilo–early Caradoc interval and is entirely of N polarity (Author, in preparation). Both polarities are represented in the succeeding Caradocian epoch (Fig.

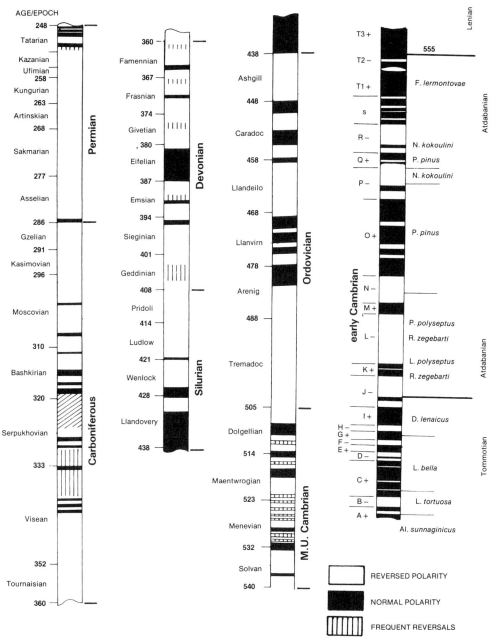

Fig. 12.3 A preliminary assessment of the polarity time scale in Palaeozoic times, based mainly on Khramov & Rodinov (1980), with additions noted in the text. The Lower Cambrian scale linked to the indicated archaeocyathid zones is based on the work of Kirschvink & Rozanov (1984). The evidence at ca. 330–320 Ma (inclined symbol) is conflicting. Note that the interpretation of pre-Lower Silurian intervals as N or R is uncertain and depends on interpretation of Cambrian and Ordovician APWPs (Sections 9.4, 10.1 and 10.3). The Westphalian (p. 338) incorporates the latter part of the Bashkirian and the Moscovian epochs.

12.3), but probably they were relatively long, because they are represented by single polarities in entire volcanic provinces of the British Caledonides (Thomas & Briden, 1976; Piper, McCook, Watkins, Brown & Morris, 1978). The Russian platform successions show a N → R inversion at the Llandovery to Wenlock boundary with the remainder of Silurian times being a period of R polarity. The real polarity history was certainly more complex than this because a R → N inversion is reported from the Llandovery–Wenlock boundary (Piper, 1975d) while an inversion is present in Middle–Upper Silurian limestones in Indiana (McCabe, Van der Voo, Wilkinson & Devaney, 1985) and both polarities are recognised in British Upper Silurian data (Fig. 10.2b). The Devonian period has a complex history of magnetic inversions (e.g. Sallomy & Piper, 1973) which is however, poorly understood, because the magnetic record is frequently a composite one residing in post-orogenic molasse-facies red sediments. The Middle Devonian–Lower Carboniferous interval appears to have been uniformly reversed; mixed polarities are first recorded in early Viséan times, and seem to have been about equally balanced between the mid-Viséan and the Viséan–Namurian boundary (Turner & Tarling, 1975).

Graham (1955) was the first to note that most Permian rocks from the U.S.A. have R polarity. A long interval of constant polarity was subsequently shown to extend back in time to the Lower Carboniferous (Mississipian)–Upper Carboniferous (Pennsylvanian) boundary. It was formerly referred to as the Kiaman interval, after the type section in South Australia, but it is now more generally known as the Permo-Carboniferous Reversed (PCR) superchron. Irving & Parry (1963) established possible lower and upper limits to the PCR in Upper Carboniferous (Westphalian) and Upper Permian sediments of the Sydney basin. Correlation with worldwide data suggested a base to the PCR within the Westphalian, or between the Westphalian and Stephanian epochs, and a top between Lower and Upper Tartarian times. However, the youngest preceding N zones appear to be within the Namurian (Roy, 1969; Roy & Morris, 1983) which, according to Khramov & Rodionov (1980) is a period of frequent inversions. It seems probable that this long period of R polarity (320–250 Ma) may have been interrupted by one or more short events, because Thompson & Mitchell (1972) identify N-polarity basalts at La Colina in Argentina, dated at 295 Ma; they may be the same age as the Paterson toscanite of late Westphalian age, which was originally used by Irving & Parry (1963) to define the base of the PCR. Helsley (1965) found a single N polarity zone in the Lower Permian Dunkard series of West Virginia, and McMahon & Strangway (1968) reported several short, normal events in the Minturn and Maroon formations of Colorado. Re-study of this latter section by Miller & Opdyke (1985) has shown that the N zones, used to place the base of the PCR in Upper Carboniferous times, are later overprints, and there is now a general concensus that the boundary lies close to the Namurian–Westphalian boundary (Irving & Pullaiah, 1976; Roy, Tanczyk & Lapointe, 1983). The PCR is the longest period of constant polarity in the Phanerozoic era, and is now well established as a global event by studies in Australia (Irving & Parry, 1963; Irving, 1966; McElhinny, 1973), North America (Roy & Morris, 1983), Russia (Khramov & Rodionov, 1980), South America (Thompson & Mitchell, 1972) and several parts of Europe (McElhinny, 1973).

12.3 The geomagnetic polarity history from the palaeomagnetic record

While the polarity history of the geomagnetic field is best recovered from detailed magnetostratigraphic studies of individual sedimentary and volcanic successions, this evidence is incomplete and often difficult to interpret. To understand the long-term behaviour of the geomagnetic field throughout geological times, it is necessary to resort to a statistical study of the palaeomagnetic data. By reference to the APWP, the palaeomagnetic poles from each plate can be classified either as having N, R, or mixed polarity. The ratio of N to R directions, from the collective data for each interval of geological time, gives a statistical indication of the time spent by the geomagnetic field in each polarity state, while the frequency with which dual polarities are observed in the palaeomagnetic study of a rock formation provides an indication of frequency of inversions. An analysis of this kind was performed by McElhinny (1971, 1973) on the world palaeomagnetic data up to 1972, and by Irving & Pullaiah (1976) on the data to 1975. The analysis is repeated here (Fig. 12.4), on the data base up to 1985, but several older results, which have not been substantiated by later studies, are excluded. Those data (mainly Palaeozoic, see Section 10.4) which are believed to have been remagnetised in later times are also excluded. The new analysis has only been conducted on the

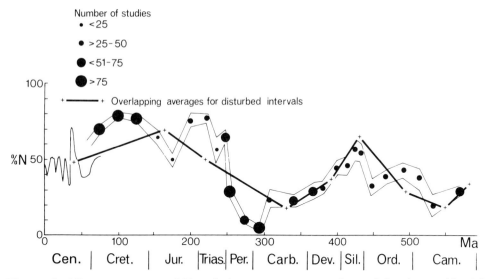

Fig. 12.4 *The percentages of N polarity measurements observed in the world wide palaeomagnetic record for Phanerozoic times. The number of observations used for the averages of the Upper, Middle and Lower divisions of each geological period are indicated by the size of the point. The post-Jurassic results, and the connected crosses, showing the overlapping averages of the polarity ratios for the disturbed intervals, are from Irving & Pullaiah (1976), and cover data up to 1975. Note that this graph shows the same cyclic change as the polarity bias. The line curve illustrating the fraction of time spent in each polarity state, averaged over 5 Ma intervals, and covering the last 75 Ma, is taken from McFadden & Merrill (1984.).*

pre-Cretaceous record, because the geomagnetic field behaviour is clearly much better understood from the oceanic magnetic record between 0 and 180 Ma.

The updated study yields similar conclusions to the analysis of McElhinny (1973). The Mesozoic is seen to be a period of predominantly normal polarity, and the frequency of inversions is highest in Middle Jurassic and Upper Triassic times, as inferred from the magnetostratigraphic record (Fig. 12.2). The Palaeozoic, in contrast, is seen to be a period of predominantly R polarity, with a low frequency of inversions during the PCR superchron, and a high frequency of inversions during the Upper Cambrian–Lower Ordovician and Lower–Middle Devonian intervals, suggesting that these were characterised by high inversion rates. The long-term cyclic variation in polarity noted by McElhinny (1971) and Irving & Pullaiah (1976) is evident in this assessment, and the even balance of N and R polarities (which has applied over the later part of Cenozoic times) is seen to be the exception through geological times. It only occurred during Permo-Triassic times at *ca.* 220 Ma, and in Silurian times at *ca.* 420 Ma. Furthermore, the long-term cyclic changes in polarity have not been caused by quiet intervals superimposed on a background of disturbed intervals, because the bias in the disturbed intervals shows the same cyclic change (Irving & Pullaiah, 1976 and Fig. 12.4). The frequency with which inversions are observed in palaeomagnetic studies is roughly proportional to the polarity bias during the time period, and it is evident that reversal frequency is least when the polarity bias is strongest (Fig. 12.5).

The quiet intervals of constant bias range from 30 to 70 Ma in length during Phanerozoic times, and are therefore nearly an order of magnitude longer than the chrons. These *superchrons* have sometimes been named after type localities (Khramov, 1967; Khramov & Rodionov, 1980) or distinguished scientists (McElhinny & Burek, 1971), but it seems preferable to adopt the system of Irving & Pullaiah (1976) and simply refer them to the time interval concerned, because the age assignment is then immediately apparent. On this basis, Harland, Cox, Llewellyn, Pickton, Smith & Walters (1982) identify six superchrons in the last 350 Ma, namely the Cretaceous-Tertiary-Quaternary (CTQ) mixed, the Cretaceous

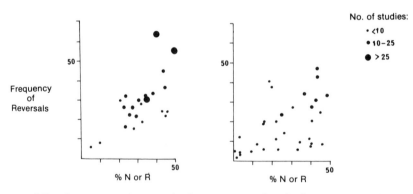

Fig. 12.5 The frequency of reversals during intervals of Phanerozoic and Precambrian times, plotted against the fraction of N or R polarity (whichever is <50%) within the time interval. These graphs illustrate the tendency for reversals to be less frequent when the polarity is heavily biased towards one polarity or the other.

Normal (CN), the Jurassic-Cretaceous (JC) mixed, the Permo-Triassic (PT) mixed, the Permo-Carboniferous (PC) reversed and the Carboniferous mixed superchrons.

Crain, Crain & Plaut (1969) performed a Fourier spectral analysis on the geomagnetic polarity bias, using an early data compilation by Simpson (1966) and found a sharp peak at 300 Ma and a smaller peak at 75 Ma. Maximum entropy spectral analysis is a more appropriate method for treating these data, and has been applied by Ulrych (1972) to the compilation of McElhinny (1971) to suggest that there are periodicities of 700 ± 100 and 250 ± 50 Ma in the data. Irving & Pullaiah (1976) also carried out a maximum entropy spectral analysis on their polarity bias curve, and identified three major components with mean periodicities of 297 ± 34, 113 ± 5 and 57 ± 1 Ma; they were unable to reproduce the 700 Ma periodicity of Ulrych (1972), and were sceptical of periods longer than the 590 Ma length of the data set. Analysis of the geomagnetic field as a Walsh function (a series of pulsed telegraph-like signals), using the data set of McElhinny (1971) also produces a comparable spectral analysis, with a major cyclicity of 285 Ma (Negi & Tiwari, 1983). Thus the *ca.* 300 Ma periodicity apparent from Fig. 12.4 has been evident from the expanding palaeomagnetic data base since the 1960s.

From a statistical study of the youngest and best-defined part of the polarity time scale (< 185 Ma), McFadden & Merrill (1984) showed that the frequency of inversions decreases in an approximately linear manner back to 165 Ma, until it reaches zero in the CN superchron. This implies that the cause of instabilities leading to inversions (and presumably located at the core–mantle boundary), gradually decreased to zero. After *ca.* 86 Ma the inversion frequency increased linearly with time; it reached a maximum *ca.* 10 Ma ago, and appears to be declining again at present. Since the stabilities of the N and R field states are not statistically different, there seems to be no *a-priori* reason why this 300 Ma cycle should build up to one polarity rather than the other. The possible control mechanisms include changes in the topography of the core–mantle boundary (Hide, 1967) or changes in electrical conductivity here. Since bumps are probably maintained by some form of convection, and electrical conductivity is exponentially dependent on temperature, both of these possible causes are temperature-dependent. This is the most important conclusion with geodynamic implications to come from analysis of long-term geomagnetic field behaviour. It requires that there be a long-term cyclic change in the temperature at the base of the mantle. This in turn, suggests that heat transfer from the lowermost mantle is an episodic process, and not a monotonic one, as has often been assumed in the past, although the temperature changes might be the result of episodic mantle convection rather than the cause of it. Some possible implications of this result are examined in Section 12.8 & 12.9.

12.4 The polarity history of Precambrian times

Hitherto it has been considered that too little was known about Precambrian APWPs to allow conclusions about the polarity time scale of these times. However, the recognition of a unified APWP, and its definition from the collective palaeomagnetic data (Chapter 7), provides a concise outline description of the field since late Archaean times. Taking the data in the figures of Chapter 7 at their face

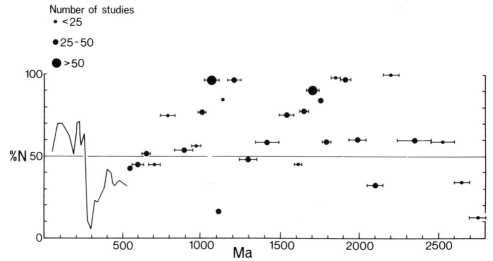

Fig. 12.6 The bias of the geomagnetic field polarity in Precambrian times, determined from the percentages of palaeomagnetic results of normal polarity falling within segments of the APWP (Chapters 7 & 9) defined between key anchor points. The size of the sample during each interval is shown by the symbol, and the width of the bar indicates the time interval considered.

value and subdividing the APWP into segments between well-dated anchor points, yields an assessment of the polarity history, illustrated in Fig. 12.6. The uncertainties in this assessment should be stressed again (see Sections 7.2 & 7.9): the quality of the date is very variable, and the positioning of those poles with broad age assignments may be ambiguous. Nevertheless, the tendency for polarities of poles from different shields to accord with one another has already been noted, and a polarity record emerges which shows that the geomagnetic field was strongly biased towards one polarity through most of Precambrian times. The frequency of reversals was also correspondingly lower during most of this interval (Fig. 12.7). Some intervals characterised by rapid reversals have been found in Precambrian times (e.g. Piper, 1977), but these appear to have been exceptional,

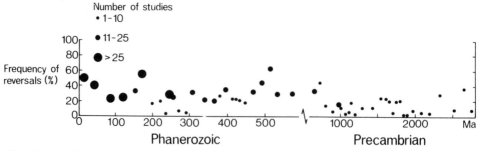

Fig. 12.7 The reversal frequency through geological time, as indicated by the percentages of palaeomagnetic studies showing reversals of magnetisation. In Phanerozoic times the data are considered within the subdivisions of the periods, and in Precambrian times they are subdivided in the same time intervals as those used in Fig. 12.6.

and some very long periods of *ca.* 100 Ma or more (see Figs. 10.4 & 10.5) were dominated by one polarity. Following the arguments of McFadden & Merrill (1984) for the CN superchron (see also Section 12.9) this geomagnetic field signature may have resulted from higher temperatures at the core–mantle interface during these times.

12.5 The strength of the geomagnetic field

It is much more difficult to recover the magnitude (palaeo-intensity) of the ancient geomagnetic field than its direction; the practical investigations are fraught with difficulties and are invalidated if appreciable remagnetisation has taken place during either geological time or laboratory treatment. Hence, the way in which field magnitude has changed through geological time is poorly understood. Archaeomagnetic studies have shown that the strength of the dipole source can fluctuate by 2–3 times over periods of the order of 10^3 years (Games, 1980), so that determinations of field intensity in the past are of very little geological use, unless they represent a time average over an interval comparable to that required to average secular variation. Most data relating to the intensity of the pre-Cenozoic field were acquired in the 1960s, using assumptions that are now regarded as unsatisfactory. Whilst little importance can be attached to individual determinations, they do give a general indication of gross changes in the magnitude of the field source (Fig. 12.8). Palaeo-intensity studies have usually been reported in terms of the equivalent intensity at the ancient equator. In Fig. 12.8 this information is converted into the equivalent Virtual Dipole Moment (VDM) which is the moment of a geocentric dipole source, required to produce the observed intensity at the sampling site, when the palaeolatitude has already been determined from directional measurements (Smith, 1967). Most of these determinations have employed the method of Thellier & Thellier (1959) which compares the NRM removed, against a laboratory TRM installed over progressively incremented temperature steps. Strictly, this method requires that the remanence be wholly SD, so that the TRM in one temperature range is independent of that residing in another temperature range (Section 3.6). Some of the more recent studies employ the method of Shaw (1974) which compares the a.f. demagnetisation of the NRM with the a.f. demagnetisation of a TRM acquired during a single heating. An ARM is imparted to the sample after each investigation, to identify the region of the coercive force spectrum which remains free of thermal alteration. Since Shaw's method requires only a single heating, the possibility of sample alteration is reduced, but the method can only be applied to a full TRM, wheras the Thelliers' method can be used to investigate PTRMs.

The data fall into two classes: firstly, there is a wide range of individual measurements made on material exhibiting a stable TRM, but otherwise with little control on the quality or reproducibility of the result. These determinations are plotted as small symbols in Fig. 12.8. Secondly, there are studies, based on a large number of determinations, with a distribution of palaeo-intensity estimates expressed in terms of a mean and standard deviation; these are plotted as large symbols. The age estimates of the rock units have been adjusted where more reliable estimates are now available.

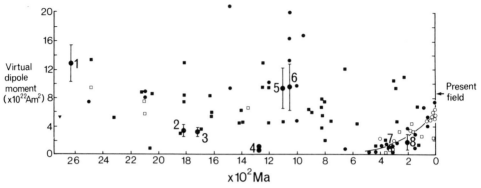

Fig. 12.8 The magnitude of the Virtual Dipole Moment over geological time, as determined from palaeo-intensity determinations. The Phanerozoic data are taken from McElhinny (1973) and do not incorporate a large number of late Cenozoic determinations; the mean value indicated by these latter data is shown by the arrow (see McFadden & McElhinny, 1982). The small symbols are based on isolated or small numbers of samples which do not permit a statistical analysis. The sources are; closed circles: Carmichael (1967, 1970); closed squares: Schwarz & Symons (1969); open squares: Briden (1966); open circles: Smith (1967); inverted triangle: Bergh (1970), and upright triangle: Kobayashi (1968). The results of Carmichael incorporate tests for stability and TRM origin of the remanence, while Schwarz & Symons used rocks with assigned radiometric ages, but did not analyse the contributions to the NRM. Note that most of these results are not corrected for the effects of cooling rate; this is probably of little importance in examples used by Carmichael, which come largely from lava flows and may have cooled over short time periods. Results based on a large number of determinations are shown by the large circles as average results, and the error bars are standard deviations. The sources are: 1, McElhinny & Evans (1968); 2, Schwarz & Symons (1970); 3, Roberts (1983); 4, G. K. Taylor & Author, unpublished data; 5, Pesonen & Halls (1983); 6, Carmichael (1967); 7, Rolph & Shaw (1985, error bar here has a length of $0\cdot 1 \times 10^{22}$ Am^2); 8, Carmichael (1970).

McFadden & McElhinny (1982) have analysed 166 VDM estimates covering the last 5 Ma, and calculate a mean value for the field strength over this period of $8.67 \pm 0.65 \times 10^{22} Am^2$. The VDM appears to have increased roughly exponentially to this value during Phanerozoic times, from a value of only *ca.* 10% of the present value at 500 Ma (McElhinny, 1973 and Fig. 12.8). This effect cannot be caused by a spontaneous time-decay of the natural remanence, because it does not extend back into Proterozoic times. Values in excess of $13 \times 10^{22} Am^2$, during this eon, come from one source only, but values higher than the present field are reported from two well-founded studies, namely of the Modipe gabbro of Botswana (McElhinny & Evans, 1968) and the Keweenawan dykes and sills (Pesonen & Halls, 1983). The acquired field strength depends on cooling rate, with slow cooling leading to enhanced magnetisation intensity (Dodson & McClelland-Brown, 1980), so that both of these results require some qualification. Pesonen & Halls reduced their palaeo-intensity values by an average of 19%, but since corrections were not applied to the Modipe result, it is likely that the VDM at 2600 Ma was significantly lower than the estimate of $12.8 \pm 1.6 \times 10^{22} Am^2$. Thorough studies of the Sudbury lopolith (Swartz & Symons, 1970, uncorrected) and the

Lunch Creek gabbro (Roberts, 1983, corrected) imply low values of the field magnitude at *ca.* 1900–1700 Ma. The *ca.* 1280 Ma Gardar lavas yield an estimate nearly an order of magnitude lower than the 150 Ma-younger Keweenawan rocks and the present field.

Hale & Dunlop (1984) report palaeofield intensities of 34 and 36 μT from the 3470 Ma Barberton lavas, and suggest that a geomagnetic field of comparable magnitude to the present was already established in early Archaean times. This observation, and the variation of VDM through geological time, place important constraints on the energy source of the geomagnetic dynamo, but isolate no single cause. The energy required to sustain the field is proportional to the square of the field magnitude, and a simple removal of the source would lead to decay of the field over a geologically-short period of *ca.* 10^5 years. It has been suggested that small amounts of radioactive elements, notably ^{40}K, are present in the core, and provide sufficient energy to power the dynamo (Verhoogen, 1980). There is, however, no simple decline in the field magnitude which could be linked to a decline in the heat output from a radiogenic heat source. Whilst the magnitudes of the parameters involved are very uncertain, it is generally believed that the latent heat of crystallisation provided by formation of the solid inner core is insufficient to contribute the required energy, although by modifying the convective pattern it could influence the field. A gravitational power source linked to the upward migration of light elements, or to the growth of the inner core by freezing out of a heavier component, is regarded by some workers as the most promising source of energy for the dynamo; it may explain why the VDM does not exhibit a long-term decline. Since the angular velocity of the Earth has been declining throughout this time, as a consequence of tidal friction, the geomagnetic field has not simply obeyed the empirical law which links the strength of planetary magnetic fields to angular velocities of rotation (Russell, 1979). Models for the dynamo source must accommodate the long-term variations in the strength of the source, which appear to include minima in Middle Proterozoic times and near the Proterozoic–Phanerozoic boundary, and maxima at *ca.* 2600, 1100 and in late Cenozoic times (Fig. 12.8). It might be anticipated that these changes reflect stages in the growth of the inner or whole core and/or heat release through the core–mantle boundary. The single feature which can be linked to a major tectonic event at the Earth's surface is the decline of the field strength, by an order of magnitude, near the Proterozoic–Phanerozoic boundary (Chapter 9 and Section 12.11). Energy release from the Earth's core at this time seems to have produced a catastrophic decline in the energy available to drive the dynamo; the source has subsequently built up again through Phanerozoic times.

12.6 Palaeosecular variation

The short-term variations of the geomagnetic field involve directional changes of $10°-20°$ over periods of 10^2-10^4 years, and comprise the secular variation. In principle, this may be caused by one or more of three possible effects (Brock, 1971). These are: (i) changes in the strength and direction of the non-dipole field component, (ii) changes in the orientation of the dipole, so that over a period of time, the average axis coincides with the axis of rotation (this is known as *dipole wobble*), and (iii) changes in the strength of the dipole (this is called *dipole oscillation*). Secular variation is generally linked to short-lived fluid eddies near the core–mantle interface, and is therefore likely to be sensitive to changes in temperature and/or topography of this interface. Its long-term variation is referred to as the *palaeosecular variation* (PSV). PSV cannot properly be studied on the time scale of geomagnetic observations, but it is amenable to investigation by archaeomagnetic and palaeomagnetic study. The significance of PSV is not simply to establish the continuity of the geomagnetic field. It is also able to highlight features of the core–mantle boundary, which may, in turn, link with processes taking place in the mantle and crust. For example, the model for geomagnetic inversions described by Cox (1968) proposes that the cause of inversions lies in the interaction of dipole and non-dipole components of varying magnitude. If the dipole field is much larger than the non-dipole field, then the polarity is stable. However, when the vector sum of axial components of the non-dipole terms exceeds the axial dipole component, but is in the opposite direction, it will swamp the dipole field; the geomagnetic field is then amplified in the opposite sense, and an inversion occurs. The probability of an inversion is high if the ratio of non-dipole to dipole components (σ) is large, and low if it is small. Hence, if secular variation is related to the non-dipole terms, inversions should be frequent when this variation is large, and infrequent when it is small.

PSV has been examined in terms of several theoretical models coded A, B, C, D, E, M and F. Only models C, D, E and M consider contributions from all three factors (i) to (iii), but they make assumptions about these variations which are now known to be incorrect. Model A, due to Irving & Ward (1964), considers that a geocentric dipole field of fixed strength 'M' is perturbed by randomly-directed non-dipole components of constant magnitude 'm'. The resultant field dispersion diminishes as latitude increases. Analysis of the dispersion of directions at a particular latitude yields an estimate of dispersion expressed by the Fisher precision parameter, k, as:

$$k \simeq 3(M/m)^2 = 3\sigma^{-2}$$

To accommodate the variation of field strength with latitude (λ), k can be related to σ_0 (the value of σ at the equator) by the equation:

$$k \simeq 3\sigma_0^{-2} (1 + 3 \sin^2\lambda)$$

PSV is most readily envisaged as an angular dispersion of field directions, and measured as an angular standard deviation, S, which is normalised to its equatorial value, S_0, according to the equation:

$$S_0 = 46 \cdot 8\sigma_0(1 - 3 \sin^2\lambda)^{-1/2}$$

This model predicts that dispersion is inversely proportional to field intensity and therefore increases by a factor of 2 from the poles to the equator. Both this prediction and the observed distribution of directions are closely followed by the present field and are better than subsequent models for PSV in this respect (Irving 1964; Brock, 1971). Models B and C for example, predict a dispersion change by a factor nearer to 3. Although model A is unlikely to have any physical reality (Merrill & McElhinny, 1983), it produces a crude but effective measure of PSV. In part, this is because the effects of dipole oscillation and the non-dipole field cannot be distinguished separately, since they both have the same variation with latitude, and a single randomly-directed vector is an effective approximation. Physically, the most plausible model is model F of McFadden & McElhinny (1984) which supposes that the non-dipole field bears some relationship to the main field; it accords remarkably well with Upper Cenozoic data, but cannot yet be effectively tested against the much older data.

Creer (1962b) examined the 1945 geomagnetic field, and found that the value of S_o declined from a value of just under $20°$ at the equator, to little over $9°$ at the poles. From a rigorous analysis of 1620 normal and 762 reversed lavas, formed over the last 5 Ma, Lee & McElhinny (results discussed in Merrill and McElhinny 1983) found that the average value of S_o decreased from $ca.\ 21\cdot0°$ at the equator to $12\cdot8°$ at the poles; the normal and reversed field states were not significantly different from one another, although the dispersion of the reversed data was somewhat greater, but could be explained by incomplete subtraction of a present field component by the cleaning techniques. Brock (1971) examined PSV over geological time, using a selected body of 83 palaeomagnetic results, which yielded a mean value of contributions to S_o by the dipole field (S_D) and the non-dipole field (S_N) of $2\cdot0°$ and $18\cdot8°$ respectively. Noting that the Cenozoic dispersions were $ca.\ 15\%$ higher than pre-Cenozoic ones, Brock derived a mean value of S_N of $20\cdot4°$ in Cenozoic times, which compares with a value of $18\cdot5°$ for the 1945 field and $17\cdot1°$ in pre-Cenozoic times. There is no significant difference in the small value of S_D ($<2°$), which suggests that dipole wobble is very small, and that PSV is due largely to the combined contribution of dipole oscillation and variation in strength and direction of the non-dipole field. However, this latter conclusion seems to be an artefact of the model, because, by refined analysis of data sets covering the intervals 110–45 Ma and 45–5 Ma, McFadden & McElhinny (1982) have found that the non-dipole field had similar average values over these periods of low and high inversion frequencies, respectively (Section 12.2), while the dipole field was much higher during the latter period. They accordingly suggest that dipole wobble is the most important influence on inversion frequency.

Using model A, Irving & Pullaiah (1976) extended Brock's analysis by using 188 selected results covering the last 300 Ma. Their results show a general increase in the magnitude of PSV up to the present, with minima during the PCR (320–250 Ma) and CN (118–83 Ma) quiet intervals, implying that the PSV (recording the sum of the non-dipole terms and dipole wobble) is reduced during these stable times of low or zero inversion frequency. Irving & Pullaiah's analysis is extended here, by incorporating the expanded data base to 1985, and covering the whole of geological time (Fig. 12.9). The same criteria for selection have been used, namely a minimum of 5 sites, and a number of results (mainly from Palaeozoic sediments, Section 10.4) have been excluded where remagnetisation is suspected. In addition, the results from Phanerozoic times are subdivided according to periods and epochs, in place of the 50 Ma windows used by Irving & Pullaiah, because many of the results are constrained by stratigraphic rather than absolute ages. Although there is the prevailing suspicion that many magnetisations in older rock specimens may be time-averages of the ambient field, this effect does

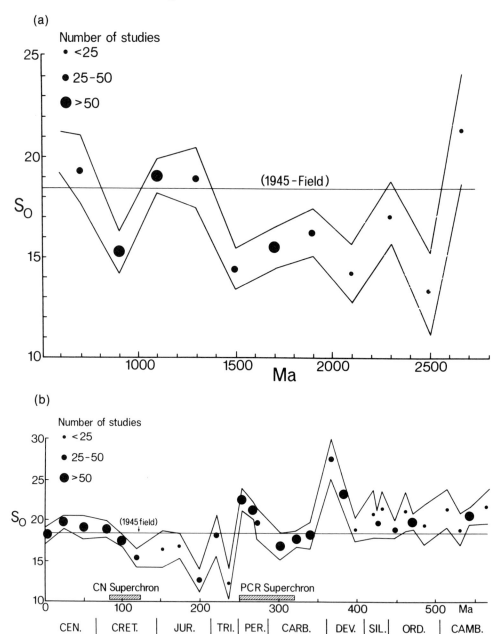

Fig. 12.9 *The mean value of palaeosecular variation (PSV) through geological time, at the palaeoequator (S_o), as estimated from palaeomagnetic data, using this model A of Irving & Ward (1964). The size of the symbols indicate the number of results passing the selection criterion, and the swathe incorporates the standard deviation of the mean. The Phanerozoic results are averaged over the Lower, Middle and Upper divisions of the geological periods. The Precambrian results are averaged over time intervals of 200 Ma, and are shown in (a). The results for Cenozoic times are taken from the analysis of Irving & Pullaiah (1976).*

not appear to have masked real variations in PSV. For example, the mean PSV during Palaeozoic times was higher (S = 20·6°, standard deviation 1·8°, based on 823 observations) than comparable values in both the preceding Precambrian (S_o = 17·0°, standard deviation 1·2° based on 416 observations) and succeeding Mesozoic S_o = 16·5°, standard deviation 1·1°, based on 546 observations) times. This was also the interval which experienced the lowest geomagnetic field strength with intensities of only 10–25% of the present field (Section 12.5). This implies that sources contributing to PSV experienced a reduction, which was proportionally less than the source of the main field. PSV was uniquely high in Devonian times. Although the inversion rate has not yet been found to have been notably high at that time, some high inversion frequencies have been noted, together with a tendency for the transitional field to become locked into intermediate (A and B) field directions, for protracted periods of time during inversions (Sallomy & Piper, 1973). These transitional fields yield comparable field strengths to the main dipolar field (Kono, 1979). Kono believes that these fields represent the effects of multipole components, which mask the dipole source, and account for the relatively dispersed distributions of directions observed during A and B events. Shaw (1977) has also found some Cenozoic examples of important transitional fields.

The quiet superchrons are confirmed to be intervals of relatively low PSV (Brock, 1971), but there is now a real indication that PSV started to drop before the onset of both the PCR and CN superchrons, and then began to build up again before the end of these intervals. The expanded data base from Triassic and Jurassic times suggests some large variations in PSV. The low result in Middle Jurassic times is based on only 5 data points and may not be reliable, but the Upper Triassic result is constrained by 27 observations. A tentative correlation is present here, with the disturbed interval in Lower Triassic times, a short and poorly-defined, normal, quiet interval in Middle-Upper Triassic times, and a disturbed interval in Lower Jurassic times. The data base is too small, and the sampling intervals (200 Ma) too large, to discern detailed variations in PSV during Precambrian times, but the low PSV in Proterozoic times corresponds with the low inversion frequency during this eon (Section 12.4). A relatively large number of inversions, which are in part asymmetrical (Section 7.9), are described from the interval 1400–1000 Ma when PSV was relatively high.

12.7 APW: Continental drift or true polar wander?

The APWPs described in preceding chapters define large, and essentially continuous, relative movements between the continents and the spin axis of the Earth. In principle, these motions could record either continental drift relative to a stationary spin axis, or shift of this axis with respect to a stationary continent. Continental drift can be isolated as the cause, when contrasting APWPs are defined by different continents over the same time period. However, if the APWPs are comparable with one another, or common to a single supercontinent, the two possibilities cannot readily be distinguished. The migration of the spin axis is described as *true polar wander* (TPW); it may be analysed in terms of a net movement of the entire lithosphere with respect to the spin axis (Model A), or in terms of the net motion of the lithosphere with respect to a lower mantle reference frame

(Model B). A choice between the two possibilities is possible if such a reference frame can be identified, and the *hot spots* can probably be used for the reference over short periods of geological time. Hot spots are the lithospheric expression of rising mantle plumes; they tend to produce chemically-primitive magmas, and are believed to have their origin in the lower mantle (Section 12.9). They are responsible for updoming the lithosphere and producing volcanic provinces remote from the plate margins; indeed, the plates are always observed to move over the hot spots. Within the oceanic crust this motion produces linear island chains, and within the continental crust, volcanic provinces show a temporal migration of volcanic activity (Morgan, 1972, 1981). The location of contemporary hot spots is illustrated in Fig. 12.10.

For the hot spots to provide a reliable reference frame, it is necessary to know that they do not move relative to one another. Morgan (1972) calculated that relative movements of no more than 0·5 cm/year could explain the divergence of Pacific island chains. Burke, Kidd & Wilson (1973) and Molnar & Francheteau (1975) believed that the hot spots could show relative movements of at least 1·5 cm/year. However, more recent studies show that relative movements are unlikely to be more than a few mm/year, and no relative movements are detectable over the past 10 Ma (Chase, 1979; Minster & Jordan, 1978). The hot spots thus appear to be the closest thing to a reference frame within the Earth, and they have been used extensively to resolve the relative movements of the tectosphere[*] and mesosphere.

TPW can be produced in at least two ways. Firstly, the net torque acting on the plates, due to push at the ridges and pull into the trenches, is balanced by shear tractions at the base of the tectosphere; any change in the driving forces leads to a change in the counteracting forces, and this could, in principle, lead to TPW. Secondly, changes in the inertial tensor of the Earth can lead to movement of the entire tectosphere with respect to the mantle. The phenomenon has been analysed in detail by Goldreich & Toomre (1969) who show that it will take place, provided that the equatorial bulge is sufficiently mobile to adjust continuously to changes in the rotation axis. This requires an average viscosity within the mantle of < *ca.* $1·5 \times 10^{24}$ Pa s, although for significant amounts of polar wander to occur on a geological time scale would require a viscosity an order less than this. The more recent evidence suggests that the mantle has a fairly uniform viscosity of *ca.* 10^{21} Pa s (Cathles, 1975), and polar wander caused by inertial changes is therefore, theoretically likely.

TWP can be distinguished from the collective effects of ocean-floor spreading and continental drift, by the identification of simultaneous intervals of rapid APW in separate continental plates. Irving & Robertson (1969) first attempted to identify translation of the entire lithosphere (Model A), by using the spreading ridges as a reference frame. However, Francheteau & Sclater (1970), subsequently showed that both ridges and trenches are in relative motion, and do not therefore provide such a frame. McElhinny (1973) used a method proposed by McKenzie (1972) to estimate the magnitude of TPW since Eocene times. This method calculates

---

[*] The *tectosphere* is regarded as the section of crust and upper mantle which moves laterally as a coherent unit over the low-viscosity layer beneath; it is thicker than the *lithosphere*, which deforms elastically under vertical crustal loading.

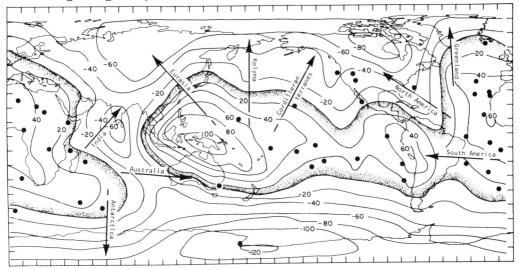

Fig. 12.10 The geoid, based on the global gravity model GEM 10B, plotted relative to a hydrostatic figure of the Earth. The uphill side of the zero-metre contour is stippled, and a 20 metre contour interval is used. The dots are the locations of 42 hot spots recognised by Crough & Jurdy (1980). Modified after Chase & Sprowl (1983). The dispersal of the present continents from their probable locations within the supercontinent of Pangaea is modified after Pal (1983, see also Section 12.10).

weighted vectors, representing the angular velocities of the plates derived from the palaeomagnetic poles. The resultant vector derived from the plate velocity vectors was shown to be small and essentially random, thus implying that no significant TPW (Model A) has taken place since Eocene times. Jurdy & Van der Voo (1974) developed a test for TPW, which employs known poles of rotation and spreading rates, and which, in principle, requires only a single reliable pole for any specific time period to position all the plates. Their method reconstructs the lithospheric plate displacements, and decomposes them into a best-fitting rigid rotation and residual motion: this is equivalent to finding the net angular momentum of the lithosphere. Calculations by this method for the Lower and Upper Cretaceous, and early Cenozoic times yield estimates for TPW of $4 \cdot 7°$, $5 \cdot 0°$ and $2 \cdot 0°$ respectively, which are within the error limits of the data, and suggest, like the method of McElhinny, that no significant motion of the combined lithosphere has occurred, with respect to the palaeomagnetic reference axis, since the latter part of Mesozoic times (Jurdy & Van der Voo 1975).

This result may not, however, be typical of geological time as a whole. Davies & Solomon (1985) have analysed the net contributions of ridge-push and trench-pull forces in the present plate framework, and conclude that these forces are at present almost equally and oppositely directed. This is a consequence of contemporary plate geometry and (in the absence of inertial changes producing the Goldreich-Toomre response), any residual forces may have been inadequate to promote significant TPW over the last part of geological time.

A different picture is obtained when the hot-spot frame is compared with the geomagnetic (and therefore the rotational) axis, according to Model B. By examining the temporal migration of volcanism in the Faeroe-Iceland sector of the

Thulean Province and the Central European Volcanic Province, Duncan, Petersen & Hargraves (1972) suggest that these hot spots have moved by some 23° relative to the rotation axis since early Cenozoic times. Hargraves & Duncan (1973) subsequently examined the traces of eight plumes over the last 50 Ma, and recognised variable rotational movements with respect to the spin axis. Jurdy (1981) was able to confirm that 10°–12° of movement of the tectosphere relative to the hot-spot framework has taken place since early Cenozoic times. Since the combined errors are ca. 4.5°, this motion is significant, and implies that net motion of the tectosphere has taken place with respect to the hot spots and host mantle. Harrison & Lindh (1983) have also compared the geomagnetic and hotspot reference frames, to show that deviation between them increases to ca. 7° at 50 Ma; at that time, the deviation diminishes somewhat, only to increase between 70 and 180 Ma to a value of 17.6°. Since the errors of this approach are little more than 4°, this appears to be a real effect. This relative motion of the mesosphere has been referred to as *mantle roll* by Hargraves & Duncan (1973), who performed a best fit on the movement vectors to estimate that the apparent rotation axis lies near 30°E, 30°N. This position is not only close to the centre of the land hemisphere, but is also the focus of a radiating system of global geosutures, which can be traced over the entire globe (Neev & Hall, 1982), thus suggesting that mantle roll influences the stress system within the tectosphere.

The distribution of the hot spots and subduction zones over the globe controls the position of the spin axis, and the principal axis of their combined effect is within a degree of this axis (Jurdy, 1983). This explains why equatorial plate velocities are higher than polar plate velocities, and hence why the Euler rotation poles of the present plates tend to be concentrated near the present geographical poles. Since mantle convection can cause horizontal displacement of density anomalies, it is anticipated that changes in the perturbing masses near the Earth's surface, in particular, the distribution of subduction zones, will control the observed mantle roll. By reconstructing the probable subduction zones of early Cenozoic times, Jurdy (1983) finds that 10° of motion since this time can be explained in terms of lateral migration of these zones.

The minimal movement between tectosphere and the spin axis, since mid-Mesozoic times, may not be typical of geological time. The recognition of rapid APW in all of the major continental plates in Jurassic times (Chapter 11 and Fig. 12.12) is a strong indication of TPW (Jones & Gartner, 1980). It is not possible to prove this because the movement is recorded only in the continental lithosphere. However, the rates of movement involved are more than 50% greater than the velocities of the fastest (and entirely oceanic) plates at the present day, and it is doubtful whether absolute velocities of this magnitude could be sustained by convective processes in the mantle. Furthermore, a characteristic signature of APW in Proterozoic (Section 7.10), and, to a lesser extent, in Phanerozoic times, is the closed loop, which implies that the plate system was periodically upset and then restored to an equilibrium position. This is the type of response to be expected from a mantle perturbation displacing the principal axis of intertia, and a restoration of this axis to the rotation axis over a period of 10^7 to 10^8 years. It is not readily explicable in terms of, for example, pulses of rapid ocean-floor spreading. The probability that APW motions include a component of TPW implies that plate velocities calculated from APW must be interpreted with caution: pulses of high velocity cannot necessarily be directly linked to more energetic ocean-floor spreading and an enhanced thermal input from the mantle. However, since thermal perturbations within the Earth are likely to upset the inertial figure of the Earth and lead to TPW, it is likely that the two phenomena are linked in some way. This point will be examined in the next section.

12.8 Geomagnetic intervals, continental velocities and geological cycles

Over the last 180 Ma of geological time, there is an impressive correlation between the reversal history of the geomagnetic field and surface phenomena (Fig. 12.11). The CN superchron coincided with a period of rapid ocean-floor spreading, while the progressive increase in inversion frequency, since late Cretaceous times, has accompanied a general reduction in the rates of sea-floor spreading. Several geological effects may be expected to accompany high rates of ocean-floor growth (Force, 1984): the buoyant ridges displace sea-water onto the continental shelves, resulting in marine transgression, which in turn decreases the Earth's albedo; more solar energy is then absorbed and a global warming results. The warm periods in turn cause the polar ice caps, with high albedos, to retreat at the same time as poleward transport of the latent heat of evaporation increases. Hence, the temperature gradients in the surface ocean waters decrease and the deep ocean currents

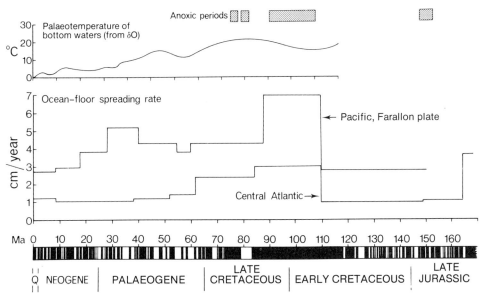

Fig. 12.11 *A comparison of the polarity time scale of Harland, Cox, Llewellyn, Picton, Smith & Walters (1982) with the average half rates of sea-floor spreading observed in the Pacific (Farallon) and the South Atlantic plates (after Larson & Pitman, 1972), and the oxygen isotope-derived bottom palaeotemperatures fom the north Pacific (Douglas & Savin, 1975). Also indicated are the intervals during which widespread anoxic facies were developed (after Arthur, 1979). Modified after Force (1984).*

of oxygenated polar waters begin to wane. As a consequence, the oxygen-starved zones in the deep ocean basins intensify and expand, so that deposition of reduced facies, such as black shales, becomes important and base metals formed in abyssal environments are selectively preserved. These stable conditions favour the stratification of waters in the ocean basins, so that plant and animal matter is preserved on sinking to lower levels, and the initial stages of hydrocarbon formation are promoted (Stoneley & Bailey, 1981). All of these effects are specifically observed during the CN superchron (Fig. 12.11). It has long been recognised, from fossil assemblages and oxygen isotope measurements, that warmer temperatures prevailed during middle to late Cretaceous times (Frakes, 1979), and this was also a time of widespread deposition of anoxic facies (Arthur, 1979).

When viewed over the whole of Phanerozoic times, this correlation between geomagnetic quiet intervals and fast plate movements is not apparent. Continental velocities were higher during Lower–Middle Jurassic times, when the geomagnetic field was probably disturbed (Section 12.2,Figs. 12.2 and 12.12), and the PCR superchron correlates with low rates of continental movement. The velocities of the continental plates are illustrated in Fig. 12.12 in terms of V_{RMS}. Bearing in mind the limitations of this method for determining true continental velocities (Second part of Section 6.6), these calculations identify three Phanerozoic episodes of rapid movement, in early Cambrian (*ca.* 590–550 Ma), Upper Ordovician to Lower Devonian (*ca.* 450–400 Ma, see Fig. 7.10) and late Triassic–early Jurassic (*ca.* 240–200 Ma) times. Owing to the problems associated with widespread magnetic overprinting (Section 10.4), it is not clear whether the second episode is applicable to all of the continental crust, while the last episode is described by a movement which is probably common to the Gondwanaland, Eurasian and North American continents (Gordon, McWilliams & Cox, 1979).

It is clear that these episodes of V_{RMS} do not directly correlate with the *ca.* 300 Ma geomagnetic polarity cycle of Fig. 12.4. Indeed it is probably wrong to seek a single correlation between geomagnetic inversions and V_{RMS}, because the former may depend on parameters other than the thermal state of the core–mantle boundary. The close similarity between the 300 Ma geomagnetic cycle and the *ca.* 280 Ma galactic year — the time taken by the solar system to revolve in an elliptical orbit about the centre of the galaxy — has prompted some workers (e.g. Negi & Tiwari, 1983) to seek a link between the two. The magnitude of tidal friction controls the torque acting at the core–mantle interface, and it seems probable that variations in this torque will influence the inversion mechanism (Brosche 1981). The magnitude of the retarding couple resulting from the oceanic tides is largely controlled by the geometry and extent of shelf seas. The oceanic and continental configurations of mid-Cretaceous and Permian times (both are 'quiet' geomagnetic intervals) predict that the magnitude of tidal friction was then smaller than the present value by a factor of two (Sundermann & Brosche, 1978). If inversions are influenced by tidal torque, they will therefore be partly controlled by marine transgressions and the formation of supercontinents. The decline in the Earth's angular velocity, as a consequence of tidal friction, can be monitored from the growth rhythms of certain organisms (Rosenberg & Runcorn, 1975) and the present data suggest a non-uniform decline in the rotation rate. As shown in Fig. 12.13, intervals of more rapid deceleration have alternated with periods of less rapid deceleration. The former appear to correlate with periods of reversed bias in

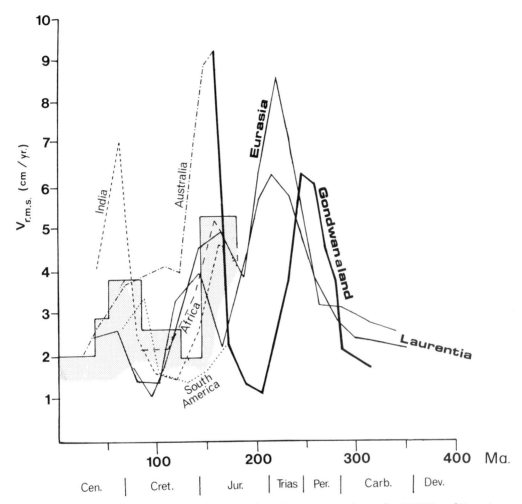

Fig. 12.12 Calculation of V_{RMS} since Carboniferous times from the APWPs of Eurasia, Laurentia and Gondwanaland. Note that the absolute values obtained in this way will depend on the manner in which the palaeomagnetic data are averaged (Section 11.1) and the method is most significant as an indicator of relative velocity changes. In this instance, the APWPs have been averaged over windows of 20 Ma, and incremented in steps of 10 Ma. The separate graphs, for fragments of Gondwanaland after 170 Ma, are calculated by averaging the data through windows of 40 Ma, in steps of 20 Ma. Note that the divergence of these paths reflects the variable quality of the APWP definitions, and is not a true measure of continental dispersal (Section 11.3). Similarly, the displacement of the peak in the V_{RMS} graph for Gondwanaland, from the peak in the Eurasian and Laurentian graphs, probably reflects uncertainties in definition and dating of the APWPs, because these blocks were moving together as the single supercontinent of Pangaea during these times. The histogram (stippled) shows the average continental velocity with respect to the hot-spots over the past 180 Ma, after Schult & Gordon (1984).

the geomagnetic cycle, and the latter with periods of normal bias. The transitions recording changes in deceleration rate appear to define spin maxima at *ca.* 65 and 375 Ma, and spin minima at 235 and 445 Ma; all four are times of no clear bias, while the transitions at 65 and 235 Ma are close to the end of quiet intervals, and suggest that the onset of these disturbed intervals may be linked to changing torques at the core–mantle interface. Neither the geomagnetic nor the palaeontological evidence is good enough to evaluate the significance of the older turning points. While extra-terrestrial causes cannot be entirely discounted, these effects are most likely to have resulted from changes in the Earth's moment of inertia, or to an internal redistribution of angular momentum (Whyte, 1977). More calculations of the kind performed by Sundermann & Brosche (1978) are required to evaluate the contribution of tidal friction to the geomagnetic cycle of Fig. 12.4. Such calculations may also clarify the apparent, but enigmatic, link between the turning points of Fig. 12.13 and times of climatic instability and faunal extinctions. The strong normal polarity bias during Precambrian times (Section 12.4), may be linked to both the lower magnitude of tidal friction during these times (Piper, 1983b) and to higher core-mantle temperatures.

In principle, the periods of high continental velocities, identified by the palaeomagnetic record in Phanerozoic times, could be caused either by episodic increases in the rate of ocean-floor spreading, or by TPW. Since no ocean-floor remains from the time periods in question, a resolution of these possibilities can only be made from the geological record. Increased rates of ocean floor spreading lead indirectly to marine transgressions; these may, in turn, result in a decreased terrigenous input into the marine sediment balance, but they may be accompanied by an *increase* in: (i) chemical/biological deposition in a more equitable and extensive oceanic environment (p. 353–354) and (ii) island arc volcanism, orogenesis and accretion.

Changes in sea-level through geological time can be estimated by a variety of methods, including seismic stratigraphy, facies analysis and mapping of the areal distribution of marine deposits. Vail, Mitchum & Thompson (1977) analysed seismic stratigraphy in terms of cycles of coastal onlap and offlap and produced a curve (Fig. 12.14) which has been widely employed. There is continuing debate about whether the sea-level changes are eustatically or tectonically controlled (e.g.

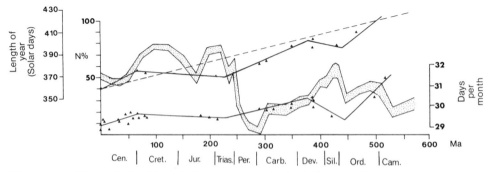

Fig. 12.13 The polarity bias of the geomagnetic field during Phanerozoic times, compared with the number of solar days per month and the number of solar days per year, derived from palaeontological data. The dashed line is extrapolated from the present rate of deceleration. Modified after Creer (1975) and Whyte (1977).

Morner, 1981), and are therefore global or regional phenomena. In addition, the significance of features in the seismic stratigraphy, interpreted as evidence for very rapid falls in sea-level, has been disputed (Brown & Fisher, 1979; Pitman, 1978) and a revised curve shows fewer instantaneous falls in sea level (Vail & Todd, 1981). The first-order changes identified by Vail, Mitchum & Thompson are in broad agreement with the eustatic sea-level curves of Egyed (1959) and Hallam (1977 and Fig. 12.4). The marine inundation curves of Hallam (1977) use palaeogeographic and facies maps of the USSR and North America to show episodes of transgression and regression, superimposed on a general trend of emergence over Phanerozoic times.

The two most important causes of sea-level changes seem to be linked to the mid-ocean ridge system. Firstly, the growth of additional new ridges follows the rifting and breakup of continents, and decreases the overall volume of the ocean basins. Secondly, when the rate of ocean-floor spreading increases, the ocean crust has less time to cool and contract, and the volume of the ocean basins is again reduced. These effects produce transgressions two or three magnitudes longer than the changes caused by fluctuating ice sheets (Hays & Pitman, 1973). The observed changes are complicated by the closure of contracting ocean basins, by the accompanying elimination of segments of the mid-ocean ridge system, and by the evolving geometry of new subsiding continental margins. Subsidence is responsible for the time delay noted in the second part of Section 9.6, where a close correlation is identified between continental rifting, high V_{RMS}, and ensuing sea-level rise in early Cambrian times. High V_{RMS} rates between 220 and 150 Ma (Fig. 12.14) also correlate with continental breakup, and a later sea-level rise of 200–300 m, over a period of *ca.* 75 Ma in Jurassic and Cretaceous times. The marked secular regression in Cenozoic times cannot be explained simply in terms of a decline in ocean-floor spreading rates (Fig. 12.11). The interaction of new ridge development in the North Atlantic, and between Australia and Antarctica only partly counteracted the destruction of ridges in Tethys and the western Pacific, to produce a complex but rapid regression (Hallam, 1977). The consequence of exceptionally high continental velocities in late Ordovician–Lower Devonian times is more obscure. Marine transgression is identified in the sea-level curve, either during, or immediately following, this episode, while the USSR curve (for which the data are probably of unequalled quality) does show the predictable delayed response (Fig. 12.14). However, it is likely that the continental collisions during this episode (which in part produced the Caledonian orogeny, second part of Section 10.3) consumed a long mid-ocean ridge system, which may, in part, have counteracted the effects of buoyant ridges elsewhere.

Other geological cycles were associated with these cycles of sea-level changes, and were therefore an indirect consequence of new ridge growth or changes in velocity. A cycle of carbonate mineralogy, for example, is in phase with the sea-level curve (Fig. 12.14 and Sandberg, 1983). This is probably because high sea levels, together with high rates of ocean-floor spreading and subduction-zone metamorphism, favour a transfer of CO_2 to the atmosphere through weathering reactions. This elevates atmospheric P_{CO_2} and favours the precipitation of low-Mg calcite. Conversely, low sea-levels and reduced metamorphism result in decreased CO_2 flux, and favour precipitation of high-Mg calcite and (less commonly) aragonite (Sandberg, 1983). Faunal diversity was at a minimum in Permo-Triassic

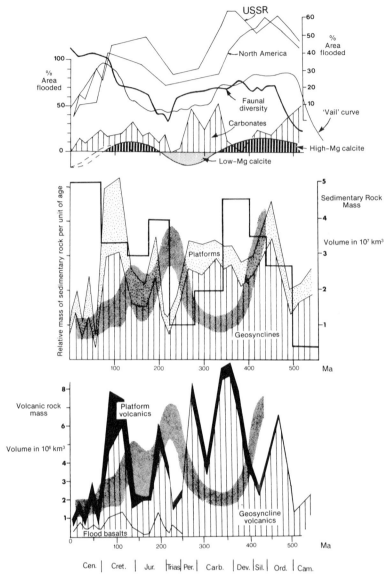

Fig. 12.14 (a) *The first-order cycles of relative sea-level changes (left hand ordinate) during Phanerozoic times (excluding second-order cycles with mean periods betweeen 35 and 155 Ma) after Vail, Mitchum & Thompson (1977), and the maximum degree of marine inundation for each geological period is indicated by the sea-level curves for the USSR and North America (right hand ordinate), after Hallam (1977). Also shown is the relative importance of carbonate development through Phanerozoic times, after Ronov, Khain, Balukhovsky & Seslavinsky (1980). The cycle of carbonate mineralogy, with the aragonite threshold separating periods of Mg-calcite and aragonite from low-Mg calcite, is after Sandberg (1983); note that this cycle closely follows the sea-level curve. (b) The total sedimentary rock mass through Phanerozoc times (right hand ordinate), after Garrels & Mackenzie (1971) and the changes in marine shelf and geosynclinal rock volumes (left hand ordinate), according to Ronov, Khain, Balukhovsky & Seslavinsky (1980). (c) The changes*

times (Valentine & Moores, 1972 and Fig. 12.14) when the assembly of the continental crust into Pangaea led to a significant reduction in diversity. Extinction was greatest among groups that were either least adapted to fluctuating conditions, or were unable to meet competition and withstand predation from the invading members of other biotas. Diversity increased again during transgression in Jurassic and Cretaceous times. The causes of the end-Ordovician faunal extinction have been discussed in Section 10.3, while the end-Cretaceous extinction stands apart from the other major extinction episodes, because it took place during an interval of continental dispersal rather than during continental suturing.

Quantitative changes in sedimentation and volcanism through geological time have been studied mainly by Ronov (1968) and colleagues (1980); their studies cover all of the continents, with the exception of Antarctica. Enhanced rates of ocean-floor spreading should lead to increased rates of calc-alkaline magmatism in island arcs, and these rocks (grouped as 'geosynclinal volcanics' by Ronov) show maxima 50–100 Ma after the pulses of high V_{RMS} (Fig. 12.14(c)). Changes in the character and style of sedimentation during Phanerozoic times are shown in Fig. 12.14(b). According to Ronov's assessment, enhanced sedimentation in both geosynclines and on passive platforms follows (in Carboniferous and Cretaceous times) the episodes of high V_{RMS} over periods of 50–100 Ma. This is the expected response from faster ocean-floor spreading in the former areas, and marine transgression in the latter areas. With the exception of Cenozoic times (from which era an abnormally high proportion of sediments are still preserved), the sedimentary rock volume approximately follows the sea-level curve (Garrels & Mackenzie, 1971 and Fig. 12.14(b)): it is high during periods of transgression, and low during periods of regression. The volume of carbonate deposits shows a distinct inverse relationship to the V_{RMS} curve, with no apparent time lag; periods of rapid continental motion were evidently unfavourable to carbonate sedimentation. Rubey (1955) showed that the CO_2 contained within the sedimentary rocks is about 80 times more than could be produced by weathering of the crystalline rocks, and proposed that the excess could only have been produced by degassing of the upper mantle. Ronov (1968) subsequently showed that the volume of the carbonate sediments is proportional to the intensity of volcanic activity and the area of the shelf seas. The excess CO_2 in carbonate deposits is mostly supplied during periods of high volcanic activity; it is fixed in the rocks by plant and animal activity, which are most important in shallow seas during periods of transgression. These effects are to be expected during episodes of rapid ocean floor spreading; hence the inverse relationship between carbonates and V_{RMS} implies that the episodes of high V_{RMS} are *not* to be explained in terms of rapid spreading.

There are uncertainties (see for example Hallam, 1977) in the interpretation of pre-Cretaceous ocean-floor spreading rates which, in part, obscure the significance of correlations suggested in Fig. 12.11. However, it does seem that spreading rates were lower prior to the CN superchron, and there is no clear indica-

---

in volumes of terrestrial, geosynclinal and flood basalts through Phanerozoic times, from the same source. The stippled swathe in (b) and (c) is the generalised pattern of V_{RMS} changes, after Figs. 7.10 and 12.12. The faunal diversity curve in (a) is after Valentine & Moores (1972).

tion that they have declined since the pulse of rapid V_{RMS} in late Triassic and early Jurassic times. This inference is supported by the geological record, which shows that peaks of orogenic and ocean-floor-related magmatism and sedimentation were actually reached in Cretaceous times, following relatively low values in Jurassic times, and *ca*. 100 Ma after the pulse of high V_{RMS}. Furthermore, high input (?) and fixing of CO_2 in the sedimentary environment occurred in Cretaceous times, and not at the time of highest continental velocity. These observations would also support the view of Jones & Gartner (1980) that the early Jurassic peak in V_{RMS} was due to TPW, and not to more rapid ocean-floor spreading. (The large amounts of sea-floor spreading in Cretaceous times were mainly latitudinal (E–W), so that the V_{RMS} calculations from the APWP paths show appreciably lower velocities than those identified from ocean-floor studies (cf. Figs. 12.11 and 12.12). In practice, TPW and ocean-floor spreading are probably linked, because a new outburst of hot-spot activity will upset the inertial figure of the Earth, to induce TPW, and will, in turn, result in continental rifting, breakup and spreading along a lengthened mid-ocean ridge system. A time delay between lower mantle events producing TPW, and their surface expression, might explain the lag between outstandingly high V_{RMS} in late Ordovician times (Fig. 7.10) and peaks in volcanism and sedimentation in Devonian times (Fig. 12.14). Similarly, although a direct link between geomagnetic reversals, V_{RMS} and surface phenomena has not been identified, it is likely that some kind of link does exist, because factors inducing TPW may relate to thermal conditions at the core–mantle interface, and because they will influence the tidal retarding couple. The observation that the long term geomagnetic polarity cycle and pulses of high V_{RMS} are of different period may reflect the point alluded to above (p. 356), namely that the inversion mechanism is controlled by more than one phenomenon.

12.9 Palaeomagnetism and mantle convection

Although plate tectonics demonstrates the reality of mantle convection, its scales and rates are still a matter of contentious debate, focused on experimental and theoretical studies of the contemporary Earth. The palaeomagnetic evidence provides a contrasting perspective on the subject by defining the past relative movements between the continents; this information records, albeit incompletely, ancient convective motions of the asthenosphere.

Much controversy concerns the central question of whether mantle convection is layered, restricted to the upper mantle, or occurs in the whole mantle. It has long been known that the mantle is chemically heterogeneous (Anderson, 1965; Gast, 1965; Armstrong, 1968), but this knowledge made little impact on the discussion until it was discovered that variations in $^{143}Nd:^{144}Nd$ could be used to monitor Sm:Nd fractionation by geological processes over most of geological history. This evidence requires that igneous rocks from continental and oceanic sources have maintained their identity for periods of 2000 Ma (De Paolo & Wasserberg 1976, 1977; De Paolo, 1979). The upper mantle and continental crust are complimentary reservoirs. The upper mantle (as characterised by mid-ocean ridge basalts (MORBs)) is systematically depleted in *REE*, incompatible with the mantle material, whereas the continental crust is correspondingly enriched. A comparison with ocean-island basalts (OIBs), generated above the hot-spots, has produced the popular geochemical interpretation in terms of a two-layer model, with a depleted upper mantle overlying a primitive lower mantle (De Paolo & Wasserburg, 1976, 1977; O'Nions, Evenson & Hamilton, 1979). The mass balance of the Sm–Nd system suggests that *ca.* 10% (the top 700 km) of the mantle has been involved in this segregation. Since this involves mantle above the 670 km discontinuity (which is also near the maximum depth from which earthquakes are recorded in the vicinity of down-going lithospheric slabs), the collective evidence has been used to infer that the upper mantle is a discrete convecting layer, with a large viscosity increase at the 670 km discontinuity between this layer and the underlying lower mantle (see for example McKenzie & Weiss, 1975). However, in recent years, evidence has suggested that the lower mantle is also (at least episodically) convecting. There are six critical points:

(i) One of the most important problems with layered convection is that it requires large viscosity contrasts between the layers. This is in disagreement with viscosity depth profiles inferred from lateral and vertical movements of the lithospheric plates (Kenyon & Turcotte, 1983). The heat transfer in layered models requires a complex exchange via a thermal boundary layer, which leads to predictions which are not in accord with observation. However, to preclude lower mantle convection would require viscosities some 10^5 times higher than in the upper mantle (Davies, 1977). This problem becomes most acute as the upper layer is made thinner. More recent evidence suggests that the lower mantle has a much lower viscosity (*ca.* 10^{21} Pa s) than previously believed (10^{22}–10^{26} Pa s) and is therefore almost certainly able to convect (Cathles, 1975; Peltier, 1983).

(ii) The layered mantle interpretation of the geochemical evidence has been criticised on several counts (e.g. Chase, 1981; White & Hoffmann, 1982). Furthermore, MORBs are not systematically depleted; instead they exhibit a range of geochemical signatures from depleted to enriched (Schilling 1973). Indeed, the increasing complexity of geochemical evidence implies more complex subdivision of the mantle, and is undermining the rationale of layered models (Davies, 1984).

(iii) The persistence and stability of the hot-spot framework over periods in 150 Ma implies that its source is in the lower mantle (Morgan, 1981). If the mantle were strictly layered there should be no such evidence of the lower layer within the upper layer.

(iv) The observation that deep earthquakes do not emanate from very far below the 650 km discontinuity is not a valid line of evidence against convection below this depth. There are other explanations for this observation, such as a transition from brittle failure to ductile flow (Toksoz, Minear & Julian, 1971), or a phase transition from olivine to basic oxides (magnesio-wüstite and perovskite); the latter occurs at conditions appropriate to the 650–700 km discontinuity, and present evidence appears to favour a phase-change mechanism (Lees, Bukowinski & Jeanloz, 1983; Loper, 1985). Furthermore, Creager & Jordan, (1984) have interpreted travel-time anomalies to indicate that the subducting slabs descend to at least 900 km.

(v) The thermodynamic efficiency of mantle convection is optimal in the whole-mantle configuration, and the very large scale of some of the lithospheric plates (10,000 km in the case of the Pacific plate) is most compatible with the dominance of a whole-mantle system (e.g. Stacey, 1977). It is difficult to see how plates with arcuate dimensions of 60°–180° could be stable over hundreds of Ma, if they were constrained only by a thin convecting layer.

(vi) The shape of the residual geoid (Fig. 12.10) is powerful evidence of contemporary lower mantle convection (Gough, 1977), while the shapes of Pangaea, and the late Proterozoic supercontinent, explored in the next section, indicate that a link between the geoid, the continental crust and mantle-wide convection has applied to former eras. Noting the high degree of organisation in the distribution pattern of continental crust throughout Phanerozoic times, Kanasewich, Havskiv & Evans (1978) came to a similar conclusion.

While these important lines of evidence argue for the reality of convection through the lower mantle, the relationship between the convection and asthenospheric movements is still unclear. It is generally believed that convection in the mantle is not driven solely by internal heat sources, because these are small, and would produce a stable temperature gradient. Negative buoyancy at the surface is produced by the continuous formation of a buoyant crust at the mid-ocean ridges, and the movements are caused by sources and sinks of buoyancy at the plate boundaries (Stacey, 1977; Fowler, 1983). Current modelling is stressing the difference between ridge-push force — a gravitational body force (Lister 1975), and trench-pull force (McKenzie, 1969; Elsasser, 1971). Collectively these forces have the resultant effect of producing the plate motions which cool the mantle. Study of contemporary plate movements has led to several deductions, which have been used to infer the type of forces driving the plates. The first deduction is that plate velocity is inversely proportional to the continental area of the plate: plates with little or no continental crust move fast, while plates incorporating large continents move slowly. This correlation suggests that continental lithosphere has greater resistance to plate motion than oceanic lithosphere (Forsyth & Uyeda, 1975). The second deduction is that plates with substantial subducting boundaries move at velocities which are functions of the relative length of the subducting perimeter, with a maximum value of 8 cm/year. This could indicate the importance of trench pull forces, which cause the subducting plate to approach a terminal velocity where gravitational body forces are balanced by viscous resistance acting on the slab (Forsyth & Uyeda, 1975). A third deduction is that plate velocities are positively correlated with their mean colatitude: plates in equatorial latitudes move faster than plates in polar latitudes. In the light of the preceding discussion, the first two conclusions are evidently not generally appliable to Phanerozoic plate tectonics. Both the analysis of V_{RMS} (Section 12.8) and of continental movements relative to the hot-spot frame (Schults & Gordon, 1984 and Figs. 12.10 & 12.12) show that large continental plates, which are now moving slowly, have been subjected to pulses of high velocity during Phanerozoic times, at rates up to an order larger than their present velocities. This implies that the surface expression of mantle convection is intermittent rather than continuous.

By investigating convection in a 2D model, equivalent to a layer heated from below, Foster (1971) was able to show that convection is intermittent at Rayleigh numbers comparable to those in the mantle. The temperature gradient of a heated layer may build up until it becomes unstable. At this point it ejects hot material and destroys itself in the process; a new layer then develops and the process is repeated. The quantitative constraints on this problem are very imperfectly understood, but a lower mantle viscosity of 10^{21} Pa s and a layer comprising a discrete entity in thermal terms with a maximum thickness of *ca.* 80 km should result in intermittent convection, with a similar period to the pulses of high V_{RMS} (Foster, 1971; Jones, 1977). A number of workers have proposed that the D" layer, which occupies the lowermost mantle between 2700 and 2900 km depth, is such a discrete layer. Here the velocity gradients of P and S waves decrease significantly, and may actually be negative (Cleary, 1974). This layer may originate from the incorporation of dense material from the core into the mantle (Bolt, 1972), or from a local increase in the temperature gradient (Jones, 1977). According to Loper (1985, see also Peltier, 1983, and Loper & Stacey, 1983), isolated plumes could result from destabilisation of the D" layer, and rise so rapidly to the base of the lithosphere that their ascent is essentially adiabatic. In addition, they may be able to transfer heat upwards through a relatively quiescent mantle without causing overturning, therefore preserving the observed chemical discreteness. Jones (1977) concludes that the time taken to destroy the D" boundary layer is small compared with the time taken for it to build up, and he argues that the effect of lateral temperature changes along the core–mantle boundary would be confined to a thin layer near the top of the core. However, Jacobs (1984) considers that the reduction in temperature, resulting from release of thermal plumes and breakdown of the layer, would increase the Rayleigh number of the core, and would in turn affect the flow pattern throughout a large part of the liquid outer core. He suggests that the increased instability would promote geomagnetic field inversions. Therefore the CN superchron should have been a period of higher temperatures at the core–mantle boundary, when the layer was building up (see also McFadden & Merrill, 1984). If this reasoning is correct, then the correlation between the CN superchron and high ocean-floor spreading rates (Fig. 12.11) could be fortuitous, and the latter phenomenon could represent a response of the asthenosphere to deep mantle and outer core events which took place in Triassic and Jurassic times (Section 12.8).

There are three possible mechanisms which might promote positive buoyancy, and hence convection, in the D" layer. These are (i) heat flux from the core, (ii) the flow of less dense material segregating from the core, and (iii) the flow of dense material segregating from the mantle into the core. The first may be the direct cause of dynamo action (e.g. Gubbins, Masters & Jacobs, 1979) and is generally reckoned to be due to the presence of small amounts of radioactive heat sources (notably ^{40}K), in the core, and/or slow cooling of the inner core with concomitant release of gravitational energy (Verhoogen, 1980). The contribution of the latter two mechanisms is conjectural.

12.10 Supercontinents and the geoid

The best indication of a lower mantle convection scheme is probably provided by the geoid (the sea-level equipotential surface of the Earth's gravity field). Since this potential field is an inverse function of depth (and not an inverse square function like the gravity field), it is more sensitive to deeper effects. The largest anomalies are linked to the dense subducting plate margins, but when these effects are subtracted, the residual geoid divides the Earth's surface into two nearly equal positive and negative zones, which interleave rather like the stripes which cover a tennis

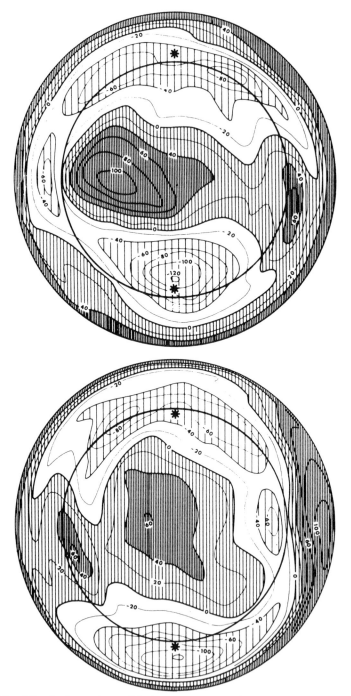

Figure 12.15 The hydrostatic geoid shown on a Lambert equal-area projection of the whole globe. In (a) the projection is centred at 184°E, 6°S, and in (b) it is centred at 4°E, 6°N. The opposite and normally hidden hemisphere is illustrated in a highly distorted fashion beyond the great circle shown by the thick line, and the present North and South poles are indicated by stars. Note the tennis-ball pattern and the discontinuities in both the polar negative and equatorial positive belts. Reproduced from Le Pichon & Huchon (1984) after Chase & Sprowl (1983) and Lerch, Klosko, Laubscher & Wagner (1979).

ball (Fig. 12.15). This residual geoid does not reflect the present-day distribution of continents and oceans, or tectonic elements such as the mid-ocean ridge system. However, the positions of the rising mantle plumes producing the hot spots correlate strongly with the geoid high: of 42 hot spots recognised by Crough & Jurdy (1980), 33 are in positive geoid regions, and only two are in strongly negative regions (Fig. 12.10). Other areas correlating with the geoid high are regions of anomalously shallow sea-floor, and sites of extensive Cretaceous volcanism.

Although there has been some dispute about the extent of lower mantle contributions to the residual geoid (see for example Parsons & Daly, 1983), the clearly recognisable global symmetry leaves little doubt that a lower mantle source is a major contributor (Gough, 1977; Le Pichon & Huchon, 1984). Dziewonski, Hager & O'Connell (1977) find a negative correlation between P-wave velocity and the geoid, implying that high-velocity regions are associated with low densities and *vice versa*. These authors conclude that temperature differences alone are unable to explain the residual geoid, and that it must result from one or more of these three effects: (i) sinking of dense eclogite-rich plates beneath the subduction zones and into the lower mantle, (ii) the distribution of chemical phases of light, high-velocity material which originated in the D" layer, and (iii) temperature anomalies and perturbations on the core–mantle interface, associated with mantle-wide convection.

There is a high degree of symmetry to the geoid, which is best illustrated by the projections of Fig. 12.15. One plane of symmetry is close to the equatorial plane and the positive belt here is interrupted by a low running through the Indian Ocean, which produces the tennis-ball effect. This is a clear demonstration of control of the geoid by the Earth's rotation, with the positive zone elongated along the equator in a response which minimises the kinetic energy of rotation (Goldreich & Toomre, 1969). When the continental crust was entirely aggregated together, as the supercontinent Pangaea, the perimeter (which was also where subduction and accretion were actively taking place) conformed closely to a great circle (Fig. 12.16). Pangaea was therefore essentially a hemisphere continent (Le Pichon & Huchon, 1984). Furthermore, the APWP for Carboniferous to Jurassic times, which defines the ancient poles of rotation throughout the lifetime of this continent, lies within 20° of the great circle defining the perimeter of the Pangaean hemisphere (Fig. 12.16). This observation shows that the equator was also an ancient line of symmetry, and it relates the shape of Pangaea to the geoid in a remarkably simple way. The present geoid pattern can be symmetrically superimposed onto the shape of Pangaea, with the geoid maximum over Tethys, and the join between Gondwanaland and Eurasia over the bridge between the hemisphere continent and the Pacific high. (Since little more than 30% of the global surface is continental crust, only a part of the negative segment of the 'tennis ball' can be occupied by this crust). This observation also implies that the present geoid shape has been a quasi-permanent feature, and shows that Pangaea represented a least-energy configuration for the subduction of rigid plates constrained by a mantle convection system and imposed from below. It also demonstrates that the transfer of continental blocks from Gondwanaland to build up the Eurasian landmass in Carboniferous–Triassic times (Section 10.6) represented a systematic reorganisation of the continental shape to balance up the two 'wings' of the Pangaean shape.

The correlation between the geoid shape and Pangaea does not imply that the sources responsible for the geoid in Carboniferous–Triassic times were identical to those operating today. Noting that the Pangaean continental assembly straddled the present location of the positive geoid anomaly and the hot spots (Fig. 12.10), Anderson (1982) proposed that Pangaea insulated large areas of mantle; he postulated that this led to thermal expansion, partial melting, and ultimately to generation of the mantle plumes, producing breakup and dispersal. This cannot have been the primary cause of dispersal however, because a direct comparison with the present location of the geoid shows that the eastern periphery of Gondwanaland (Australia, India and part of Antarctica) actually occupied a position

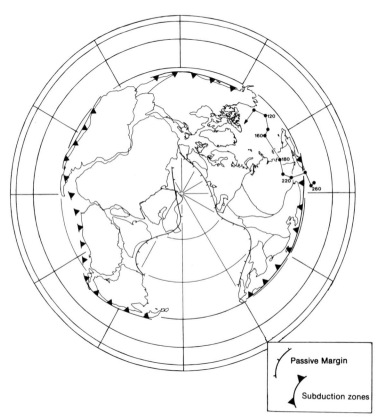

Fig. 12.16 *The reconstruction of Pangaea, plotted on a pole-centred Lambert equal-area projection of the whole globe calibrated in divisions of 30° and located at 40°E, 40°N with respect to Africa in present-day coordinates. The supercontinent is reconstructed using the operations of Bullard, Everitt & Smith (1965), Smith & Hallam (1970) and Smith, Briden & Drewry (1973); note that it does not accomodate the probable Cenozoic motions of the blocks of eastern Asia discussed in Section 11.4. Also plotted is the APWP for the Carboniferous–Jurassic interval, as defined by North American data and discussed in Section 11.2. Note that the supercontinent is constrained within the equator of the projection, and that the APWP lay close to this perimeter throughout the lifetime of the supercontinent (i.e. the supercontinent occupied one hemisphere with the rotational poles remaining at the continental edges). Modified after Le Pichon & Huchon (1984).*

The geomagnetic field, continental movements since Archaean times   367

which is now part of the geoid low, whilst the present Pacific positive was only partially covered by continental crust (Pal, 1983). Nevertheless, a definite correlation does exist between the present position of geoid highs and regions of continental rupture and dispersal, and between geoid lows and regions of continental convergence and assembly (Fig. 12.10).

The Eurasian wing of Pangaea was still accreting in Triassic times (second part of Section 10.5) and seems to have lain across the western part of the present Pacific high. Subsequently, this block moved northwards, while a number of smaller terranes of uncertain affinity moved northwards by larger amounts: Japan has moved nothwards by up to 35° since Permian times, and some of the suspect terranes of the North American Cordillera originated here, far south of their present position (third part, and concluding part of Section 11.4). These are part of the collective northward movement of the crust, which is the most distinctive feature of much

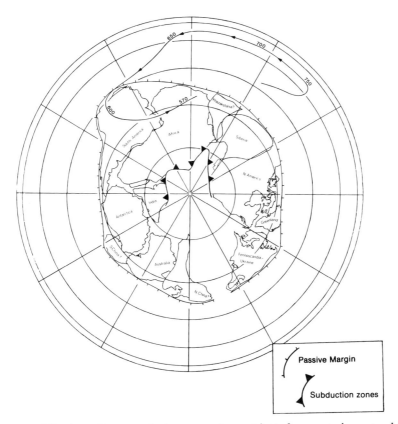

Figure 12.17 The Late Proterozoic Supercontinent plotted on a pole-centred Lambert equal area projection of the whole globe and centred at 270°E, 15°N with respect to the Laurentian Shield in present day coordinates. The reconstruction is based on the operations listed in Table 9.1 and discussed in Chapters 7 and 9. The arrowed line with dates is the APWP for the continental crust throughout the lifetime of this supercontinent and is discussed in detail in sections 9.2 and 9.3. The symbols for subduction zones and passive margins are the same as those used in Figure 12.16; they are based on evidence discussed in sections 8.5 and 9.6; the projection is calibrated in divisions of 30°.

of the post-Jurassic palaeomagnetic record (Chapter 11). Their original locations now contain submarine plateaux, rises, and sea-mount chains, which have also been carried far to the NW of their points of origin. They include the Hess, Line Islands and Necker Rises, which were derived from the Polynesian sea-mount province to the SE, and the Bering Sea plateau which was carried northwards on the Kula plate (Anderson, 1982). The area from the NW Pacific to the East Pacific rise is a region of anomalously shallow bathymetry (the Darwin Rise) and represents predominantly the product of voluminous Cretaceous volcanism which originated 50°–80° to the SE. The larger Gondwanaland–North Atlantic sector of Pangaea originally lay across the present Africa-Atlantic geoid high and has dispersed outwards since Jurassic times, with only Africa near the centre, moving relatively little. This dispersal followed the enormous Karroo-Ferrar volcanism of early Jurassic times (Fig. 11.3) which developed in a tensional environment not far from the contemporaneous (and compressive) cordilleran margin of Gondwanaland (Fig. 12.16). Subsequent large-scale volcanism took place in early Cretaceous times in Brazil and India, and in early Cenozoic times in India and the North Atlantic. In each case this magmatism defined a new outbreak of hot spots and was linked to the continental breakup and dispersal from locations within the present geoid high (Fig. 12.10).

The clear correlation of sites of Mesozoic breakup with the geoid high, and of zones of subsequent accretion and assembly with the geoid low, is a strong indication that the deep mantle events which produced the present geoid were ultimately responsible for this dispersal. According to one model (Chase, 1979; Pal, 1983), the geoid positive is the result of a rising equatorial belt of hot thermal plumes, which cause thermal expansion in the upper mantle; these plumes are responsible for updoming and rupturing the lithosphere, and they then initiate the divergent motions of the plates at the locations of maximum deviatoric tensile stress. The contribution of the lithosphere to the asthenospheric expansion must be an ancillary one, but (especially if it is continental) it provides thermal insulation and enhanced heat production: the effect of a continental cap at one location for 100–200 Ma gives a highly unstable thermal situation, and promotes circulation beneath the continents, so accelerating heat transfer to the oceanic areas (Christensen, 1983). This model implies that the residual geoid negative is related to the complementary polar descending convection of the cold, contracted mantle. The negative buoyancy and down-welling of the lithosphere accelerates the convergent motions of the plates, and causes the geoid lows to become zones of convergent tectonism in the stable Pangaean configuration of Fig. 12.16. (It can, therefore, be predicted that translation of continental blocks towards the present geoid lows will ultimately result in formation of another supercontinent, with a similar form to Pangaea).

The time of establishment of the present hot spot framework (Morgan, 1981) correlates approximately with the episode of rapid V_{RMS} in late Triassic–early Jurassic times, and one consequence of this event appears to have been a shift of *ca.* 30° between the pole of the Pangaean great circle and the pole of the geoid maximum at *ca.* 200 Ma (Le Pichon & Huchon, 1984). This offset was somewhat reduced by 125 Ma, although the geometrical adjustments to the reconstruction caused the continental crust to remain within the continental hemisphere, because crustal separation in the Central Atlantic and Indian Oceans was made at the expense of the area of Tethys. Between 125 and 80 Ma, this circle was breached, and the continents were dispersed to move towards the polar geoid low; the exception was Africa, which lay near the centre of the geoid high, and was hence the focus of dispersion. By calculating the average shift of the continents along the Pangaean great circle, from the present configurations to the 80, 125 and 200 Ma reconstructions, Le Pichon & Huchon (1984) have estimated that the magnitude of TPW (Model A in section 12.7) was only significantly different from zero (by 17°–20°) during the initiation of dispersal at 200 Ma. This would support an interpretation of the contemporaneous episode of high V_{RMS} (Fig. 12.12) in terms of global adjustments to lower mantle events. In contrast, breaching of the hemisphere in Cretaceous times correlates with the outbreak of most of the present hot spot activity; it also marks the initiation of a pulse of rapid ocean-floor spreading, and was ultimately responsible for the range of geological events discussed above and in Section 12.8.

The fundamental significance of the geoid shape to the form of the continental crust over the latter part of geological time is illustrated by the late Proterozoic Supercontinent, described in detail in Chapter 9. When this continental crust is centred at 270°E, 15°N with respect to the Laurentian Shield, in present-day coordinates, it is apparent that this supercontinent too, was a hemisphere continent, with a gross shape almost identical to the later supercontinent of Pangaea (Fig. 12.17). The APWP of this supercontinent shows that it was not tightly constrained to the equatorial line of symmetry between 800 and 650 Ma although it did become so constrained (like Pangaea) by ca. 650 Ma, some 80–100 Ma prior to breakup in Lower Cambrian times. This observation again defines the equatorial plane as an ancient symmetry axis, and implies that the tennis-ball shape of the geoid was the controlling force responsible for distributing and locating the continental crust in the latter part of Precambrian times. This important conclusion enables further deductions on the nature of mantle convection over the latter part of geological time, because, whilst the sizes and shapes of the late Proterozoic and Pangaean supercontinents were nearly identical, the polarity of their tectonic elements was reversed: the active subduction zones in the Late Proterozoic Supercontinent were within the indentation in the reconstruction, and the passive margins were around the perimeter (Fig. 12.17 & Section 9.6). On Pangaea this situation was exactly reversed (Fig. 12.16 & Le Pichon & Huchon, 1984). This may imply that the balance of deep and shallow forces in the mantle responsible for contraining the continental crust in this way, changed between late Precambrian and late Palaeozoic times.

Unfortunately, the understanding of the relationship between deep mantle convection and these observed motions of the asthenosphere is still very imperfect. In the context of the geochemical evidence and the observation that contemporary plate positions show little resemblance to the geoid, most workers have favoured the view that they comprise two separate systems. Busse (1983) suggested that the equatorial positive belt is the ascending limb of a cell, and the polar negative belt is a descending limb. Several authors have addressed the difficulties implied by two superimposed ascending and descending limbs to convection cells in the mantle. Gough (1977) and Le Pichon & Huchon (1984) postulate an ascending limb in the lower mantle, beneath the geoid polar low, with a descending limb below the equatorial high; this would imply that the hot spots are underlain by *descending* limbs of lower mantle convection cells. These interpretations seek to retain a two-level convection scheme in the mantle. Alternative solutions are possible (see for example, Loper, 1985), if a non-steady-state solution is pursued. If lower mantle convection is a pulsed phenomenon, as supported by the geophysical evidence of Section 12.9, and suggested by the geological evidence of Section 12.8, it may occur on a major scale only during breakdown of the D'' layer. This intepretation would imply that the present geoid records relict thermal and density inhomogeneities established during an earlier episode of convection. (It is relevant to note here that changes in polarity length, going into and out of the CN superchron, were gradual, and not sudden, as expected from direct application of the model of Jones (1977); this suggests that the D'' layer does not collapse in a catastrophic way (McFadden & Merrill, 1984)). There was clearly a much closer correlation between the geoid and the pre-Cretaceous plate boundaries (Chase & Sprowl, 1983; Pal, 1983), implying that coupling between lower mantle and asthenospheric convection was then more intimate. This observation prompts a consideration of the continental insulation effect of Anderson (1982). When applied for periods of *ca*. 250 Ma in the case of both the Late Proterozoic and Pangaean supercontinents, it appears to be sufficient to produce high-level thermal anomalies, which may ultimately produce complex surface motions and largely uncouple mesospheric and asthenospheric circulations (Christensen, 1983); this unstable situation would accompany heat release at the surface and persist until equilibrium was again achieved.

A time scale for these effects can be gauged from the growth and destruction of Pangaea. This supercontinent grew into a hemisphere continent from the series of suturing episodes described in Chapters 10 and 11 which began in late Ordovician times (*ca*. 430 Ma) and continued until Permo-Triassic times (*ca*. 230 Ma). Rifting began at *ca*. 200 Ma, and the hemisphere circle was breached in mid-Cretaceous times at *ca*. 120 Ma. This burst of high-level activity had largely waned 50 Ma later, at the end of Cretaceous times (Figs. 12.11 & 12.12). It would, therefore, appear to take some 100 Ma to release the bulk of the accumulated heat at the surface; the dominance of body forces in the present plate régime suggests that thermal transfer from mantle to surface as a result of the Cretaceous episode of rapid ocean-floor spreading is now essentially complete (Section 12.8). Considering now the Late Precambrian history of the continental crust: it first began to break up at *ca*. 1000 Ma (Section 7.8) and became fully constrained to the geoid shape by *ca*. 800 Ma (Section 9.2). The Late Proterozoic Supercontinent persisted for *ca*. 250 Ma, before breaking up in catastrophic events which began in Lower Cambrian times (*ca*. 570–550 Ma, Section 9.6), and were largely concluded by Middle Cambrian times (*ca*. 520 Ma). Hence, the time periods occupied by the growth, duration and destruction of this supercontinent were remarkably similar to Pangaea. The palaeomagnetic evidence has therefore, demonstrated that a long-term 400–500 Ma cycle of continental development has been executed twice since the demise of the Proterozoic supercontinent *ca*. 1000 Ma ago.

12.11 The palaeomagnetic signature of global behaviour in Archaean and Proterozoic times

The primary continental crust also had a simple form (Fig. 7.1(b)). This was unrelated to the shape of the present geoid, although a developing influence of

The geomagnetic field, continental movements since Archaean times 371

this geoid shape, and hence of deep mantle convection, is suggested by the concentration of intermittent subduction and accretion along the concave margin of the crust after 1800 Ma (Fig. 12.18). From the foregoing analysis, it might be anticipated that this subduction occurred only when the supercontinent moved into the geoid-constrained position, but the age constraints on this activity are still too poor to test this possibility (Section 8.5). In addition, it would be unwise to make a close comparison with the present geoid shape, because mantle conditions were certainly more mobile in Proterozoic times, and it is possible that the Earth's heat loss was then concentrated within the oceanic lithosphere (Tarling, 1980). The thinner continental lithosphere could have been constrained to a polar negative zone, because a zone of rising convection beneath the oceanic lithosphere probably produced a quasi-permanent equatorial positive zone. This inference is supported by the observation that the stable configuration of the primary crust contrasts with the post-1000 Ma one: between 2150 and 1000 Ma this primary crust was pole-centred, with the geometrical centre of the crust lying near to the (probably north) geographical pole (Fig. 7.9). This situation was repeatedly upset at *ca.* 200 Ma intervals by events which translated the crust away from this position. Each time it moved back into the equilibrium position, so that the APWP executed the radiating loops characteristic of the Proterozoic eon (Fig. 7.9). It is certain that the

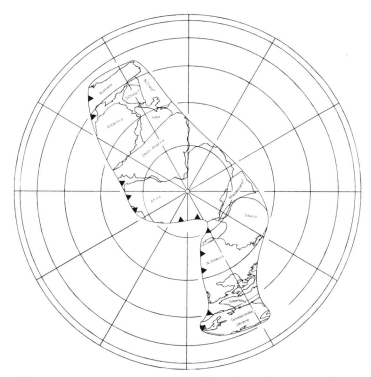

Fig. 12.18 The Proterozoic Supercontinent, plotted on the same pole-centred Lambert equal-area projection of the whole globe as Figs. 12.16 and 12.17, with the geometrical centre of the continent placed at the pole. The periphery of the supercontinent, affected by intermittent subduction-related accretion events after 1800 Ma (see Section 8.4) is indicated by the black triangles.

approximately pole-centred configuration was occupied for much longer periods of Proterozoic time than the positions defined by the extremities of the loops, for two reasons: firstly, the former position is repeatedly defined by the palaeomagnetic record (Morris, Schmidt & Roy, 1979) and seems to have been occupied for quite long periods at *ca.* 1900 Ma (Fig. 7.4) and *ca.* 1350 Ma (Fig. 7.7); secondly, the extremities of the loops are typically poorly-defined by existing data (see for example Figs. 7.3 & 7.5).

The cyclic shifts away from this position, recorded by the APW loops, are applicable to the continental crust *en masse*, and strongly suggest that TPW, caused by the Goldreich-Toomre mechanisms occurred here (Section 12.7 and Piper, 1983b). If mantle disturbances were created by effects such as breakdown of the D″ layer (or a more complete convective overturn) then a convective translation of the continental crust might have been followed by migration of the thermal upwelling towards the equatorial bulge, and a restoration of equilibrium. Thus the episodic models of Foster (1971), Jones (1977) and Loper (1985) have more direct application to Precambrian crustal tectonics, because mantle conditions were then more mobile, and because the continental crust was never stationary for long enough for the insulation effect to generate complex asthenospheric motions and uncouple the lithosphere from the mesophere. Calculations incorporating pressure-dependent physical parameters, with a thermal history incorporating the radiogenic heat sources in a three-dimensional mantle, predict an episodic model for mantle convection (Arkani-Hamed, Mafi-Toksoz & Itsui, 1981); the periodicity of mantle convection is predicted to have been 50–250 Ma, or similar to the periodicity of the Proterozoic APW loops (Fig. 12.19). The features of the model depend on the viscosity assumed for the mantle: models with lower viscosity boundaries cool faster, so that convection slows down more rapidly. Arkani-Hamed, Mafi-Toksoz & Itsui (1981) regard the model shown in Fig. 12.19, with likely viscosity bounds of $10^{20}-10^{22}$ Pa s as the most realistic of their calculations. It does not accommodate core heat sources (an important reservation, see Section 12.9) but it shows a complicated time-dependence of mantle velocity, due to 'beats' arising from the coupling of upper-mantle and whole-mantle convection. There is a possible analogy with the Archaean–Proterozoic transition present here, and also a close correlation with the periodicity of the Proterozoic APW loops. Although the interval of Proterozoic-type behaviour is not yet precisely matched, this model shows a very promising approach to observed Earth behaviour, and predicts that the magnitude of the disturbances producing the Proterozoic APW loops has declined appreciably over the latter part of geological time.

An important feature of the Precambrian crust, which ceased to operate when the crust broke up (and so became constrained by the present geoid form), was the long-term continuity of the global stress field. This has already been noted in terms of the alignment of the younger greenstone belts and straight belts (2700–2200 Ma and Section 8.4 and Fig. 8.2), and it also persisted in most of the subsequent mobile belts formed through to 1100 Ma (Fig. 8.4). The dyke swarms formed prior to 1100 Ma also exhibit a simple relationship to the axial grain of the Precambrian crust, identified in Figs. 8.2 & 8.4. In practice, they seem to have been intruded either parallel, or at right angles, to this fabric. The earliest post-straight belt dykes, intruded at *ca.* 2600 Ma, are orthogonal to the structural grain of the host granite-greenstone terranes (Halls, 1982 and Fig. 8.5). Dykes formed

The geomagnetic field, continental movements since Archaean times 373

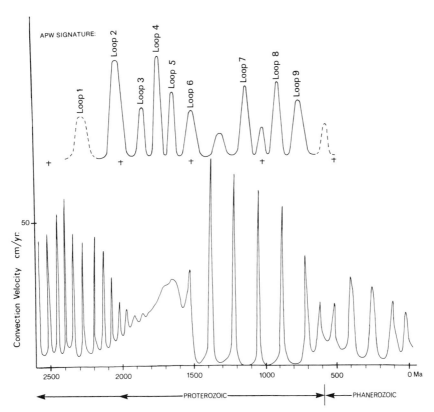

Fig. 12.19 The radial velocity of mantle convection at 900 km depth, using a model, with the mantle viscosity constrained by bounds of 10^{20} and 10^{22} Pa s, and the lateral variations of temperature and velocity expressed in terms of spherical harmonics up to the fifth order. After Arkani-Hamed, Mafi-Toksoz & Itsui (1981). The upper curve shows the movements of the continental crust, represented as a cyclic function, with the amplitude of the cycles proportional to the magnitude of the APW loops, and with minima at the stable (pole-centred) configuration. This function is derived from the APWPs of Figs. 7.3–7.8 and 9.3–9.5. Modified after Piper (1983b).

somewhat later, at *ca.* 2400 Ma in Zimbabwe and Western Australia (which include the longest and thickest examples on Earth) were emplaced parallel to older Archaean trends (Fig. 10.5); later giant dykes in Sweden (1550 Ma) and S. Greenland (1170 Ma) were emplaced approximately orthogonal to older tectonic trends. Highly oblique intersections between Proterozoic dyke swarms and tectonic fabrics are much less commonly observed. This preferential dyke orientation is also recorded by the palaeomagnetic evidence: Precambrian dyke swarms show a prominent statistical preference for intrusion close to a N–S orientation (Morris & Tanzyk 1978). This implies that regional deviatoric tensional stresses were oriented close to lines of latitude during these times. It is strikingly demonstrated, for example, by the Gardar dyke swarms of South Greenland, which were intruded between 1275 and 1165 Ma in a regional stress field which changed progressively from NNW to ESE. The pole positions fall sequentially along the earlier limb of

Loop 7 in Fig. 7.7, in conformity with this change in the regional stress field (Piper & Stearn, 1977). Subsequent analysis by Tanczyk, Ranalli & Morris (1985) has shown that this preferred orientation of dyke swarms did not persist after the crust began to break up at *ca*. 1100 Ma, and subsequent stress fields (with the possible exception of the last 190 Ma) have no preferred azimuth.

The central question of why the continental crust failed to break up in a brittle fashion until more than 2000 Ma after it had first started to grow and survive at the Earth's surface has not been addressed here. A possible cause has been noted in Section 8.5, but it also seems likely that the increasing thickness of the lithosphere beneath the continents, and opportunities for contributions by the insulation effect, must form part of an integrated explanation. Nevertheless, it has been demonstrated that continental reconstructions derived from palaeomagnetism are not random events. Rather, they are ordered features, systematically related to large-scale geophysical anomalies, on the one hand, and to small-scale geological features on the other. Probably the two most important conclusions of the palaeomagnetic evidence are; firstly, that supercontinents represent the longest, and include the most stable phases of, Earth history, and secondly, that continental movements have not been steady but have been concentrated in pulses of high velocity. These observations should change our understanding of both the palaeogeography of the ancient continents (maps illustrating dispersed continental fragments are a very incidental view of what actually happened), and of tectonics (which are ultimately controlled by these pulses). As this chapter has attempted to show, palaeomagnetism has now reached a stage of development at which it can provide most important constraints to interpretation of phenomena as diverse as regional tectonics, marine transgressions and regressions, and the nature of mantle convection.

References

ADDISON, F. T., (1980) 'A magnetic study of diagenesis in the Pendleside Limestone Group (abstract)'. *Geophys. J. R. Astr. Soc.*, 69, p 291.

ADDISON, F. T., TURNER, P. & TARLING, D. H., (1985) 'Magnetic studies of the Pendleside Limestone: evidence for remagnetisation and late-diagenetic dolomitisation during a post-Absian normal event'. *Jour. Geol. Soc. Lond.*, 142, pp 983–994.

ADE-HALL, J. M., KHAN, M. A., DAGLEY, P. & WILSON, R. L., (1968a) 'A detailed opaque petrological and magnetic investigation of a single Tertiary lava flow from Skye, Scotland — I. Iron-Titanium oxide petrology.' *Geophys. J. R. Astr. Soc.* 16, pp 375–388.

ADE-HALL, J. M., KHAN, M. A., DAGLEY, P & WILSON, R. L., (1968b) 'A detailed opaque petrological and magnetic investigation of a single Tertiary lava flow from Skye, Scotland — II. Spatial variations of magnetic properties and selected relationships between magnetic and opaque petrological properties'. *Geophys. J. R. Astr. Soc.*, 16, pp 389–399.

ADE-HALL, J. M. & LAWLEY, A. E., (1970) 'An unsolved problem — opaque petrological differences between Tertiary basaltic dykes and lavas', in: *Mechanisms of Igneous Intrusion* (eds. G,. NEWALL & N. RAST), Galley Press, Liverpool, pp 217–230.

ADE-HALL, J. M., PALMER, H. C. & HUBBARD, T. P., (1971) 'The magnetic and opaque petrological response of basalts to regional hydrothermal alteration'. *Geophys. J. R. Astr. Soc.*, 24, pp 137–174.

ADE-HALL, J. M. & WATKINS, N. D., (1970) Absence of correlations between opaque petrology and natural remanence polarity in Canary Island lavas, *Geophys. J. R. Astr. Soc.*, 19, pp 351–360.

ADE-HALL, J. M. & WILSON, R. L., (1963) 'Petrology and natural remanence of the Mull lavas'. *Nature*, 198, pp 659–660.

ADE-HALL, J. M. & WILSON, R. L., (1969) 'opaque petrology and natural polarity in Mull (Scotland) dykes'. *Geophys. J. R. Astr. Soc.* 18, pp 333–352.

AGER, D. V., (1975). 'The geological evolution of Europe'. *Proc. Geol. Assoc.*, 86, pp 127–154.

ALVAREZ, W., KENT, D. V., SILVA, I. P., SCHWEICKERT, R. A. & LARSON, R. A., (1980) 'Franciscan complex limstone deposited at 70° south palaeolatitude' *Bull. Geol. Soc. Amer.*, 91, pp 476–487.

ANDERSON, A. T., (1966) 'Mineralogy of the Labrieville anorthosite, Quebec'. *Amer. Mineralogist*, 51, pp 1671–1711.

ANDERSON, D. L., (1965) 'Recent evidence concerning the structure and composition of the Earth's mantle'. *Phys. Chem. Earth*, 6, pp 1–131.

ANDERSON, D. L., (1982) 'Hot spots, polar wander, Mesozoic convection and the geoid'. *Nature*, 297, pp 391–393.
ANDERSON, J. M. & SCHWYZER, R. U., (1977) 'The biostratigraphy of the Permian and Triassic, Part 4 — Palaeomagnetic evidence for large scale intra-Gondwanan plate movements during the Carboniferous to Jurassic', *Trans. Geol. Soc. South Afr.*, 80, pp 211–234.
ANDERTON, R., (1980) 'Did Iapetus start to open during the Cambrian?' *Nature*, 286, pp 706–708.
ANHAEUSSER, C. R., (1978) 'The geological evolution of the primitive earth — evidence from the Barberton Mountain Land,' in: *Evolution of the Earth's Crust*, (ed. D. H. TARLING), Academic Press, London, pp 71–106.
ANON., (1979) 'Magnetostratigraphic polarity units, a supplementary chapter of the International Subcommission on Stratigraphic Classification. International Stratigraphic Guide'. *Geology*, 7, pp 578–583.
ARKANI-HAMID, J., MAFI-TOKSOZ, M. & ITSUI, A. T., (1981), 'Thermal evolution of the Earth'. *Tectonophysics*, 75, pp 19–30.
ARMSTRONG, R. L., (1968) 'A model for the evolution of strontium and lead isotopes in a dynamic Earth'. *Rev. Geophys. Space Phys.*, 6, pp 175–199.
AMSTRONG, R. L., (1981) 'Radiogenic isotopes: the case for crustal re-cycling on a near-steady-state no-continental-growth earth'. *Phil. Trans. R. Soc. Lond.*, A301, pp 443–472.
ARTHAUD, F. and MATTE, P., (1977) 'Détermination de la position initiale de la Corse et de la Sardiagne à la fin de l'orogenèse Hercynienne grâce aux marqueurs géologiques anté-mesozoïques'. *Bull. Geol. Soc. Fr.*, 19, pp 833–840.
ARTHUR M. A., (1979) 'Palaeoceanographic events: recognition, resolution and reconsideration'. *Rev. Geophys. Space Phys.*, 17, pp 1474–1494.
AS, J. A., (1960) 'Instruments and measuring methods in palaeomagnetic research'. *Med. Verh., Kon. Ned, Meteorrol. Instit.* 78, 56 pp.
ASHWAL, L. D. & WOODEN, J. L., (1983) 'Isotopic evidence from the eastern Canadian Shield for geochemical discontinuity in the Proterozoic mantle'. *Nature*, 306, pp 679–680.
AZIZ-UR-RAHMAN, GOUGH, D. I. & EVANS, M. E., (1975) 'Anisotropy of magnetic susceptibility of the Martin Formation, Saskatchewan, and its sedimentological implications'. *Can. Jour. Earth. Sci.*, 12, pp 1465–1473.
BAAG, C. G & HELSLEY, C. E., (1974) 'Evidence for penecontemporaneous magnetisation of the Moenkopi Formation'. *Jour. Geophys. Res.*, 79, pp 3308–3320.
BACHTADSE, V., HELLER, F. and KRÖNER, A., (1983) 'Palaeomagnetic investigations in the Hercynian mountain belt of Central Europe'. *Tectonophysics*, 91, pp 285–299.
BACKUS, G. E., (1958). 'A Class of self-sustaining dissipative spherical dynamos'. *Ann. Phys.*, 4, 372–447.
BADHAM, J. P. N., (1981). 'The origins of ore deposits in sedimentary rocks', in: Economic Geology and Geotectonics (ed. D. H. Tarling) Blackwell, Oxford, 149–192.
BADHAM, J. P. N., (1982) 'Strike-slip orogens — an explanation for the Hercynides'. *Jour. Geol. Soc. Lond.*, 139, pp 493–505.
BADHAM, J. P. N. & HALLS, C., (1975) 'Microplate tectonics, oblique collisions, and the evolution of the Hercynian orogenic systems'. *Geology*, 3, pp 373–376.
BAER, A. J., (1976) 'The Grenville Province in Helikian times: a possible model of evolution'. *Phil. Trans. R. Soc. Lond.*, A280, pp 499–516.
BAILEY, M. E. & DUNLOP, D. J., (1983) 'Alternating field characteristics of pseudo single-domain (2–14 μm) and multi-domain magnetite', *Earth Planet. Sci. Lett.*, 63, pp 335–352.
BAK, J., SØRENSEN, K., GROCOTT, J., KORTSGAARD, J. A., NASH, D. & WATTERSON, J., (1975) 'Tectonic implications of Precambrian shear belts in Western Greenland'. *Nature*, 254, pp 566–569.
BAKSI, A. K., (1982) '$^{40}Ar-^{39}Ar$ incremental heating studies on the Tudor Gabbro, Grenville Province, Ontario: its bearing on the North American apparent polar wander path in late Proterozoic times'. *Geophys. J. R. Astr. Soc.*, 70, pp 545–562.

BALSLEY, J. R. & BUDDINGTON, A. F., (1958) 'Iron-titanium oxide minerals, rocks and aeromagnetic anomalies of the Adirondack area, New York'. *Econ. Geol.*, 53, pp 777–805.

BANERJEE, S. K., (1980) 'Magnetism of the oceanic crust: evidence from ophiolite complexes'. *Jour. Geophys. Res.*, 85, pp 3557–3566.

BARAGAR, W. R. A. & ROBERTSON, W. A., (1973) 'Fault rotation of palaeomagnetic directions in Coppermine River lavas and their revised pole'. *Can. J. Earth Sci.*, 10, pp 1519–1532.

BARAGAR, W. R. A. & SCROATES, R. F. J., (1980) 'The Circum-Superior belt: a Proterozoic plate margin'? in: *Precambrian Plate Tectonics* (ed. A. Kröner), Elsevier Publ. Co., Amsterdam, pp 261–296.

BARBETTI, M. F. & MCELHINNY, M. W., (1976) 'The Lake Mungo geomagnetic excursion'. *Phil. Trans. R. Soc., Lond.*, A281, pp 515–542.

BARR, S. M., MACDONALD, A. S. & HAILE, N. S., (1978) 'Reconnaissance palaeomagnetic measurements on Triassic and Jurassic sedimentary rocks from Thailand'. *Geol. Soc. Malaysia Bull.*, 10, pp 53–63.

BARRON, E. J., HARRISON, C. G. A. & HAY, W. W., (1978) 'A revised reconstruction of the southern continents', *EOS, Trans. Amer. Geophys. Un.*, 59, pp 436–450.

BARTON, C. E., MCELHINNY, M. W. & EDWARDS, D. J., (1980) 'Laboratory studies of depositional DRM'. *Geophys. J. R. Astr. Soc.*, 61, pp 355–377.

BATTEY, M. H., (1967). 'The identification of the opaque oxide minerals by optical and X-ray methods'. in: *Methods in Palaeomagnetism* (eds. D. W. COLLINSON, K. M. CREER & S. K. RUNCORN) Elsevier Publ. Co., Amsterdam, pp 485–495.

BEACH, A., (1976) 'The interrelations of fluid transport, deformation, geochemistry and heat flow in early Proterozoic shear zones in the Lewisian complex'. *Phil. Trans. R. Soc. Lond.*, A280 pp 569–604.

BECK, M. E., (1980) 'Palaeomagnetic record of plate-margin tectonic processes along the western edge of North America'. *J. Geophys. Res.*, 85, pp 7115–7131.

BECKMANN, G. E. J., OLSEN, N. O. & SØRENSEN, K., (1978) 'A palaeomagnetic experiment of crustal uplift in West Greenland', *Earth Planet. Sci. Lett.*, 36, pp 269–279.

BELLON, H., COULON, C. & EDEL, J-B., (1977) 'Le déplacement de la Sardaigne; Synthèse des données géochronologiques, magmatiques et paléomagnétiques'. *Bull. Soc. Geol. Fr.*, 19, pp 825–831.

BERGER, G. W. & YORK, D., (1979) '$^{40}Ar^{39}Ar$ of multicomponent magnetisations in the Archaean Shelley Lake granite, north western Ontario'. *Can. Jour. Earth Sci.*, 16, pp 1983–1941.

BERGER, G. W. & YORK, D. & DUNLOP, D. J., (1979) 'Calibration of Grenvillian paleopoles by $^{40}Ar/^{39}Ar$ dating'. *Nature*, 277, pp 46–47.

BERGER, G. W. & YORK, D., (1981) '$^{40}Ar/^{39}Ar$ dating of the Thanet gabbro, Ontario: looking through the Grenvillian metamorphic veil and implications for palaeomagnetism'. *Can. Jour. Earth. Sci.*, 18, pp 266–275.

BERGH, H. W., (1970) 'Palaeomagnetism of the Stillwater Complex, Montana', in: *Palaeogeophysics'* (ed. S. K. RUNCORN), Academic Press, London, pp 143–158.

BERKNER, L. V. & MARSHALL, L. C., (1967) 'The rise of oxygen in the Earth's atmosphere with notes on the Martian atmosphere'. *Adv. Geophys.*, 12, pp 309–331.

BERNER, R. A., (1969). 'Geothite stability and the origin of red beds'. *Geochim. Cosmochim. Acta*, 33, pp 267–273.

BERRY, G. R., (1980) *World Atlas of Geology and Mineral Deposits*. Mining Journal Books, London, 110 pp.

BERTHELSEN, A., (1976). 'Himalayan tectonics: a key to the understanding of Precambrian shield patterns', in: *Coll. Int. CNRS*, 268, 'Ecologie et Géologie de l'Himalaya, pp 61–67.

BEUF, S., BIJU-DUVAL, B., DE CHARPAL, O., ROGNON, P., GARIEL, O. & BENNACEF, A., (1971) 'Les grés du paleozoique inférieur au Sahara', *Publ. Inst. Fr., Pet., Coll. Sci. Tech. Pet.*, 18, 464 pp.

BIJU-DUVAL, B., DERCOURT, J. & LE PICHON, X., (1977) 'From the Tethys Ocean to the Mediterranean Seas: a plate tectonic model of the evolution of the western Alpine System', in: *Interna-*

tional Symposium on the Structural History of the Mediterranean Basins (eds. B. BIJU-DUVAL and L. MONTADERT), Editions Technip, Paris, pp 143–164.

BINGHAM, C., (1964) *Distributions on the sphere and the projective plane*, Ph.D. thesis, Yale University.

BINGHAM, D. K. & EVANS, M. E., (1975) 'Precambrian geomagnetic field reversal'. *Nature*, 253 pp 332–333.

BINGHAM, D. K. & EVANS, M. E., (1976) 'Palaeomagnetism of the Great Slave Supergroup, Northwest Territories, Canada: the Stark Formation'. *Can. J. Earth Sci.*, 13, pp 563–578.

BINGHAM, D. K. & KLOOTWIJK, C. T., (1980) 'Palaeomagnetic constraints on Greater India's underthrusting of the Tibetan Plateau', *Nature*, 284, pp 336–338.

BIRD, C. F. & PIPER, J. D. A., (1981) 'Opaque petrology, magnetic polarity and thermomagnetic properties in the Reydarfjordur dyke swarm, eastern Iceland', *Jokull*, 30, pp 34–42.

BLACKETT, P. M. S., (1952) 'A negative experiment relating to magnetism and the Earth's rotation'. *Phil. Trans. R. Soc., Lond.*, A245, pp 309–370.

BLACKETT, P. M. S., (1961) 'Comparisons of ancient climate with the ancient latitude deduced from rock magnetic measurements'. *Proc. R. Soc., Lond.*, A263, pp 1–30.

BLAKEMORE, R., (1975) 'Magnetotactic bacteria', *Science*, 190, pp 377–379.

BLAKEMORE, R. & FRANKEL, R. B., (1981) 'Magnetic navigation in bacteria'. *Sci. Amer.*, 245, pp 42–48.

BLEIL, U., HALL, J. M., JOHNSON, H. P., LEVI, S. & SCHONHARTING, G., (1982) 'The natural magnetisation of a 3-kilometre section of Icelandic crust'. *Jour. Geophys. Res.*, 87, pp 6569–6589.

BOLT, B. A., (1972) 'The density distribution near the base of the mantle and near the Earth's centre'. *Phys. Earth Planet. Int.*, 5, pp 301–311.

BONATTI, E., LAWRENCE, J. R. & MORANDI, N., (1984) 'Serpentinisation of oceanic peridotites: temperature dependence of mineralogy and boron content'. *Earth Planet. Sci. Lett.* 70, pp 88–94.

BOND, G. C., CHRISTIE-BLICK, N., KOMINZ, M. A. & DEVLIN, W. J., (1985) 'An early Cambrian rift to post-rift transition in the Cordillera of western North America'. *Nature*, 315, pp 742–745.

BOND, G. C., NICKESON, P. A. & KOMINZ, M. A., (1984) 'Break-up of a supercontinent between 625 Ma and 555 Ma: new evidence and implications for continental histories'. *Earth Planet. Sci. Lett.*, 70, pp 325–345.

BOUCOT, A. J., JOHNSON, J. G. & TALENT, J. A., (1969) 'Early Devonian brachiopod zoogeography.' *Geol. Soc. Amer., Spec. Pap.* 119, 111 pp.

BOULIN, J., (1981) 'Afghanistan structure, greater India concept and eastern Tethys evolution', *Tectonophysics* 72, pp 261–287.

BRAZIER, M. D., (1979) 'The Cambrian radiation event, in: *The Origin of Major Invertebrate Groups* (ed. M. R. House), Academic Press, London, pp 103–159.

BRENCHLEY, P. J., (1985) 'Late Ordovician extinctions and their relationship to the Gondwana Glaciation', in: *Fossils and Climate* (ed. P. J. BRENCHLEY), John Wiley, Chichester, pp 291–315.

BRIDEN, J. C., (1966). 'Variation of intensity of the palaeomagnetic field through geological time'. *Nature*, 212, pp 246–247.

BRIDEN, J. C., (1972) 'A stability index of remanent magnetism'. *J. Geophys. Res.*, 77, pp 1401–1405.

BRIDEN, J. C. & IRVING, E., (1964) 'Palaeolatitude spectra of sedimentary palaeoclimatic indicators', in: *Problems in Palaeoclimatology* (ed. A. E. M. NAIRN) Interscience, New York, pp 199–244.

BRIDEN, J. C. & MORRIS, W. A., (1973) 'Palaeomagnetic studies in the British Caledonides — III. Igneous rocks of the Northern Lake District, England'. *Geophys. J. R. Astr. Soc.*, 34, pp 27–46.

BRIDEN, J. C., MORRIS, W. A. & PIPER, J. D. A., (1973) 'Palaeomagnetic studies in the British Caledonides — VI. Regional and global implications'. *Geophys. J. R. Astr. Soc.*, 34, pp 107–134.

BRIDEN, J. C. & MULLAN, A. J., (1984) 'Superimposed Recent, Permo-Carboniferous and Ordovician palaeomagnetic remanence in the Builth Volcanic Series, Wales'. *Earth Planet. Sci. Lett.*, 69, pp 413–421.
BRIDEN, J. C., SMITH, A. G. & SALLOMY, J. T., (1971) 'The geomagnetic field in Permo-Triassic time'. *Geophys. J. R. Astr. Soc.*, 23, pp 101–117.
BRIDEN, J. C. & WARD, M. A., (1966) 'Analysis of magnetic inclination in bore cores'. *Pure Appl. Geophys.*, 63, pp 133–152.
BRIDEN, J. C., TURNELL, H. B. & WATTS, D. R., (1984) 'British palaeomagnetism, Iapetus Ocean, and the Great Glen Fault'. *Geology* 12, pp 428–431.
BRIDGWATER, D. & WINDLEY, B. F., (1973) 'Anorthosites, post-orogenic granites, acid volcanic rocks, and crustal development in the North Atlantic Shield during the Proterozoic', in: *Symposium on Granites and Related Rocks.* (ed. L. A. LISTER), Geol. Soc. S. Afr. Sp. Publ., 3, pp 307–318.
BRIDGWATER, D., SUTTON, J. & WATTERSON, J., (1974) 'Crustal downfolding associated with igneous activity'. *Tectonophysics*, 21, pp 57–77.
BROCK, A., (1971) 'An experimental study of palaeosecular variation'. *Geophys. J. R. Astr. Soc.*, 24, pp 303–312.
BROSCHE, P., (1981) 'Geomagnetic reversals and tidal friction'. *Naturwissenchaften*, 68, pp 139–140.
BROWN, G. C.., (1977) 'Mantle origin of Cordilleran granites.' *Nature*, 265, pp 21–24.
BROWN, G. C. & HENNESSEY, J., (1978) 'The initiation and thermal diversity of granite magmatism'. *Phil. Trans. R. Soc. Lond.*, A288, pp 631–643.
BROWN JR, L. E. & FISHER, W. L., (1979) *Principles of Seismic Stratigraphic Interpretation*, AAPG-ASEG Stratigraphic interpretation of seismic data, School Notes, Amer. Assoc., Pet. Geol. Ed. Dept., Austin, Texas.
BROWN, P. M. & VAN DER VOO, R., (1983) 'A palaeomagnetic study of Piedmont metamorphic rocks in northern Delaware'. *Bull. Geol. Soc. Amer.*, 94, pp 815–822.
BUCHA, V. V., (1965) 'Results of the palaeomagnetism research on rocks of Precambrian and Lower Palaeozoic age in Czechoslovakia'. *J. Geomag. Geoelect*, 17, pp 435–444.
BUCHAN, K. L., (1978) 'Magnetic overprinting in the Thanet gabbro complex Ontario'. *Can. J. Earth Sci.*, 15, pp 1407–1421.
BUCHAN, K. L. & DUNLOP, D. J., (1976) 'Palaeomagnetism in the Haliburton intrusions: superimposed magnetisations, metamorphisms and tectonics in the late Precambrian'. *Jour. Geophys. Res.*, 81, pp 2951–2967.
BUDDINGTON, A. F. & LINDSLEY, D. H., (1964) 'Iron-titanium oxide minerals and synthetic equivalents'. *Jour. Petrol.*, 5(2), pp 310–357.
BULLARD, E. C., (1949) 'The magnetic field within the earth', *Proc. Roy. Soc.*, Lond., A197, pp 433–453.
BULLARD, E. C., EVERITT, J. E. and SMITH A. G., 1965. 'A symposium on continental drift-IV. The fit of the continents around the Atlantic'. *Phil. Trans. R. soc. Lond.*, A258 41–51.
BUREK, P. J. H., (1964) 'Korrelation revers magnetisierter Gesteinfolgen in Oberen Buntsandstein SW-Deutschland'. *Geologisches Jahrbuch* 84, pp 591–616.
BUREK, P. J. H. (1970) 'Magnetic reversals: their application to stratigraphic problems'. *Amer. Assoc. Pet. Geol., Bull.* 54, 1120–1139.
BURKE, K. & DEWEY, J. F., (1973) 'Plume generated triple junctions: key indicators in applying plate tectonics to old rocks'. *Jour. Geol.*, 81, pp 406–433.
BURKE, K., DEWEY, J. F. & KIDD, W. S. F., (1977) 'Precambrian palaeomagnetic results compatible with contemporary operation of the Wilson cycle'. *Tectonophysics*, 33, pp 287–299.
BURKE, K., KIDD, W. S. F. & WILSON, J. T., (1973) 'Relative and latitudinal motion of Atlantic hotspots'. *Nature*, 245, pp 133–137.
BURMESTER, R. F., (1977) 'Origin and stability of drilling induced remanence'. *Geophys. J. R. Astr. Soc.*, 48, pp 1–14.

BURNS, K. L., RICKARD, M. J., BELBIN, L. & CHAMALAUN, F., (1980) 'Further palaeomagnetic confirmation of the Magellanes Orocline.' *Tectonophysics* 63, pp 75–90.
BURRETT, C. F., (1972) 'Plate tectonics and the Hercynian Orogeny'. *Nature* 239, pp 155–157.
BURRETT, C. F., (1973) 'Ordovician biogeography and continental drift'. *Palaeogeo., Palaeoclim,, Palaeoecol.*, 13, pp 161–201.
BURRETT, C. F., (1974) 'Plate tectonics and the fusion of Asia'. *Earth Planet. Sci. Lett.*, 21, pp 181–189.
BURRETT, C. F., (1983) 'Palaeomagnetism and the Mid-European Ocean – an alternative interpretation of Lower Palaeozoic apparent polar wander'. *Geophys. J. R. Astr. Soc.*, 72, pp 523–534.
BURRETT, C. & STAIT, B., (1985). 'South-east Asia as a part of an Ordovician Gondwanaland — a palaeo-biogeographic test of a tectonic hypothesis'. *Earth Planet. Sci. Lett.*, 75, pp 184–190.
BUSSE, F. H., (1976) 'Generation of planetary magnetism by convection', *Phys. Earth Planet. Int.*, 12, pp 350–358.
BUSSE, F. H., (1983) 'Quadrupole convection in the lower mantle?' *Geophys. Res. Lett.*, 10, pp 285–288.
BUTAKOVA, E. L., (1975) 'Regional distribution and tectonic relations of the alkaline rocks of Siberia'. in: *The Alkaline Rocks* (ed. H. Sørensen), Wiley, London, pp 172–182.
BUTLER, R. F. & BANERJEE, S. K., (1975) 'Theoretical single domain size range in magnetite and titanomagnetite', *Jour. Geophys. Res.*, 80, pp 4049–4058.
BYLUND, G., (1985) 'Palaeomagnetism of Middle Proterozoic basic intrusives in central Sweden and the Fennoscandian apparent polar wander path'. *Precambrian Research*, 28, pp 283–310.
CABY, R., BERTRAND, J. R. L. & BLACK, R., (1981) 'Pan-African ocean closure and continental collision in the Hoggar-Iforas segment, Central Sahara', in: *Precambrian Plate Tectonics*, (ed. A. KRÖNER) Elsevier Publ. Co., Amsterdam, pp 407–434.
CAMPBELL, I. H. & TAYLOR, S. R., (1983) 'No water, no granites – no oceans, no continents'. *Geophys. Res. Lett.*, 10, pp 1061–1064.
CANDE, S. C., LARSON, R. L. & LABREQUE, J. L., (1978) 'Magnetic lineations in the Pacific Jurassic Quiet Zone'. *Earth Planet. Sci. Lett.*, 41, pp 434–440.
CAREY, S. W., (1958) 'A tectonic approach to continental drift'. in: *Continental Drift, A Symposium* (ed. S. W. CAREY) Univ. Tasmania, Hobart, pp 177–355.
CARMICHAEL, C. M., (1961) 'The magnetic properties of ilmenite-hematite crystals'. *Proc. Roy. Soc. Lond.*, A263, pp 508–530.
CARMICHAEL, C. M. (1967) 'An outline of the intensity of the palaeomagnetic field of the Earth'. *Earth Planet. Sci. Lett.*, 3, pp 351–354.
CARMICHAEL, C. M., (1970) 'The intensity of the Earth's palaeomagnetic field from 2.5×10^9 years ago to the present'. in: *Palaeogeophysics* (ed. S. K. RUNCORN). Academic Press, London, pp 73–77.
CATHLES III, L. M., (1975) *The viscosity of the Earth's mantle*. Princeton University Press, Princeton, N. J., 371 pp.
CHAMALAUN, F. H., (1964) 'Origin of the secondary magnetisation of the Old Red Sandstone of the Anglo-Welsh Cuvette'. *Jour. Geophys. Res.*, 69, pp 4327–4337.
CHAMALAUN, F. H., (1977) 'Palaeomagnetic evidence for the relative position of Timor and Australia in the Permian'. *Earth Planet. Sci. Lett.*, 34, pp 107–112.
CHAMALAUN, F. H. & CREER, K. M., (1964) 'Thermal demagnetisation studies of the Old Red Sandstone of the Anglo-Welsh cuvette'. *J. Geophys. Res.*, 69, pp 1607–1616.
CHAMBERLAIN, V. E. & LAMBERT, R. St. J., (1985) 'Cordilleria, a newly defined Canadian microcontinent'. *Nature*, 314, pp 707–713.
CHAMPION, D. E., HOWELL, D. G. & GROMMÉ, C. S., (1984) 'Palaeomagnetic and geologic data indicating 2500 km of northward displacement for the Salinian and related terranes, California'. *J. Geophys. Res.* 89, pp 7736–7752.
CHANNELL, J. E. T., CATALANO, R. & D'ARGENIO, B., (1980) 'Palaeomagnetism and the deformation of the Mesozoic continental margin in Sicily'. *Tectonophysics*, 61, pp 391–407.

CHANNELL, J. E. T. & D'ARGENIO, B., (1980) 'The evolution of the Calabrian Arc'. *Instit. Geograf. Nacional, Madrid, Spec. Publ.* 201, pp 129–147.

CHANNELL, J. E. T., D'ARGENIO, B. & HORVATH, F., (1980) Adria, the African promontory, in: Mesozoic Mediterranean Palaeography'. *Earth Sci. Rev.*, 15, pp 213–292.

CHANNELL, J. E. T. & HORVATH, F., (1976) 'The African/Adriatic promontory as a palaeogeographic premise for Alpine orogeny and plate movements in the Carpatho-Balkan region'. *Tectonophysics*, 35, pp 71–101.

CHANNELL, J. E. T., LOWRIE, W., MEDIZZA, F. & ALVAREZ, W., (1980) 'Palaeomagnetism and tectonics in Umbria, Italy'. *Earth Planet. Sci. Lett.*, 39, pp 199–210.

CHANNELL, J. E. T., LOWRIE, W., PIALLI, P. & VENTURI, F., (1984) 'Jurassic magnetic stratigraphy from Umbrian (Italian) land sections'. *Earth Planet. Sci. Lett.*, 68, pp 309–325.

CHASE, C. G., (1979) 'Subduction, the geoid and lower mantle convection', *Nature*, 282, pp 464–468.

CHASE, C. G., (1981) 'Oceanic Island Pb: two stage histories and mantle evolution'. *Earth Planet. Sci. Lett.*, 52, pp 277–284.

CHASE, C. G. & SPROWL, D. R., (1983) 'The modern geoid and ancient plate boundaries'. *Earth Planet, Sci. Lett.*, 62, pp 314–320.

CHAUDHURI, S., (1976) 'The significance of the Rubidium Strontium age of sedimentary rock'. *Contrib. Mineral. Petrol.*, 59, pp 161–170.

CHAUVEL, J. J. & SCHOPF, J. W., (1978) 'Late Precambrian microfossils from Brioverian cherts and limestones of Brittany and Normandy, France,' 275, pp 640–642.

CHEVALLIER, R. & MATHIEU, S., (1943) 'Propriétés magnétiques des poudres d'hématites-influence des dimensions des grains'. *Ann. Phys.*, 18, pp 258–288.

CHRISTENSEN, U., (1983) 'A numerical model of coupled subcontinental and oceanic convection'. *Tectonophysics*, 95, pp 1–23.

CHUMAKOV, N. M., (1981) 'Upper Proterozoic glaciogenic rocks and their stratigraphic significance', *Precambrian Research* 15, pp 73–395.

CISNE, J. L., (1984) 'A basin model for massive banded iron-formations and its geophysical applications'. *Jour. Geology*, 92, pp 471–488.

CISOWSKI, S. M., (1984) 'Evidence for early Tertiary remagnetisation of Devonian rocks from the Orcadian Basin, northern Scotland, and associated transcurrent fault motion'. *Geology*, 12, pp 369–372.

CLAESSON, C., (1978) 'Swedish Ordovician limestones: problems in clarifying their directions of magnetisation'. *Phys. Earth Planet. Int.*, 16, pp 65–72.

CLEARY, J. R., (1974) 'The D″ region'. *Phys. Earth Planet. Int.*, 9, pp 13–27.

CLIFFORD, T. N., (1966) 'Tectono-metallogenic units and metallogenic provinces of Africa'. *Earth Planet. Sci. Lett.*, 1, pp 421–434.

CLOUD JR, P. J., (1968). 'Atmospheric and hydrospheric evolution on the primitive earth'. *Science*, 160, pp 729–736.

COGNÉ, J. (1971) 'Le Massif Armoricain', in: *Géologie de la France*, ed. J. DEBELMAS, vol. 1, Doin Editeurs, Paris, pp 106–161.

COLLINSON, D. W., (1965a) 'Origin of remanent magnetisation and initial susceptibility of certain red sandstones'. *Geophys. J. R. Astr. Soc.*, 9, pp 203–217.

COLLINSON, D. W., (1965b) 'Depositional remanent magnetisation in red sediments'. *J. Geophys. Res.*, 70, pp 4663–4668.

COLLINSON, D. W., (1974) 'The role of pigment and specularite in the remanent magnetism of red sandstones', *Geophys. J. R. Astr. Soc.*, 38, pp 253–264.

COLLINSON, D. W., (1983) *Methods in Palaeomagnetism and Rock Magnetism*, Chapman & Hall, London, 503 pp.

COLLINSON, D. W. & RUNCORN, S. K., (1960) 'Polar wandering and continental drift: evidence of palaeomagnetic observations in the United States'. *Geol. Soc. Amer. Bull.*, 71, pp 915–958.

CONDIE, K. C., (1976) *Tectonics and Crustal Evolution*. Pergamon, 288 pp.

CONDIE, K. C., (1982) 'Plate tectonic model for Proterozoic continental accretion in the south western United States'. *Geology*, 10, pp 37–42.

CONEY, P. J., (1978) 'Mesozoic–Cenozoic Cordilleran plate tectonics'. *Geol. Soc. Am. Mem.*, 152, pp 33–50.

CONEY, P. J., JONES, D. L. & MONGER, J. W. H., (1980) 'Cordilleran suspect terranes'. *Nature*, 288, pp 329–333.

COOK, P. J. & MCELHINNY M. W., (1979) 'A re-evaluation of the spatial and temporal distribution of sedimentary phosphate deposits in the light of plate tectonics'. *Econ. Geol.*, 74, pp 315–330.

COOPER, J. A., NESBITT, R. W., PLATT, J. P. & MORTEMER, G. E., (1978) 'Crustal development in the Agnew region, Western Australia, as shown by Rb/Sr isotopic and chemical studies'. *Precambrian Research*, 7, pp 31–59.

COWARD, M. P. & DALY, M. C., (1984) 'Crustal lineaments and shear zones in Africa: their relationship to plate movements'. *Precambrian Research*, 24, pp 27–45.

COWIE, J. W., (1971) 'Lower Cambrian faunal provinces', in: *Faunal Provinces in Space and Time* (ed. F. A. MIDDLEMISS, P. F. RAWSON & G. NEWHALL), Seel House Press, Liverpool, pp 31–46.

COX, A., (1957) 'Remanent magnetism of lower to middle Eocene basalt flows from Oregon', *Nature*, 179, pp 685–686.

COX, A., (1968) 'Lengths of geomagnetic polarity intervals'. *J. Geophys. Res.*, 73, pp 3247–3260.

COX, A., (1969) 'Geomagnetic reversals', *Science*, 163, pp 237–245.

COX, A. & DOELL, R. R., (1960) 'Review of palaeomagnetism, ' *Geo. Soc. Amer. Bull.*, 71 pp 645–768.

CRAIN, I. K., CRAIN, P. L. & PLAUT, M. G., (1969) 'Long period Fourier spectrum of geomagnetic reversals'. *Nature*, 223, pp 283–284.

CRAWFORD, A. R. & DAILY, B., (1971) 'Probable non-synchroneity of late Precambrian glaciations'. *Nature*, 230, pp 111–112.

CREAGER, K. C. & JORDAN, T. H., (1984) 'Slab penetration into the lower mantle', *J. Geophys. Res.*, 83, pp 3031–3050.

CREER, K. M., (1962a) 'An analysis of the geomagnetic field using palaeomagnetic methods'. *J. Geomag. Geoelect.*, 13, pp 113–119.

CREER, K. M., (1962b) 'The dispersion of the geomagnetic field due to secular variation and its determination for remote times from palaeomagnetic data'. *J. Geophys. Res.*, 67, pp 3461–3476.

CREER, K. M., (1968a) 'Arrangement of the continents during the Palaeozoic era'. *Nature*, 219, pp 41–44.

CREER, K. M., (1968b) 'Palaeozoic palaeomagnetism'. *Nature*, 219, pp 246–250.

CREER, K. M., (1975) 'On a tentative correlation between changes in the geomagnetic polarity bias and reversal frequency and the Earth's rotation through Phanerozoic time', in: *Rhythms and the history of the Earth's rotation* (eds. G. D. ROSENBERG & S. K. RUNCORN), John Wiley, London, pp 293–318.

CREER, K. M., (1978) 'Geomagnetic secular variations during the last 25,000 years: an interpretation of data obtained from rapidly deposited sediments', *Geophys. J. R. Astr. Soc.*, 48 pp 91–110.

CREER, K. M., IRVING, E. & RUNCORN, S. K., (1957) 'Geophysical interpretation of palaeomagnetic directions from Great Britain'. *Phil. Trans. R. Soc., Lond.*, A250 pp 144–155.

CREER, K. M., IRVING, E. & RUNCORN, S. K., (1957) 'Palaeomagnetic investigations in Great Britain'. *Phil. Trans. R. Soc., Lond.*, A250, pp 144–156.

CREER, K. M. & SANVER, M., (1967) 'The use of the sun compass', in: *Methods in palaeomagnetism* (eds. D. W. COLLINSON, K. M. CREER & S. K. RUNCORN), Elsevier, Publ. Co., Amsterdam, pp 11–15.

CRIMES, T. P., (1976) 'Sand fans, turbidites, slumps and the origin of the Bay of Biscay: a facies analysis of the Guipuzcoan Flysch'. *Palaeogeo., Palaeoclim. and Palaeoecol.*, 19, pp 1–15.

CRIMES, T. P. & OLDERSHAW, M. A., (1967) 'Palaeocurrent determinations by magnetic fabric

measurements on the Cambrian rocks of St. Tudwal's peninsula, North Wales'. *J. Geol.*, 75, pp 217–232.
CROUGH, S. T. & JURDY, D. M., (1980) 'Subducted lithosphere, hot spots and the geoid', *Earth Planet Sci. Lett.*, 48, pp 15–22.
CROWELL, J. C. & FRAKES, L. A., (1970) 'Phanerozoic glaciation and the causes of ice ages'. *Amer. J. Sci.*, 268, pp 193–224.
DALLMEYER, R. D. & SUTTER, J. F., (1980) 'Acquisition chronology of remanent magnetisation along the 'Grenville Polar Paths': evidence from ^{40}Ar/^{39}Ar ages of hornblende and biotite from the Whitestone Diorite, Ontario'. *J. Geophys. Res.*, 85, pp 3177–3186.
DALRYMPLE, G. B., (1979) 'Critical tables for conversion of K–Ar ages from old to new constants'. *Geology*, 7, pp 558–560.
DALY, M. C., (1986) 'The Irumide belt of Zambia and its bearing on collision orogeny during the Proterozoic of Africa', in: *Collision Tectonics* (eds. M. P. COWARD & A. C. RIES), *Spec. Publ. Geol. Soc. Lond.*, 19, pp 321–328.
DALZIEL, I. W. D., (1982) 'The early (Pre-Middle Jurassic) history of the Scotia Arc region: a review and progress report', in: *Antarctic Geoscience* (ed. C. CRADDOCK), University of Wisconsin Press, Madison, Wisconsin, pp 111–126.
DANKERS, P., (1982) 'Implications of early Devonian poles from the Canadian Arctic Archipelago for the North American apparent polar wander path'. *Can. J. Earth Sci.*, 19, pp 1802–1809.
DANKERS, P. & LAPOINTE, P. L., (1981) 'Palaeomagnetism of Lower Cambrian volcanics and cross-cutting Cambro-Ordovician diabase dyke from Buckingham (Quebec)', *Can. J. Earth Sci.*, 18, pp 1174–1186.
DAVIES, D. M. & SOLOMON, S. C., (1985) 'True polar wander and plate-driving forces', *J. Geophys. Res.*, 90, pp 1837–1841.
DAVIES, G. F., (1977) 'Whole mantle convection and plate tectonics', *Geophys. J. R. Astr. Soc.*, 49, pp 459–486.
DAVIES, G. F., (1984) 'Geophysical and isotopic constraints on mantle convection: an interim synthesis'. *J. Geophys. Res.*, 89, pp 6017–6040.
DAVIES, P. A. & RUNCORN, S. K., (eds) (1981) *Mechanisms of continental drift and plate tectonics*. Academic Press, London, 362 pp.
DAVIS, K. D., (1981) 'Magnetite rods in plagioclase as the primary carrier of stable NRM in ocean floor gabbros', *Earth Planet, Sci. Lett.*, 55, pp 190–198.
DAY, R., (1975) 'Some curious thermomagnetic curves and their interpretation'. *Earth Planet. Sci. Lett.*, 27, pp 95–100.
DAY, R., (1977) 'TRM and its variation with grain size: a review'. *J. Geomag. Geoelect.* 29, pp 223–266.
DEARNLEY, R., (1966) 'Orogenic fold-belts and a hypothesis of earth evolution'. *Phys. Chem. Earth*, 7, pp 1–114.
DE PAOLO, D. J., (1979) 'Implications of correlated Nd and Sr isotopic variations for the chemical evolution of the crust and mantle'. *Earth Planet. Sci. Lett.*, 43, pp 201–211.
DE PAOLO, D. J., (1981) 'Neodymium isotopes in the Colorado Front Range and crust-mantle evolution in the Proterozoic'. *Nature*, 291, pp 193–196.
DE PAOLO, D. J. & WASSERBURG, G. J., (1976) 'Inferences about magma sources and mantle structure from variations of ^{143}Nd/^{144}Nd', *Geophys. Res. Lett.*, 3, pp 743–746.
DE PAOLO, D. J. & WASSERBURG, G. J., (1977) 'The sources of island arcs as indicated by Nd and Sr isotopic studies', *Geophys. Res. Lett.*, 4, pp 465–468.
DEUTSCH, E. R., (1980) 'Magnetism of the Mid-Ordovician Tramore Volcanics, SE Ireland, and the question of a wide Proto-Atlantic Ocean'. *J. Geomag. Geoelectr.*, 32, Suppl. III, SIII, pp 77–98.
DEUTSCH, E. R., RAO, K. V., LAURET, R. & SEGUIN, M. K., (1972) 'New evidence and possible origin of native iron in ophiolites of eastern Canada'. *Nature*, 269, pp 684–685.
DEWEY, J. F., (1969) 'Evolution of the Appalachian/Caledonian orogen'. *Nature*, 222, pp 124–129.

DEWEY, J. F. & BIRD, J. M., (1970) 'Mountain belts and the new global tectonics', *J. Geophys. Res.*, 75, pp 2625–2647.

DEWEY, J. F. & BURKE, K., (1973) 'Tibetan, Variscan and Precambrian basement reactivation: products of continental collision'. *J. Geol.*, 81, pp 683–692.

DEWEY, J. F. & PANKHURST, R. J., (1970) 'The evolution of the Scottish Caledonides in relation to their isotopic age pattern', *Trans. R. Soc. Edin. ; Earth Sciences.*, 68, pp 351–389.

DICKINSON, B. B. & WATSON, J., (1976) 'Variations in crustal level and geothermal gradient during the evolution of the Lewisian complex of northwest Scotland'. *Precambrian Research*, 3, pp 363–374.

DICKSON, G. O., EVERITT, C. W. F., PARRY, L. G. & STACEY, F. D., (1966) 'Origin of the thermoremanent magnetisation', *Earth Planet. Sci. Lett.*, 1, pp 222–224.

DIEHL, J. F. & SHIVE P. N., (1981) 'Palaeomagnetic results from the late Carboniferous/early Permian Casper Formation: implications for northern Appalachian tectonics'. *Earth Planet. Sci. Lett.*, 54, pp 281–292.

DODSON, M. A., (1973) 'Closure temperature in cooling geochronological and petrological systems'. *Contr. Mineral. Petrol.*, 40, pp 259–274.

DODSON, M. A. & MCCLELLAND-BROWN, E., (1980) 'Magnetic blocking temperatures of single-domain grains during slow cooling', *J. Geophys. Res.*, 85, pp 2625–2637.

DOELL, R. R. & COX, A., (1963) 'The accuracy of palaeomagnetic method as evaluated from historic Hawaiian lava flows'. *J. Geophys. Res.*, 68, pp 1997–2009.

DOELL, R. R. & COX, A., (1967) 'Palaeomagnetic sampling with a portable coring drill', in: *Methods in Palaeomagnetism* (ed.: D. W. COLLINSON, K. M. CREER & S. K. RUNCORN), Elsevier Publ. Co., Amsterdam, 21–25.

DOIG, R., (1970) 'An alkaline rock province linking Europe and North America'. *Can. J. Earth Sci.*, 7, pp 22–28.

DONOVAN, R. N., ARCHER, R., TURNER, P. & TARLING, D. H., (1976) 'Devonian palaeogeography of the Orcadian Basin and the Great Glen Fault'. *Nature*, 259, pp 550–551.

DONOVAN, T. J., FORGEY, R. L. & ROBERTS, A. A., (1979) 'Aeromagnetic detection of diagenetic magnetite over oil fields'. *Bull. Amer. Assoc. Petrol. Geol.*, 63, pp 245–248.

DOUGLAS, R. G. & SAVIN, S. M., (1975) 'Oxygen and Carbon isotopic analyses of Tertiary and Cretaceous microfossils from Shatsky Rise and other sites in the North Pacific', *Initial Rep. Deep Sea Drill. Proj.*, 32, 509–520.

DOWNING, K. N. & COWARD, M. P., (1981) 'The Okahandja lineament and its significance in Damaran tectonics in Namibia'. *Geologische Rundschau*, 70, pp 972–1000.

DREWRY, G. E., RAMSAY, A. T. S. & SMITH, A. G., (1974) 'Climatically-controlled sediments, the geomagnetic field and trade windbelts in Phanerozoic time', *J. Geol.*, 82 pp 531–533.

DUBOIS, P. M., (1962) 'Palaeomagnetism and correlation of Keweenawan rocks'. *Bull. Geol. Surv. Can.*, 71, 75 pp.

DUFF, B. A., (1979a) 'Peaked thermomagnetic curves for hematite-bearing rocks and concentrates', *Phys. Earth Planet. Int.*, 19, pp P1–P4.

DUFF, B. A., (1979b) 'The palaeomagnetism of Cambro-Ordovician red beds, the Erquy Spilite Series and the Trégastel-Ploumanach granite complex, Armorican Massif (France and the Channel Islands)'. *Geophys. J. R. Astr. Soc.*, 59, pp 345–365.

DUFF, B. A., (1980) 'The palaeomagnetism of Jersey volcanics and dykes, and the Lower Palaeozoic apparent polar wander path for Europe'. *Geophys. J. R. Astr. Soc.*, 60, pp 355–375.

DUNCAN, R. A., PETERSEN, N. & HARGRAVES, R. B., (1972) 'Mantle plumes, movement of the European Plate and polar wandering'. *Nature*, 239, pp 82–86.

DUNLOP, D. J., (1968) 'Monodomain theory: experimental verification', *Science*, 162, pp 256–258.

DUNLOP, D. J., (1971) 'Magnetic properties of fine-particle hematite'. *Ann. Geophys.*, 27, pp 269–293.

DUNLOP, D. J., (1972) 'Magnetic mineralogy of unheated and heated rock sediments by coercivity spectrum analysis', *Geophys. J. R. Astr. Soc.*, 27, pp 37–55.

DUNLOP, D. J., (1977) 'Rocks as high-fidelity tape recoders', *IEEE Transactions on Magnetics*, Vol. MAG-13, 5, pp 1267–1271.

DUNLOP, D. J., (1979a) 'On the use of Zijderveld vector diagrams in multicomponent palaeomagnetic studies'. *Phys. Earth Planet. Int.*, 20, pp 12–24.

DUNLOP, D. J., (1979b). 'A regional palaeomagnetic study of Archaean rocks from the Superior Geotraverse area, northwestern Ontario'. *Can. Jour. Earth Sci.*, 16, pp 1906–1919.

DUNLOP, D. J., (1981) 'The rock magnetism of fine particles'. *Phys. Earth Planet. Int.*, 26, pp 1–26.

DUNLOP, D. J., (1983) 'Determination of domain structure in igneous rocks by alternating field and other methods'. *Earth Planet Sci. Lett.*, 63, pp 353–367.

DUNLOP, D. J., HANES, J. A. & BUCHAN, K. L., (1983) 'Indices of multidomain magnetic behaviour in basic igneous rocks: alternating field demagnetisation hysteresis and oxide petrology'. *Jour. Geophys. Res.*, 70, pp 1387–1393.

DUNLOP, D. J. & PRÉVOT, M., (1982) 'Magnetic properties and opaque mineralogy of drilled submarine intrusive rocks'. *Geophys. J. R. Astr. Soc.*, 69, pp 763–802.

DUNLOP, D. J. & STIRLING, J. M., (1977) 'Hard viscous remanent magnetisation (VRM) in fine-grain hematite' *Geophys. Res. Lett.*, 4, pp 163–166.

DUNLOP, D. J. & WADDINGTON, E. D., (1975) 'The field dependence of thermoremanent magnetisation in igneous rocks'. *Earth Planet. Sci. Lett.*, 25, pp 11–25.

DUNN, J. R., FULLER, M. D., ITO, H. & SCHMIDT, V. A., (1975) 'Palaeomagnetic study of a reversal of the Earth's magnetic field', *Science*, 172, pp 840–845.

DUNN, P. R., THOMPSON, B. P. & RANKAMA, K., (1971) 'Late Precambrian glaciations in Australia as stratigraphic boundaries'. *Nature*, 231, pp 498–502.

DUNNETT, D., (1976) 'Some aspects of the Panantarctic-cratonic margin in Australia'. *Phil. Trans. R. Soc. Lond.*, A280, pp 641–654.

DU TOIT, A. L., (1937) *Our Wandering Continents*, Oliver & Boyd, London, 366 pp.

DZIEWONSKI, A. M., HAGER, B. H. & O'CONNELL, R. J., (1977) 'Large-scale heterogeneities in the lower mantle'. *J. Geophys. Res.*, 82, pp 239–255.

EDWARDS, J., (1980) 'Comparisons between the generation and properties of rotational remanent magnetisation and anhysteritic remanent magnetisation'. *Geophys. J. R. Astr. Soc.*, 62, pp 379–392.

EGYED, L., (1959) 'The expansion of the Earth in connection with its origin and evolution'. *Geophysica*, 7, pp 13–22.

EINARSSON, T., (1957) 'Magneto-geological mapping in Iceland with the use of a compass'. *Phil. Mag., Suppl.* 6, 22, pp 232–238.

ELLWOOD, B. B., (1978) 'Flow and emplacement directions determined for selected basaltic bodies using magnetic anisotropy measurements'. *Earth Planet. Sci. Lett.*, 41, pp 254–264.

ELLWOOD, B. B., (1982) 'Palaeomagnetic evidence for the continuity and independent movement of a distinct major crustal block in the southern Appalachians'. *Jour. Geophys. Res.*, 87, pp 5339–5350.

ELLWOOD, B. B., (1984) 'Bioturbation: some effects on remanent magnetisation acquisition'. *Geophys. Res. Lett.*, 11, pp 653–655.

ELSASSER, W. M., (1946) 'Induction effects in terrestrial magnetism'. *Phys. Rev.*, 69, pp 106–116.

ELSASSER, W. M., (1971) 'Sea floor spreading as thermal convection'. *J. Geophys. Res.*, 76, pp 1101–1112.

ELSTON, D. P. & BRESSLER, S. L., (1977) 'Palaeomagnetic poles and polarity zonation from Cambrian and Devonian strata of Arizona'. *Earth Planet. Sci. Lett..*, 36, pp 423–433.

ELSTON, D. P. & GROMMÉ, C. S., (1974) 'Precambrian polar wandering from the Unkar Group and Nankoweap Formation, Eastern Grand Canyon, Arizona', in: *Geology of North Arizona with notes on archaeology and palaeoclimate, Pt. I — Regional Studies* (ed. T. N. V. KARLSTROM, G. A. SWANN & R. L. EASTWOOD), Geol. Soc. Amer. Rocky Mountain Section guidebook, 27th Ann. Meeting, Flagstaff, Arizona, pp 97–117.

ELSTON, D. P. & PURUCKER, M., (1979) 'Detrital magnetisation in redbeds of the Moenkopi Formation'. *J. Geophys. Res.*, 84, pp 1653-1655.
EMBLETON, B. J. J. & MCELHINNY, M. W., (1975) 'The palaeoposition of Madagascar: palaeomagnetic evidence from the Isolo Group', *Earth Planet. Sci. Lett.*, 27, pp 329-341.
EMBLETON, B. J. J., MCELHINNY, M. W., CRAWFORD, A. R. & LUCK, G. R., (1974) 'Palaeomagnetism and the tectonic evolution of the Tasman orogenic zone'. *J. Geogr. Soc. Aust.*, 21, pp 187-194.
EMBLETON, B. J. J. & MCELHINNY, M. W., (1982) 'Marine magnetic anomalies, palaeomagnetism and the drift history of Gondwanaland'. *Earth Planet. Sci. Lett.*, 58, pp 141-150.
EMBLETON, B. J. J. & SCHMIDT, P. W., (1977) 'Revised palaeomagnetic data for the Australian Mesagoic and a synthesis of late Palaeozoic-Mesozoic results from Gondwanaland'. *Tectonophysics*, 38, pp 355-364.
EMBLETON, B. J. J., VEEVERS, J. J., JOHNSON, & POWELL, C. MCA. (1980) 'Palaeomagnetic comparison of a new fit of east and west Gondwanaland with the Smith and Hallam fit'. *Tectonophysics*, 61, pp 381-390.
ENGEL, A. E. J. & KELM, D. L., (1972) 'Pre-Permian global tectonics: a tectonic test'. *Geol. Soc. Amer. Bull.*, 83, pp 2325-2340.
ENGEL, A. E. J., ITSON, S. P., ENGEL, C. G., STICKNEY, D. M. & CRAY, E. J., (1974) 'Crustal evolution and global tectonics: a petrogenic view'. *Bull. Geol. Soc. Amer.*, 85, pp 843-858.
ENGLAND, P. C. & MCKENZIE, D. P., (1982). 'A thin viscous sheet model for continental deformation'. *Geophys. J. R. Astr. Soc.*, 70, pp 295-321.
ESANG, C. B. & PIPER, J. D. A., (1984) 'Palaeomagnetism of Caledonian intrusive suites in the Northern Highlands of Scotland: constraints to tectonic movements within the Caledonian orogenic belt'. *Tectonophysics*, 104, pp 1-34.
EVANS, M. E., (1976) 'Test of the dipolar nature of the geomagnetic field throughout Phanerozoic time'. *Nature*, 262, pp 676-678.
EVANS, M. E. & BINGHAM, D. K., (1976) 'Palaeomagnetism of the Great Slave Supergroup, Northwest Territories, Canada: the Tochatwi Formation'. *Can. J. Earth Sci.*, 13, pp 555-562.
EVANS, M. E. & BINGHAM, D. K., (1973) 'Palaeomagnetism of the Precambrian Martin Formation, Saskatchewan'. *Can. J. Earth Sci.*, 10, pp 1485-1495.
EVANS, M. E., BINGHAM, D. K. & MCMURRY, E. W., (1975) 'New palaeomagnetic results from the Upper Belt-Purcell Supergroup of Alberta'. *Can. J. Earth Sci.*, 12, pp 52-61.
EVANS, M. E. & MCELHINNY, M. W., (1969). 'An investigation of the origin of stable remanence in magnetite-bearing igneous rocks.' *J. Geomag. Geoelect.*, 21, pp 757-773.
EVANS, M. E. & WAYMAN, M. L., (1970) 'An investigation of small magnetic particles by means of electron microscopy'. *Earth Planet. Sci. Lett.*, 9, pp 365-370.
EVERITT, C. W. F., (1962) 'Thermoremanent magnetisation II. Experiments on multidomain grains'. *Philos. Mag.*, 7, pp 583-597.
EVERITT, C. W. F. & CLEGG, J. A., (1962) 'A field test of palaeomagnetic stability'. *Geophys. J. R. Astr. Soc.*, 6, pp 312-319.
EUGSTER, H. P., (1972) 'Reduction and oxidation in metamorphism (II)'. *Int. Geol. Cong.*, 24th sess., Montreal Sect. 10, pp 3-11.
FACER, R. A., (1983) 'Folding, strain and Graham's fold test in palaeomagnetic investigations'. *Geophys. J. R. Astr. Soc.*, 72, pp 165-172.
FAHRIG, W. F., IRVING, E. & JACKSON, G. D., (1971) 'Palaeomagnetism of the Franklin diabases'. *Can. J. Earth Sci.*, 8, pp 455-467.
FALLER, A. M., (1975) 'Palaeomagnetism of the oldest Tertiary basalts in the Kangerdlugssuag area of East Greenland'. *Meddelser fra Dansk Geologisk Forening*, 24, pp 173-178.
FALLER, A. M., BRIDEN, J. C. & MORRIS, W. A., (1977) 'Palaeomagnetic results from the Borrowdale Volcanic Group, English Lake District'. *Geophys. J. Astr. Soc.*, 48, pp 111-121.
FIELD, D. & RAHEIM, A., (1981) 'Age relationships in the Proterozoic high grade gneiss regions of southern Norway'. *Precambrian Research* 14, pp 261-275.

FINLAYSON, D. M. & MATHUR, S. P., (1984) 'Seismic refraction and reflection features of the lithosphere in northern and eastern Australia and continental growth'. *Annales Geophysicae*, 2, pp 711–722.

FISHER, Sir R. A., (1953) 'Dispersion on a sphere'. *Proc. R. Soc., Lond.*, A217, pp 295–305.

FITCH, T. J., (1972) 'Plate convergence, transcurrent faults and internal deformation adjacent to southeast Asia and the western Pacific', *J. Geophys. Res.*, 77, pp 4432–4461.

FLOYD, P. A., (1972) 'Geochemistry, origin and tectonic environment of the basic and acidic rocks of Cornubia, England'. *Proc. Geol. Assoc.*, 83, pp 385–404.

FORCE, E. R., (1984) 'A relation among geomagnetic reversals, seafloor spreading rate, palaeoclimate and black shales, EOS,. *Trans. Amer. Geophys. Un.* 65, pp 18–19.

FORSYTH, D. W. & UYEDA, S., (1975) 'On the relative importance of the driving forces of plate motions'. *Geophys. J. R. Astr. Soc.*, 43, pp 163–200.

FOSS, C., (1981) 'Graphical methods for rapid vector analysis of demagnetisation data'. *Geophys. J. R. Astr. Soc.*, 65, pp 217–221.

FOSTER, T. D., (1971) 'Intermittent convection'. *Geophys. Fluid Dyn.*, 2, pp 201–217.

FOWLER, A. C., (1983) 'On the thermal state of the Earth's mantle'. *J. Geophys.*, 53, pp 42–51.

FOX, K. F. & BECK, M. E., (1985) 'Palaeomagnetic results for Eocene volcanic rocks from NE Washington and the Tertiary tectonics of the Pacific NW'. *Tectonics*, 4, pp 323–341.

FOX, P. J. & OPDYKE, N. D., (1973) 'Geology of the oceanic crust: magnetic properties of oceanic rocks'. *J. Geophys. Res.*, 78, pp 5139–5154.

FRAKES, L. A., (1979) *Climates through Geologic Time*, Elsevier, Amsterdam, 310 pp.

FRANCHETEAU, J. & SCLATER, J. G., (1970) 'Comment on paper by E. Irving and W. A. Robertson: Test for polar wandering and some possible implications.' *J. Geophys. Res.*, 75, pp 1023–1026.

FREUND, R. & TARLING, D. H., (1979) 'Preliminary Mesozoic palaeomagnetic results from Israel and inferences for a microplate structure in the Lebanon'. *Tectonophysics*, 60, pp 189–205.

FROST, C. D. & O'NIONS, R. K., (1984) 'Nd evidence for Proterozoic crustal development in the Belt-Purcell Supergroup'. *Nature*, 312, pp 53–56.

FUDAO, X. and GUANGHONG, X., (1978) 'The age of anorthosite event and its geological implications.' (in Chinese). *Geochimica*, 3, pp 202–208.

FUJIWARA, Y. & MORINAGA, Y., (1984) 'Cretaceous remagnetisation of the Palaeozoic rocks in the south Kitakami mountains, NE Honshu, Japan'. *Rock Mag. Palaeogeophys.*, 10, pp 85–86.

FULLER, M. D., (1960) 'Anisotropy of susceptibility and the natural remanent magnetisation of some Welsh slates'. *Nature*, 186, pp 791–792.

FUNNELL, B. M. & SMITH, A. G., (1968) 'Opening of the Atlantic Ocean'. *Nature*, 219, pp 1328–1333.

GAMES, K. P., (1977) 'The magnitude of the palaeomagnetic field: a new non-thermal, non-detrital method using sun-dried bricks'. *Geophys. J. R. Astr. Soc.*, 48, pp 315–329.

GAMES, K. P., (1980) 'The magnitude of the archaeomagnetic field in Egypt between 3000 and 0 BC.' *Geophys. J. R. Astr. Soc.*, 63, pp 45–56.

GARRELS, R. M. & CHRIST, C. L., (1965) *Solutions, Minerals and Equilibria*, Harper & Row, New York, 450 pp.

GARRELS, R. M. & MACKENZIE, F. T., (1971) *Evolution of sedimentary rocks*, W. W. NORTON, New York, 397 pp.

GAST, P. W., (1965) 'Terrestrial ratio of potassium to rubidium and the composition of the Earth's mantle'. *Science*, 147, pp 858–860.

GASTIL, G., (1960) 'The distribution of mineral dates in time and space'. *Amer. J. Sci.*, 258, pp 1–35.

GIBSON, I. L. & PIPER, J. D. A., (1972) 'Structure of the Icelandic basalt plateau and the process of drift'. *Phil. Trans. R. Soc., Lond.*, A271, pp 141–150.

GIDDINGS, J. W. & MCELHINNY, M. W., (1976) 'A new index of palaeomagnetic stability for magnetite-bearing igneous rocks'. *Geophys. J. R. Astr. Soc.*, 44, pp 239–251.

GILLETT, S. L. & VAN ALSTINE, D. R., (1979) 'Palaeomagnetism of Lower and Middle Cambrian sedimentary rocks from the Desert Range, Nevada'. *J. Geophys. Res.*, 84, pp 4475–4489.

GILKSON, A. Y., (1983) 'Geochemical, isotopic and palaeomagnetic tests of early sial-sima patterns: the Precambrian crustal enigma revisited', in: *Proterozoic Geology: selected papers from an International Proterozoic Symposium.* (eds. L. G. MEDARIS, C. W. BYERS, D. H. MICKELSON & W. C. SHANKS)., *Geol. Soc. Amer. Mem.,* 161, pp 95–117.

GOLDICH, S. S., (1973) 'Ages of Precambrian banded iron formation'. *Econ. Geol.*, 68, pp 1126–1134.

GOLDREICH, P. & TOOMRE, A., (1969) 'Some remarks on polar wandering'. *Jour. Geophys. Res.*, 74, pp 2555–2567.

GOLEBY, B. R., (1980) 'Early Palaeozoic palaeomagnetism in south east Australia'. *J. Geomag. Geoelect.*, 32, Suppl. III, SIII-11-SIII-21.

GOODWIN, A. M., (1973) 'Archaean iron formations and tectonic basins of the Canadian Shield'. *Econ. Geol.*, 68, pp 915–933.

GOODWIN, A. M., (1977) 'Archaean volcanism in the Superior Province, Canadian Shield', in: *Volcanic Régimes in Canada* (ed. W. R. A. BARAGAR, L. C. COLEMAN & J. M. HALL), *Geol. Assoc. Can., Spec. Pap.* 16, pp 205–224.

GOODWIN, A, M., (1981) 'Archaean plates and greenstone belts,' in: *Precambrian Plate Tectonics*, (ed. A. KRÖNER), Elsevier Publ. Co., Amsterdam, pp 105–135.

GOODWIN, A. M. & RIDLER, R. H., (1970) 'The Abitibi orogenic belt'. *Geol. Surv. Can. Paper* 70-40, pp 1–24.

GORDON, R. G., MCWILLIAMS, M. O. & COX, A., (1979) 'Pre-Tertiary velocities of the continents: a lower bound from palaeomagnetic data'. *J. Geophys. Res.*, 84, pp 5480–5486.

GOSE, W. A. & SCHWARTZ, D. K., (1977) 'Palaeomagnetic results from Cretaceous sediments in Honduras: tectonic implications'. *Geology*, 5, pp 505–508.

GOSE, W. A., BELCHER, R. C. & SCOTT, G. R., (1982). 'Palaeomagnetic results from north eastern Mexico: evidence for large Mesozoic rotations.' *Geology*, 10, pp 50–54.

GOUGH, D. I., (1977) 'The geoid and single cell mantle convection'. *Earth Planet. Sci. Lett.*, 34, pp 360–364.

GOUGH, D. I., RAHMAN, A. & EVANS, M. E., (1977) 'Magnetic anisotropy and fabric of red beds of the Great Slave Supergroup of Canada'. *Geophys. J. R. Astr. Soc.*, 50, pp 685–697.

GRAHAM, J. W., (1949) 'The stability and significance of magnetism in sedimentary rocks'. *J. Geophys. Res.*, 54, pp 131–167.

GRAHAM, J. W., (1955) 'Evidence of polar shift since Triassic time'. *J. Geophys. Res.*, 60, pp 329–349.

GRANAR, L. (1958) 'Magnetic measurements on Swedish varved sediments'. Arkiv for Geofysik, Svenska Vetenskap Akademiens, Stockholm 3, pp 1–40.

GREENWOOD, W. R., HADLEY, D. G., ANDERSON, R. E., FLECK, R. J. & SCHMIDT, D. L., (1976) 'Late Proterozoic cratonisation in south western Saudi Arabia'. *Phil. Trans. R. Soc. Lond.* A280 pp 517–527.

GREGOR, C. B. & ZIJDERVELD, J. D. A., (1964) 'The magnetism of some Permian red sandstones from north-western Turkey', *Tectonophysics*, 1, pp 289–306.

GRIFFITHS, D. H., KING, R. F., REES, A. J. & WRIGHT, A. E., (1960) 'The remanent magnetism of some recent varved sediments'. *Proc. R. Soc., Lond.*, A256, pp 359–383.

GROSS, G. A., (1970) 'Nature and occurrence of iron ore deposits', in: *Survey of World Iron Ore Resources*, United Nations, New York.

GUBBINS, D., MASTERS, T. G. & JACOBS, J. A., (1979) 'Thermal evolution of the Earth's core'. *Geophys. J. R. Astr. Soc.*, 59, pp 57–99.

GUOLIANG, Y. C., QINGGE, S., YUHANG, S., YONFAN, L., YUJE, D. & DONGJIANG, S., (1985) 'Late Palaeozoic polar wandering path for the Tarim Block and its tectonic significance'. *Abstracts, IAGA Meeting, Prague,* 1985, p 189.

HAGGERTY, S. E., (1976) 'Opaque mineral oxides in terrestrial igneous rocks', in: *Oxide Minerals* (ed. D. RUMBLE). *Mineralogical Soc. America, Short Course Notes*, III, pp 101–176.

HAGGERTY, S. E., (1978) 'The aeromagnetic mineralogy of igneous rocks'. *Can. J. Earth Sci.*, 16, pp 1281–1293.

HAGSTRUM, J., VAN DER VOO, R., AUVRAY, B. & BONHOMMET, N., (1980) 'Eocambrian-Cambrian palaeomagnetism of the Armorican Massif, France'. *Geophys. J. R. Astr. Soc.*, 61, pp 489–513.

HAILE, N. S., (1979) 'Palaeomagnetic evidence for rotation and northward drift of Sumatra'. *J. Geol. Soc. Lond.*, 136, pp 541–545.

HAILE, N. S., (1980) 'Palaeomagnetic evidence from the Ordovician and Silurian of northwest Peninsular Malaysia'. *Earth Planet. Sci. Lett.*, 48, pp 233–236.

HAILE, N. S., (1981) 'Palaeomagnetism of southeast and east Asia', in: *Palaeoreconstruction of the continents. Geodynamics Series* 2, Amer. Geophys. Un., pp 129–135.

HAILWOOD, E. A., (1974) 'Palaeomagnetism of the Msissi norite (Morocco) and the Palaeozoic reconstruction of Gondwanaland', *Earth Planet. Sci. Lett.*, 23 pp 376–386.

HALE, C. J., & DUNLOP, D. J., (1984) 'Evidence for an early Archaean geomagnetic field: a palaeomagnetic study of the Komati Formation, Barberton Greenstone Belt, South Africa. *Geophys. Res. Lett.*, 11, pp 97–100.

HALLAM, A., (1973) *Atlas of Palaeobiogeography*. Elsevier Publ. Co., Amsterdam, 531 pp.

HALLAM, A., (1977) 'Secular changes in marine inundation of USSR and North America through the Phanerozoic'. *Nature*, 269, pp 769–772.

HALLAM, A., (1983) 'Supposed Permo-Triassic megashear between Laurasia and Gondwana', *Nature*, 301, pp 499–502.

HALLER, J., (1970) *Tectonic map of East Greenland (1:500,000). An account of tectonism, plutonism and volcanism in East Greenland*, Meddelelser om, Grønland Geoscience 171, 286 pp.

HALLS, H. C., (1975) 'Shock induced remanent magnetisation in late Precambrian rocks from Lake Superior'. *Nature*, 255, pp 692–695.

HALLS, H. C., (1976) 'A least squares method to find a remanence direction from converging remagnetisation circles'. *Geophys. J. R. Astr. Soc.*, 45, pp 297–304.

HALLS, H. C., (1978) 'The use of converging remagnetisation circles in palaeomagnetism'. *Phys. Earth Planet. Int.*, 16, pp 1–11.

HALLS, H. C., (1982) 'The importance and potential of mafic dyke swarms in studies of geodynamic processes'. *Geoscience Canada*, 9, pp 145–184.

HALLS, H. C. & PESONEN, L. J., (1983) 'Palaeomagnetism of Keweenawan rocks'. *Geol. Soc. Amer., Mem.* 156, pp 173–201.

HAMANO, Y., (1980) 'An experiment on the post-depositional remanent magnetisation in artificial and natural sediments'. *Earth Planet. Sci. Lett.*, 51, pp 221–232.

HAMILTON, N & REES, A. I., (1970) 'The use of magnetic fabric in palaeocurrent estimation', in: *Palaeogeophysics*, (ed. RUNCORN S. K.) Academic Press, New York, pp 445–464.

HAMILTON, W., (1970) 'The Uralides and the motion of the Russian and Siberian Platform'. *Bull. Geol. Soc. Amer.*, 81, pp 2553–2567.

HANSON, G. N., (1975) '$^{40}Ar:^{39}Ar$ spectrum ages on Logan intrusions, a Lower Keweenawan flow and mafic dykes in north eastern Minnesota–northwestern Ontario'. *Can. J. Earth Sci.*, 12, pp 821–835.

HARGRAVES, R. B. & DUNCAN, R. A., (1973) 'Does the mantle roll?' *Nature*, 245, pp 361–363.

HARGRAVES, R. B. & FISH, J. R., (1972). 'Palaeomagnetism of anorthosite in southern Norway and comparison with an equivalent in Quebec'. *Geol. Soc. Amer. Mem.*, 132 141–148.

HARGRAVES, R. B. & YOUNG, W. M. (1969) Source of stable remanent magnetism in Lambertville diabase, '*Amer. J. Sci.*, 267, pp 1161–1177.

HARLAND, W. B. (1964). 'Critical evidence for a great infra-Cambrian glaciation'. *Geologische Rundschau.*, 54 45–61.

HARLAND, W. B., (1971) 'Tectonic transpression in Caledonian Spitsbergen', *Geol. Mag.*, 108(1), pp 27–42.
HARLAND, W. B., (1972) 'The Ordovician ice age', *Geol. Mag.*, 109(5), pp 451–456.
HARLAND, W. B., (1978) 'The Caledonides of Svalbard', in: *Caledonian-Appalachian orogen of the North Atlantic region* (IGCP Project 27), Geol. Surv. Can. Pap., 78-13, pp 3–11.
HARLAND, W. B., (1983) 'The Proterozic glacial record', in: *Proterozic Geology: selected papers from an International Proterozoic Symposium* (ed. L. G. MEDARIS, C. W. BYERS, D. M. MICKELSON & W. C. SHANKS). Geol. Soc. Amer., Mem. 161, pp 279–288.
HARLAND, W. B. & HEROD, K. N., (1974) 'Glaciations through time, in: *Ice Ages: Ancient and Modern*', Proc. 21st Inter-University Geol. Congress, Birmingham, Geol. J. Spec. Issue 6, Seel House Press, pp 189–216.
HARLAND, W. B., COX, A. V., LLEWELLYN, P. G., PICKTON, C. A. G., SMITH, A. G. & WALTERS, R., (1982) *A geologic time scale*, Cambridge University Press, Cambridge, 131 pp.
HARPER, C. T., (1967) 'On the interpretation of potassium argon ages from Precambrian shields and Phanerozoic orogens', *Earth Planet. Sci. Lett.*, 3, pp 128–132.
HARRIS, N. B. W., & GASS, I. G., (1981) 'Significance of contrasting magmatism in North East Africa and Saudi Arabia'. *Nature*, 289, pp 394–396.
HARRISON, C. G. A. & LINDH, T., (1982) 'Comparison between the hot spot and geomagnetic field reference frames', *Nature*, 300, pp 251–252.
HARRISON, C. G. A., MCDOUGALL, I. & WATKINS, N. D., (1979) 'A geomagnetic field reversal time scale back to $13\cdot0$ million years before present'. *Earth Planet. Sci. Lett.*, 42, pp 143–152.
HART, S. R., (1964) 'The petrology and isotopic mineral age relations of a contact zone in the Front Range, Colorado'. *J. Geophys. Res.*, 72, pp 493–525.
HATCHER, R. D., (1978) 'Tectonics of the Western Piedmont and Blue Ridge, southern Appalachians: review and speculations', *Amer. Jour. Sci.*, 278, pp 276–304.
HAYS, J. D. & PITMAN, III W. C. (1973). 'Lithospheric plate motion, sea level changes and climatic and ecological consequences; *Nature*, 246, 18–22.
HEIRTZLER, J. R., DICKSON, G. O., HERRON, E. M., PITMAN III, W. C. & LE PICHON, X., (1968) 'Marine magnetic anomalies, geomagnetic field reversals and motions of the ocean floor and continents'. *J. Geophys. Res.*, 73, pp 2119–2136.
HEKI, K., HAMANO, Y., KINOSHITA, H., TAIRA, A. & KONO, M., (1984) 'Palaeomagnetic study of Cretaceous rocks of Peru, South America: evidence for rotation of the Andes'. *Tectonophysics*, 108, pp 267–281.
HELSLEY, C. E., (1965). 'Palaeomagnetic results from the Lower Permian Dunkard Series of West Virginia'. *J. Geophys. Res.*, 70, pp 413–424.
HELSLEY, C. E., (1969) 'Magnetic reversal stratigraphy of the Lower Triassic Meonkopi Formation of western Colorado'. *Geol. Soc. Amer. Bull.*, 80, pp 2431–2450.
HELSLEY, C. E. & STEINER, M. B., (1969) 'Evidence for long intervals of normal polarity during the Cretaceous period'. *Earth Planet. Sci. Lett.*, 5, pp 325–332.
HELSELY, C. E. & STEINER, M. B., (1974) 'Palaeomagnetism of the Lower Triassic Moenkopi Formation'. *Geol. Soc. Amer. Bull.*, 85, pp 457–464.
HENRY, S. G., (1979) 'Chemical demagnetisation methods, procedures and applications through vector analysis'. *Can. J. Earth Sci.*, 16, pp 1832–1841.
HENRY, S. G., MAUK, F. J. & VAN DER VOO, R., (1977) 'Palaeomagnetism of the Upper Keweenawan sediments: the Nonesuch Shale and Freda Sandstone'. *Can. J. Earth Sci.*, 14, pp 1128–1138.
HENSHAW, P. C. & MERRILL, R. T., (1980) 'Magnetic and chemical changes in marine sediments'. *Rev. Geophys. Space Phys.*, 18, pp 483–504.
HENTHORN, D. I., (1972) *Palaeomagnetism of the Witwatersrand Triad, Republic of South Africa, and related topics*, Ph.D thesis, Univ. of Leeds.
HERERO-BERVERA, E. & HELSLEY, C. E., (1983) 'Palaeomagnetism of a polarity transition in the Lower (?) Triassic Chugwater Formation Wyoming'. *Jour. Geophys. Res.*, 88, pp 3506–3522.

HERZ, N., (1969) 'Anorthosite belts, continental drift, and the anorthosite event', *Science*, 164, pp 944–947.
HERZENBERG, A., (1958) 'Geomagnetic dynamos'. *Phil. Trans. R. Soc., Lond.*, A250 pp 543–585.
HICKMAN, M. H., (1979) 'An Rb–Sr age and isotope study of the Ikertôq, Nordre Strømfjord and Evighedsfjord shear belts, West Greenland — outline and preliminary results'. *Grønlands Geologiske Undersøgelse Rapport*, 89, pp 125–128.
HIDE, R., (1967) 'Motion of the Earth's core and mantle, and variations of the main geomagnetic field'. *Science*, 157, pp 55–59.
HIETANEN, A., (1975) 'Generation of potassium-poor magmatism in the north Sierra Nevada and the Svecofennian of Finland'. *Journal of Research, U.S. Geological Survey* 3, pp 631–645.
HILLHOUSE, J. W., (1977) 'Palaeomagnetism of the Triassic Nikolai greenstone, McCarthy Quadrangle, Alaska'. *Can. J. Earth Sci.*, 14, pp 2578–2592.
HILLHOUSE, J. W., GROMMÉ, C. S. & VALLIER, T. L., (1982) 'Palaeomagnetism and Mesozoic tectonics of the Seven Devils volcanic arc in northeastern Oregon'. *J. Geophys. Res.*, 87, pp 3777–3794.
HOFFMAN, K. A. & DAY, R. (1978) 'Separation of multi-component NRM: a general method'. *Earth Planet. Sci. Lett.*, 40, pp 433–438.
HOFFMAN, P., (1980) 'Wopmay orogen: a Wilson-cycle of early Proterozoic age in the northwest of the Canadian Shield'. *Geol. Assoc. Can. Spec. Pap.*, 20, pp 523–549.
HOLLAND, C. H., (1974) (ed). *Cambrian of the British Isles, Norden and Spitsbergen*, Wiley, London, 300 pp.
HOLSER, W. T., (1977) 'Catastrophic chemical events in the history of the oceans'. *Nature*, 267, pp 403–408.
HORNER, F. & FREEMAN, R., (1983) 'Palaeomagnetic evidence from pelagic limestones for clockwise rotation in the Ionian Zone, western Greece', *Tectonophysics* 98, pp 11–27.
HORNER, F. & HELLER, F., (1983) 'Lower Jurassic magnetostratigraphy at the Breggia Gorge (Ticino, Switzerland) and Alpe Turati (Como, Italy)'. *Geophys. J. R. Astr. Soc.*, 73, pp 705–718.
HORTON, R. A. & GEISSMAN, J. W., (1984) 'Palaeomagnetism and rock magnetism of the Mississippian leadville (Carbonate) Formation and implication for the age of sub-regional dolomitisation'. *Geophys. Res. Lett.*, 11, pp 649–652.
HOSPERS, J & VAN ANDEL, S. I., (1968) 'Palaeomagnetism and tectonics: a review', *Earth Sci. Rev.*, 5, pp 5–44.
HROUDA, F. & JANAK, F., (1976) 'The changes in the shape of the magnetic susceptibility ellipsoid during progressive metamorphism and deformation'. *Tectonophysics*, 34, pp 135–148.
HROUDA, F., JANAK, F. & REJL, L., (1978) 'Magnetic anistropy and ductile deformation of rocks in zones of progressive regional metamorphism'. *Gerlands Beitrage zur Geophysik* (Leipzig), 87, pp 126–134.
HUNTER, D. R., (1973) 'The localisation of tin mineralisation with reference to southern Africa'. *Minerals Sci Engng.*, 5, pp 53–77.
HURLEY, P. M., (1973) 'On the origin of 450 ± 200 m.y. orogenic belts', in: *Implications of Continental Drift to the Earth Sciences*, vol. 2, (eds. D. H. TARLING & S. K. RUNCORN) Academic Press, London, pp 1083–1089.
HURLEY, P. M. & Rand, J. R., (1969) 'Predrift continental nuclei'. *Science* 164, pp 1229–1242.
HUTCHINS, A., (1967) 'Computations of the behaviour of two- and three-axis rotation systems', in: *Methods in Palaeomagnetism* (ed. D. W. COLLINSON, K. M. CREER & S. K. RUNCORN), Elsevier Publ. Co., Amsterdam, pp 224–236.
HUTCHINSON, R. W., (1981) 'Mineral deposits as guides to supracrustal evolution' in: *Evolution of the Earth* (ed. R. J. O'CONNELL & W. S. FYFE). Geodynamics Series vol. 5, Amer. Geophys. Un., pp 120–140.
IRVING, E., (1957) 'The origin of the palaeomagnetism of the Torridonian Sandstones of north-west Scotland'. *Phil. Trans. R. Soc., Lond.*, A250, pp 100–110.
IRVING, E. (1960–1962) 'Palaeomagnetic directions and pole positions,' Parts I–VII. *Geophys. J. R.*

Astr. Soc., 3: pp 97–111, 3: pp 444–449, 4: pp 72–79, 6: pp 263–267, 7: pp 263–374, 8: pp 249–257 (with P. M. STOTT), & 9: pp 185–194.

IRVING, E., (1964) *Palaeomagnetism and its application to geological and geophysical problems*, Wiley, New York, 399 pp.

IRVING, E., (1966) 'Palaeomagnetism of some Carboniferous rocks from New South Wales and its relation to geological events'. *J. Geophys. Res.*, 71, pp 6025–6051.

IRVING, E. (1977) 'Drift of the major continental blocks since the Devonian'. *Nature*, 270, pp 304–309.

IRVING, E. & BROWN, D. A., (1964). 'Abundance and diversity of the labyrinthodonts as a function of palaeolatitude'. *Amer. J. Sci.*, 262, 689–708.

IRVING, E. & BRIDEN, J. C., (1964) 'Palaeolatitude of evaporite deposits'. *Nature*, 196, pp 425–428.

IRVING, E. & COUILLARD, R. W., (1973) 'Cretaceous normal polarity interval'. *Nature*, 244, pp 10–11.

IRVING, E., EMSLIE, R. F. & UENO, H., (1974) 'Upper Proterozoic palaeomagnetic poles from Laurentia and the history of the Grenville structural province'. *Jour. Geophys. Res.*, 79, pp 5491–5502.

IRVING, E. & GASKELL, T. F., (1962) 'The palaeogeographic latitude of oilfields', *Geophys. J. R. astr. Soc.*, 7, pp 54–64.

IRVING, E. & HASTIE, J., (1975) *Catalogue of palaeomagnetic directions and poles, second issue: Precambrian results 1957–1974*. Ottawa: *Publ. Earth Physics Branch, Geomagnetism Series 3*, pp 1–42.

IRVING, E. & IRVING, G. A., (1982) 'Apparent polar wander paths, Carboniferous through Cenozoic and the assembly of Gondwana', *Geophys. Surv.*, 5, pp 141–188.

IRVING, E. & MAJOR, A., (1964) 'Post depositional detrital remanent magnetisation in a synthetic sediment'. *Sedimentology*, 3, pp 135–143.

IRVING, E. & MCGLYNN, J. C., (1976a) 'Polyphase magnetisation of the Big Spruce Complex, North West Territories'. *Can. J. Earth Sci.*, 13, pp 476–489.

IRVING, E. & MCGLYNN, J. C., (1976b) 'Proterozoic magnetostratigraphy and the tectonic evolution of Laurentia', *Phil. Trans. R. Soc. Lond.*, A280, pp 243–265.

IRVING, E. & MCGLYNN, J. C., (1979) 'Palaeomagnetism in the Coronation Geosyncline and arrangement of continents in the middle Proterozoic'. *Geophys. J. R. Astr. Soc.*, 58, pp 309–336.

IRVING, E. & MCGLYNN, J. C., (1981) 'On the coherence, relation and palaeolatitudes of Laurentia in the Proterozoic', in: *Precambrian Plate Tectonics* (ed. A. KRÖNER), Elsevier Publ. Co., Amsterdam, pp 561–598.

IRVING, E., MOLYNEUX, L. & RUNCORN, S. K., (1966) 'The analysis of remanent magnetisation intensities and susceptibilities of rocks'. *Geophys. J. R. Astr. Soc.*, 10, pp 451–464.

IRVING, E. & NALDRETT, A. J., (1977) 'Palaeomagnetism in Abititi and Matachewan diabase dykes: evidence of the Archaean geomagnetic field'. *Jour. Geol.*, 85, pp 157–176.

IRVING, E., NORTH, F. K. & COUILLARD, R., (1974) 'Oil, climate and tectonics'. *Can. J. Earth Sci.*, 11, pp 1–17.

IRVING, E. & PARK, J. K., (1972a) 'Palaeomagnetism of metamorphic terrains: errors owing to intrinsic anisotropy', *Geophys. J. R. astr. Soc.*, 34, 489–493.

IRVING, E. & PARK, J. K. (1972b), 'Hairpins and superintervals'. *Can. J. Earth Sci.*, 9, pp 1318–1324.

IRVING, E., PARK, J. K. & EMSLIE, R. F., (1974) 'Palaeomagnetism of the Morin complex'. *Jour. Geophys. Res.*, 79, pp 5482–5490.

IRVING, E. & PARRY, L. G., (1963) 'The magnetism of some Permian rocks from New South Wales'. *Geophys. J. R. Astr. Soc.*, 7, pp 395–411.

IRVING, E. & PULLAIAH, G., (1976) 'Reversals of the geomagnetic field, magnetostratigraphy and relative magnitude of palaeosecular variation in the Phanerozoic'. *Earth Sci. Revs.*, 12, pp 35–64.

IRVING, E. & ROBERTSON W. A., (1969) 'Test for polar wandering and some possible implications'. *J. Geophys. Res.*, 74, pp 1026–1036.

IRVING, E. & RUNCORN, S. K., (1957) 'Analysis of the palaeomagnetism of the Torridonian Sandstone Series of north-west Scotland'. *Phil. Trans. R. Soc.*, A250, pp 83–99.

IRVING, E. & STRONG D. F., (1984) 'Palaeomagnetism of the Early Carboniferous Deer Lake Group, western Newfoundland: no evidence for mid-Carboniferous displacement of Acadia'. *Earth Planet. Sci. Lett.*, 69, pp 379–390.

IRVING, E. & STRONG, D. F., (1985) 'Palaeomagnetism of rocks from Burin Peninsula, Newfoundland: hypothesis of late Palaeozoic displacement of Acadia criticised'. *J. Geophys. Res.*, 90, pp 1949–1962.

IRVING, E., TANCZYK, E., & HASTIE, J., (1976a) *Catalogue of palaeomagnetic directions and poles, third issue: Palaeozoic results 1949–1975.* Ottawa: *Pub. Earth Physics Branch, Geomagnetism*, Series 5, pp 1–98.

IRVING, E., TANCZYK, E. & HASTIE, J., (1976b) *Catalogue of palaeomagnetic directions and poles, fourth issue: Mesozoic results 1954–1975 and results from seamounts*, Ottawa: *Publ. Earth Physics Branch, Geomagnetism*, Series 6, pp 1–70.

IRVING, E., TANCZYK, E. & HASTIE, J., (1976c) *Catalogue of palaeomagnetic directions and poles, fifth issue: Cenozoic results 1927–1965*, Ottawa: *Publ. Earth Physics Branch, Geomagnetism*, Series 10, pp 1–87.

IRVING, E. & WARD, M. A., (1964) 'A statistical model of the geomagnetic field.' *Pure Appl. Geophys.*, 57, pp 47–52.

ISACHSEN, Y. W., (1969) *Origin of Anorthosite and Related Rocks*, 1, New York Museum and Science Serivce Memoir 18 N.Y., 466 pp.

ITO, H. & TOKIEDA, K., (1980) 'An interpretation of palaeomagnetic results from Cretaceous granites in South Korea'. *J. Geomag. Geoelectr.*, 32, pp 275–284.

JACOBS, J. A., (1984) *Reversals of the Earth's magnetic field.* Adam Hilger, Bristol, 230 pp.

JIANG-ZHIWEN, W., (1984) 'Global distribution of the earliest shelly metazoans'. *Geol. Mag.*, 121, pp 185–188.

JOHNSON, H. P. & HALL, J. M., (1978) 'A detailed rock magnetic and opaque mineralogy study of the basalt of the Nazca plate'. *Geophys. J. R. Astr. Soc.*, 52, pp 45–64.

JOHNSON, H. P., LOWRIE, W. & KENT, D. V., (1975) 'Stability of anhysteritic remanent magnetisation in fine and coarse magnetite and maghemite particles'. *Geophys. J. R. Astr. Soc.*, 41, pp 1–10.

JOHNSON, R. J. E., VAN DER VOO, R. & LOWRIE, W. (1984) 'Palaeomagnetism and late diagenesis of Jurassic carbonates from the Jura Mountains, Switzerland and France'. *Geol. Soc. Amer. Bull.*, 95, pp 478–488.

JONES, D. L., COX, A., CONEY, P. & BECK, M. E., (1982) 'The growth of western North America'. *Sci. Amer.*, 247, pp 70–84.

JONES, D. L. & MCELHINNY, M. W., (1967) 'Stratigraphic interpretation of palaeomagnetic measurements on the Waterberg red beds of South Africa'. *J. Geophys. Res.*, 72, pp 4171–4179.

JONES, D. L., ROBERTSON, I. D. M. & MCFADDEN, P. L., (1975) 'A palaeomagnetic study of Precambrian dyke swarms associated with the Great Dyke of Rhodesia'. *Trans. Geol. Soc. S. Afr.*, 78, pp 57–65.

JONES, D. L., SILBERLING, N. J. & HILLHOUSE, J., (1977) 'Wrangellia — a displaced terrane in north western North America', *Can. J. Earth Sci.*, 14, pp 2565–2577.

JONES, G. M., (1977) 'Thermal interaction of the core and the mantle and long-term behaviour of the geomagnetic field'. *J. Geophys. Res.*, 82, pp 1703–1709.

JONES, G. M. & GARTNER, S., (1980) 'Commentary on 'Pre-Tertiary velocities of the continents: A lower bound from Palaeomagnetic data', by R. G. GORDON, M. O. MCWILLIAMS & A. COX, *Jour. Geophys. Res.*, 85, pp 4431–4432.

JONES, M., VAN DER VOO, R. & BONHOMMET, N., (1979) 'Late Devonian to early Carboniferous palaeomagnetic poles from the Armorican Massif, France'. *Geophys. J. R. Astr. Soc.*, 58, pp 287–308.

JURDY, D. M., (1981) 'True polar wander'. *Tectonophysics*, 74, pp 1–16.

JURDY, D. M. & VAN DER VOO, R., (1974) 'A method for the separation of true polar wander and continental drift, including results for the last 55 Ma'. *Jour. Geophys. Res.*, 79, pp 2945–2952.

JURDY, D. M. & VAN DER VOO, R., (1975) 'True polar wander since the Early Cretaceous'. *Science*, 187, pp 1193–1196.

KANESEWICH, E. R., HAVSKOV, J. & EVANS, M. E., (1978) 'Plate tectonics in the Phanerozoic'. *Can. J. Earth Sci.*, 15, pp 919–955.

KAULA, W. M., (1972) 'Global gravity and mantle convection'. *Tectonophysics*, 13, pp 341–359.

KAULA, W. M., (1975) 'Absolute plate motions by boundary velocity minimisation'. *J. Geophys. Res.*, 80, pp 244–248.

KAWAI, N., ITO, H. & KUME, S., (1961) 'Deformation of the Japanese Islands as inferred from rock magnetism'. *Geophys. J. R. Astr. Soc.*, 16, pp 124–130.

KAWAI, N., ITO, H. & KUME S., (1962) 'Study of magnetisation of the Japanese rocks'. *J. Geomag. Geoelect.*, 13, pp 150–153.

KELLOG, K. S. & REYNOLDS, R. L., (1978) 'Palaeomagnetic results from the Lassiter coast, Antarctic, and a test for oroclinal bending of the Antarctic Peninsula'. *J. Geophys. Res.*, 83, pp 2293–2299.

KENT, D. V., DIA, O. & SOUGY, J. M. A., (1984) 'Palaeomagnetism of Lower–Middle Devonian and Upper Proterozoic–Cambrian (?) rocks from Mejeria (Mauretania, West Africa)'. in: *Plate reconstruction from Palaeozoic palaeomagnetism, Geodynamics Series, vol. 12*, American Geophysical Union, pp 99–115.

KENT, D. V. & LOWRIE, W., (1974) 'Origin of magnetic instability in sediment cores from the central North Pacific'. *Jour. Geophys. Res.*, 79, pp 2987–2999.

KENT, D. V. & OPDYKE, N. D., (1978) 'Palaeomagnetism of the Devonian Catskill red beds: evidence for motion of coastal New England — Canadian Maritime region relative to cratonic North America'. *J. Geophys. Res.*, 83, pp 4441–4450.

KENT, J. T., BRIDEN, J. C. & MARDIA, K. V., (1983) 'Linear and planar structure in ordered multivariate data as applied to progressive demagnetisation of palaeomagnetic remanence'. *Geophys. J. R. Astr. Soc.*, 75, pp 593–622.

KENYON, P. M. & TURCOTTE, D. L., (1983) 'Convection in a two-layer mantle with strongly temperature dependent viscosity'. *Jour. Geophys. Res.*, 88, pp 6403–6414.

KHRAMOV, A. N., (1958) *Palaeomagnetism and stratigraphic correlation*. Gostoptechizdat, Leningrad (translated by A. J. LOJKINE and ed. E. IRVING), Australian National University, 204 pp.

KHRAMOV, A. N., (1967) 'The Earth's magnetic field in the late Palaeozoic. *Izvestiya Akademii Nauk SSSR, Earth Phys. Ser.* pp 86–108.

KHRAMOV, A. N., (1973) *Palaeozoic Palaeomagnetism* (in Russian), Nedra Press, Leningrad, 238 pp.

KHRAMOV, A. N. & RODIONOV, V. P., (1980) 'The geomagnetic field during Palaeozoic times', in: *Advances in Earth and Planetary Sciences 10* (ed. M. W. MCELHINNY, A. N. KHRAMOV, M. OZIMA & D. A. VALENCIO) Academic Publications, Tokyo, Japan, pp 99–116.

KHRAMOV, A. N., RODIONOV, V. P. & KOMISSAROVA, R. A., (1965) 'New data on the Palaeozoic history of the geomagnetic field in the USSR', in: *The Present and past of the geomagnetic field* (in Russian), Nauka press, Moscow, pp 206–213.

KING, R. F., (1955) 'The remanent magnetism of artificially deposited sediments'. *Mon. Not. R. Astr. Soc.*, 7, pp 115–134.

KING, R. F., (1967) 'Errors in anisotropy measurements with the torsion balance', in: *Methods in Palaeomagnetism* (ed. D. W. COLLINSON, K. M. CREER & S. K. RUNCORN), Elsevier Publ. Co., Amsterdam, pp 387–398.

KING, R. F. & REES, A. I., (1962) 'The measurement of the anisotropy of magnetic susceptibility of rocks by the torque method'. *Jour. Geophys., Res.*, 67, pp 1565–1572.

KIRBY, G. A. (1979) 'The Lizard Complex as an ophiolite'. *Nature* 282 pp 58–61.
KIRSCHVINK, J. L., (1978) 'The Precambrian–Cambrian boundary problem: palaeomagnetic directions from the Amadeus Basin, Central Australia'. *Earth Planet. Sci. Lett.*, 40, pp 91–100.
KIRSCHVINK, J. L., (1980) 'The least-squares line and plane and the analysis of palaeomagnetic data'. *Geophys. J. R. Astr. Soc.*, 62, pp 699–718.
KIRSCHVINK, J,. L. & LOWENSTAM, H. A., (1979) 'Mineralisation and magnetisation of chiton teeth: palaeomagnetic, sedimentologic and biologic implications of organic magnetite'. *Earth Planet. Sci. Lett.*, 44, pp 193–204.
KIRSCHVINK, J. L. & ROZANOV A.Yu., (1984) 'Magnetostratigraphy of Lower Cambrian Strata from the Siberian Platform: a palaeomagnetic pole and preliminary polarity time-scale '. *Geol. Mag.*, 121, (3) pp 189–203.
KITTEL, C., (1949) 'Ferromagnetic domain theory'. *Rev. Mod. Phys.*, 21, pp 451–583.
KLIGFIELD, R. & CHANNELL, J. E. T., (1981) 'Widespread remagnetisation in Helvetic Limestones'. *Jour. Geophys. Res.*, 86, pp 1888–1900.
KLOOTWIJK, C. T., (1979) 'A review of palaeomagnetic data from the Indo-Pakistani fragment of Gondawanaland'. in: *Geodynamics of Pakistan*, (ed. A. FARAH & K. DE JONG) Geological Survey of Pakistan, Quetta, pp 41–80.
KLOOTWIJK, C. T., (1980) 'Early Palaeozoic palaeomagnetism in Australia'. *Tectonophysics*, 64, pp 249–332.
KLOOTWIJK, C. T. & BINGHAM, D. K., (1980) 'The extent of Greater India. III. Palaeomagnetic data from the Tibetan sedimentary series, Thakkola Region, Nepal Himalaya'. *Earth Planet. Sci. Lett.*, 51, pp 381–405.
KLOOTWIJK, C. T., CONAGHAN, R. & POWELL, C. (1985) 'Himalayan Arc: large-scale continental subduction, oroclinal bending and back-arc spreading', *Earth Planet. Sci. Lett.*, 75, 167–183.
KLOOTWIJK, C. J., NAZIRULLA, R., DE JONG, K. A. & AHMED, H., (1981) 'A palaeomagnetic reconnaissance of northern Baluchistan, Pakistan'. *J. Geophys. Res.*, 86, pp 289–306.
KLOOTWIJK, C. T. & PIERCE, J. W., (1979) 'India and Australia's pole path since the late Mesozoic and the India–Asia collision'. *Nature*, 282, pp 605–607.
KLOOTWIJK, C. T. & RADHAKRISHNAMURTY, C., (1981) 'Phanerozoic palaeomagnetism of the Indian Plate and India–Asia collision', in: *Palaeoreconstruction of the continents* (eds. M. W. MCELHINNY & D. A. VALENCIO); *Geodynamics Series, vol. 2, Amer. Geophys. Un.*, pp 93–105.
KNEEN, S. J., (1973) 'The palaeomagnetism of the Foyers Plutonic Complex, Inverness-shire', *Geophys. J. R. Astr. Soc.*, 32, pp 53–63.
KNELLER, E. & LUBORSKY, F. E., (1963) 'Particle size dependence of coercivity and remanence of single domain particles'. *J. App. Phys.*, 34, pp 656–658.
KOBAYASHI, K., (1959) 'Chemical remanent magnetisation of ferromagnetic minerals and its application to rock magnetism'. *J. Geomag. Geoelect.*, 10 pp 99–117.
KOBAYASHI, K., (1968) 'Palaeomagnetic determination for the intensity of the geomagnetic field in the Precambrian'. *Phys. Earth Planet. Int.*, 1, pp 387–395.
KOBAYASHI, K. & NAMURA, M., (1974) 'Ferromagnetic minerals in the sediment cores collected from the Pacific Basin'. *Jour. Geophys. Res.*, 40, pp 501–517.
KODAMA, K. P., (1984) 'Palaeomagnetism of granitic intrusives from the Precambrian basement under eastern Kansas: orienting drill cores using secondary magnetisation components'. *Geophys. J. R. Astr. Soc.*, 76, pp 273–288.
KODAMA, K. A., TAIRA, A., OKAMURA, M. & SAITO, Y., (1983) 'Palaeomagnetism of the Shimanto belt in Shikoku, southwest Japan', in: *Accretion Tectonics in the Circum-Pacific Regions* (ed. M. HASHIMOTO & S. UYEDA); Terrapub, Tokyo, pp 231–241.
KONO, M., (1979) 'Palaeomagnetism and palaeointensity studies of Scottish Devonian volcanic rocks'. *Geophys. J. R. Astr. Soc.*, 56, pp 385–396.
KONO, M., (1980) 'Statistics of palaeomagnetic inclination data'. *J. Geophys. Res.*, 85, pp 3878–3882.

KRATZ, M. B., (1974) 'Paired metamorphic belts in Precambrian granulite rocks in Gondwanaland'. *Geology*, 2, pp 237–241.

KRATZ, K. A., GERLING, E. K. & LOBACH-ZHUCHENKO, S. B., (1968) 'The isotope geology of the Precambrian of the Baltic Shield'. *Can. J. Earth Sci.*, 5, pp 657–660.

KREBS, W. & WACHENDORF, H., (1973) 'Proterozoic–Palaeozoic geosynclinal and orogenic evolution of central Europe'. *Bull. Geol. Soc. Amer.*, 84, pp 2611–2630.

KRÖNER, A., (1977) 'Precambrian mobile belts of southern and western Africa — ancient sutures or sites of ensialic mobility? A case for crust evolution towards plate tectonics'. *Tectonophysics*, 40, pp 101–135.

KRÖNER, A., (1981) (ed.) *Precambrian Plate Tectonics*. Elsevier Publ. Co., Amsterdam, 759 pp.

KRÖNER, A., MCWILLIAMS, M. O. & LAYER, P. W. (1982) 'A provisional Archaean APWP for the Kaapvaal Craton'. *EOS, Trans. Amer. Geophys. Un.*, 63, 921 pp.

KRÖNER, A., MCWILLIAMS, M. O., GERMS, G. J. B., SCHALK, K. L. & REID, A., (1980) 'Palaeomagnetism of late Precambrian to early Palaeozoic mixtite bearing formations in Namibia (South West Africa): the Nama Group and Blaubecker Formation'. *Amer. J. Sci.*, 280, pp 942–968.

KROONENBERG, S. B., (1982) 'A Grenvillian granulite belt in the Columbian Andes and its relation to the Guiana Shield'. *Geologie en Mijnbouw*, 61 pp 325–333.

KROGH, T. C. & DAVIS, G. L., (1972) 'Zircon U–Pb ages of Archaen metavolcanic rocks in the Canadian Shield'. Carnegie Institute of Washington Yearbook, 70, pp 241–242.

KROUGH, T. H., MCNUTT, R. H. & DAVIS, G. L., (1982). 'Two high precision U/Pb zircon ages for the Sudbury Nickel Irruptive'. *Can. J. Earth Sci.*, 19, pp 723–728.

KRS, M. & VLASIMSKY, P., (1976) 'Palaeomagnetic study of Cambrian rocks of the Barrandian (Bohemian Massif)', *Geofysikalni Sbornik*, 24, pp 263–280.

KRS, M., MUSKA, P. & PAGAC, P., (1982) 'Review of palaeomagnetic investigations in the West Carpathians of Czechoslovakia', *Geologické Práce, Správy*, 78, 39–58.

KRUGLYAKOVA, G. I., (1961) 'Results of palaeomagnetic research on the Ukrainian crystalline massif and adjacent regions'. *Izvestiya Akademii Nauk SSSR, Earth Phys. Ser.*, pp 150–153.

KUMAR, A. & BHALLA, M. S., (1983) 'Palaeomagnetics and igneous activity of the area adjoining the south western margin of the Cuddapah Basin, India.' *Geophys. J. R. Astr. Soc.*, 73, pp 27–38.

KUMAR, A. & BHALLA, M. S., (1984) 'Source of stable remanence in chromite ores'. *Geophys. Res. Lett.*, 11(3) pp 177–180.

LABREQUE, J. L., KENT, D. V. & CANDE, S. C., (1977) 'Revised magnetic polarity time-scale for Late Cretaceous and Cenozoic time'. *Geology*, 5(6), pp 330–335.

LAJ, C., JAMET, M., SOREL, D. & VALENTE, J. P., (1982) 'First palaeomagnetic results from Mio-Pliocene Series of the Hellenic sedimentary arc.' *Tectonophysics*, 86, pp 45–67.

LANGMUIR, D., (1971) 'Particle size effect on the reaction goethite = hematite + water.' *Amer. J. Sci.*, 271, pp 147–156.

LAPOINTE, P. L., ROY, J. L. & MORRIS, W. A., (1978) 'What happened to the high latitude palaeomagnetic poles?' *Nature*, 273, pp 655–657.

LARSON, E. E. & WALKER, T. R., (1975) 'Development of chemical and remanent magnetisation during early stages of red bed formation in late Cenozoic sediments, Baja, California.' *Geol. Soc. Amer. Bull.*, 86, pp 639–650.

LARSON, E. E. & WALKER, T. R., (1985) 'Comment on Palaeomagnetism of a polarity transition in the Lower (?) Triassic Chugwater Formation Wyoming' by E. HERERO-BERVERA & C. E. HELSLEY, *Jour. Geophys. Res.*, 90, pp 2060–2062.

LARSON, E. E., WALKER, T. R., PATTERSON, P. E., HOBLITT, R. P. & ROSENBAUM, J. G., (1982) 'Palaeomagnetism of the Moenkopi Formation, Colorado Plateau: basis for long-term model of acquisition of chemical remanent magnetism in red beds.' *Jour. Geophys. Res.*, 87, pp 1081–1106.

LARSON, R. L., GOLOVCHENKO, X. & PITMAN III, W. C. (1981)'Geomagnetic polarity time-scale, in:

Plate tectonic map of the circum-Pacific region, northeast quadrant (ed. C. K. J. DRUMMOND). Amer. Assoc. Petrol. Geol., Tulsa.

LARSON, R. L. & HELSLEY, C. E., (1975) 'Mesozoic reversal sequence.' *Rev. Geophys. Space Phys.*, 13, pp 174–176.

LARSON, R. L. & HILDE, T. W. C., (1975) 'A revised timescale of magnetic reversals for the early Cretaceous and late Jurassic.' *J. Geophys. Res.*, 80, pp 2586–2594.

LARSON, R. L. & PITMAN III, W. C., (1972) 'Worldwide correlation of Mesozoic magnetic anomalies and its implications,' *Bull. Geol. Soc. Amer.*, 83, pp 3642–3662.

LAWLEY, E. A., (1970) 'The intensity of the geomagnetic field in Iceland during Neogene polarity transitions and systematic deviations.' *Earth Planet. Sci. Lett.*, 10, pp 145–149.

LAYER P. W., KRÖNER, A. & McWILLIAMS, M. O., (1982) 'Palaeomagnetism of the 2813 Ma Usushwana Complex, southern Africa.' *EOS, Trans. Am. Geophys. Un.* 63, pp 912.

LEBLANC, M., (1981) 'The late Proterozoic ophiolites of Bou Azzer (Morocco): evidence for Pan-African Plate Tectonics,' in: *Precambrian Plate Tectonics*, (ed. A. KRÖNER) Elsevier Publ. Co. Amsterdam, pp 435–451.

LEBORGNE, E., LE MOUEL, J. L. & LE PICHON, X., (1971) 'Aeromagnetic survey of southwestern Europe.' *Earth Planet. Sci. Lett.*, 12, pp 287–299.

LEES, A. C., BUKOWINSKI, M. S. T. & JEANLOZ, R., (1983) 'Reflection properties of phase transition and compositional change models of the 670-km discontinuity.' *J. Geophys. Res.*, 88, pp 8145–8159.

LEFORT, J. P. & VAN DER VOO, R., (1981) 'A kinetic model for the collision and complete suturing between Gondwanaland and Laurussia in the Carboniferous.' *Jour. Geol.*, 89, pp 537–550.

LERCH, F. J., KLOSKO, S. M., LAUBSCHER, R. E. & WAGNER, C. A., (1979) 'Gravity model improvement using GEOS-3 (GEM9 and 10).' *J. Geophys. Res.*, 86, pp 3897–3916.

LE PICHON, X. & HUCHON, P., (1984) 'Geoid, Pangea and convection.' *Earth Planet. Sci. Lett.*, 67, pp 123–135

LE PICHON, X. & SIBUET, J. C., (1971) 'Western extension of the boundary between the European and Iberian plates during the Pyrenean orogeny.' *Earth Planet. Sci. Lett.*, 12, pp 83–88.

LIENERT, B. R. & WASILEWSKI, P. J., (1979) 'A magnetic study of the serpentisation process at Burro Mountain, California.' *Earth Planet. Sci. Lett.*, 43, pp 406–416.

LIN, J. L., FULLER, M. D. & ZHANG, W.-Y. (1985) 'Preliminary Phanerozoic polar wander paths for the North and South China blocks.' *Nature*, 313. pp 444–449.

LINDSLEY, D. J., (1973) Delimitation of the hematite–ilmenite miscibility gap.' *Geol. Soc. Amer. Bull.*, 84, pp 657–662.

LISTER, C. R. B., (1975) 'Gravitational drive on oceanic plates caused by thermal contraction.' *Nature*, 257, pp 663–665.

LIU, C. & FENG, H. (1965) 'Alternating field demagnetisation study on a Lower Sinean sandstones in Xuining district, Anhui Province,' *Acta Geophys. Sinica*, 14, pp 173–180 (translated by E. R. HOPE, Directorate of Scientific Information Services, DRB Canada, T7C, 1966).

LONGSHAW, S. K. & GRIFFITHS, D. H., (1983) 'A palaeomagnetic study of Jurassic rocks from the Antarctic Peninsula and its implications.' *J. Geol. Soc. Lond.*, 140, pp 945–954.

LOPER, D. E., (1985) 'A simple model of whole-mantle convection.' *J. Geophys. Res.*, 90, pp 1809–1836.

LOPER, D. E. & STACEY, F. D., (1983) 'The dynamical and thermal structure of deep mantle plumes.' *Phys. Earth Planet. Int.*, 33, pp 304–317.

LOVLIE, R., (1974) 'Post-depositional remanent magnetisation in a re-deposited deep-sea sediment.'*Earth Planet. Sci. Lett.*, 21, pp 315–320.

LOVLIE, R., LOWRIE, W. & JACOBS, M., (1971) 'Magnetic properties and mineralogy of four deep-sea cores.' *Earth Planet. Sci. Lett.*, 15, pp 157–168.

LOVLIE, R., TORSVIK, T., JELENSKA, M. & LEVANDOWSKI, M., (1984) 'Evidence for detrital remanent magnetisation carried by hematite in Devonian red beds from Spitzbergen: palaeomagnetic implications.' *Geophys. J. R. Astr. Soc.*, 79, pp 573–588.

LOWRIE, W. & ALVAREZ, W., (1981) 'One hundred million years of geomagnetic history.' *Geology*, 9, pp 392–397.

LOWRIE, W., CHANNELL, J. E. T. & ALVAREZ, W., (1980) 'A review of magnetic stratigraphy investigations in Cretaceous pelagic carbonate rocks.' *J. Geophys. Res.*, 85, pp 3597–3605.

LOWRIE, W. & FULLER, M. D., (1971) 'On the alternating field demagnetisation characteristics of multi-domain thermoremanent magnetisation in magnetite.' *J. Geophys. Res.*, 76, pp 6339–6349.

LOWRIE, W. & HELLER, F. (1982) 'Magnetic properties of marine limestones.' *Rev. Geophys. Space Phys.*, 20, pp 171–192.

LUYENDYK, B. P., KAMMERLING, M. J. & TERRES, R. (1980) 'Geometric model for Neogene crustal rotations in southern California.' *Bull. Geol. Soc. Amer.* 91, pp 211–217.

LYNAS, B. D. T., RUNDLE, C. C. & SANDERSON, R. W. (1986) 'A note on the age and pyroxene chemistry of the igneous rocks of the Shelve Inlier, Welsh Borderlands.' *Geol. Mag.*, 122. pp 641–647.

MACDONALD, W. D., (1980) 'Net tectonic rotation, apparent tectonic rotation and the structural tilt correction in palaeomagnetic studies,' *Jour. Geophys. Res.*, 85, pp 3659–3669.

MACDONALD, W. D. & OPDYKE, N. D., (1972) 'Tectonic rotations suggested by palaeomagnetic results from Northern Colombia, South America.' *J. Geophys. Res.*, 77, pp 5720–5730.

MALFAIT, B. T. & DINKELMAN, M. G., (1972) 'Circum-Caribbean tectonic and igneous activity and the evolution of the Caribbean Plate.' *Geol. Soc. Amer. Bull.* 83, pp 251–272.

MANKINEN, E. A. & DALRYMPLE, G. B., (1979) 'Revised geomagnetic polarity time scale for the interval 0–5 m.y. B. P.', *J. Geophys. Res.*, 84, pp 615–676.

MANNING, P. G., WILLIAMS, G. D. H., CHARLTON, M. N., ASH, L. A. & BIRCHALL, T., (1979) 'Mössbauer spectral studies of the diagenesis of iron in a sulphide-rich sediment core.' *Nature*, 280, pp 134–136.

MARSHALL, J. F. & COOK, P. J. (1980) 'Petrology of iron and phosphorus-rich nodules from the E. Australian continental shelf,' *J. Geol. Soc. Lond.*, 137, pp 765–771.

MARTIN, D. L., (1976) 'A palaeomagnetic polarity transition in the Devonian Columbus Limestone of Ohio: a possible stratigraphic tool.' *Tectonophysics*, 28, pp 125–134.

MARTIN, D. L., NAIRN, A. E. M., NOLTIMER, H. C., PETTY, M. H. & SCHMIDT, H.C., (1978) 'Palaeozoic and Mesozoic palaeomagnetic results from Morocco.' *Tectonophysics*, 44, pp 91–114.

MARTON, E. & MARTON, P., (1981) 'Mesozoic palaeomagnetism of the Transanubian central mountains and its tectonic implications.' *Tectophysics*, 72, pp 129–140.

MARTON, E. & MARTON, P., (1983) 'A refined apparent polar wander curve for the Transdanubian central mountains and its bearing on the Mediterranean tectonic history. *Tectonophysics*, 98, pp 43–57.

MARTON, E. and VELJOVIC, D., (1983) 'Palaeomagnetism of the Istria Peninsula, Jugoslavia. *Tectonophysics*, 91, pp 73–87.

MASSEY, N. W. D., (1979) 'Keweenawan palaeomagnetic reversals at Mamainse Point, Ontario: fault repetition or three reversals?' *Can. Jour. Earth Sci.* 16, pp 373–375.

MATTAUER, M., (1969) 'Sur la rotation de l'Espagne,' *Earth Planet. Sci. Lett.*, 7, pp 87–93.

MAXWELL, A. E., VON HERZEN, R. P., HSU, K. J., ANDREWS, J. E., SAITO, T., PERCIVAL, S. F., MITLOW, E. O. & BOYCE, R. E., (1970) 'Deep-sea drilling in the South Atlantic.' *Science*, 168, pp 1047–1059.

MCCABE, C., VAN DER VOO, R. & BALLARD, M. M., (1984) 'Late Palaeozoic remagnetisation of the Trenton Limestone.' *Geophys. Res. Lett.*, 94, pp 979–982.

MCCABE, C., VAN DER VOO, R., PEACOR, D. R., SCOTESE, C. R. & FREEMAN, R. (1983) 'Diagenetic magnetite carries ancient yet secondary remanence in some Palaeozoic sedimentary carbonates.' *Geology*, 11, pp 221–223.

MCCABE, C., VAN DER VOO, R., WILKINSON, B. H. & DEVANEY, K., (1985) 'A Middle/Late Silurian palaeomagnetic pole from limestone reefs of the Wabash Formation, Indiana, U.S.A.,' *Jour. Geophys. Res.*, 90, pp 2959–2965.

MCCLELLAND-BROWN, E., (1983) 'Palaeomagnetic studies of fold development and propagation in the Pembrokeshire Old Red Sandstone.' *Tectonophysics*, 98, pp 131–149.

MCDOUGALL, I., SAEMUNDSSON, K., JOHANNESSON, H., WATKINS, N. D. & KRISTJANSSON, L., (1977) 'Extension of the geomagnetic polarity time-scale to 6·5 Myr: K–Ar dating, geological and palaeomagnetic study of a 3500 m lava succession in western Iceland.' *Bull. Geol. Soc. Amer.*, 88, pp 1–15.

MCDOUGALL, I., WATKINS, N. D., WALKER, G. P. L. & KRISTJANSSON, L., (1976) 'Potassium–argon and palaeomagnetic analysis of Icelandic lava flows: limits on the age of anomaly 5.' *J. Geophys. Res.*, 81, pp 1505–1512.

MCDOUGALL, I., WATKINS, N. D. & KRISTJANSSON, L., (1976) 'Geochronology and palaeomagnetism of a Miocene–Pliocene lava sequence at Bessastadaa, Eastern Iceland.' *Amer. J. Sci.* 276, pp 1078–1095.

MCELHINNY, M. W., (1964) 'Statistical significance of the fold test in palaeomagnetism.' *Geophys. J. R. Astr. Soc.*, 8, pp 338–340.

MCELHINNY, M. W., (1966) 'An improved method for demagnetising rocks in alternating fields.' *Geophys. J. R. Astr. Soc.*, 10, pp 369–374.

MCELHINNY, M. W., (1967) 'Statistics of a spherical distribution,' in: *Methods in Palaeomagnetism* (ed. D. W. COLLINSON, K. M. RUNCORN & S. K. RUNCORN) Elsevier Publ. Co., Amsterdam, pp 313–321.

MCELHINNY, M. W., (1968a) 'Palaeomagnetic directions and pole positions — VIII.' *Geophys. J. R. Astr. Soc.*, 15, pp 409–430.

MCELHINNY, M. W., (1968b) 'Palaeomagnetic directions and pole positions — IX.' *Geophys. J. R. Astr. Soc.*, 16, pp 207–224.

MCELHINNY, M. W., (1969) 'Palaeomagnetic directions and pole positions — X.' *Geophys. J. R. Astr. Soc.*, 19, pp 305–327.

MCELHINNY, M. W., (1970),'Palaeomagnetic directions and pole positions — XI.' Geophys. J. R. Astr. Soc., 20, pp 417–429.

MCELHINNY, M. W., (1971) 'Geomagnetic reversals during the Phanerozoic.' *Science*, 172, pp 157–159.

MCELHINNY, M. W., (1972a) 'Palaeomagnetic directions and pole positions — XII.' *Geophys. J. R. Astr. Soc.*, 27, pp 237–257,

MCELHINNY, M. W., (1972b) 'Palaeomagnetic directions and pole positions — XIII.' *Geophys. J. R. Astr. Soc.*, 30, pp 281–293.

MCELHINNY, M.W., (1973) *Palaeomagnetism and Plate Tectonics*. Cambridge University Press, Cambridge, 358 pp.

MCELHINNY, M. W. & BUREK, P. J., (1971) 'Mesozoic palaeomagnetic stratigraphy.' *Nature*, 232,, pp 98–102.

MCELHINNY, M. W. & COWLEY, J. A., (1977) 'Palaeomagnetic directions and pole positions — XIV. Pole numbers 14/1 to 14/574.' *Geophys. J. R. Astr. Soc.*, 49, pp 313–356.

MCELHINNY, M. W. & COWLEY, J. A., (1978) 'Palaeomagnetic directions and pole positions — XV. Pole numbers 15/1 to 15/232.' *Geophys. J. R. Astr. Soc.*, 52, pp 259–276.

MCELHINNY, M. W. & COWLEY, J. A., (1980) 'Palaeomagnetic directions and pole positions — XVI. Pole numbers 16/1 to 16/296.' *Geophys. J. R. Astr. Soc.*, 61, pp 549–571.

MCELHINNY, M. W., COWLEY, J. A., BROWN, D. A. & WIRUBOV, N., (1977) *Palaeomagnetic results from the USSR.*, Australian National University, Research School of Earth Sciences, Publ. No. 1268, 82 pp.

MCELHINNY, M. W., COWLEY, J. A. & BROWN, D. A., (1979) *Palaeomagnetic results from the USSR, Supplement No. 1*, Australian National University, Research School of Earth Sciences, Publ. No. 1377, 23 pp.

MCELHINNY, M. W. & EMBLETON, B. J. J., (1974) 'Australian palaeomagnetism and the Phanerozoic plate tectonics of eastern Gondwanaland.' *Tectonophysics*, 22, pp 1–29.

MCELHINNY, M. W. & EMBLETON, B. J. J., (1976) 'Precambrian and early Palaeozoic palaeomagnetism in Australia.' *Phil. Trans. R. Soc. Lond.*, A280, pp 417–432.

McElhinny, M. W., Embleton, B. J. J., Ma, X. H. & Zhang, X. K. (1981) 'Fragmentation of Asia in the Permian.' *Nature*, 293, pp 212–216.

McElhinny, M. W. & Evans, M. E., (1968) 'An investigation of the strength of the magnetic field in the early Precambrian.' *Phys. Earth Planet. Int.*, 1, pp 485–497.

McElhinny, M. W., Giddings, J. W. & Embleton, B. J. J., (1974) 'Palaeomagnetic results and late Precambrian glaciations.' *Nature*, 248, pp 557–561.

McElhinny, M. W., Haile, N. S. & Crawford, A. R., (1974) 'Palaeomagnetic evidence shows Malay Peninsula was not part of Gondwanaland.' *Nature*, 252, pp 641–645.

McElhinny, M. W. & McWilliams, M. O., (1977) 'Precambrian geodynamics — a palaeomagnetic view.' *Tectonophysics*, 40, pp 137–159.

McElhinny, M. W. & Opdyke, N. D., (1973) 'Remagnetisation hypothesis discounted: a palaeomagnetic study of the Trenton Limestone, New York State.' *Bull. Geol. Soc. Amer.*, 84, pp 3697–3708.

McElhinny, M. W. & Senanayake, W. E., (1980) 'Palaeomagnetic evidence for the existence of the magnetic field 3·5 Ga ago.' *Jour. Geophys. Res.*, 85, pp 3523–3528.

McFadden, P. L., (1980a) 'Determination of the angle in a Fisher distribution which will be exceeded with a given probability.' *Goephys. J. R. Astr. Soc.*, 60, pp 391–396.

McFadden, P. L., (1980b) 'The best estimate of Fisher's precision parameter x.' *Geophys. J. R. Astr. Soc.*, 60, pp 397–407.

McFadden, P. L. & Jones, D. L., (1981) 'The fold test in palaeomagnetism.' *Geophys. J. R. Astr. Soc.*, 67, pp 53–58.

McFadden, P. L. & Lowes, F. J., (1981) 'The discrimination of mean directions drawn from Fisher distributions.' *Geophys. J. R. Astr. Soc.*, 67, pp 19–33.

McFadden, P. L. & Merrill, R. T., (1984) 'Lower mantle convection and geomagnetism.' *J. Geophys. Res.*, 89, pp 3354–3362.

McFadden, P. L. & McElhinny, M. W., (1982) 'Variations in the geomagnetic dipole 2: statistical analysis of VDMs for the past 5 million years.' *J. Geomag. Geoelect.*, 34, pp 163–178.

McGlynn, J. C. & Irving, E., (1978) 'Multicomponent magnetisation of the Pearson Formation (Great Slave Supergroup, N. W. T.) and the Coronation Group.' *Can. J. Earth Sci.*, 15, pp 642–654.

McKenzie, D. P., (1966) 'The viscosity of the lower mantle.' *J. Geophys. Res.* 71, pp 3995–4010.

McKenzie, D. P., (1969) 'Speculations on the consequences and causes of plate motions.' *Geophys. J. R. Astr. Soc.*, 18, pp 1–32.

McKenzie, D. P., (1972) 'Plate tectonics,' in: *The nature of the solid earth* (ed. E. Robertson), McGraw Hill, New York, pp 323–360.

McKenzie, D. P., (1978) 'Some remarks on the development of sedimentary basins.' *Earth Planet. Sci. Lett.*, 40, pp 25–32.

McKenzie, D. P. & Richter, F. M., (1976) 'Convection currents in the Earth's mantle.' *Scientific Amer.*, 242, pp 72–89.

McKenzie, D. P. & Sclater, J. G., (1971) 'The evolution of the Indian Ocean since the Late Cretaceous.' *Geophys. J. R. Astr. Soc.*, 25, pp 437–528.

McKenzie, D. P. & Weiss, N., (1975) 'Speculations on the thermal and tectonic history of the Earth.' *Geophys. J. R. Astr. Soc.*, 42, pp 131–174.

McLennan, S. M. & Taylor, S. R., (1982) 'Geochemical constraints on the growth of the continental crust.' *J. Geol.*, 90, pp 347–361.

McLennan, S. M. & Taylor, S. R., (1983) 'Continental freeboard, sedimentation rates and growth of continental crust.' *Nature*, 306, pp 169–172.

McMahon, B. E. & Strangway, D. W., (1968) 'Investigation of Kiaman magnetic division in Colorado red beds.' *Geophys. J. R. Astr. Soc.*, 15, pp 265–285.

McMenamin, M., (1982) 'A case for two late Proterozoic–earliest Cambrian faunal province loci.' *Geology*, 10, pp 290–292.

McWilliams, M. O., (1981) 'Palaeomagnetism and Precambrian tectonic evolution of Gondwana,' in: *Precambrian Plate Tectonics* (ed. A. Kröner), Elsevier Publ. Co. Amsterdam, pp. 649–687.

MCWILLIAMS, M. O. & DUNLOP, D. J., (1978) 'Grenville palaeomagnetism and tectonics.' *Can. Jour. Earth Sci.*, 15, pp 687–695.

MCWILLIAMS, M. O. & KRÖNER, A., (1981) 'Palaeomagnetism and tectonic evolution of the Pan-African Damara Belt, southern Africa.' *J. Geophys. Res.*, 86, pp 5147–5162.

MCWILLIAMS, M. O. & MCELHINNY, M. W., (1980) 'Late Precambrian palaeomagnetism of Australia: the Adelaide Geosyncline.' *Jour. Geol.*, 88, pp 1–26.

MERRILL, R. T., (1970) 'Low temperature treatment of magnetite and magnetite-bearing rocks' *Jour. Geophys. Res.*, 75, pp 3343–3349.

MERRILL, R. T., (1975) 'Magnetic effects associated with chemical changes in igneous rocks.' *Geophys. Surv.*, 2, pp 227–243.

MERRILL, R. T. & MCELHINNY, M. W., (1983) *The Earth's Magnetic Field, International Geophysics Series, vol. 32*, Academic Press, London, 401 pp.

MIDDLEMISS, F. A., RAWSON, P. F. & NEWALL, G. (ed.), (1971) *Faunal Provinces in Space and Time.* (Proc. 17th Inter-Univ. Geol. Cong.), Seel House Press, Liverpool, 236 pp.

MILLER, R. G. & O'NIONS, R. K., (1985) 'Source of Precambrian chemical and clastic sediments.' *Nature*, 314, pp 325–330.

MILLER, J. D. & OPDYKE, N. D., (1985) 'Magnetostratigraphy of the Red Sandstone Creek Section, Vail, Colorado.' *Geophys. Res. Lett.*, 12, pp 133–136.

MINSTER, J. B. & JORDAN, T. H., (1978) 'Present day plate motions'. *J. Geophys. Res.*, 83, pp 5331–5354.

MITCHELL, C., TAYLOR, G. K., COX, K. G. & SHAW, J., (1985) 'The Falkland Islands — a rotated microplate?' *Nature*, 319, pp 131–134.

MOLNAR, P. & FRANCHETEAU, J., (1975) 'The relative motion of 'hotspots' in the Atlantic and Indian Oceans during the Cenozoic.' *Geophys. J. R. Astr. Soc.*, 43, pp 763–774.

MOLNAR, P. & TAPPONNIER, P., (1977) 'Relation of the tectonics of eastern China to the India–Eurasian collision: application of strike-line field theory to large-scale continental tectonics.' *Geophys.*, 5, pp 212–216.

MOLYNEUX, L., (1971) 'A complete results magnetometer for measuring the remanent magnetisation of rocks.' *Geophys. J. R. Astr. Soc.*, 24, pp 429–433.

MONGER, J. W. H., PRICE, R. A. & TEMPELMAN-KLUIT, D. J. (1982) 'Tectonic accretion and the origin of the two major metamorphic and plutonic belts in the Canadian Cordillera.' *Geology*, 10, pp 70–75.

MONTIGNY, R., EDEL, J. B. & THVIZAT, R., (1981) 'Oligo-Miocene rotation of Sardinia: K–Ar ages and palaeomagnetic data of Tertiary volcanics.' *Earth Planet. Sci. Lett.*, 54, pp 261–271.

MOORBATH, S., (1967) 'Recent advances in the application and interpretation of radiometric age data.' *Earth Sci. Rev.*, 3, pp 113–133.

MOORBATH, S., (1969) 'Evidence for the age of deposition of the Torridonian sediments of north-west Scotland.' *Scott. J. Geol.*, 5, pp 154–170.

MOORBATH, S., (1977) 'Age isotopes and evolution of Precambrian continental crust.' *Chemical Geol.* 201, pp 151–187.

MORALEV, V. M., (1981) 'Tectonics and petrogenesis of early Precambrian complexes of the Aldan Shield, Siberia.' in *Precambrian Plate Tectonics*, (ed. A. KRÖNER) Elsevier, Amsterdam, pp 237–260.

MORGAN, G. E., (1976) 'Palaeomagnetism of a slowly-cooled plutonic terrain in western Greenland.' *Nature*, 259, pp 383–385.

MORGAN, G. E. & BRIDEN, J. C., (1981) 'Aspects of Precambrian palaeomagnetism with new data from the Limpopo mobile belt and Kaap Vaal craton in southern Africa. *Phys. Earth Planet. Int.*, 24, pp 142–168.

MORGAN, G. E. & SMITH, P. P. K., (1981) 'Transmission electron microscope and rock magnetic investigations of remanence carriers in a Precambrian metadolerite.' *Earth Planet. Sci. Lett.*, 53, pp 226–240.

MORGAN, W. J., (1968) 'Rises trenches, great faults and crustal blocks.' *J. Geophys. Res.*, 73, pp 1959–1982.

MORGAN, W. J., (1972) 'Deep mantle convection plumes and plate motions.' *Am. Assoc. Pet. Geol. Bull.*, 56, pp 203-213.
MORGAN, W. J., (1981) 'Hotspot tracks and the opening of the Atlantic and Indian Oceans.' in: *The Sea*, 7 (ed. C. EMILIANI), J. Wiley & Sons, New York, pp 443-487.
MORNER, N. A., (1981) 'Revolution in Cretaceous sea-level analysis.' *Geology*, 9, pp 344-346.
MORRIS, W. A., (1976) 'Palaeomagnetic results from the lower Palaeozoic of Ireland.' *Can. J. Earth Sci.* 13, pp 294-304.
MORRIS, W. A. (1977a) 'Palaeomagnetism of the Gowganda and Chibougamau Formations: evidence for 2,200 m.y. old folding and remagnetisation event of the southern province.' *Geology*, 5, pp 137-140.
MORRIS, W. A., (1977b) 'Palaeolatitude of glaciogenic upper Precambrian Rapitan Group and the use of tillites as chronostratigraphic marker horizons.' *Geology*, 5, pp 85-88.
MORRIS, W. A., (1979) 'A positive contact test between Nipissing diabase and Gowganda argillites.' *Can. Jour. Earth Sci.* 16, pp 607-611.
MORRIS, W. A. (1980) 'Tectonic and metamorphic history of the Sudbury Norite: the evidence from palaeomagnetism.' *Econ. Geology*, 75, pp 260-277.
MORRIS, W. A., (1981a) 'A positive fold test from Nipissing diabase.' *Can. J. Earth Sci.*, 18, pp 591-598.
MORRIS, W. A., (1981b) 'Fault block rotations in the Southern Province defined by palaeomagnetism of the Nipissing diabase.' *Can. J. Earth Sci.*, 18, pp 1755-1757.
MORRIS, W. A. & ROY, J. L., (1977) 'Discovery of the Hadrynian Track and further study of the Grenville problem.' *Nature*, 266, pp 689-692.
MORRIS, W. A., SCHMIDT, P. W. & ROY, J. L., (1979) 'A graphical approach to polar paths: palaeomagnetic cycles and global tectonics.' *Phys. Earth Planet. Int.*, 19, pp 85-99.
MORRIS, W. A. & TANCZYK, E. I., (1978) 'A preferred orientation of intrusion of Precambrian dykes.' *Nature*, 275, pp 120-121.
MORRISH, A. H. & WATT, L. A. K., (1958) 'Coercive force of iron oxide powders at low temperatures.' *J. App. Phys.*, 29, pp 1029-1033.
MORRISH, A. H. & YU, S. P., (1955) 'Dependence of the coercive force on the grain size and the density of some iron oxide powders.' *J. Appl. Phys.*, 26, pp 1049-1055.
MULLINS, C. E., (1978) 'Magnetic susceptibility of the soil and its significance in soil science — a review.' *J. Soil Sci.*, 28, pp 223-246.
MULLINS, C. E. & TITE, M. S., (1973) 'Magnetic viscosity, quadrature susceptibility and frequency dependence of susceptibility in single-domain assemblies of magnetite and maghemite.' *Jour. Geophys. Res.*, 78, pp 804-809.
MURPHY, G. S., EVANS, M. E. & GOUGH, D. I., (1971) 'Evidence of single-domain magnetite in the Michikamau anorthosite.' *Can. J. Earth Sci.*, 8, pp 361-370.
MUSSAKOVSKY, A. A., (1973) 'Palaeozoic orogenic volcanism of Eurasia (principal complexes and the tectonic factors in their distribution),' in: *Developmental stages of folded belts and the problem of ophiolites*, III, Institute of Geology, Academy of Sciences of the USSR, pp 140-161.
NAGATA, T., (1961) *Rock Magnetism*, 2nd Ed., Maruzen Press, Tokyo, 350 pp.
NAGATA, T., AKIMOTO, S. & UYEDA, S. (1952) 'Reverse thermo-remanent magnetism.' *Proc. Jap. Acad.*, 27., 27, pp 643-645.
NAGATA, T., UYEDA, S. & AKIMOTO, S., (1952) 'Self reversal of thermoremanent magnetism of igneous rocks.' *J. Geomag. Geoelect.*, 4, pp 22-38.
NAIRN, A. E. M., (ed.), (1974) *Problems in palaeoclimatology*, Interscience Publishers, London, 720 pp.
NAIRN, A. E. M. & WESTPHAL, M., (1968) 'Possible implications of the palaeomagnetic study of late Palaeozoic igneous rocks of north western Corsica.' *Palaeogeo., Palaeoclim. Palaeoecol.*, 5, pp 179-204.
NAQVI, S. M., DIVAKARA RAO, V. & NARAIN, H., (1974) 'The protocontinental growth of the Indian Shield and the antiquity of its rift valleys.' *Precambrian Research*, 1, pp 345-398.

NÉEL, L., (1949), 'Théorie due traînage magnétiques des ferromagnétiques en grains fins avec applications aux terres cuites.' *Ann. Geophys.*, 5, pp 99–136.

NÉEL, L., (1955) 'Some theoretical aspects of rock magnetism.' *Adv. Phys.*, 4, pp 191–243.

NEEV, D. & HALL, J. K., (1982) 'A global system of spiralling geosutures.' *J. Geophys. Res.*, 87, pp 10689–10708.

NEGI, J. G. & TIWARI, R. K., (1983) 'Matching long term periodicities of geomagnetic reversals and galactic motions of the solar system.' *Geophys. Res. Lett.*, 10, pp 713–716.

NESS, G., LEVI, S. & COUCH, R., (1980) 'Marine magnetic anomaly time-scales for the Cenozoic and Late Cretaceous: a précis, critique and synthesis.' *Rev. Geophys. Space Phys.*, 18, pp 753–770.

NEVANLINNA, H. & PESONEN, L. J., (1983) 'Late Precambrian Keweenawan asymmetric polarities as analysed by axial offset dipole geomagnetic models.' *J. Geophys. Res.*, 88, pp 647–658.

NORMAN, T. N., (1960) 'Azimuths of primary linear structures in folded strata.' *Geol. Mag.*, 97, pp 338–343.

NORTHOLT, A. J. G., (1980) 'Economic phosphatic sediments: mode of occurrence and stratigraphical distribution.' *J. Geol. Soc. Lond.*, 137, pp 793–805.

NORTON, I. O. & SCLATER, J. (1979) 'A model for the evolution of the Indian Ocean and the break-up of Gondwanaland.' *J. Geophys. Res.*, 84, pp 6803–6829.

NOZHAROV, P., PETKOV, N., YANEV, S., KROPACEK, V., KRS, M. & PURNER, P., (1980) 'A palaeomagnetic and petromagnetic study of Upper-Carboniferous, Permian and Triassic sediments, NW Bulgaria,' *Studia Geophysica et Geodaetica (Prague)*, 24, pp 252–284.

NUNES, P. D., (1981) 'The age of the Stillwater complex — a comparison of U–Pb zircon and Sm–Nd isochron systematics.' *Geochim. Cosmochim. Acta.*, 45, pp 1961–1963.

NUNES, P. D. & JENSEN, L. S., (1980) 'Geochronology of the Abitibi Metavolcanic Belt, Kirkland Lake area — Progress Report.' *Ontario Geological Survey, Miscellaneous Paper* 92, pp 252–284.

O'HARA, M. J., (1977) 'Thermal history of excavation of Archaean gneisses from the base of the continental crust.' *J. Geol. Soc. Lond.*, 134, pp 185–200.

O'NIONS, R. K., EVENSON, N. M. & HAMILTON, P. J., (1979) 'Geochemical modelling of mantle differentiation and crustal growth.' *J. Geophys. Res.*, 84, pp 6091–6101.

O'NIONS, R. K., SMITH, D. G. W., BAADSGARD, H. & MORTON, R. D., (1969) 'Influence of chemistry and compositions on argon retentivity in metamorphic calcic amphiboles from South Norway.' *Earth Planet. Sci. Lett.*, 5, pp 339–345.

ONSTOTT, T. C., (1980) 'Application of the Bingham distribution function in palaeomagnetic studies.' *J. Geophys. Res.*, 85, pp 1500–1510

ONSTOTT, T. C. & HARGRAVES, R. B., (1981) 'Proterozoic transcurrent tectonics: palaeomagnetic evidence from Venezuela and Africa.' *Nature*, 289, pp 131–136.

ONSTOTT, T. C., HARGRAVES, R. B., YORK, D. & HALL, C., (1984) 'Constraints of the motions of South American and African Shields during the Proterozoic: I. $^{40}Ar/^{39}Ar$ and palaeomagnetic correlations between Venezuela and Liberia.' *Bull. Geol. Soc. Amer.* 95, pp 1045–1054.

OPDYKE, N. D., (1962) 'Palaeoclimatology and continental drift,' in: *Continental drift* (ed. S. K. RUNCORN), Academic Press, New York, pp 41–65.

OPDYKE, N. D., GLASS, B., HAYS, J. D. & FOSTER, J., (1966) 'Palaeomagnetic study of Antarctic deep-sea cores.' *Science*, 154, pp 349–357.

OPDYKE, N. D., KENT, D. V. & LOWRIE, W., (1973) 'Details of magnetic polarity transitions recorded in a high deposition rate deep-sea core'. *Earth Planet. Sci. Lett.*, 20, pp 315–324.

OPDYKE, N. D. & RUNCORN, S. K., (1960) 'Wind direction in the western United States in the late Palaeozoic.' *Geol. Soc. Amer. Bull.*, 71, pp 959–972.

O'REILLY, W., (1983) *Rock And Mineral Magnetism*. Blackie, Glasgow, 220 pp.

OTOFUJI, Y. & MATSUDA T., (1983) 'Palaeomagnetic evidence for the clockwise rotation of southwest Japan.' *Earth Planet. Sci. Lett.*, 62, pp 349–359.

OTOFUJI, Y., MATSUDA, T. & NOHDA, S., (1985) 'Palaeomagnetic evidence for the Miocene counter-clockwise rotation of northeast Japan — rifting process of the Japan Arc.' *Earth Planet. Sci. Lett.*, 75, pp 265–277.

OZDEMIR, O. & BANERJEE, S. K., (1984) 'High temperature stability of maghemite (γFe_2O_3).' *Geophys. Res. Lett.*, 11(3), pp 161–164.

OZIMA, M., OZIMA, M. & NAGATA, T., (1964) 'Low temperature treatment as an effective means of 'magnetic cleaning' of natural remanent magnetisation'. *J. Geomag. Geoelect.*, 16, pp 37–40.

PAL, P.C., (1983) 'Palaeocontinental configurations and geoid anomalies'. *Nature*, 303, pp 513–516.

PALMER, H. C., (1970) 'Palaeomagnetism and correlation of some Middle Keweenawan rocks, Lake Superior', *Can. J. Earth Sci.*, 7, pp 1410–1436.

PALMER, H. C., BARAGER, W. R. A., FORTIER, M. & FOSTER, J. H., (1983) 'Palaeomagnetism of Late Proterozoic rocks, Victoria Island, Northwest Territories, Canada'. *Can. J. Earth Sci.*, 20, pp 1456–1469.

PALMER, H.C. & CARMICHAEL, C. M., (1973). 'Palaeomagnetism of some Grenville Province rocks,' *Can. J. Earth Sci.* 10, 1175–1190.

PALMER, H. C. & HALLS H. C., (1985), 'The palaeomagnetism of the Powder Mill Group: its relevance to correlation with other Keweenawan sequences and to tectonic development of the South Range,' *31st Ann. meeting, 1st Lake Superior Geology*, (abstract).

PALMER, H. C., HALLS, H. C. & PESONEN, L. J., (1981) 'Remagnetisation in Keweenawan rocks. Part I: Conglomerates.' *Can. J. Earth Sci.*, 18, pp 599–618.

PALMER, H. C. & HAYATSU, A., (1975) 'Palaeomagnetism and K–Ar dating of some Franklin lavas and diabases, Victoria Island'. *Can. J. Earth Sci.*, 12, pp 1439–1447.

PALMER, H. C., HAYATSU, A. & MACDONALD, W. D., (1980) 'The Middle Jurassic Camaraca Formation, Arica, Chile: palaeomagnetism, K–Ar dating and tectonic implications'. *Geophys. J. R. Astr. Soc.*, 62, pp 155–172.

PARK, A. F. (1983) 'Sequential development of metamorphic fabric and structural elements in polyphase deformed serpentinites in the Svecokarelian of eastern Finland'. *Trans. R. Soc. Edinburgh, Earth Sciences*, 74, pp 33–60.

PARK, J. K., (1974) 'Palaeomagnetism of miscellaneous Franklin and Mackenzie diabases of the Canadian Shield and their adjacent country rocks'. *Can. J. Earth Sci.* 11, pp 1012–1017.

PARK, J. K., (1975) 'Palaeomagnetism of the Flin-Flon–Snow Lake Greenstone Belt, Manitoba and Saskatchewan'. *Can. J. Earth Sci.*, 12, pp 1272–1290.

PARK, R. G., (1981) 'Shear-zone deformation and bulk strain in granite–greenstone terrain of the western Superior Province, Canada'. *Precambrian Research*, 14, pp 31–47.

PARKER, R. L. & DENHAM, C. R., (1979) 'Interpretation of unit vectors'. *Geophys. J. R. Astr. Soc.*, 58, pp 685–687.

PARRY, L. G., (1965) 'Magnetic properties of dispersed magnetic powder'. *Phil. Mag.*, 11, pp 303–312.

PARRY, L. G., (1979) 'Magnetisation of multi-domain particles of magnetite'. *Phys. Earth Planet. Int.*, 19, pp 21–30.

PARSONS, B., (1982) 'Causes and consequences of the relations between area and age of the ocean floor'. *J. Geophys. Res.*, 87, pp 289–302.

PARSONS, B. & DALY, S., (1983) 'The relationship between surface topography, gravity anomalies and temperature structure of convection'. *J. Geophys. Res.*, 88, pp 1129–1144.

PATCHETT, P. J., (1978) 'Rb/Sr ages of Precambrian dolerites and syenites in southern and central Sweden'. *Sveriges Geologiska Undersökning Serie C*, 747, 63 pp.

PATCHETT, P. J. & BYLUND, G., (1977) 'Age of Grenville Belt magnetisations: Rb–Sr and palaeomagnetic evidence from Swedish dolerites'. *Earth Planet. Sci. Lett.*, 35, pp 92–104.

PATCHETT, P. J., BYLUND, G. & UPTON, B. G. J., (1978) 'Palaeomagnetism and the Grenville orogeny, new Rb–Sr ages from dolerites in Canada and Greenland', *Earth Planet. Sci. Lett.*, 40, pp 349–364.

PATCHETT, P. J., GALE, N. H., GOODWIN, R. & HUMM, M. J., (1980) 'Rb–Sr whole rock isochron ages of late Precambrian to Cambrian igneous rocks from southern Britain.' *J. Geol. Soc. Lond.*, 137, pp 649–656.

PELTIER, W. R., (1981) 'Surface plates and thermal plumes: separate scales of the mantle convective circulation'. in: *Evolution of the Earth*, (eds. R. J. O'CONNELL and W. S. FYFE) Geodynamics Series, vol. 5, pp 229–248.

PELTIER, W. R., (1983) 'Constraint on deep mantle viscosity from LAGEOS acceleration data'. *Nature*, 304, pp 434–436.

PERIGO, R., VAN DER VOO, R., AUVRAY, B. & BONHOMMET, N., (1983) 'Palaeomagnetism of late Precambrian–Cambrian volcanics and intrusives from the Armorican massif, France'. *Geophys. J. R. Astr. Soc.*, 75, pp 235–260.

PERROUD, H., BONHOMMET, N., & ROBARDET, M., (1982) 'Comment on 'A palaeomagnetic study of Cambrian red beds from Carteret, Normandy, France', by W. A, Morris'. *Geophys. J. R. Astr. Soc.*, 69, pp 573–578.

PERROUD, H. & VAN DER VOO R., (1985) 'Palaeomagnetism of the late Ordovician Thouars Massif, Vendee Province, France,' *J. Geophys. Res.*, 90, pp 4611–4625.

PERROUD, H., VAN DER VOO, R. & BONHOMMET, N., (1984) 'Palaeozoic evolution of the Armorican plate on the basis of palaeomagnetic data'. *Geology*, 12, pp 579–582.

PERRY, E. A. & TUREKIAN K. K., (1974) 'The effects of diagenesis on the redistribution of strontium isotopes in shales'. *Geochim. Cosmochim. Acta.*, 38, pp 929–935.

PESONEN, L. J. & HALLS, H. C., (1983) 'Geomagnetic field intensity and reversal asymmetry in Late Precambrian Keweenawan rocks'. *Geophys. J. R. Astr. Soc.,*, 73, pp 241–270.

PESONEN, L. K. & NEUVONEN K. J., (1981) 'Palaeomagnetism of the Baltic Shield — implications for Precambrian tectonics'. in: *Precambrian Plate Tectonics* (ed. A. KRÖNER), Elsevier Publ. Co. Amsterdam, pp 623–648.

PETRASCHECK, W. E., (1973) 'Some aspects of the relations between continental drift and metallogenic provinces', in *Implications of Continental Drift to the Earth Sciences*, (ed.D. H. TARLING & S. K. RUNCORN), vol. 1, Academic Press, London, pp 567–572.

PHILLIPS, W. E. A., STILLMAN, C. J. & Murphy, T. (1976). 'A Caledonian plate tectonic model'. *J. Geol. Soc. Lond.*, 132, pp 579–609.

PICARD, M. D, (1964) 'Palaeomagnetic correlation of units within the Chugwater (Triassic) Formation, west-central Wyoming'. *Amer. Assoc. Pet. Geol. Bull.*, 48, pp 269–291.

PIPER, J. D. A., (1972) 'A palaeomagnetic study of the Bukoban System, Tanzania'. *Geophys J. R. Astr. Soc.*, 28, pp 111–127.

PIPER, J. D. A., (1974) 'Magnetic properties of the Cunene anorthosite complex, Angola. *Phys. Earth Planet. Int.*, 9, pp 353–363.

PIPER, J. D. A., (1975a) 'A palaeomagnetic study of the coast-parallel Jurassic dyke swarm in southern Greenland.' *Phys. Earth Planet. Int.*, 11, pp 36–42.

PIPER, J. D. A., (1975b) 'The palaeomagnetism of Precambrian igneous and sedimentary rocks of the Orange River Belt in South Africa and South West Africa'. *Geophys. J. R. Astr. Soc.*, 40, pp 313–344.

PIPER, J. D. A., (1975c)'Palaeomagnetic correlations of Precambrian formations of east-central Africa and their tectonic implications'. *Tectonophysics*, 26, pp 135–161.

PIPER, J. D. A., (1975d) 'Palaeomagnetism of Silurian lavas of Somerset and Gloucestershire, England'. *Earth Planet. Sci. Lett.*, 25, pp 355–360.

PIPER, J. D. A., (1976a) 'Palaeomagnetic evidence for a Proterozoic Supercontinent.' *Phil. Trans. R. Soc.*, Lond., A280, pp 469–490.

PIPER, J. D. A., (1976b) 'Definition of pre-2000 m.y. apparent polar wander movements'. *Earth Planet. Sci. Lett.*, 28, pp 470–478.

PIPER, J. D. A. (1977) 'Magnetic stratigraphy and magnetic petrologic properties of Precambrian Gardar lavas, South Greenland.' *Earth Planet. Sci. Lett.*, 34, pp 247–263.

PIPER, J. D. A., (1978) 'Palaeomagnetic survey of the (Palaeozoic) Shelve Inlier and Berwyn Hills, Welsh Borderlands'. *Geophys. J. R. Astr. Soc.*, 53, pp 355–371.

PIPER, J. D. A., (1979a) 'Palaeomagnetic study of late Precambrian rocks of the Midland Craton of England and Wales'. *Phys. Earth Planet. Int.*, 19, pp 59–72.

PIPER, J. D. A., (1979b) 'Aspects of Caledonian palaeomagnetism and their tectonic implications'. *Earth Planet. Sci. Lett.*, 46, pp 443–461.

PIPER, J. D. A., (1980), 'Palaeomagnetic study of the Swedish Rapakivi suite: Proterozoic tectonics of the Baltic Shield.' *Earth Planet. Sci. Lett.*, 46, pp 443–461.

PIPER, J. D. A., (1981a) 'The altitude dependence of magnetic remanence in the slowly-cooled Precambrian plutonic terrain of West Greenland'. *Earth Planet. Sci. Lett.*, 54, pp 449–466.

PIPER, J. D. A. (1981b) 'Palaeomagnetic study of the (Late Precambrian) West Greenland kimberlite-lamprophyre suite: definition of the Hadrynian Track,' *Phys. Earth Planet. Int.*, 27, pp 164–186.

PIPER, J. D. A., (1982a) 'The Precambrian palaeomagnetic record: the case for the Proterozoic Supercontinent'. *Earth Planet. Sci. Lett.*, 59, pp 61–89.

PIPER, J. D. A. (1982b) 'A palaeomagnetic investigation of the Malverian and Old Radnor Precambrian, Welsh Borderlands'. *Geol. Jour.*, 17, pp 69–88.

PIPER, J. D. A., (1983a) Dynamics of the continental crust in Proterozoic times, in: (eds. L. G., MEDARIS, C. W, BYERS, D. M. Mickelson and W. C. SHANKS), *Proterozoic Geology*. Geol. Soc. Amer. Mem., 161, 11–34.

PIPER, J. D. A., (1983b) 'Proterozoic palaeomagnetism and single continent plate tectonics' *Geophys. J. R. Astr. Soc.*, 74, pp 163–197.

PIPER, J. D. A., (1985a) 'Continental movements and break-up in late Precambrian–Cambrian times: prelude to Caledonian orogenesis'. in: *The Caledonide Orogen: Scandinavia and Related Areas* (ed. D. G. GEE & B. A. STURT), John Wiley, London, pp 19–34.

PIPER, J. D. A., (1985b) 'Palaeomagnetic study of the Nagssugtoqidian mobile belt in central-west Greenland,' *Precambrian Research*, 28, 75–110.

PIPER, J. D. A., (1987) '*A Directory of Global Palaeomagnetic Data.*' Open University Press, Milton Keynes.

PIPER, J. D. A., BRIDEN, J. C. & LOMAX, K., (1973) 'Precambrian Africa and South America as a single continent'. *Nature*, 245, pp 244–248.

PIPER J. D. A., MCCOOK, A. S., WATKINS, K. P., BROWN, G. C. & MORRIS, W. A., (1978) 'Palaeomagnetism and chronology of Caledonian igneous episodes in the Cross Fell Inlier and Northern Lake District'. *Geol. J.*, 13, pp 73–92.

PIPER, J. D. A. & STEARN, J. E. F., (1977) 'Palaeomagnetism of the dyke swarms of the Gardar Igneous Province, South Greenland'. *Phys. Earth Planet. Int.*, 14, pp 345–358.

PITCHER, W., (1969) 'Northeast-trending faults of Scotland and Ireland, and chronology of displacements'. in: *North Atlantic Geology and Continental Drift, a symposium* (ed. M. KAY), *Amer. Assoc. Pet. Geol.*, Mem. 12, pp 724–733.

PITMAN III, W. C., (1978) 'Relationship between eustasy and stratigraphic sequences of passive margins'. *Geol. Soc. Amer. Bull.*, 89, pp. 1389–1403.

PITMAN III, W. C., HERRON, E. M. & HEIRTZLER, J. R., (1968) 'Magnetic anomalies in the Pacific, and sea floor spreading'. *J. Geophys. Res.*, 73, pp 2069–2085.

PORATH, H. & CHAMALAUN, F. H., (1968) 'Palaeomagnetism of Australian hematite ore bodies — II Western Australia.' *Geophys. J. R. Astr. Soc.*, 15, pp 253–264.

POWELL, C., JOHNSON, B. D. & VEEVERS, J. J., (1980) 'A revised fit of east and west Gondwanaland.' *Tectonophysics*, 63, pp 313–29.

PRÉVOT, M., LE CAILLE, A. & MANKINEN, E. A., (1981) 'Magnetic effects of maghemitisation of ocean crust'. *Jour. Geophys. Res.*, 86, pp 4009–4020.

PULLAIAH, G., IRVING, E., BUCHAN, K. L. & DUNLOP, D. J., (1975) 'Magnetisation changes caused by burial and uplift'. *Earth Planet. Sci. Lett.*, 28, pp 133–143.

QUENNELL, A. M., (1983) 'On the evolution of the Dead Sea Rift — a review', in: *Proc. 1st Jordanian Geol. Conf. Amman* (ed. A. M. ABED & H. M. KHALID), Amman, pp 460–483.

RADHAKRISHNAMURTY, C., LIKHITE, S. D., DEUTSCH, E. R. & MURPHY, G. S., (1981) 'A comparison of the magnetic properties of synthetic titanomagnetites and basalts'. *Phys. Earth Planet. Int.*, 26, pp 37–46.

RADHAKRISHNAMURTY, C., LIKHITE, S. D. & SAHASTABUDE, P. W. (1977) 'Nature of magnetic grains and their effect on the remanent magnetisation of basalts'. *Phys. Earth Planet. Int.*, 13, pp 289–300.

RANKIN, D. W., (1976) 'Appalachian salients and recesses: Late Precambrian continental breakup and the opening of the Iapetus Ocean.' *J. Geophys. Res.*, 81, pp 5606–5619.

RAO, K. V., SEGUIN, M. K. & DEUTSCH, E. R. (1981) 'Palaeomagnetism of Siluro-Devonian and Cambrian granitic rocks from the Avalon zone in Cape Breton Island, Nova Scotia'. *Can. J. Earth Sci.*, 18, pp 1187–1210.

RAO, K. V. & VAN DER VOO, R., (1980) 'Palaeomagnetism of a Palaeozoic anorthosite from the Appalachian Piedmont, Northern Delaware: possible tectonic implications'. *Earth Planet. Sci. Lett.*, 47, pp 113–120.

RATHORE, J. S., (1979) 'Magnetic susceptibility anisotropy in the Cambrian Slate Belt of North Wales and correlation with strain'. *Tectonophysics*, 53, pp 83–97.

READ, H. H., (1949) 'A Contemplation of time in plutonism'. *Quart. J. Geol. Soc.*, Lond., 105, pp 101–156.

READMAN, P. W. & O'REILLY, W., (1972) 'Magnetic properties of oxidised titanomagnetites'. *J. Geomag. Geoelect.*, 24, pp 69–90.

REES, A. I., (1961) 'The effect of water currents on the magnetic remanence and anistropy of susceptibility of some sediments'. *Geophys. J. R. Astr. Soc.*, 5, pp 235–251.

REIS, A. C., RICHARDSON, A. & SHACKLETON, R. M., (1980) 'Rotation for the Iberian arc: palaeomagnetic results from north Spain'. *Earth Planet. Sci. Lett.*, 50, pp 301–310.

RICE, P. D., HALL, J. M. & OPDYKE, N. D., (1980) 'Deep drill 1972: a palaeomagnetic study of the Bermuda Seamount'. *Can. Earth Sci.*, 17, pp 232–243.

RICHTER, F. M. & PARSONS, B., (1976) 'On the interaction of two scales of convection in the mantle'. *J. Geophys. Res.*, 80, pp 2529–2541.

RIDING, R., (1974) 'Model of the Hercynian fold belt'. *Earth Planet. Sci. Lett.*, 24, pp 125–135.

RINGWOOD, A. E., (1975) *Composition and Petrology of the Earth's Mantle*, McGraw-Hill, New York, 618 pp.

RIKITAKE, T., (1966) *Electromagnetism and the Earth's Interior*, Devel. Solid Earth Geophys., 2, Elsevier Publ. Co., Amsterdam, 308 pp.

ROBERTS, D. & GALE, G. H., (1978) 'The Caledonian–Appalachian Iapetus Ocean' in: *Evolution of the Earth's Crust* (ed. D. H. TARLING) Academic Press, pp 255–342.

ROBERTS, N., (1983) *The Earth's magnetic field during field reversals*, Ph.D. thesis, University of Cardiff.

ROBERTSON, W. A. (1973) 'Pole positions from the Mamainse Point lavas and the bearing on a Keweenawan pole path and polarity sequence'. *Can. Jour. Earth Sci.*, 10, pp 1541–1555.

ROBERTSON, W. A. & FAHRIG, W. F., (1971) 'The great Logan Palaeomagnetic Loop — the polar wandering path from Canadian Shield rocks during the Neohelikian Era'. *Can. Jour. Earth Sci.* 8, pp 1355–1372.

ROLPHE, T. C. & SHAW, J., (1985) 'A new method of palaeofield magnitude correction for thermally altered samples and its application to lower Carboniferous lavas', *Geophys. J. R. Astr,. Soc.*, 80, pp 773–781.

ROBINSON, P. L. (1973) 'Palaeoclimatology and continental drift', in: *Atlas of palaeobiogeography* (ed. A. HALLAM), Elsevier Publ. Co., Amsterdam, pp 451–476.

ROBINSON, R. A. & ROWELL, A. J., (ed.), (1976) *Paleontology and depositional environments: Cambrian of Western North America*, Geol. Studies, Provo, Utah, 23, 227 pp.

RONOV, A. B., (1968) 'Probable changes in the composition of sea water during the course of geological time'. *Sedimentology*, 10, pp 25–43.

RONOV, A. B., KAIN, V. E., BALUKHOVSKY, A. W. & SESLAVINSKY, K. B., (1980). 'Quantitative analysis of Phanerozoic sedimentation'. *Sedimentary Geology*, 25, pp 311–325.

ROSENBERG, G. D., & RUNCORN, S. K., (ed.) (1975) *Rhythms and the history of the Earth's rotation*, John Wiley, London, 559 pp.

ROSS, C. A., (1974) 'Palaeogeographic provinces and provinciality'. *Soc. Econ. Palaeont. Mineral., Spec. Publ.* 21.

ROY, J. L., (1969) 'Palaeomagnetism of the Cumberland Group and other Palaeozoic formations'. *Can. J. Earth Sci.*, 6, pp 663–669.

ROY, J. L. & ANDERSON, P., (1981) 'An investigation of the remanence characteristics of three sedimentary units in the Silurian Mascarene Group of New Brunswick, Canada'. *Jour. Geophys. Res.*, 86, pp 6357–6368.

ROY, J. L. & FAHRIG, W. F., (1973) 'The palaeomagnetism of Seal and Croteau rocks from the Grenville Front, Labrador: polar wandering and tectonic implications'. *Can. J. Earth Sci.*, 10, pp 1279–1301.

ROY, J. L. & LAPOINTE, P. L., (1976) 'The palaeomagnetism of Huronian red beds and Nipissing diabase; Post-Huronian igneous events and apparent polar path for the interval -2300 to -1500 for Laurentia'. *Can. Jour. Earth Sci.* 13, pp 749–773.

ROY, J. L. & LAPOINTE, P. L., (1978) 'Multiphase magnetisations: problems and implications.' *Phys. Earth Planet. Int.*, 16, pp 20–39.

ROY, J. L., LAPOINTE, P. L. & ANDERSON, P., (1975) 'Palaeomagnetism of the oldest red beds and the direction of Late Aphebian polar wander relative to Laurentia'. *Geophys. Res. Lett.*, 2, pp 537–540.

ROY, J. L. & MORRIS, W. A., (1983) 'A review of palaeomagnetic results from the Carboniferous of North America; the concept of Carboniferous geomagnetic field horizon markers'. *Earth Planet. Sci. Lett.*, 65, pp 167–181.

ROY, J. L. & Park J. K., (1974) 'The magnetisation process of certain red beds: vector analysis of chemical and thermal results'. *Can. Jour. Earth. Sci.*, 11, pp 437–471.

ROY, J. L. & ROBERTSON, W. A., (1978) 'Palaeomagnetism of the Jacobsville Formation and the apparent polar path for the interval -1100 to -670 Ma for North America'. *J. Geophys. Res.*, 83, pp 1289–1304.

ROY, J. L., TANCZYK, E. & LAPOINTE, P., (1983) 'The palaeomagnetic record of the Appalachians', in: *Regional Trends in the Geology of the Appalachian–Caledonian–Hercynian–Mauritanide Orogen* (ed. P. E. SCHENK), D. Reidel Publ. Co., pp 11–26

RUBEY, W. W., (1955) 'Developments in the hydrosphere with special reference to the probable composition of the early atmosphere'. *Geol. Soc. Amer., Spec. Pap.*, 62, pp 631–650.

RUMBLE, D., (1976) 'Oxide minerals in metamorphic rocks', in: *Oxide Minerals* (ed. D. RUMBLE) Mineral. Soc. Amer. Short Course Notes, vol. 3, R1–24.

RUNCORN, S. K., (1956) 'Palaeomagnetic comparisons between Europe and North America' *Proc. Geol. Assoc. Canada*, 8, pp 77–85.

RUNCORN S. K., (1961) 'Climatic change through geological time in the light of the palaeomagnetic evidence for polar wandering and continental drift'. *Quart. J. Roy. Met. Soc.*, 87, pp 282–313.

RUSSELL, C. T., (1979) 'Scaling law test and two predictions of planetary magnetic moments'. *Nature*, 281, pp 552–553.

RUTLAND, R. W. R. (1973) 'Tectonic evolution of the continental crust in Australia', in: *Implications of Continental Drift to the Earth Sciences* (ed. D. H. TARLING, & S. K. RUNCORN), vol. 2, Academic Press, London, pp 1011–1033.

RYALL, P. J. C., HALL, J. M., CLARK, J. & MILLIGAN, T., (1977) 'Magnetisation of oceanic crustal layer 2 — results and thoughts after DSDP leg 37'. *Can. Jour. Earth Sci.*, 14, pp 684–706.

SAAD, A. H., (1969) 'Palaeomagnetism of Franciscan ultramafic rocks from Red Mountain, California'. *Jour. Geophys. Res.*, 74, pp 6567–6578.

SALLOMY, J. T., & Piper, J. D. A., (1973) 'Palaeomagnetic studies in the British Caledonides IV.

Lower Devonian lavas of the Strathmore region, Scotland'. *Geophys. J. R. Astr. Soc.*, 34, pp 49–68.

SANDBERG, P. A., (1983) 'An oscillating trend in Phanerozoic non-skeletal carbonate mineralogy'. *Nature*, 305, pp 19–22.

SASAJIMA, S. & SHIMADA, M., (1964) 'The palaeomagnetic study upon the geotectonics of the Ryukyu Islands'. *Ann. Rep. Rock. Mag. Res. Grp.*, Tokyo, pp 67–70.

SASAJIMA, S. & SHIMADA, M., (1966) 'Palaeomagnetic studies of the Cretaceous volcanic rocks of SW Japan'. *J. Geol. Soc. Japan*, 72, pp 503–514.

SASAJIMA, S., OTOFUJI, Y., KIROOKA, K., SUPARKA, S. & HEHUWAT, F., (1978) 'Palaeomagnetic studies on Sumatra being part of Gondwanaland'. *Rock Magnet. Palaeogeophys.*, 5, pp 104–110.

SCHEMERHORN, L. J. G., (1983) 'Late Proterozoic glaciation in the light of CO_2 depletion in the atmosphere', in: *Proterozoic Geology* (eds. L. G. MEDARIS, C. W. BYERS, D. M. MICKELSON & W. C. SHANKS) *Geol. Soc. Amer. Mem.* 161, pp 309–315.

SCHILLING, J. G., (1973) 'Icelandic mantle plume, geochemical evidence along the Reykjanes Ridge'. *Nature*, 242, pp 565–578.

SCHMIDT, P. W. (1980) 'Palaeomagnetism of igneous rocks from the Belcher Islands, N. W. T., Canada,' *Can. J. Earth Sci.*, 17, pp 807–822.

SCHMIDT, P. W., (1985) 'Bias in converging great circle method.' *Earth Planet. Sci. Lett.*, 72, pp 427–432.

SCHMIDT, P. W. & EMBLETON, B. J. J., (1981) 'Magnetic overprinting in south eastern Australia and the thermal history of its rifted margin'. *J. Geophys. Res.*, 86, pp 3998–4008.

SCHMIDT, P. W. & EMBLETON, B. J. J., (1985) 'Prefolding and overprint magnetic signatures in Precambrian (2·9–2·7 Ga) igneous rocks from the Pilbara Craton and Hammersley basin, NW Australia'. *Jour. Geophys. Res.*, 90, pp 2967–2984.

SCHMIDT, P. W. & MORRIS, W. A., (1977) 'An alternative view of the Gondwana Palaeozoic apparent polar path'. *Can. J. Earth Sci.*, 14, pp 2674–2678.

SCHUILLING, R. D., (1967) 'Tin belts on the continents around the Atlantic ocean.' *Econ. Geol.*, 62, pp 540–550.

SCHULT, F. R., & GORDON, R. G., (1984) 'Velocities of the continents'. *J. Geophys. Res.*, 89, pp 1767–1801.

SCHUTTS L. D. & DUNLOP, D. J., (1981) 'Proterozoic magnetic overprinting of Archaean rocks in the Canadian Superior Province'. *Nature*, 291, pp 642–644.

SCHWARTZ, E. J. & FUJIWARA, Y., (1981) 'Palaeomagnetism of the Circum-Ungava Belt II: Proterozoic rocks of Richmond Gulf and Manitounuk Islands', in: *Proterozoic Basins of Canada*, (ed. F. H. A. CAMPBELL), *Geol. Surv. Can. Pap.* 81-10 pp 255–267.

SCHWARZ, E. J. & SYMONS, D. T. A., (1969) 'Geomagnetic intensity between 100 million and 2500 million years ago'. *Phys. Earth Planet. Int.*, 2, pp 11–18.

SCHWARZ, E. J. & SYMONS, D. T. A., (1970) 'Palaeomagnetic field intensity during cooling of the Sudbury irruptive 1700 million years ago'. *J. Geophys. Res.*, 75, pp 6631–6640.

SCLATER, J. G., JARRARD, R. D., MCGOWRAN, B. & GARTNER, S., (1974) 'Comparison of the magnetic and biostratigraphic scales since the late Cretaceous'. *Init. Reps. Deep Sea Drilling Project*, 22, pp 381–386.

SCOTESE, C. R., BAMBACH, R. K., BARTON, C., VAN DER VOO, R. & ZIEGLER, A. M., (1979) 'Palaeozoic base maps. *Jour. geol.*, 87, pp 217–277.

SEARS, J. W. & PRICE, R. A., (1978) 'The Siberian Connection: a case for Precambrian separation of the North American and Siberian cratons'. *Geology*, 6, pp 267–270.

SEGUIN, M. K., (1981) 'Reconnaissance palaeomagnetic investigation in the Spider Lake-Woburn area'. *J. Geomag. Geoelectr.*, 33, pp 205–224.

SEGUIN, J. K., RAO, K. V. & PINEAULT, R., (1982) 'Palaeomagnetic study of Devonian rocks from Ste. Cecile-St. Sebastien Region, Quebec Appalachians'. *J. Geophys. Res.*, 87, pp 7853–7864.

SELLEY, R. C., (1976) *An Introduction to Sedimentology*. Academic Press, London, 408 pp.
SEMENENKO, N. P., SCHERBAK, A. P., VINOGRADOV, A. P., TOUGARINOV, A. I., ELISEEVA, G. B., COTLOVSKAY, F. J. & DEMIDENKO, S. G., (1968) 'Geochronology of the Ukrainian Precambrian'. *Can. J. Earth Sci.*, 5, pp 661–671.
SENANAYAKE, W. E. & MCELHINNY, M. W., (1981) 'Hysteresis and susceptibility characteristics of magnetite and titanomagnetites: interpretation of results from basaltic rocks'. *Phys. Earth Planet. Int.*, 26, pp 47–55.
SENANAYAKE, W. E. & MCELHINNY, M. W. (1982) 'The effects of heating on low temperature susceptibility and hysteresis properties of basalts'. *Phys. Earth Planet. Int.*, 30, pp 317–321.
SENGOR, A. M. C., (1979) 'Mid-Mesozoic closure of Permo-Triassic Tethys and its implications'. *Nature*, 279, pp 590–593.
SEYFERT, C. K. & SIRKIN, L. A., (1973) *Earth History and Plate Tectonics*, Harper & Row, New York, 504 pp.
SHACKLETON, R. M., (1973) 'Correlation of structures across Precambrian orogenic belts in Africa', in: (ed. D. H. TARLING & S. K. RUNCORN), *Implications of Continental Drift to the Earth Sciences*, vol. 2, Academic Press, London, pp 317–321.
SHACKLETON, R. M., RIES, A. C., COWARD, M. P. & COBBOLD, P. R., (1979) 'Structure, metamorphism and geochronology of the Arequipa Massif of coastal Peru'. *J. Geol. Soc. Lond.*, 136, pp 195–214.
SHAW, J., (1974) 'A new method of determining the magnitude of the palaeomagnetic field. Application to five historic lavas and five archaeological samples'. *Geophys. J. R. Astr. Soc.*, 39, pp 133–141.
SHAW, J., (1977) 'Further evidence for a strong intermediate state of the palaeomagnetic field'. *Geophys. J. R. Astr. Soc.*, 48, pp 263–270.
SHAW, J., SHARE, J. A. & ROGERS, J., (1984) 'An automated superconducting magnetometer and demagnetising system'. *Geophys. J. R. Astr. Soc.*, 78, pp 209–218.
SHEGELSKI, R. J., (1979) 'Geologic evolution of the Shebandowan–Thunder Bay greenstone terrain'. *Precambrian Research*, 12, pp 331–348.
SHIMRON, A. E., (1980) 'Proterozoic island arc volcanism and sedimentation in Sinai'. *Precambrian Res.*, 12, pp 437–458.
SILVER, L. T. & GREEN, J. C., (1975) 'Time constants for Keweenawan igneous activity'. *Geol. Soc. Amer. Abstr. Programs*, 4, p 665.
SIMPSON, J. F., (1966) 'Evolutionary pulsations and geomagnetic polarity'. *Bull. Geol. Soc. Amer.*, 77, pp 197–203.
SIMPSON, R. W. & COX, A. V., (1977) 'Palaeomagnetic evidence for tectonic rotation of the Oregon Coast Range'. *Geology*, 5, pp 585–589.
SKERLEC, G. M. & HARGRAVES, R. B., (1980) 'Tectonic significance of palaeomagnetic data from northern Venezuela'. *J. Geophys. Res.*, 85, pp 5303–5315.
SMITH, A. G., BRIDEN, J. C., & DREWRY, G. E., (1973) 'Phanerozoic World Maps, *Spec. Pap. Palaeontol.* 12, pp 1–42.
SMITH, A. G. & HALLAM, A., (1970) 'The fit of the southern continents'. *Nature*, 225, pp 139–144.
SMITH, A. G., HURLEY, A. M. & BRIDEN, J. C., (1980) *Phanerozoic Palaeocontinental World Maps*, Cambridge University Press, Cambridge, 102 pp.
SMITH, P. J., (1967) 'The intensity of the ancient geomagnetic field: a review and analysis'. *Geophys. J. R. Astr. Soc.*, 12, pp 213–362.
SMITH, R. L. (1979) 'A high presure cell for chemical demagnetisation of sediments'. *Geophys. J. R. Astr. Soc.*, 59, pp 605–608.
SMITH, R. L. & PIPER, J. D. A., (1982) 'Palaeomagnetism of the Southern Zone of the Lewisian (Precambrian) Foreland, NW Scotland'. *Geophys. J. R. Astr. Soc.*, 68, pp 325–347.
SMITH, R. L., STEARN, J. E. F. & PIPER, J. D. A., (1983) 'Palaeomagnetic studies of the Torridonian sediments, NW Scotland'. *Scott. J. Geol.*, 19, pp 29–45.

SMITH, R. W. & FULLER, M. D., (1967) 'Alpha-hematite: stable remanence and memory'. *Science*, 156, pp 1130–1133.

SOFFEL, H. C., POHL, W. & BUSER, S., (1983) 'Palaeomagnetism of Permo-Triassic rocks from northern Slovenia, Yugoslavia, and the eastern margin of the Adriatic Plate'. *Tectonophysics*, 91, pp 301–320.

SPALL, H., (1971) 'Precambrian apparent polar wandering evidence from North America'. *Earth Planet. Sci. Lett.*, 10, pp 273–280.

SPALL, H., (1972) 'Palaeomagnetism and Precambrian continental drift.' *24th Int. Geol. Cong, Section 3*, pp 172–179.

SPARIOSU, D. J. & KENT, D. V., (1983) 'Palaeomagnetism of the Lower Devonian Traveller Felsite and the Acadian Orogeny in the New England Appalachians'. *Bull. Geol. Soc. Amer.*, 94, pp 1319–1328.

STACEY, F. D., (1960) 'Magnetic anisotropy of igneous rocks'. *Jour. Geophys. Res.*, 65, pp 2429–2442.

STACEY, F. D., (1963) 'The physical theory of rock magnetism'. *Phil. Mag. Supp. Adv. Phys.*, 12, pp 46–133.

STACEY, F. D., (1977) 'A thermal model of the Earth'. *Phys. Earth Planet. Int.*, 18, pp 341–348.

STACEY, F. D. & BANERJEE, S. K., (1974) *The Physical Principles of Rock Magnetism, Developments in Solid Earth Geophysics* 5, Elsevier, Publ. Co. Amsterdam, 195 pp.

STARKEY, J. & PALMER, H. C., (1971) 'The sensitivity of the conglomerate test in palaeomagnetism'. *Geophys. J. R. Astr. Soc.*, 71, pp 235–240.

STAUFFER, P. H., (1974) 'Malaya and southeast Asia in the pattern of continental drift'. *Bull. Geol. Soc. Malaysia*, 7, pp 89–138.

STEARN, J. E. F. & PIPER, J. D. A., (1984) 'Palaeomagnetism of the Sveconorwegian mobile belt of the Fennoscandian Shield'. *Precambrian Research*, 23, pp 201–246.

STEARNS, C., MONKS, F. J. & VAN DER VOO, R., (1982) 'Late Cretaceous–early Tertiary palaeomagnetism of Aruba and Bonaire (Netherlands Leeward Antilles)'. *J. Geophys. Res.*, 87, pp 1127–1141.

STEIGER, R. H. & JÄGER, E., (1977) 'Subcommission of geochronology: convention on the use of decay constants in geo- and cosmochronology'. *Earth Planet. Sci. Lett.*, 36, pp 359–362.

STEINER, M. B., (1978) 'Magnetic polarity during the Middle Jurassic as recorded in the Summerville and Cartis Formations'. *Earth Planet. Sci. Lett.*, 38, pp 331–345.

STEINER, M. B. (1983) 'Detrital remanent magnetisation in hematite'. *Jour. Geophys. Res.*, 88, pp 6523–6539.

STEINER, M. B. & HELSLEY, C. E., (1974) 'Magnetic polarity sequence of the Upper Triassic Kayenta Formation'. *Geology*, 2, pp 191–194.

STEPHENSON, A., (1980) 'Rotational remanent magnetisation and the torque exerted on a rotating rock in an alternating magnetic field'. *Geophys. J. R. Astr. Soc.*, 62, pp 113–132.

STEWART, A. D. & IRVING, E., (1974) 'Palaeomagnetism of Precambrian sedimentary rocks from NW Scotland and the apparent polar wandering path of Laurentia'. *Geophys. J. R. Astr. Soc.*, 37, pp 51–72.

STEWART, J. H., (1976) 'Late Precambrian evolution of North America: plate tectonic implications'. *Geology*, 4, pp 11–15.

STOCKWELL, C. H., (1973) 'Revised Precambrian time scale for the Canadian Shield'. *Geol. Surv. Can. Pap.* 72-52, pp 1–4.

STONE, D. B., (1967a) 'Torsion-balance method of measuring anisotropic susceptibility', in: *Methods in Palaeomagnetism* (ed. D. W. COLLINSON, K. M. CREER & S. K. RUNCORN) Elsevier Publ. Co., Amsterdam, pp 381–386.

STONE, D. B., (1967b) 'An anisotrophy meter', in: *Methods in palaeomagnetism* (ed. D. W. COLLINSON, K. M. CREER, & S. K. RUNCORN), Elsevier Publ. Co., Amsterdam, pp 373–380.

STONE, D. B., PANUSKA, B. C. & PACKER, D. R., (1982) 'Palaeolatitudes versus time for southern Alaska', *J. Geophys. Res.*, 87, pp 3697–3707.

STONELEY, R. & BAILEY, R. J., (1981) 'Petroleum: introduction and the formation and migration of hydrocarbons', in: *Economic Geology and Geotectonics* (ed. D. H. TARLING), Blackwell, Oxford, pp 31–50.

STORETVEDT, K. M., (1972) 'Crustal evolution in the Bay of Biscay'. *Earth Planet Sci. Lett.*, 17, pp 135–141.

STORETVEDT, K. M., (1974) 'A possible large-scale sinistral displacement along the Great Glen Fault in Scotland'. *Geol. Mag.*, 111, pp 23–30.

STORETVEDT, K. M. & HALVORSEN, E., (1968) 'On the palaeomagnetic reliability of the Scottish Devonian lavas'. *Tectonophysics*, 5, pp 447–457.

STORETVEDT, K. M., HALVORSEN, E. & GJELLESTAD, G., (1968) 'Thermal analysis of the natural remanent magnetism of some Upper Silurian red sandstones in the Oslo region'. *Tectonophysics*, 5, pp 413–426.

STRANGWAY, D. W., (1970) *History of the Earth's Magnetic Field*, McGraw-Hill, New York, 168 pp.

STRANGWAY, D. W., LARSON, E. E. & GOLDSTEIN, M., (1968) 'A possible cause of high magnetic stability in volcanic rocks'. *J. Geophys. Res.*, 73, pp 3787–3795.

STUBBS, P. H. S., (1958) *Continental drift and polar wandering: a palaeomagnetic study of British and European Trias and of the British Old Red Sandstones*, Ph.D thesis, University of London.

STUPAVSKY, M. & SYMONS, D. T. A., (1978), 'Separation of magnetic components from a.f. step demagnetisation data by least squares computer methods'. *Jour. Geophys. Res.*, 83, pp 4925–4931.:

STUPAVSKY, M. & SYMONS, D. T. A., (1982) 'Isolation of early Paleohelikian remanence in Grenville anorthosites of the French River area, Ontario'. *Can. J. Earth Sci.*, 19, pp 819–828.

STUPAVSKY, M., SYMONS, D. T. A. & GRAVENOR, C. P., (1982) 'Evidence for metamorphic remagnetisation of Upper Precambrian tillite in the Dalradian Supergroup of Scotland'. *Trans. R. Soc. Edin; Earth Sciences*, 73, pp 59–65.

SUNDERMANN, J. & BROSCHE, P., (1978) 'Numerical computation of tidal friction for present and ancient oceans', in: *Tidal Friction and the Earth's Rotation* (ed. P. BROSCHE & J. SUNDERMANN), Springer, Berlin, pp 125–144.

SUTTILL, R. J., (1980) 'Post-depositional remanent magnetisation in recent tidal-flat sediments'. *Earth Planet. Sci. Lett.*, 49, pp 132–140.

SUTTON, J., (1971) 'Some developments in the crust'. *Spec. Publ. Geol. Soc. Aust.* 3, pp 1–10.

SUTTON, J., (1978) 'Proterozoic of the North Atlantic', in: *Evolution of the Earth's Crust* (ed. D. H. TARLING), Academic Press, London, pp 239–254.

SUTTON, J. & WATSON, J. V., (1974) 'Tectonic evolution of continents in early Proterozoic times'. *Nature*, 247, pp 433–435.

SYMONS, D. T. A., (1971) 'A palaeomagnetic study of the Nipissing diabase, Blind River, Elliot Lake area, Ontario'. *Geol. Surv. Can. Pap.*, 70-63, pp 18–30.

SYMONS, D. T. A., (1973) 'Unit correlations and tectonic rotation from palaeomagnetism of the Triassic Copper Mountain intrusions, British Columbia'. *Geol. Surv. Can. Pap.* 73-19, pp 11–28.

SYMONS, D. T. A. & STUPAVSKY, M. M., (1974) 'A rational paleomagnetic stability index'. *Jour. Geophys. Res.*, 79, pp 1717–1720.

SYMONS, D. T. A., (1983) 'New palaeomagnetic data for the Triassic Guichon batholith of south-central British Columbia and their bearing on Terrane 1 tectonics'. *Can. J. Earth Sci.*, 20, pp 1340–1344.

SYMONS, D. T. A. & STUPAVSKY, M., (1979) 'Magnetic characteristics of the iron formation near Temagami, Ontario'. *Ontario Geological Survey, Miscellaneous Paper* 87, pp 133–147.

SYMONS, D. T. A. & STUPAVASKY, M., (1980) 'Grant 5, Part 2. Magnetic anomaly type curves and palaeomagnetism of the Algoman-type iron formation near Temagami, Ontario'. *Ontario Geological Survey*, Miscellaneous Paper 93, pp 215–219.

SYMONS, D. T. A., QUICK, A. W. & STUPAVSKY, M., (1982) 'Magnetic and palaeomagnetic characteristics of the Archaean iron formation and host rocks at the Adams Mine, Ontario'. *Ontario Geological Survey, Miscellaneous Paper* 98, pp 293–307.

TAIRA, A. & SCHOLLE, P. A., (1979) 'Deposition of resedimented sandstone bed in the Pico Formation, Ventura Basin, California, as interpreted from magnetic fabric measurements.' *Geol. Soc.. Amer. Bull.*, 90, pp 952–962.

TALBOT, C., (1973) 'A Plate Tectonic model for the Archaean crust'. *Phil. Trans. R. Soc. Lond.*, A273, pp 413–428.

TANCZYK, E. I., RANALLI, G. & MORRIS, W. A., (1985) 'Original strike and latitude of dykes and their bearing on the state of stress in the lithosphere', *Abstracts, Int. Conf. on Mafic Dyke Swarms, Univ of Toronto*, p 180.

TANKARD, A. J., JACKSON, M. P. A., ERIKSSON, K. A., HOBDAY, D. K., HUNTER, D. R. & MINTER, W. E. L., (1982). *Crustal Evolution of Southern Africa*. Springer-Verlag, Berlin, 523 pp.

TAPPONIER, P., (1977) 'Evolution tectonique du système Alpin en Méditerranée: poinçonnement et écrasement rigide-plastique'. *Bull. Soc. Geol. Fr.*, 19, pp 437–460.

TAPPONIER, P. & MOLNAR, P., (1976) 'Slip-line theory and large-scale continental tectonics,' *Nature*, 264, pp 319–324.

TARLING, D. H., (1974) 'A palaeomagnetic study of Eocambrian tillites in Scotland,' *J. geol. Soc. Lond.*, 130, pp 163–177.

TARLING, D. H. (1979) 'Palaeomagnetic reconstructions and the Variscan Orogeny'. *Proc. Ussher Soc.*, 4, pp 233–261.

TARLING, D. H., (1980) 'Lithosphere evolution and changing tectonic regimes', *J. Geol. Soc., Lond.*, 137, pp 459–467.

TARLING, D. H., (1983). *Palaeomagnetism, Principles and Applications in Geology, Geophysics and Archaeology*, Chapman and Hall, London, 379 pp.

TARLING, D. H. & MITCHELL, J. G., (1967) 'Revised Cenozoic polarity time-scale.' *Geology*, 4, pp 133–136.

TARLING, D. H. & SYMONS, D. T. A., (1967). 'A stability index of remanence in palaeomagnetism.' *Geophys. J. R. astr. Soc.*, 12 443–448.

TAYLOR, G. K., (1984) *Geophysical investigations of a Caledonian ophiolite complex, NE Shetland*, Ph.D thesis, University of Liverpool.

TARNEY, J., (1976) 'Geochemistry of Archaean high grade gneisses with implications as to the origin and evolution of the Precambrian crust', in: *The Early History of the Earth*. (ed. B. F. WINDLEY), John Wiley, London, pp 405–417.

TAYLOR, S. R., (1979) 'Chemical composition and evolution of the continental crust: the rare earth element evidence', in: *The Earth, its origin, structure and evolution* (ed. M. W. MCELHINNY), Academic Press, London, pp 353–376.

TAYLOR, S. R. & MCLENNAN, S. M., (1981) 'The rare earth element evidence in Precambrian sedimentary rocks: implications for crustal evolution', in: *Precambrian plate tectonics* (ed. A. Kröner), Elsevier Publ. Co., Amsterdam, pp 527–548.

THELLIER, E., (1951) 'Propietés magnétiques des terres cuites et des roches'. *J. de Phys. et Radium*, 12, pp 205–218.

THELLIER, E. & THELLIER, O., (1959) Sur l'intensité du champ magnétique terestre dans le passé historique et géologique'. *Ann. Geophys.*, 15, pp 285–376.

THOMAS, C. & BRIDEN, J. C., (1976) 'Anomalous geomagnetic field during the late Ordovician'. *Nature*, 259, pp 380–382.

THOMPSON, J. B., (1972) 'Oxides and sulphides in regional metamorphism of pelitic schists.' *Int. Geol. Congr.*, 29th Sess., Montreal, Sect. 110, pp 27–35.

THOMPSON, R., (1972) 'Palaeomagnetic results for the Paganzo Basin of northwest Argentina.' *Earth Planet. Sci. Lett.*, 15, pp 145–156.

THOMPSON, R. & CLARK, R. M. (1981) 'Fitting polar wander paths.' *Phys. Earth Planet. Int.*, 27, pp 1–7.

THOMPSON, R. & KELTS, K., (1974) 'Holocene sediments as magnetic stratigraphy from Lakes Zug and Zurich, Switzerland.' *Sedimentology*, 21, pp 588-596

THOMPSON, R. & MITCHELL, J. G. (1972) 'Palaeomagnetic and radiometric evidence for the age of the lower boundary of the Kaiman magnetic interval in South America. *Geophys. J. R. Astr. Soc.*, 27, pp 207-214.

THORPE, R. S., BECKINSALE, R. D., PATCHETT, P. J., PIPER, J. D. A., DAVIES, G. R. & EVANS, J. A., (1984) 'Crustal growth and late Precambrian-early Palaeozoic plate tectonic evolution of England and Wales'. *J. Geol. Soc. Lond.*, 141, pp 521-530.

TOKSOZ, M. N., MINEAR, J. W. & JULIAN, B. R., (1971) 'Temperature field and geophysical effects of a down-going slab'. *J. Geophys. Res.*, 76, pp 1113-1138.

TORSKE, T., (1977) 'The south Norway Precambrian region — a Proterozoic Cordilleran-type orogenic segment'. *Norsk Geologisk Tidsskrift*, 57, pp 97-150.

TORSVIK, T. H., (1985a) 'Palaeomagnetic results from the Peterhead granite, Scotland; implication for regional late Caledonian magnetic overprinting.' *Phys. Earth Planet. Int.*, 39, pp 108-117.

TORSVIK, T. H., (1985b) 'Magnetic properties of the Lower Old Red Sandstone lavas in the Midland Valley, Scotland; palaeomagnetic and tectonic considerations'. *Phys. Earth Planet. Int.*, 39, pp 194-207.

TORSVIK, T. H., LOVLIE, R. & STURT, B. A., (1975) 'Palaeomagnetic argument for a stationary Spitsbergen relative to the British Isles (Western Europe) since late Devonian and its bearing on North Atlantic reconstruction'. *Earth Planet. Sci. Lett.*, 75, pp 278-288.

TOSHA, T. & KONO, M., (1984) 'Palaeomagnetism and accretion tectonics of Japanese Islands', in: *Accretion Tectonics in the Circum-Pacific Regions* (eds. M. Hashimoto and S. Uyeda), Terrapub, Tokyo, pp 65-72.

TUCKER, P., (1980) 'A grain mobility model of post-depositional realignment'. *Geophys. J. R. Astr. Soc.*, 63, pp 149-163.

TUNNELL, H. B. & BRIDEN, J. C., (1983) 'Palaeomagnetism of NW Scotland syenites in relation to local and regional tectonics'. *Geophys. J. R. Astr. Soc.*, 67, pp 217-234.

TURNER, J. S., (1949) 'The deeper structure of central and northern England,' *Proc. Yorks. Geol. Soc.*, 27, pp 280-297.

TURNER, P., (1975) 'Palaeozoic secular variation recorded in Pendleside Limestone'. *Nature*, 257, pp 207-208.

TURNER, P. (1980) *Continental Red Beds*. Developments in sedimentology No. 29, Elsevier Publ. Co., Amsterdam, 562 pp.

TURNER, P., (1981) 'Relationship between magnetic components and diagenetic features in reddened Triassic alluvium (St. Bees Sandstone, Cumbria, U.K.)'. *Geophys. J. R. Astr. Soc.*, 67, pp 395-413.

TURNER, P. & ARCHER, R., (1977) 'The role of biotite in the diagenesis of red beds from the Devonian of Northern Scotland'. *Sedimentary Geology*, 19, pp 241-251.

TURNER, P., CLEMMEY, H. & FLINT, S., (1984) 'Palaeomagnetic studies of a Cretaceous molasse sequence in the central Andes (Coloso Formation, Northern Chile)'. *J. Geol. Soc. Lond.*, 141, pp 869-876.

TURNER, P & IXER, R. A., (1976) 'Diagenetic development of unstable and stable magnetisation in the St. Bees Sandstone (Triassic) of Northern England'. *Earth Planet. Sci. Lett.*, 34, pp 113-124.

TURNER, P., METCALFE, I. & TARLING, D. H., (1979) 'Palaeomagnetic studies of some Dinantian limestones from the Craven Basin and a contribution to Asbian magnetostratigraphy'. *Proc. Yorks. Geol. Soc.*, 42, pp 371-395.

TURNER, P. & TARLING, D. H., (1975) 'Implications of new palaeomagnetic results from the Carboniferous System in Britain'. *Jour. Geol. Soc. Lond.*, 131, pp 469-488.

TURNER, P., VAUGHAN, D. J. & TARLING, D. H., (1978) 'Palaeomagnetic and mineralogical studies of Devonian lacustrine sediments from Caithness, Scotland'. *Phys. Earth Planet. Int.*, 16, pp 73-83.

TURNER, S. & TARLING, D. H., (1982) 'Thelodont and other agnathan distributions as tests of Lower Palaeozoic continental reconstructions'. *Palaeogeog., Palaeoclim., Palaeoecol.*, 39, pp 295-311.

ULRYCH, T., (1972) 'Maximum entropy power spectrum of long period geomagnetic reversals'. *Nature*, 235, pp 218-219.

URRUTIA-FUCUGAUCHI, J., (1981) 'Palaeomagnetic evidence for tectonic rotation of northern Mexico and the continuity of the Cordilleran orogenic belt between Nevada and Chilhuahua'. *Geology*, 9, pp 178-183.

URRUTIA-FUCUGAUCHI, J., (1982) 'Reconnaissance palaeomagnetic investigation of Cretaceous limestones from southern Mexico'. *Geol. Intern.*, 20, pp 203-217.

URRUTIA-FUCUGAUCHI, J., (1983) 'On the tectonic evolution of Mexico: palaeomagnetic constraints'. *Geodynamics Series, vol. 12, Amer. Geophys. Un.*, pp 29-47.

UYEDA, S., (1958) 'Thermoremanent magnetism as a medium of palaeomagnetism with special reference to reverse thermoremanent magnetism'. *Japan. J. Geophys.*, 25, pp 1-121.

VAIL, P. R., MITCHUM, R. M. & THOMPSON, S., (1977) 'Seismic stratigraphy and global changes of sea level, Part 4: Global cycles of relative changes of sea level', in: *Seismic Stratigraphy — Applications to Hydrocarbon Exploration* (ed. C. W. PAYTON). *Amer. Assoc. Petro. Geol., Mem.*, 26, pp 83-97.

VAIL, P. R., & TODD, R. G., (1981) 'Northern North Sea Jurassic unconformities, chronostratigraphy and sea level changes from seismic stratigraphy', in: *Petroleum Geology of the Continental shelf of North-west Europe* (ed. J. V. CLING & C. D. HOBSON) Institute of Petroleum, London, pp 216-235.

VALENCIO, D. A., VILAS, J. F. & MENDIA, J. E., (1977) 'Palaeomagnetism of a sequence of red beds of the Middle and Upper Sections of Tagnazo Group (Argentina) and the correlation of Upper Palaeozoic-Lower Mesozoic rocks'. *Geophys. J. R. Astr. Soc.*, 51, pp 59-74.

VALENTINE, J. W. & MOORES, E. M., (1972) 'Global tectonics and the fossil record'. *J. Geol.*, 80, pp 167-184.

VANDENBERG, J., (1979) 'Palaeomagnetic data from the Western Mediterranean: a review'. *Geologie en Mijnbouw*, 58, pp 161-174.

VANDENBERG, J., (1980) 'New palaeomagnetic data from the Iberian peninsula' *Geologie en Mijnbouw*, 59, pp 49-60.

VANDENBERG, J., (1983) 'Reappraisal of palaeomagnetic data from Gargano (South Italy)'. *Tectonophysics*, 98, pp 29-41.

VAN DER LINDEN, W. J. M., (1985) 'Looping the loop; geotectonics of the Alpine-Mediterranean region'. *Geologie en Mijnbouw*, 64, pp 281-295.

VAN DER VOO, R., (1968) 'Palaeomagnetism and the Alpine tectonics of Eurasia — IV, Jurassic, Cretaceous and Eocene pole positions from north east Turkey'. *Tectonophysics*, 6, pp 252-269.

VAN DER VOO, R., (1979) 'Age of the Alleghenian folding in the Central Appalachians'. *Geology*, 7, pp 297-298.

VAN DER VOO, R. & CHANNELL, J. E. T., (1980) 'Palaeomagnetism in orogenic belts'. *Rev. Geophys. Space Phys.*, 18, pp 455-482.

VAN DER VOO, R. & FRENCH, R. B., (1977) 'Palaeomagnetism of the late Ordovician Juniata Formation and the remagnetisation hypothesis'. *J. Geophys. Res.*, 82, pp 5796-5802.

VAN DER VOO, R., FRENCH, A. N. & FRENCH, R. B., (1979) 'A palaeomagnetic pole position from the folded Upper Devonian Catskill red beds and its tectonic implications'. *Geology*, 7, pp 345-348.

VAN DER VOO, R., HENRY, S. G. & POLLACK, H. N., (1978) 'On the significance and utilisation of secondary magnetisations in red beds'. *Phys. Earth Planet. Int.*, 16, pp 12-19.

VAN DER VOO, R., JONES, M., GROMMÉ, C. S., EBERLEIN, G. D. & CHURKIN, M., (1980) 'Palaeozoic palaeomagnetism and the northward drift of the Alexander terrane, southeastern Alaska'. *J. Geophys. Res.*, 85, pp 5281-5296.

VAN DER VOO, R., PEINADO, J. & SCOTESE, C. (1984) 'A palaeomagnetic re-evaluation of Pangaea reconstructions', in: *Reconstructions from Palaeozoic Palaeomagnetism*, Geodynamics Series, vol. 12, American Geophysical Union, pp 11–26.

VAN DER VOO, R. & SCOTESE, C., (1981) 'Palaeomagnetic evidence for a large (~ 2000 km) sinistral offset along the Great Glen Fault during Carboniferous times'. *Geology*, 9, pp 583–589.

VAN DER VOO, R. & ZIJDERVELD, J. D. A., (1971) 'Renewed palaeomagnetic study of the Lisbon Volcanics and implications for the rotation of the Iberian peninsula'. *J. Geophys. Res.*, 76, pp 3913–3921.

VAN DONGEN, P. G., VAN DER VOO, R. & RAVEN, Th., (1967) 'Palaeomagnetism and the Alpine tectonics of Eurasia III. Palaeomagnetic research in the Central Lebanon Mountains and the Tartous area of Syria'. *Tectonophysics*, 4, pp 35–53.

VAN HOUTEN, F. B., (1968) 'Iron oxides in red beds'. *Bull. Geol. Soc. Amer.*, 79, pp 399–416.

VAN HOUTEN, F. B., (1973) 'Origin of red beds: a review, 1961–1972'. *Ann. Rev. Earth Planet. Sci.*, 1, pp 39–61.

VAN SCHMUS, W. R., (1976) 'Early and Middle Proterozoic history of the Great Lake area, North America'. *Phil. Trans. R. Soc. Lond.*, A280, pp 605–628.

VAN ZIJL, J. S., GRAHAM, K. W. T. & HALES, A. L., (1962) 'The palaeomagnetism of the Stormberg lavas II. The behaviour of the magnetic field during a reversal'. *Geophys. J. R. Astr. Soc.*, 7, pp 169–182.

VEARNCOMBE, J. R., (1983) 'A proposed continental margin in the Precambrian of Western Kenya'. *Geologische Rundschau* 72, pp 663–670.

VEEVERS, J. J. & MCELHINNY, M. W., (1976) 'The separation of Australia from other continents'. *Earth Sci. Rev.*, 12, pp 139–159.

VERHOOGEN, J., (1959) 'The origin of thermoremanent magnetisation' *J. Geophys. Res.*, 64, pp 2441–2449.

VERHOOGEN, J., (1980) *Energetics of the Earth*, National Academy Press, Washington D.C., 139 pp.

VEROSUB, K. L., (1977) 'Depositional and post-depositional processes in the magnetisation of sediments'. *Rev. Geophys. Space Phys.*, 15, pp 129–143.

VEROSUB, K. L., (1979) 'Palaeomagnetism of varved sediments from western New England; variability of the palaeomagnetic recorder'. *Geophys. Res. Lett.*, 6, pp 241–244.

VIDAL, G., (1977) 'Late Precambrian microfossils.' *Geol. Mag.*, 114, pp 292–294.

VIDAL, G., (1979) 'Acritarchs and the correlation of the Upper Proterozoic.' *Pub. Inst. Mineral. Palaeont. Quart. Geol. Univ. Lund*, 219, pp 1–22.

VIDAL, G., & KNOLL, A. H., (1983) 'Radiations and extinctions of plankton in the late Proterozoic and early Cambrian'. *Nature*, 297, pp 57–60.

VIEZER, J. & COMPSTON, W., (1976) '^{87}Sr:^{86}Sr in Precambrian carbonates as an index of crustal evolution'. *Geochim. Cosmochim. Acta.* 40, pp 905–914.

VIEZER, J. & JANSEN, S. L., (1979) 'Basement and sedimentary recycling and continental evolution'. *J. Geol.*, 87, pp 341–370.

VINCENZ, S. A. & DASGUPTA, S. N., (1978) 'Palaeomagnetic study of some Cretaceous and Tertiary rocks on Hispaniola'. *Pure Appl. Geophys.*, 116, pp 1200–1210.

VILAS, J. F. & VALENCIO, D. A., (1978) 'Palaeomagnetism and K-Ar age of the Upper Ordovician Alcaparrosa Formation, Argentina'. *Geophys. J. R. Astr. Soc.*, 55, pp 143–154.

VINE, F. J., (1966) 'Spreading of the ocean floor: new evidence'. *Science*, 154, pp 1405–1415.

VINE, F. J. & MATTHEWS, D. H., (1963) 'Magnetic anomalies over oceanic ridges'. *Nature*, 199, pp 947–949.

VISWANATHA, M. N., (1972) 'The geology and geochemistry of the high-grade Sargus schist complex in southern Karnataka with special reference to the early history of crustal development of the Southern Peninsula Shield', in: *Abstrs. Archaean Geochemistry Symposium, Hyderabad, India*, pp 7–9.

VLASOV, A. A. & POPOVA, A. V., (1968) 'Palaeomagnetism of Precambrian deposits of the Yenisey Ridge'; *Izvestiya Akademii Nauk SSR Earth Physics Section*, 2, pp 63–70.

WALKER, G. P. L., (1960) 'Zeolite zones and dyke distribution in relation to the structure of the basalts of Eastern Iceland'. *Jour. Geol.*, 68, pp 515–528.

WALKER, T. R., (1976) 'Diagenetic origin of continental red beds', in: *The Continental Permian in Central, West and South Europe* (ed. H. FALKE), D. Reidel Publ. Co., Dordrecht, pp 240–282.

WALKER, T. R., LARSON, E. E. & HOBLITT, R. P., (1981) 'Nature and origin of hematite in the Moenkopi Formation (Triassic), Colorado Plateau: a contribution to the origin of magnetism in red beds'. *Jour. Geophys. Res.*, 86, pp 317–333.

WATKINS, N. D., (1969) 'Non-dipole behaviour during an Upper Miocene geomagnetic polarity transition in Oregon'. *Geophys. J. R. Astr. Soc.*, 17, pp 121–149.

WATKINS, N. D. & PASTER, T. P. (1971) 'The magnetic properties of igneous rocks from the ocean floor.' *Phil. Trans. R. Soc., Lond.*, A268, pp 507–550.

WATKINS, N. D. & RICHARDSON, A., (1968) 'Palaeomagnetism of the Lisbon Volcanics'. *Geophys. J. R. Astr. Soc.*, 15, pp 287–304.

WATSON, G. S., (1956) 'A test for randomness of directions'. *Mon. Not. R. Astr. Soc.*, Geophys. Suppl. 7, pp 160–161.

WATSON, G. S. & IRVING, E. (1957) 'Statistical methods in rock magnetism'. *Mon. Not. Roy. Astr. Soc., Geophys.*, Suppl 7, pp 289–300.

WATSON, J. V., (1973) 'Effects of reworking on high-grade gneiss complexes'. *Phil. Trans. R. Soc. Lond.*, A273, pp 433–456.

WATSON, J. V., (1976) 'Vetical movements in Proterozoic structural provinces'. *Phil. Trans. R. Soc., Lond.*, A280, pp 629–646.

WATSON, J. V., (1978) 'Ore-deposition through geological time'. *Proc. R. Soc. Lond.*, A362, pp 305–328.

WATSON, J. V., (1980) 'The origin and history of the Kapuskasing structural zone, Ontario, Canada'. *Can. Jour. Earth Sci.*, 17, pp 866–875.

WATSON, J. V., (1984) 'The ending of the Caledonian Orogeny in Scotland'. *J. Geol. Soc. Lond.*, 141, pp 193–214.

WATTERSON, J., (1978) 'Proterozoic intra-plate deformation in the light of S. E. Asian neotectonics'. *Nature*, 273, pp 636–640.

WATTS, D. R., (1981) 'Palaeomagnetism of the Fond du Lac Formation and the Eileen and Middle River sections with implications for Keweenawan tectonics and the Grenville problem'. *Can. J. Earth Sci.*, 78, pp 829–841.

WATTS, D. R., (1982) 'A multicomponent dual polarity palaeomagnetic regional overprint from the Moine of Northwest Scotland'. *Earth Planet,. Sci. Lett.*, 61, pp 190–198.

WATTS, D. R. & BRAMALL, A. M., (1981) 'Palaeomagnetic evidence for a displaced terrane in western Antarctica'. *Nature*, 293, pp 638–640.

WATTS, D. R. & BRIDEN, J. C., (1984) 'Palaeomagnetic signature of slow postorogenic cooling of the north-west Highlands of Scotland recorded in the Newer Gabbros of Aberdeenshire'. *Geophys. J. R. Astr. Soc.*, 77, pp 775–778.

WATTS, D. R., VAN DER VOO, R. & FRENCH, R. B., (1980) 'Palaeomagnetic investigations of the Cambrian Waynesboro and Rome Formations of the Valley and Ridge Province of the Appalachian Mountains'. *J. Geophys,. Res.*, 85, pp 5331–5343.

WATTS, D. R., VAN DER VOO, R. & REEVE, S. C., (1980) 'Cambrian palaeomagnetism of the Llano Uplift, Texas. *Jour. Geophys. Res.*, 85, pp 5316–5330.

WEBB, G. W., (1969) 'Palaeozoic wrench faults in Canadian Appalachians', in: *North Atlantic Geology and Continental Drift* (ed. M. KAY), Amer. Assoc. Petrol. Geol. Mem. 12, pp 754–788.

WEGENER, A., (1967). *The Origin of Continents and Oceans*, English translation of 4th Edition (1929), Methuen, London, (First edition published in 1923).

WELLS, R. E. & COE, R. S., (1985) 'Palaeomagnetism and geology of Eocene Volcanic rocks of southwest Washington, implications for mechanisms of tectonic rotation'. *J. Geophys. Res.*, 90, pp 1925–1948.

WENNER, D. B. & TAYLOR, H. P., (1971) 'Temperatures of serpentinisation of ultramafic rocks based on O^{18}/O^{16} fractionation between coexisting serpentine and magnetite'. *Contrib. Mineral Petrol.*, 32, pp 165–185.

WENSINK, H., (1979) 'The implications of some palaeomagnetic data from Iran for its structural hisory'. *Geologie en Mijnbouw*, 58, pp 175–185.

WENSINK, H. (1983) 'Palaeomagnetism of red beds of early Devonian age from central Iran.' *Earth Planet. Sci. Lett.*, 63, pp 325–334.

WESTPHAL, M., (1973) 'Etudes paléomagnétiques de quelques formations Permiennes et Triasiques dans les Alpes Occidentales (France)'. *Tectonophysics*, 17, pp 323–335.

WHITE, G. W., (1980) 'Permian–Triassic continental reconstruction of the Gulf of Mexico–Caribbean area', *Nature*, 283, pp 823–826.

WHITE, W. M. & HOFFMANN, W. A., (1982) 'Sr and Nd isotopic geochemistry of oceanic and mantle evolution'. *Nature*, 296, pp 821–825.

WHITTINGTON, H. B. & HUGHES, C. P., (1973) 'Ordovician trilobite distribution and geography,' *Spec. Pap. Palaeontol.*, 12, pp 235–240.

WHITTINGTON, H. B. & HUGHES, C. P., (1974) 'Geography and faunal provinces in the Tremadoc Epoch,' in: *Palaeogeographic provinces and provinciality* (ed. C. A. ROSS), Soc. Econ. Palaeontol. Min., Spec. Publ. 21, pp 203–218.

WHYTE, M. A., (1977) 'Turning points in Phanerozoic history'. *Nature*, 267, pp 679–682.

WILKINSON, B. H., OWEN, R. M. & CARROLL, A. R., (1985) 'Submarine hydrothermal weathering, global eustasy, and carbonate polymorphism in Phanerozoic marine oolites'. *Jour. Sed. Petrol.*, 55, pp 171–183.

WILLIAMS, A., (1973) 'Distribution of brachiopod assemblages in relation to Ordovician palaeogeography'. *Spec. Pap. Palaeontol.*, 12, pp 241–270.

WILLIAMS, C. A., (1975) 'Sea-floor spreading in the Bay of Biscay and its relationship to the North Atlantic'. *Earth Planet. Sci. Lett.*, 24, pp 440–456.

WILLIAMS, D. M., (1980) 'Evidence for glaciation in the Ordovician rocks of Western Ireland'. *Geol. Mag.*, 117, p 81.

WILLIAMS, G. E., (1975) 'Late Precambrian glacial climate and the Earth's obliquity'. *Geol. Mag.*, 112, pp 441–544.

WILLIAMS, G. E., (ed.), (1981). *Megacycles: Long-term Episodicity in the Earth and Planetary History*, Benchmark Papers in Geology No. 57, Hutchinson Ross Publ. Co., Stroudsbury, Penn., 434 pp.

WILLIAMS, H., (1978) 'Geological development of the northern Appalachians; its bearing on the evolution of the British Isles', in: *Crustal evolution of north western Britain and adjacent regions* (eds., D. R. BOWES & B. E. LEAKE), Seel House Press, Liverpool, pp 1–22.

WILLS, L. J., (1978) 'A palaeogeological map of the lower Palaeozoic floor below the cover of Upper Devonian, Carboniferous and later formations', *Mem. Geol. Soc., Lond.* 8, 32 pp.

WILSON, J. T., (1966) 'Did the Atlantic close and then reopen?' *Nature*, 211, pp 676–679.

WILSON, J. T., (1968) 'Static or mobile Earth: the current scientific revolution', in: *Gondwanaland revisited: New evidence for continental drift. Proc. Amer. Philos. Soc.*, 112, pp 309–320.

WILSON, R. L., (1961) 'Palaeomagnetism in Northern Ireland. Part I. The thermal demagnetisation of natural magnetic movements in rocks'. *Geophys. J. R. Astr. Soc.*, 5, pp 45–58.

WILSON, R. L., (1962) 'The palaeomagnetic history of a doubly baked rock'. *Geophys. J. R. Astr. Soc.*, 6, pp 397–399.

WILSON, R. L., (1964) 'Magnetic properties and normal and reversed natural magnetisation in the Mull lavas'. *Geophys. J. R. Astr. Soc.*, 8, pp 424–439.

WILSON, R. L., (1970) 'Permanent aspects of the Earth's non-dipole magnetic field over Upper Tertiary times'. *Geophys. J. R. Astr. Soc.*, 19, pp 417–437.

WILSON, R. L., (1971). 'Dipole offset — the time average palaeomagnetic field over the past 25 million years'. *Geophys. J. R. Astr. Soc.*, 22, pp 491–504.

WILSON, R. L. & HAGGERTY, S. E., (1966) 'Reversals of the Earth's magnetic field'. *Endeavour*, 25, pp 104–109.

WILSON, R. L., HAGGERTY, S. E. & WATKINS, N. D., (1968) 'Variations of palaeomagnetic stability and other parameters in a vertical traverse of a single Icelandic lava.' *Geophys. J. R. Astr. Soc.*, 16, pp 79–96.

WILSON, R. L. & LOMAX, R., (1972) 'Magnetic remanence related to slow rotation of ferromagnetic material in alternating magnetic fields'. *Geophys. J. R. Astr. Soc.*, 30, pp 295–303.

WILSON, R. L. & MCELHINNY, M. W., (1974). 'Investigation of the large scale palaeomagnetic field over the past 25 million years. Eastward shift of the Icelandic spreading ridge, *Geophys. J. R. astr. Soc.*, 39, 570–586.

WILSON, R. L. & WATKINS, N. D., (1967) 'Correlation of petrology and natural magnetic polarity in Columbia Plateau basalts'. *Geophys. J. R. Astr. Soc.*, 12, pp 405–424.

WINDLEY B. F., (1984) *The Evolving Continents*, 2nd Ed., John Wiley, London, 399 pp.

WINDLEY, B. F. & BRIDGWATER, D., (1971) 'The evolution of Archean low- and high-grade terrains'. *Geol. Soc. Australia Spec. Publ.*, 3, pp 33–46.

WISE, D. O., (1974) 'Continental margins, freeboard and volumes of continents and oceans through time', in: *The Geology of Continental Margins* (eds. C. A. BURK & C. L. DRAKE), Springer-Verlag, Berlin, pp 45–58.

YOLE, R. W. & IRVING, E., (1980) 'Displacement of Vancouver Island, palaeomagnetic evidence from the Karmutsen Formation'. *Can. J. Earth Sci.*, 17, pp 1210–1288.

YORK, D., (1978) 'A formula describing both magnetic and isotopic blocking temperatures'. *Earth Planet. Sci. Lett.*, 39, pp 89–93.

YORK, D. & FARQHAR, R. M., (1972) *The Earth's Age and Geochronology*, Pergamon Press, Oxford, 178 pp.

ZICHAO, Z., GUOGAN, M. & HUAQUIN, L., (1984) 'The chronometric age of the Sinian–Cambrian boundary in the Yangtse Platform, China'. *Geol. Mag.*, 121, pp 175–178.

ZIEGLER, A. M., HANSEN, K. S., JOHNSON, M. E. & KELLY, M. A., (1977). 'Silurian continental distributions, palaeogeography, climatology and biogeography'. *Tectonophysics*, 40, pp 13–51.

ZIEGLER, A. M., SCOTESE, C. R., MCKERROW, W. S., JOHNSON, M. E. & BAMBACH, R. K., (1979) 'Palaeozoic palaeogeography'. *Ann. Rév. Earth Planet. Sci.*, 7, pp 473–502.

ZIEGLER, P. A., (1984) 'Caledonian and Hercynian crustal consolidation of western and central Europe — a working hypothesis'. *Geologie en Mijnbouw*, 63, pp 93–108.

ZIJDERVELD, J. D. A., (1967) 'A. C. demagnetisation of rocks: analysis of results', in: *Methods in Palaeomagnetism* (ed. D. W. COLLINSON, K. M. CREER, & S. K. RUNCORN), Elsevier Publ. Co., Amsterdam, pp 254–286.

ZIJDERVELD, J. D. A., DEJONG, K. A. & VAN DER VOO, R., (1977) 'Rotation of Sardinia: palaeomagnetic evidence from Permian rocks', *Nature* 226, pp 933–934.

ZWART, H. J., (1967) 'The duality of orogenic belts', *Geologie en Mijnbouw*, 46, pp 283–309.

APPENDIX

THE GEOLOGICAL TIME SCALE

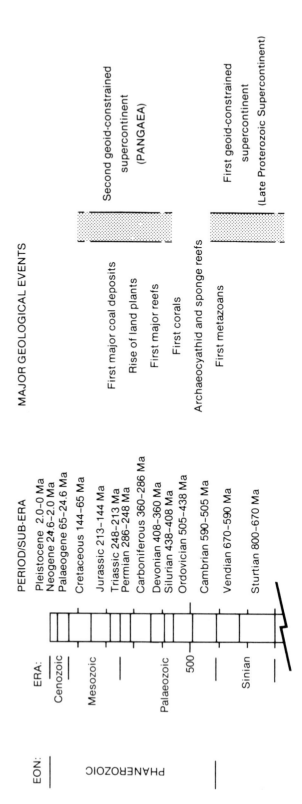

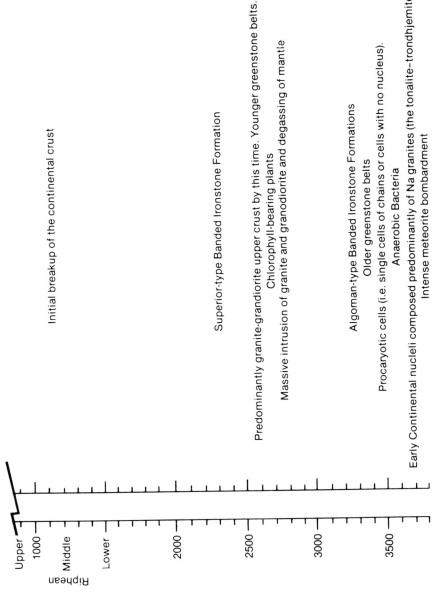

Index

Aberdeenshire gabbros 262–263, 267
Abitibi greenstone belt 155–157
abyssal environments 29, 61, 354
Acadian orogeny 273–276, 279, 281–283
accretion, continental: 141–143, 233, 312–328
acidity (pH) 22, 26, 28, 63
acid treatment 89–90
acritarchs 242, 247
activation energy 50
additivity law of PTRM 41
Adelaide geosyncline 206, 220, 222–223, 240, 242
Adirondacks 48
Adriatic 312–315; promontory: 288, 311–315
Aegean 314–315
Afghanistan 295, 315
Africa, Phanerozoic: 276, 280, 288, 290, 305–308, 311–315, 351, 368–369; Precambrian, geology: 140–143, 145, 203–204, 206–207, 209, 211, 213, 216–218, 229–230, 240–242, 245, 250–252; palaeomagnetism of: 136, 151–159, 162–167, 170–171, 173–178, 220–225, 228, 236
Afro-Arabian arc 206, 238–239, 246
agglomerate 125–126
agnathids 294
akagenite 28
Alaska 322, 324–328
albedo 353
Alcaparossa lavas 255
Alexander terrane 324–325

algae 293, 354
algorithm 120
alkaline magmatism 215–216, 220, 231, **245–246**
allochthonous terranes 146–147, 183, 233–239, 254, 259, 269, 311–328
alluvial environments 58–63, 67–69
Alnö complex 228–229, 233, 244
alpha 95, 105
Alpine belts 140–143, 204, 271, 276, 289, **311–315**
Altai Sayan Zone 286–287
alternating field (a.f.) demagnetisation 39, **85–87**, 343
aluminium 54
Amadeus basin 222–223
Amana formation 256–258
amphiboles 7
amphibolite facies 56, 67, 73–74, 121–123, 208
Anabarites 250
analcite 56
Anatolian fault, northern, 288
Andes 142, 258, 319–321
andesite 6; -see also calc-alkaline, island arc, subduction
Angaran flora 286, 296–297
Anglesey 237, 239
Anglo-Welsh, microplate: 237–239, 271; cuvette: 276
angular velocity of the Earth 345, 356
anhydrite 7
anhysteritic remanent magnetism (ARM) 86–87, 96, 343
anisotropy, constants of: 32, 44, 88, 94; delineator 84; effect on remanence

direction: 47–49; (magnetocrystalline) energy of: 12, 36–37, 49, 88; magnetic: 22, 31, 35; shape 35–36, 126; of susceptibility: **46–49**, 62–63, 68, 71; tensor: 46–47.
anomalies, magnetic: 17, 66, 76
anorogenic magmatism 172–173
anorthosites 199, 206, 209, 212
anoxic periods 353–354
Antarctica, East Antarctica shield: 148, 152–153, 180, 182, 185, 203, 217–218, 228, 250–252, 255–257, 305–310, 319, 357, 366; West Antarctica peninsula: 305, 319–320, 359
Anti Atlas belt 258, 266, 275
antiferromagnetism **10**, 25; canted **11**, 25
antigorite 57
Antilles, Lesser and Greater, 320
Antrim plateau basalts (Australia) 224–225, 246
apatite - see phosphorites
Apennines 312–315
Appalachian belt 140–142, 237, 252, 264, 271, 276–277, 280–281, 294; Piedmont of: 264, 280–281, 283
apparent polar wander (APW) **135–147**, 151–154, 188–192, 349–353; calculation of continental velocity from: 137–140; calculation of by 'moving window' approach: 299–310,

371–373; continental drift or true polar wander? 349–353; style of : 189–191; tectonic models for: 143–147
Aptian times 334–335
Apulian zone 313–314
Arabia 248, 252, 288, 315
Arabian shield 153, 217–218, 224–225, 248, 252
aragonite 357–358
Aravalli belt 206
archaeocyathids 236, 249–250, 252, 336
archaeomagnetism 1, 3, 129, 343
Archaean eon 1, 72, **151–161**, 320, 345, 372; palaeomagnetic data: 148, **161**; geology of: 194–202, 209–213, 216
Arenig times 259–261, 294
Arequipa massif 320
arenites -see sandstones
Argentina 223, 246, 252, 258, 338
argon 121–122
argon 39: argon 40 dating 121–123, 149
aridity 66
arkose 67
Armorican, microplate: 233–237, 250; massif: 260, 264–265, 267, 270–271, 273–274, 294, 312
arsenopyrite 6
Ashgill times 260–261, 264–265, 269 –see also Hirnantian
Asia 276, 283–289, 317–319; Minor: 315; -see also Eurasia
assymetry, of geomagnetic field: 185–187
astatic magnetometers 78–79
asthenosphere 192, 201–203, 349–353, 360, 363, 366–369, 370
Athapuscow aulacogen 164, 168–169
Atlas, mountains 258; fault zone: 259
Atlantic Ocean 302, 332, 353, 357, 368
atmosphere 68–69, 292
atomic dipole moments -see dipole moments
atomic structure 5, 9–12, 22–23, 25, 29, 33, 54
aulacogens 164, 168, 178, 206, 215–216, 243, 245–246, 248
aureole -see contact test, metamorphism

authigenic mineral growth 61–62, 66–72, 159
Australia, Precambrian, geology: 198, 207, 213, 240–242, 246, 373, 250–252, 254–258; palaeomagnetic data: 151–159, 169, 173–178, 217–218, 220–225, 228; Phanerozoic: 305–309, 338, 351, 357, 366, -see also Adelaide geosyncline, Sydney basin, Tasman geosyncline
Avalonian terrane 237, 279–281
Avzyan supergroup 174–175
axial geocentric dipole -see geomagnetic field

bacteria 71, 248
Baffin island 221, 246
Baikal, lake, 285, 287
Balaeric basin 312, 314–315
Ballantrae ophiolite 260
Baltica 145, 227–229, 233, 249, 258, 265, 267, 271–274, 289, 295
Baltic Shield -see Fennoscandian-Ukrainian Shield
bamble sector 180–181, 184
Banda arc 319
banded ironstone formations **72**, 198, 201; Algoman type: 72, 210; Superior type: 72, 198, 210.
Barberton, belt: 199; lava: 160–161, 345
Barrandian Massif -see Bohemia
basalts 6, 54–57; submarine 6 -see also pillow lavas
Basin and Range province 322
Basinsk group 219
basic igneous rocks 48, 54–58 -see also basalts, gabbros
Bay of Biscay 260, 270, 311–312, 315
Bear Tooth terrane 155–159
bedding, 'error': 60; correction: 123–125
Belt Supergroup 173–175, 184
Bering Sea plateau 368
between site scatter 107
Bhander sediments, India 222
bias, in polarity: 340–343, 355–356; intervals: 330
Big Spruce Complex 149
Bingham distribution and statistics 109

bioherms -see reefs
biotite 7, 56, 67, 121–122
bioturbation 48, 60–62
Birrimian Belt 199
Black Sea 288
black shales 266, 354
Blake subchron 331
Bloch wall 34
blocking, of magnetic remanence: 15, radiogenic isotopes: 121–122; temperature: 36, **40–41**, 49–52, 67, 84–91; temperature spectrum: **40**, 49–52, 110, 112, 120; volume: 43.
blueschist metamorphic facies 142, 195, 204
Bohemian Massif and microplate 234, 236–237, 260, 265, 270–274, 294, 312
bore cores, palaeomagneticm of: 106–107
Borneo -see Kalimantan
Borrowdale volcanic series 261, 336
Botwood group 282–283
brachiopods 250
Bradore formation 228, 229, 232
Brazil 203, 242, 368; shield: 320
Breidden inlier 261
Breggia gorge 334
brines -see evaporites
Britain 250, 336; British microplate: 254, 259–270, 274, 276, 281, 283; -see also Anglo-Welsh, Scotland, Wales.
British Columbia 242, 246
Brittany 235, 236, 260, 274; -see also Armorica
Brownian motions 58
Brunhes chron 135, 331
Bukoban System 184
Bunter sandstones 336
Burin peninsula 279
Burma 288
Bushveld Complex 162–164

Cader Idirs basalts 260, 264
Cadomian orogeny 233, 235–236
Calabrian arc 313–314
calc-alkaline magmatism **142**, 239, 205–208, 246, 271, 310, 359
calcite 7, 357; Mg-calcite: 357–358 -see also limestones, carbonates

Caledonides 69, 140, 142, 145, 236–237, 240, 246, 259–260, 263, **268–274**, 276, 338, 357
California 322, 324–325, 327
Callovian times 334
Cambrian times 3, 66, 128, 142, 190, 217, 222, 223, 226–231, 255–256, 267, 319, 354, 357, 369–370, 336–337, 340
Campeche microplate 290
Canada 221, 242, 279; -see also Cordilleran, Laurentia, Laurentian Shield
Canary islands 290
Cape System 259
Caradocian times 260–261, 264–265, 336
carbonates 6, 66, 69, 72, 195, 198, 247, 249–252, 292–298, 312–313, 357–360, 365; see also dolomites
carbon dioxide 240, 357, 359
Carboniferous times 68, 142, 235, 243, 253, 255–286, 288, 293, 295–298, 300, 322, 337–338, 365
Cardenas lavas 178
Caribbean 320–322
Carnian times 288
Carpathians 271, 314
cartesian coordinates 101, 117
Cascades 322–325
Caspar formation 280, 282
cassiterite 6, 30
Cathaysian flora 296–297
Catskill formation 279, 282
Cauacasus 283–284, 288, 315
Celebes -see Sulawesi
cell dimensions, magnetite: 22–23; hematite: 25
Celt-Iberia 266
cementation, of sediments: 63–72
Cenozoic times 1, 3, 76, 119, 131–133, 140–141, 148, 286, 293, 299, 310–328, 330–333, 335, 340, 347–349, 352, 359, 368; sediments: 182; geomagnetic field: 186
Central Asian fold belt 285–287, 318
Central European, geosyncline: 206; volcanic province: 352
C.G.S. units 7
chabazite 56
chalcophile elements 19
chalcopyrite 6
chalk 293
chamosite 7

Channel Islands 234–235
characteristic magnetisation 16
charnockites 165
Charnoid trend 271
chemical demagnetisation 88–90
chemical heterogeneity of mantle 360–361
chemical remanent magnetisation (CRM) 16, 28, 30, **43–44**, 55, 61, 63–69, 85, 88–90, 120–121, 128, 336
cherts 71
Chile 320
China, Precambrian Shields: 152, 153, 203, 206, 217–218, 220–221, 241–242, 250–252; plate, north: 285, 287, 289, 295, 317–319; plate, south: 285, 287–289, 295, 317–319
chitons 71
chlorite 56–57, 72, 93
chromite 6, 30
chromiun 54, 213
chronozones 330, 333
chrons 330; numbering system: 334–335
'chrontours' 206, 209, 267
Chugwater formation 336
Churchill terrane, Canada 169, 203
Chutline terrane 324, 326–328
Cimmerian mosaic 288
circle; of confidence: 105
Circum-Pacific 140–142
clasts -see conglomerates, sandstones
clays 60–61, 67
climate 133, 276; -see also palaeoclimates
clinopyroxene 56–75
closure, of domains 11, 14–15; to radiogenic isotopes -see blocking
Cloudina 250
coal 6, 71, 133, 293, 295
Coastal plutonic complex, Canada, 324
cobalt 8
coccoliths 333
Cocos plate 321
coercive force: 9, 32, 35; spectra: 85–86, 110–119.
coercivity 14, 23, 37, **39–40**, 45, 53, 55, 85–87, 90–92, 95–96, 110
collisions, continental: 141, 274–276, 309–311; -see also continent names
Colombia 320

Colorado 279, 336, 338; alkaline province: 231
combining directions of magnetisation 106–109
comparison of directions 105–106
compaction of sediments 60–63
components of magnetisation 15; separation of: 110–112; dating of: 120–128
Compton metasediments 281
confidence limits 103, 131
conglomerates 115; conglomerate test: 125–126, 277
Connemara terrane 260, 269, 281
contact test 126–128, 277
continental breakup episodes 3, 215, 216, 226–251, 357, 360, 367–368, 370
continental drift 140, 349
convection, core: 341, 363; mantle: 341, 349–352, **360–363**, 368–370, 372
convergent remagnetisation circles -see remagnetisation circles
cooling rates, effects on remance 49–52
Copper Harbour conglomerate 126, 179–182, 186
Coppermine group 176–177, 216
corals -see reefs
cordierite 7
Cordillera Darwin 305, 319
cordilleran environments 142–147, 199, 207, 280
Cordilleran margin of North America 279, 300, **322–328**, 351, 367
core of the Earth 134–135, 330, 341, **345–346**, 354, 363; mantle interface: 186, 341, 343, 345, 354, 356, 359, 362–363, 365; energy sources in: 345
Coriolis force 293, 298
Coronation, loop: 163, 186; geosyncline: 206 (see also Wopmay Orogeny); Gulf 246
Corsica 312, 314
cratons 141, 143, **151**, 205
Cretaceous Normal (CN) superchron **334–335**, 339–340, 343, 353–354, 359, 363, 370
Cretaceous times 249, 293, 300, 305–328, 332–333, 351, 353–354, 357–359, 368–369

crust, continental, growth of: 194–197
crust, oceanic 195 -see also ophiolites, serpentinisation
cryogenic magnetometers 80–82
crystalline anisotrophy -see anisotropy, crystal lattice
Cuba 320
Cuddapah group and traps 176–177, 180–182, 216
Cunene complex, Angola 58, 162–164
cuprite 6
Curie, Law: 8, 93; points: 11, 15–16, 22, 28–30, 41, 49, 50, 55, **92–93**, 126–127 -see also thermomagnetic spectra
cyclic behaviour, Earth: 4; geological: 356–360; geomagnetic: 339–341
Czechoslovakia 236–237, 272–273

Dalradian rocks 240, 260, 269
Damaran mobile belt 207–208, 226, 229–230
Darwin rise 368
Dead Sea fault zone 315
Deccan traps 305, 310
declination (D) **99–105**, 129–130, 147
décollement 146, 313
deep ocean sediments 331–332; -see also abyssal, DSDP
Deer lake group 280, 282
defect magnetism 25, 90
degassing, mantle 194
dehydration 61
Delamerian orogeny 223–225
demagnetisation, analysis of: 110–120; representation of: 101–102; techniques of: 84–90
demagnetising factor 35, 37, 45, 47 -see also anisotropy, shape effect
depositional fabrics 48
deserts 293, 295–298
Desert Ranges, Nevada 231, 336; -see also Basin and Range
detrital remanent magnetisation (DRM) 16, **58–63**, 67–71, 125, 128, 135
detrital particles 16, 21
deuteric oxidation 20–22, 54–57
Devonian times 42, 124–125, 192, 235, 237, 254, 262–263, 267–276, 280–285, 293–296, 320–321, 337–338, 349, 354, 357, 360
dewatering -see lithification, diagenesis
diagenesis 28–29, 63–72, 161, 279–280
diamagnetism 8–9, 11
diamictites -see tillites
Digico magnetometers 79–80
Dinarides 312–315
diorite 6
dipole, geomagnetic: 128–135; oscillation: 346; wobble: 346–347
dipole, moments, atomic 5, 8, 11 -see also geomagnetic dipole, orbital dipole moments, spin dipole moments
direct exchange -see exchange forces
directions of magnetisation **99–105**, 129–130; dispersion of: 132
discontinuities, mantle 361–363
dislocations 14, 34, 38
dispersion of geomagnetic field 346–347
distributions of palaeomagnetic directions 103–109
'disturbed' intervals and zones 330, 339–340
Dharwar belts, India 198, 199
D" layer 262–263, 365, 370
dolerites 6, 55, 57–58, 74 -see also hypabyssal environment
dolomites 6, 72, 197, 240, 243; dolomitisation: 71
domains, multi (MD): **14–15**, 23, 28, 31–34, 37–40, 42–45, 55, 58, 74, 77, 85; pseudo-single (PSD): **15**, 23, 31–34, 41–42, 88; single (SD): **14–15**, 23, 26, 28, 31–40, 55, 58, 85, 88, 90; structure: 9, 11, 14, 34–35; tests for: 93–97; transition: 37–40; walls: 9, 11–12, 14, 34, 42.
Donbass basin 283
Dorsal Eglabs 162–164, 165
drills 77
drilling induced remanence (DIR) 77, 107
DSDP results 308, 310, 334
dune beddingg 293
Duffer formation 160, 161
dunite 6
Dunkard series 338
Du Toit, reconstruction of Gondwanaland, 289–291
Dwyka glacial episode 258
dykes swarms 48, 55, 76, 159, 164, 169, 177, 175, 200, 206, 215–216, 219, 221, 344, 372–374; giant dykes 373–374; global stress fields: 206, 373–374; front-parallel swarms 216, 219
dynamo source 134, 345, 363

earthquakes 361
Earth's magnetic field 16; -see also geomagnetic
East Indies -see South East Asia
"easy" directions of magnetisation 11–12, **23**, 31–32, 35
ecliptic, obliquity of, 240
eclogite 365
Ecuador 320
Ediacaran fauna 250–251
eigenvalues 119–120
Eights Coast-Thurston block 305, 319
electron, magnetic moments: 58; shells: 8
Eleonora Bay group 242
Ellsworth mountains 305, 319
Elsonian terrane 173, 209, 206
Egersund, dykes: 219; anorthosite: 58
energies in crystal lattice 11–13
English Channel -see Channel Islands
ensialic (intracrustal) tectonics 143–145, 151–152, 187–188, **204–208**, 274–276
Eocene times 310–315, 318–314, 322–323, 327, 333, 350–351
epidote 7, 56–57
equal area projection 99–100
equatorial projection 100
Eromanga basin 210
Euler poles of rotation 117, **137–139**, 145, 153, 189, 191, 202, 218, 310, 312, 318, 352
euphotic zone 292
Euramerican flora 296–297
Europe, palaeomagnetism of: 276–278, 280–284; -see also state and regional names
Eurasia 132, 142–143, 265, 268, 286–291, 295–298, 300–304, 310–311, 314, 317–318, 351, 365; growth of in Pangaea: 283–289, 365–367

europium (Eu) 197
eustatic changes 354–360
evaporites 133, 210, 248, 250–252, 292–298, 312
exchange, forces: 10; energy of: 10–11, 14, 35
excursions -see geomagnetic field
expanding Earth hypothesis 4
exsolution 20–22, 54–58
extinctions -see faunal extinctions
extrusive environments -see lava flows

fabric, magnetic: **46–49**, 60, 62–63, 68, 71, 277
Falkland islands 291
Farallon plate 327–328, 353
far sided poles 132–133
faulting 146–147, 269, 271, 276, 279–281, 320, 315, 317–318, 321, 324–328
faunal, diversity: 243, 357–359; extinctions: 266, 356–359; provinces: 251, 259, 267, 292, 294–298, 327–328; -see also metazoans
faunas, basal Cambrian: 247 -see also Ediacaran
fayalite 7
feldspars 58, 121, 195
Fennoscandian (Baltic) shield, palaeomagnetic data: 156–185, 219–221, 224–226, 228–229, 236; geology of: 203, 205, 209, 213, 215–218.
Fen complex 228–229, 245–233
Ferrar (-Beacon) volcanic province, Antarctica: 305–306, 308, 368
ferrimagnetism 11, 22, 30
ferromagnetism 8–11
ferromanganese oxides 30
feroxyhyte 28
Finland 157, 159, 165–167 -see also Fennoscandian shield
Finnemark 241–242, 247
Fisher's statistics **103–105**, 109, 120, 123, 131, 147
fish faunas 294
Flinders Range 221, 223
Flin Flon-Snow lake belt 169
flips 15, 25, 33
floral provinces 292, 295–298
fluorapatite -see phosphorites
flysch 142 -see also turbidites
folding 11–112, 115, 146–147, 269, 323; concentric: 124–125; fold test: 120,

123–125, 277; shear: 124–125
foraminifera 328, 333
Fortesque volcanics 160, 161
Fossa Magna zone 316
fossils 133 -see also faunas, faunal diversity
France 270, 272–273, 312 -see also Armorican Massif, Brittany
Franciscan terrane, California: 324, 327
Franconian forest 270
Franklin igneous province 148, 219, 221, 224, 231, 245–246
Fraser-Musgrave belt 203
Freda sediments 179–181
frequency of inversions (reversals) 342–343, 353
freeboard, continental 194
Frontenac dykes 219
fusilinids 328

gabbros 6, 57, -see also basic rocks
galactic year 354
galena 6
gamma (nanoTesla) 7
Gardar igneous province 148, 175–181, 186, 215–216, 344–345, 373–374
Gargantua lavas 179–181
garnet 7, 56, 73
Garzon terrane 203
Gaspé peninsula 266
Gauss 7; chron: 331–333; gaussian distribution: 104–105
geocentric dipole -see geomagnetic field
geochronology, methods, 120–123, 154–155, 182, 183, 240
geoid 351, 361, 363–371
Geological Time Scale 3, 331–338, Appendix 1
geomagnetic field 75, **128–135**; axial geocentric dipole: 129–136; axis: 351–352; dipole source: 186; dipole field: 129–130; excursions of: 134–135; intensity of: 129, 135, **343–345**, non dipole components: 129; origin of 134 -see also dynamo; other sources: 186–187; pole position: 130; strength of: 343 -see also palaeo-intensity; tests for validity of 131–134
geomagnetic sources -see

geomagnetic dipole source, non dipole components, multipole components.
geomagnetism 1, 3, 128–135
Geophysical Journal of the Royal Astronomical Society 148
geosutures, global, 352
geosynclines 358–359
geothermal gradients 201, 208, 209
Germany 272–273 -see also Hercynian, Eurasia etc.
Gilbert chron 331–333
Gilsa subchron 331
glacial deposits 133, 292 -see also glaciations, late Precambrian: 239–243, Cambro-Ordovician: 227; Hirnantian: 265–266; Palaeozoic: 227, 258, 266, 294–298; Pleistocene: 243, 266
glauconite 243, 249
Glossopteris fauna 295, 296–297
Gneiguira supergroup 256–257, 259
gneisses 7, 198–200 -see also metamorphism
geothite 6, 28, 47, 63, 65; properties of: 24; treatment of: 88–89, 95.
gold 6, 122–213
Goldreich-Toomre mechanism 350–351, 365, 372
Gondwanaland 217, 223, 226–230, 232, 234–236, 245, 250–252, 266–267, 274, 276–277, 286, 288–291, 295, 305–311, 319, 365, 368; reconstructions of: 217, 227, 229–230
Gowganda formation 69 -see also Huronian
grain shape, magnetic effects: **37–40**, 55, 58–59, 62–63, 85
Grampian, orogeny: 294; region: 259
granites 6, 198–200, 203–206, 209–212, 267–268, 271, 274–276
granodiorites 194, 201
granulation 56
granulite 7, 51, 73, 164, 205, 208; facies: **56**, 73–74
graphite 6
graptolites 259
gravitational energy 363
gravity field, Earth, 1, 363 -see also geoid

Great dyke, Zimbabwe, 156–157, 159, 206
Great Glen fault 263, 267, 269, 279–280
Great Logan palaeomagnetic loop 178
Greece 314–315
Greenland 148, 300, 333, 373; Precambrian of: 153–155, 172, 176–181, 186–187, 199–202, 205, 218, 221, 224, 231, 242, 245, 247; Caledonides: 271
greenschist facies 7, **56–57**, 72–73, 142 -see also chlorite
greenstone belts, geology of: 3, 194–195, **198–203**, 206, 210–213, 372; palaeomagnetism of: 155–161
Grenville province and mobile belt 148, 152, 169, 173, **182–185**, 203, 206, 209, 215–216, 218, 219, 321; front zone: 182–183, 185, 204, 209, 216; dykes: 219, 220
growth, rings: 293; rhythms: 354
Gubbio section 333
Gulf of Aden 315
Gunflint formation 156–157, 159
Guyana shield 162–165, 169, 203
Gwalior traps, India, 165
gypsum 7
gyromagnetic remanence 86

Hadrynian track 222
"hairpins" **191–192**, 227, 235, 237, 372–373 -see also apparent polar wander
Haliburton complex 149, 180–183, 218
"hard" directions of magnetisation 12, **23**, 32, 35
harzburgites 57
Hammersley range 160–164
Harp complex 173–175
Harz mountains 270, 272–273
Hellenic-Cyprian arc 314
Hellenides 314–315
hematite 6, 11, 16, 20–21, **25–29**, 33, 35, 27, 44, 46–48, 50, 54, 60, 62–72, 85–96, 120, 125; properties of **24**; remanences; remanences: 169, 270
hematite-ilmenite series (hemo-ilmenites) **25–27**, 48, 73–74, 92, 149

Hercynian orogeny 235–237, **269–276**, 297–281, 285, 312; remagnetisation: 259, 270–273
high grade gneiss terranes 198–200
Himalayan orogenic belt 140–143, 276, 289, **310–311**
Hirnantian stage and extinction 266, 295
Hoffman-Day method 111, 113, 117
Hokkaido 316
Holmia stage (Cambrian) 240, 247
Holy Cross mountains, Poland, 236, 273
Honshu island 316
hornblende -see amphiboles
"hotspots" **350–352**, 360–362, 365–366, 368
Hudsonian mobile episode 148, 166–169, 172, 183 -see also Trans-Hudsonian belt
Hungary 313–315, 334
Hunnedalen dykes 219
Huronian sediments 156–157, 159
hydraulic currents 58–60
hydrocarbons 133, 293–294, 354
hydrochloric acid 89
hydrogen sulphide 66
hydrosphere 292 -see also oceans, oceanic crust
hydrothermal alteration 16, 22, 54, 56–57, 72
hydroxides 6, 28; of iron: 63–67
hypabyssal environments 57, 96
hyperzones 330
hysterisis 8–9, 11, 33–34; loop: 9

Iapetus (Proto Atlantic), ocean: 237, 243, 268–270, 294–296; suture: 261–265, 269, 294 -see also Caledonides
Iberian peninsula 260, 264–265, 270–272, 274; rotation of: 311–313 -see also Pyrenees
Iblean zone 313–314
ice 7; caps: 239–243, 266, 292, 294–298, 353 -see also glaciations

Kampuchea 287–28
Kaolinite 7
Karmoy ophiolite 265
Karroo igneous province 305–307, 368

Kasai (Congo) craton 162–164, 230
Katav group 219
Kenoran mobile episode 148, 155–158
Kenya 158, 217, 291 -see also Africa, Mozambique belt
Ketilidian mobile belt 205–206
Keweenawan rift, geology: 215–216; duration of igneous activity: 178; palaeomagnetic resluts: 148, **178–185**, 218, 344–345; asymmetric reversals: 185–189
Khazakhstania 152–153, 217–218, 236, 250, 285–287, 289, 295
Khapchan series 156, 159
Kiaman -see Permo-Carboniferous superchron
Kibaran belt 203–204
kimberlites 180–182, 185, 245–256
Klintla formation 173–175
Köenigsberger ration (Q) **17**, 28, 37, 45–46, 76, 82
Kola peninsula 199, 245
Kolyma plate 286–288, 300, 304, 317–318
Komati formation 161
Koolyanobbing hematite ores 156–157, 159
Korea 317–318
Kula plate 368
Kumlinge dykes 173
Kun Lun zone 287, 311
Kuskokwim terrane 324, 326–327
kyanite 56
Kyushu 316

Labrador 155, 185, 209, 245; Trough: 168, 170–171
La Colina rocks 258, 338
Lake District 260, 259–261
Lake Froome group 223–225
lake sediments 62, 135
Lake Superior 114, 178, 183 -see also Keweenawan
Laos 288
Laschamp subchron 331
Late Proterozoic Supercontinent 216–252, 367–370
laterite 28, 293
laumontite 56–57
Laurentia 145, 227–232, 249–253, 258, 266–267, 276–283, 289, 295–298, 300–304

Laurentian Shield, geology: 198–203, 209, 213–216, 218; palaeomagnetic results: 148–140, 155–157, 159, 172–185, 219–220, 222, 224, 369
lava flows 21, 48, 53–57, 72, 75–76, 93, 135; pillow lavas: 21, 35, 113
Laxfordian belt 203; -see also Lewisian, Scotland
layered mantle models 360–363
lead isotopes 194
Lena river 228–229, 232, 336
Lenian times 336
lepidocrocite 28
Lewisian terrane 171–172 -see also Scotland
Ligerian orogenic belt 272–274
lightning strikes 16
limestones, 6, 29–30, **69–71**, 95–96, 268, 333, 334, 338 -see also carbonates
"limonite" 71 -see also goethite
Limpopo mobile belt 159, 168
lineaments, Precambrian, 152, 198–213
Lisbon volcanics 311
lithification 48, 60–63, 123 -see also diagenesis
lithophile elements 19
lithosphere 201–203, 209, 349–352, 360–363, 368, 372
littoral -see intertidal
Llandeilo times 259–265, 269, 294
Llandovery times 262–263, 267, 337–338
Llanvirn times 260–265, 294, 336–337
Llano uplift, Texas, 231, 228–229
lodestone 28
loops, APW, 156–185, 189–191, 208, 232, 371–373
Lower Congo glaciation 240
Lowrie-Fuller test 96–97
low temperature, demagnetisation 88; susceptibility behaviour: 93–95
Lufilian arc 230
lunar rocks 19
Lunch Creek lopolith 169–171, 344–345

Mackenzie, igneous province: 148, 175–177, 215–216; Mountains: 324, 327
Madagascar 2, 200, 209, 217–218, 256–257, 291, 305
maghemite 6, **27–28**, 35, 46–47, 90–91, 93; properties of: 24; maghemitisation: 28

magmas 19
magnesioferrite 30
magnesite 7
magnesio-wüstite 361
magnetic, anomalies 3; anomalies, marine: 330, 332–335 flux: 5; identification of minerals: 90–97; induction: 5, 7; intensity -see intensity of magnetisation; susceptipility -see susceptibility
magnetisation, induced 17 isothermal 9, 16; remanent 9, 10, 15–17; saturation: 9, 16, 32, 42–43, 46, (magnetite) 23–24, (hematite) 24–25; types of **15–17**
magnetite 6, 11, 16, **21–24**, 30, 33, 35, 37–39, 41–42, 44–47, 49–50, 53, 56, 58, 62–64, 71–74, 85–89, 125, 149; properties of: 24; tests for: 90–96; magnetite-ulvöspinel series: 20–24; remanences: 169, 279
magnetometers 77–82
magnetocrystalline energy 11–13, 35–36, 49
magnetostatic energy 11–12, 14, 34–35
magnetostratigraphy 76, 330–338, 340 -see also geomagnetic, polarity time scale
magnetostriction 13–14, 35–36
magnetozones 128
Malani rhyolites 226
Malaysia 287–288, 318
Mamainse lavas 179–181, 186
manganese 8,54; nodules: 30, 62
mantle 341, 349–354; convection of: 201, **360–363**, 368–370; plumes: 350–353, 360–363, 365; roll: 352; viscosity of: 350, 361–363, 372–373
marcasite 29
Marie Byrd land 305, 319
marine environments 60–66, 69–72
Maringoan formation 90
Marinoan glaciation 242
Maritime Provinces, North America, 276–280
Martin formation 68
"martite" 67
Mascarene group 281
Mashonaland dolerites 166–168

Matachewan dykes 148, 155–159, 200, 202
Matatzal belt 206
maturation 293
Matuyama chron 331–332
Mauretainides 275
Mbozi complex 226
Mealy Mountain 218
Mediterranean region 311–316
megacycles 4
Merijina tillite 241
mesophere 350, 370, 372
Mesozoic times 2–3, 141, 148, 248, 279–280, 286, 297–328, 332–336 -see also Cretaceous, Jurassic, Triassic
metallogenesis -see mineralisation
metamorphic rocks 26, 29, 48, 56, 108, 121
metamorphism 46, 48, 51, 55–57, 72–74, 126, 142
metasomatism 55–57, 200
metazoan faunas 244, 248, 250–251; provinces: 257
meteorites 19
Mexico 321–322, 328
mica -see biotite, muscovite
Michikamau intrusion 58, 174
microfossils, late Precambrian 236, 240, 242–243, 247, 249
microplates 233–239
Mid Continent Rift, North America 178, -see also Keweenawan
Middleback range ores 171
Middle America Trench 320–321
Middle East 245 -see also Arabia, Iran, Israel etc.
Midland Valley of scotland 276–277
mid-ocean ridges 247–248, 332, 357, 360–363, 365, 368; basalts 360–361; push forces 362
M.K.S. units 7
migration, of hydrocarbons: 293–294; of faunas: 292, 294
Mildunna complex 160–161
mineralisation 194, 209–213
Minturn and Maroon formations 338
Miocene times 133, 312, 315–316, 319, 320
miogeoclines 243 -see also passive margins
mobile belts 202–208, 216
model sedimentationstudies 59–63

Modipe gabbro 58, 156–158, 344
Moenkopi formation 69, 336
Moine rocks, Scotland, 262–263, 268
molasse 142, 286
molluscs -see chitons
montmorillonite 7, 61
Moonlight Valley tillites 242
Morin complex 58, 180–183, 218
Morin transition 25–26, 94
Morocco 246, 258–259, 266, 275
Moscow basin 283
Mount Isa 206, 211
Mozambique, 291; belt: 204, 229
Mssissi norite 258–259
Mt Peyton 281
muscovite 56, 121–122
multicomponent magnetisations 40, 110–120, 122
multidomain structure -see domains
multipole components 186, 349

multivariant analysis 119–120
muscovite 7
Muskox intrusion 176–177, 216
Mwembeshi zone 207, 230
mylonite zones 146, 200

Nagssugtoqidian mobile belt 166–169, 203
Nain province, Canada 169
Nama group 224–226, 240–242
Namqualand belt 182, 203, 206, 226
Namibia 219, 229, 242, 249
Namurian times 337–338
nanoTesla (gamma) 7
nappe tectonics 208, 274, 310–311, 313
Naskaupi fold belt 185
natural remanent magnetisation (NRM) **15**, 17, 30, 40, 53, 75–76, 80, 83–90, 99, 102
Nazca plate 320
near sided poles 147
Necker Rise 368
Néel, temperature: 11, 25–26, 88; theory of remanence acquisition: **31–37**, 39–43, 49
neodymium isotopes 360
Neo Tethys 288, 297 -see also Tethys
Nepal 310
Nevada 231, 336

New Brunswick 281 -see also Maritime Provinces
New England 273, 281
Newfoundland 242, 246, 250, 269, 279–281
New Guinea 308
New Zealand 308
nickel 10, 210–213; nickel-iron alloys: 30
Nipissing ingneous episode 148, 159, 161–164, 169
nodules -see manganese nodules
Nonesuch formation 179–181
non-dipole, components: 346–349; field: 129
Nordingra complex 173–175
norite 57
North America 151–154, 245, 247, 134, 147, 321–328, 334, 336, 338, 357–358 -see also Laurentia, Laurentian Shield, Canada
Northern Calcareous Alps 313
Northern Highlands of Scotland 259, 263, 267–268
North Sea 236, 271–273
North Shore lavas 179–181
Norway 159, 205, 245, 269, 272–274, 276, 294 -see also Caledonides Fennoscandian Shield
Nosib tillite 241–242
Nova Scotia 263, 280, 281 -see also Maritime Provinces
Numees tillite 241–242

obduction 204–205, 327
Obuasi greenstone belt, Ghana, 164
ocean floor spreading 350–353, 356, 359, 369; rates of: 353–354
oceanic, circulation: 293; crust: 28, 57–58, 73, 141–142, 147; island basalts: 360; -see also maghemitisation, ophiolites, serpentinisation
oceans, chemistry of: 247–249; palaeotemperatures of: 353–354
offset dipole models 132–133, 186
oil -see hydrocarbons
Okinawa 316
Old Red Sandstone 276 -see also Devonian, Silurian
Olduvai subchron 331
Oligocene times 310, 312, 315
olivine 57, 361 -see also fayalite, dunite

ophiolites **57–58**, 142, 204, 205, 260, 265, 285, 287, 289
optical properties of ore minerals **24**, 90
Ora region, E. Mediterranean, 314
orbital moments 5, 8
Ordovician times 142, 190, 192, 223–235, 247, 253–257, 259–267, 269–270, 272–273, 275–277, 282–285, 288, 294, 296, 320, 325, 336–337, 340, 354, 357, 359–360
Oregon 322, 325
organic matter 64–66, 73; magnetism of: 71; -see also hydrocarbons
orientation of palaeomagnetic cores 77
oroclinal bending 270, 310–316, 319–321
orogenic, belts 140–147; episodes: 213; -see also locality names
orthogonal projections 101–102, 110, 115–117
orthopyroxenes 56–57 -see also granulite
Osler lavas 179–181
Otavi group 219
Oti tillite 241–242
Ouachita-Marathon belt 290
overprint magnetisations 153, 155, 163, 168–169, 179, 185, 188, 195, 229, 231, 233, 259, 268, 276–279, 312, 316, 347, 354
Oxfordian times 332, 334
oxidation, indices of, 20, 54–57 -see also deuteric oxidation, redox potential
oxidation-polarity paradox 134
oxides 6, 19 -see also ilmenite, maghemite, magnetite, rutile, ulvöspinel

Pacific Ocean 328, 350, 357, 367–368
paired metamorphic belts 142
Pakistan 248, 310, 318
Palaeocene times 308, 310, 318, 333
palaeoclimates 64, 258, 291, 294–298; indicators of: 251, 292–294 -see also carbonates, evaporites, red sediments
palaeogeography 250–252, 291–298
palaeogeophysics 1
palaeo-intensities 61, 343–345

palaeolatitudes 130–133, 137–139, 253 -see also palaeogeography
palaeomagnetic data, analysis of 16, **103–150**; data base: 148–149
palaeomagnetic poles, calculation of: 130; quality of: 153; interpretation of: 2–3, **135–150**; causes for dispersion of: 187–188.
palaeosecular variation (PSV) 345–349
Palaeo-Tethys 288, 297 -see also Tethys
palaeowinds 133,
Palaeozoic times 3, 140, 190, 192, 198, 208, 221–298; palaeomagnetism of: **251–298**, 336, 340, 347
Pan African belts 208, 227, 229–230, see also Damaran, Lufilian Mozambique, Mwembeshi and Zambezi
Pangaea 143, 152, 217, 274–277, 280, 321, 351, 357, 362, 365–370; variants of reconstruction: 289–291
Pannonian basin 313–315
paramagnetism 8–9, 11, 22, 46, 93, 95–96
Parana, volcanic province: 305; basin: 258
parastaic magnetometers 78–79
partial thermal remanent magnetism (PTRM) 16, 29, **41**, 50, 87–88, 120–121, 343
passive margins **243–245**, 247–248, 366–367, 369
Paterson toscanite 338
Peary Land 242
peat 71, 293
pegmatites 213
Pembrokeshire 237
Peninsula and Chugach terrane 324, 326–327
Penokean orogeny 163, 205, 206
Pensacola mountains 319
Pentaravian terrane 239
Peri-Adriatic belt 312–314
peridotite 6
periodicity, in geomagnetic polarity: 341; in APW: in geological events:
Permian times 67, 143, 243, 253–258, 276–286, 288–291, 293, 295–298, 300,
311–313, 319, 336–337, 340, 354, 357, 367
Permo-Carboniferous (PCR) superchron 268, 270, 277, 327, 334, **337–338**, 354
perovskite 361
Peru 249, 320
petrology -see basalts, igneous rocks, metamorphism, ophilites etc
Phanerozoic times 68–70, 132–133, 140–145, 148 -see also Cambrian, Ordovician, Silurian etc.
phosphates 66, 133
phosphorites 249
phyllite 7
Piedmont -see Appalachians
pigments -see red sediments
Pilbara carton 160, 161, 195
pillow lavas -see lavas
plagioclase minerals -see feldspars
planetary magnetic fields 134
planktonic faunas -see microfossils
plates 136, 141–145
plate tectonics 140–143
Pleistocene times 243, 266, 330–331
Pliensbachian times 334
Pliocene times 133
plunge of folds, correction for, 124
plutonic rocks and terranes 49, 53, 58, 72–74, 96; -see also locality names
Poland 272–273
polar ice caps -see ice caps
polarity of geomagnetic field, Precambrian: 154, 159, 163, 173, 177, 179, 186
polarity time scale, Phanerozoic: 330–338, 353; Precambrian: 341–343
polar projection 100
polar wander -see apparent polar wander, true polar wander
Polynesian seamount chains 368
populations, test for significance of: 103–106
porosity 60–61, 63
Portage lake lavas 179–181
Portugal -see Iberian peninsula
post depositional detrital magnetisation (PDDRM) **60–63**, 66–71, 89–90, 120, 125, 128, 135, 336

potassium, -40: 345, 363; -argon method and results: 120–122, 183, 267, 332–333
Precambrian -see Archaean, Proterozoic
precision parameter (k) **104–105**, 108–109, 123
Priessac dykes 148, 162–164
primary magnetisation 15
projections -see equal area, equatorial, polar, stereographic projections
Proterozoic, geology of: 3, 51, 66–70, 132, 136, 140–143, 147–148, 194–198, 200–213, 215–217, 267; palaeomagnetic results: 74, **151–187**, 217–226 Supercontinent and tectonics of: 1, 152–153, 156, 158–185, 187–192, 210, 341–342, 344–345, 348–349, 362, 367, 369–374 -see also Late Proterozoic Supercontinent
protohematite 28
Protohertzina 250–251
protozoa -see archaeocyathids
pseudobrookite 21, 54; properties of: 24
pseudo single domains (PSD) -see domains
pseudotachylytes 53, 231
Puerto Rico 320
Purcell lavas and sediments 173–174
Pyrenees 312
pyrite 6, 29, 64–66, 71–72; properties of: 24
pyroxenes 7, 67 -see also clinopyroxenes, orthopyroxenes
pyroxenite 6
pyrrhotite 6, **29–30**, 35, 47, 66, 72, 90–91, 93; properties of: 24

quartz 6–7, 60
quartzites 72–73
Quaternary 132–133
Queen Maud block 200
Quebec 266, 281
"quiet" intervals and zones 330, 340

radiogenic isotopes 195–195, 345; -see also element names
radiometric ages, frequency over geological time: 190–192

radiometric methods -see geochronology
randomness, tests for: 103–104
Rapakivi granites 170–175, 206, 209
Rapitan group 240
rare earth elements (REEs) 194–197, 360–361
rate of cooling, -see cooling
Rayleigh number 201–202, 262–263
red sediments 6, 26, **66–70**, 89–90, 243, 251, 271, 293–298
redox potential (Eh) 22, 26, 63–69
Red Sea 315 -see also Arabia, Dead Sea Fault zone
reefs 133, 249, 292–298
reflectivities, mineral 24
regressions 266, 295, 356–359
relaxation time 39
remagnetisation circles 111–115
remagnetisation hypothesis 276–280
remagnetisations -see overprints
remanence -see magnetisations
resultant vector 103–106, 123, 125
reversals of the geomagnetic field, asymmetric: 109, 128, 134–135, 185–187; -see also inversions
Rewa sediments, India, 222, 224–225
Rhine rift 245
ridges, mid ocean, push-pull forces: 351; geometry of: 353–354, 357
rifting 215–217, 231, 243–251; -see also aulacogens
Riphean times 123, 149, 175, 219, 222, 242, 247
Rocky Mountains 324–325; troughs: 324
Rogaland complex 180–181, 184, 219
Rome formation 228–229, 231
Roraima rocks 169, 170–171
Rose Hill formation 123
rotational remanent magnetisation (RRM) 86–87
rotation of the Earth, rate of: 298, 354, 356, 362, 365
rotations, tectonic: 163–164, 177, 185, 311–328 -see also Euler poles
rubidium-strontium ages 190–192; method 121–123
Russia -see U.S.S.R.
rutile 6, 20–21, 26, 54–55, 73, 89; properties of: 24

Ryukyu arc 316

Saamo-Karelian terrane, Finland, 159, 164–165
Sahara 259
Salinian terrane 327
salt 6 -see also evaporites
Salt Range 223
samarium-neodymium methods 154, 156, 196, 198; results: 360–361
samples, palaeomagnetic 75–77
sampling, palaeomagnetic 76–77
San Andraes Fault zone 327–328
sands 58–63; sandstones 6 -see also red sediments
sanidine 121
Santonian times 334
Sardinia 312, 316
saturation magnetisation (J_S) -see magnetisation
Scandinavia -see Caledonides, Fennoscandia, Norway, Sweden
scatter of directions 105; reasons for: 107
schists 73–74
scolecite 56
Scotland, NW Scotland Precambrian: 148, 153–154, 180–182, 200, 203, 218, 222, 240; Caledonian orogenic belt: 240, 243
Scourie dykes 170–172, 200
sea floor spreading 330, 332–335
sea level 266, 365; -see also regressions, transgressions
secular variation 71, 75, 115, 129, **345–349**
sedimentary and sedimentation 6, 47–48, 58–72, 88–90, 108, 135, 195–198, 205, 216, 231–233, 243–245, 247–248, 277–279, 356–358;
sedimentary rock mass (over geological time) 190–192, 196–197, 358
sedimentation, rates of: 60–62, 332, 357–359; theory of: 48, **58–63**
seismic stratigraphy 356–358
selection criteria 149–150
self reversal of remanence 26, 134
Senonian times 311, 313
Seram 319
serpentinites 7, 16, 205, 254;

serpentinisation: 16, 30, 57–58
shales 6
shape effects, on palaeomagnetic measurements: 77–78, -see also anisotropy, fabric, grain shapes
Shaw (palaeo-intensity) method 343
shear -see stike slip
shear remanent magnetisation (SRM) 62
Shelve inlier 260
Shetland (Unst) ophiolite 265, 269
shields 141, 158; -see also Africa, Laurentia, Fennoscandia etc.
Shimanto belt 316
Siberia, Precambrian, geology: 143, 198, 203, 209, 217, 240–242, 245, 250–252; Precambrian palaeomagnetism of: 153–154, 156–157, 162–164, 180, 182, 184–185, 187, 218–222, 224, 228–229; Phanerozoic palaeomagnetism of: 232–233, 267, 276, 283–284, 286–287, 289, 318, 336
Sibumasi plate 287–288
Sicily 312–314
siderite 7, 64, 72
siderophile elements 19
Sierra Madre Oriental -see Mexico
significance tests, directions, 103
Sikote Alin 287, 300, 316–318
silicates 6–7, 19, 58, 67, 69, 72–74
sillimanite 56, 73
silts 60–63
Silurian times 66, 142, 254–260, 262–275, 280–285, 288, 293, 336–338, 340
silver 6, 211–212
Sinean 219, 221–222, 242
single domain (SD) -see domains
Sino-Korean plate 148 -see also China plates, Korea
site, palaeomagnetic 76
Siyeh formation 173–175
slates 6, 48
Slave craton 149, 157, 164, 168
slumping 62, 68, 127
small scale tectonics, palaeomagnetism of 146–147

smeared distributions 17, 111–115
Smith-Hallam reconstruction (Gondwanaland) 217, 227, 257–258, 289–291, 305–310, 366
soil 28
Song Ma/Song Da zone 287–28
South Africa 156–157, 162, 166–167, 170–171, 174–175, 180–181, 200; -see also Kaapvaal craton
South America, Precambrian: 143, 145, 153, 203, 209, 217–221, 224–225, 228, 245, 255–258; Phanerozoic: 291, 305–308, 320–322, 338; -see also country names
South Atlantic 332
South China fold belt 252
South East Asia 288–289, 292, 295, 308, 318–319
South Shetlands 320
Southern Uplands 259–260
Spain 266, 311–312 -see also Iberian peninsula
spectral content of geomagnetic polarity time scale 341
specularite 26, 67–69, 89–90
sphalerite 6
sphene 57
spin, dipole moments: 5, 10–14, 33; axis of Earth: 240, 349–352
spinner magnetometers 79–80
Spitsbergen 243, 271
Spokane group 173–174
SQUID 81–82
stability of remanence 28, 37; indices of: 119
staurolite 56, 73
St Bees sandstone 69, 89
Stephanian times 337–338
stereographic projections 99–100
Stikine terrance 324
Stillwater complex 156–158
St Lawrence graben 245
straight belts 200–203, 372
strains, crystallographic -see magnetostriction
strain tensor, tectonic 48–49
Strathmore lavas 276
strength of the geomagnetic field -see geomagnetic field, palaeo-intensities
Sturtian glaciation 240, 242
stress, effects in hematite: 25; fields, global: 198–203, 372–374
striations, glacial: 292, 294–298
strike slip, in straight belts: 200–201;
in mobile belts: 145–147, 165, 189, 207, 227, 229–230; in orogenic belts: 268–270, 275–276, 279–280, 286–288, 317–328; -see also faults
strike slip movements 227, 229–230
strontium isotopes 194–195, 247–248, 324–325
subchron 330
subduction zones 140–147, 194, 206, 212–213, 334, 352, 361–363, 365–376, 369, 371
subsidence, continental margins 243–245, 247–248
subtracted vectors **110–111**, 113, 115, 118–119
Sudbury dyke swarm 176–177, 185, 215–216; lopolith 165–168, 181, 185, 344
Sulawesi 318–319
sulphates 29
sulphides 6, 19, 29–30, 64–66, 71–72, 210–213
sulphur 7, 19, 55, 65–66; isotope (34): 248–249
Sumatra 288, 316, 318–319
Sun compass 77
Sunda Strait 318
superchrons **330**, 340, 349
supercontinents 2–4, 363–374; accretion of: 363–368; cycle of formation and destruction of: 370; -see also Proterozoic, Late Proterozoic, Pangaea, Gondwanaland
superexchange -see exchange forces
Superior craton 155, 164, 167, 168–169, 183, 198, 202–203, 205
superparamagnetic (SP) state **15**, 23, 33, 35, 38–39, 43, 46; threshold: 15, 17
susceptibility **5**, 7–8, 42, 44–49, 57, 79, 82–84, 93–96; meters 82–83; -see also anisotropy
suspect terranes **147**, 233, 269, 305, 311–328, 351, 367
sutures 140–143
Svalbard -see Spitsbergen
Svecofennian terrane 172, 184, 203, 205–206
Svecokarelian terrane 165, 172, 203, 205–206
Sveconorwegion mobile belt 180–181, 184, 203, 215–216; front zone:184, 204, 206, 209, 216, 219

Sweden 159, 172, 177, 205, 242, 265, 272–273, 373; -see also Fennoscandian Shield
switching field 33
Sydney basin 305–308, 338
Systéme International (SI) units 7

Taconic orogeny 145, 273, 281, 283, 294
Tanganyika carton 230
Tanzania 184, 204, 217, 291
Tapeats formation 232, 228–229
Tarim 285–286, 289
Tarkwaian belt 164
Tasman, Geosyncline: 227, 241, 254; Sea: 308
Tasmanian dolerites 305–306
tectosphere 350–306
tectosphere 350, 352
tectonic, models: **140–147**, 279; corrections: 123–125, 146–147
ternary diagrams 19–20
terranes, suspect, 147, 311–328
Terranes 1, 2 and 3 (Canadian cordillera) 324–325
Tesla 7
Tethyian ocean 288, 295, 311–312, 315, 328, 357, 365, 368
Texas lineament 321
Thailand 287–288, 318
Thellier method 343
thelodonts 294
thermal demagnetisation 87–90
thermal vibrations, atomic, 32
thermochemical remanence (TCRM) **16**, 51, 122, 267
thermomagnetic spectra 55, 57, 92–93
thermoremanent magnetisation (TRM) 16, **41–44**, 46, 49, 121–122, 135, 343; rate of cooling, effect on: 43, 49–52; strength of: 4243, 55, 85
thomsonite 56
Thouars massif 260
thrusts 146 -see also nappe tectonics
Thulian province 352
Thuringia 266
Tibet 140, 287–289, 295, 310–311
tidal friction 345, 354, 356, 359
Tien Shan zone 286–287
tillites 239–242, 246, 258, 246, 292, 294–298
tilloids 292

tilt correction 13–125
time, effects on remanence: 49–52, 344
Timor 319
tin 212–213
titanium, oxides of, 19–28, 54–55, 58; -see also ilmenite
titanohematites 19, **25–26**, 35, 57
titanomagnetites **19–24**, 35, 54–58, 93–94; compositions of: 21
Toby formation 242
Tommotian times 250, 336
tonalite-trondjhemite suite 194–195
Tornquist lineament 271
torquemeters -see torsion balance
Torridonian sediments 60, 68–69, 125, 179–182, 184, 216, 222
torsion balance 83–84
Tournasian 235
Trade winds 293, 295
Tramore volcanics 260
Transdanubian central mountains 313–314
Trans-Hudsonian belt 207
transgressions, marine, 243–245, 247–248, 266, 353–354, 356–359
transitions in magnetic minerals 25–29, 88, 94–95
transpression 146
Transylvanian Alps 314
Traveller felsite 281
tree growth rings 293
trench pull forces 362
Trenton limestone 272, 279
trevorite 30
Triassic times 67, 69, 245, 262–263, 268, 283–286, 288–291, 293, 295, 297–298, 311, 315, 317–319, 325–327, 334–336, 340, 349, 354, 357, 359, 363, 365, 367
trilobites 250, 252, 294
triple junctions 209
troilite 29
true polar wander (TPW) **349–353**, 360, 369, 372
Tsinling orogenic belt 286–288
tumbling 86–87
turbidites 61, 292
Turkey 244–245, 315
two-tier analysis 108
Tyrrhenian Sea 313–315

Ukraine, Precambrian of, 143, 152, 164, 166–167, 174–177, 218, 220–221, 224–225, 247

ultramafic rocks 57–58, 198, 207, 210
ulvöspinel 20–24; properties of: 24, 54
Umbrian region 303–314, 334
unblocking of magnetic remanence 39, 85–91
unconformity test 120, 128
unconformity-related ores 210–212
unit cell, -see cell dimensions
Unkar group 187
uplift and cooling remanences **49–52**, 169–171, 183–184, 267
upwelling, oceanic 247, 249
uranium-lead method 121
uranium ores -see mineralisation
Urals orogenic belt 140–141, 143, 279, 283–286
U.S.A. -see Laurentia, Laurentian Shield, North America and state names
U.S.S.R. 159, 334, 283–285, 336–338, 357–358; -see also Eurasia, Siberia
Ust'Maya group 228–229, 232
Ust'Kutsk group 228–229, 232
Usushwana complex 156–158

valancy electrons 8, 12
vanadium 54
Vancouver Island 325
Varangian glaciation 240–242
varves 62, 292
Vasparaisjarvi diorites 156–157, 159
vectors 110; analysis of: 110–120
vector plots -see orthogonal projections
vector subtraction 110–114
velocity of continental crust (V_{RMS}) **138–140**, 190–192, 208, 300, 329, 354–360, 362, 368–369
Vendian times 123, 149, 220–222, 224–226, 232, 235, 237, 240–251
Venezuela 320
Ventersdorp lavas 156–159
Verkholensk group 228–229, 232
Verkoyansk orogenic belt 286–287, 318
Verwey transition 22
Vietnam 287–288
Vindhyan supergroup 180, 182, 185, 216, 242
virtual dipole moment (VDM) 129, 343–345

virtual geomagnetic poles (VGPs) 130–132; dispersions of 132; paths during reversals 134–135
viscosity of the mantle 350, 360–363, 372–373
viscous remanent magnetisation (VRM) **16**, 41–42, 76, 85, 107, 119
Viséan times 337–338
volcanic environments 54–57, 96; rock mass: 358–359; -see also lavas
volcanogenic ores -see mineralisation
Vosges 270, 272–273

Wales 48, 128, 260; -see also Anglo-Welsh microplate, Caledonides, Welsh igneous province
Wallace's line 292
Walsh function 341
Washington (state) 242, 322–323, 327
water 7 -see also diagenesis, red sediments, redox potential
Waterberg redbeds 166–168
weathering 28, 211
Welsh igneous province 260, 264–265, 338
Wenlock times 262–263, 267, 337–338
Westphalian times 338
West Virginia 338
Wichita granites 228–229, 231
Willamette terrane 322
Wilson cycle 145, 207
Windermere supergroup 218
winds -see palaeowinds, Trade winds
within-site scatter 107
Wisconsin 205
White rock formation 266
Wood Canyon formation 228–229, 232
Wooltana volcanics 220–221, 223, 240
Wopmay orogeny 205, 206, 208; see also Coronation geosyncline
Wrangellia, mountains 325; terranes 324–327
Wyoming 205, 336

Yangste plate 148; -see also China plates
Yenesei Ridge 180, 182, 184–185, 187
Yilgarn craton 161, 198, 226

Yucatan microplate 290
Yugoslavia 313–315
Yukon-Tanana terrane 324
Yunnan province 288

Zagros thrust zone 315

Zambezi belt 204, 230
Zambia 184, 230
zeolites 55–56, 73
Zijderveld plots -see orthogonal projections
Zil'Merdak group 174–175, 219

Zimbabwe 373; craton: 159, 168, 198, 200, 204
zircon 30, 121

R00725 07484

REF QE   Piper, J. D. A.
    501.4
    .P35   Palaeomagnetism and
    P54   the continental
    1987   crust

$67.95

DATE			

© THE BAKER & TAYLOR CO.